2 - 6, 7, 8, 9, 10, 16, 18

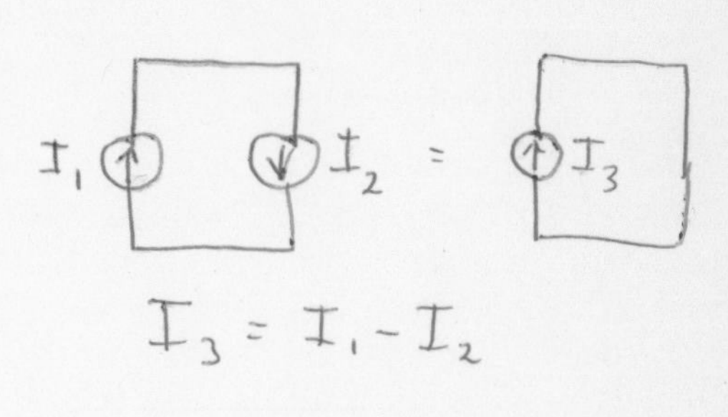

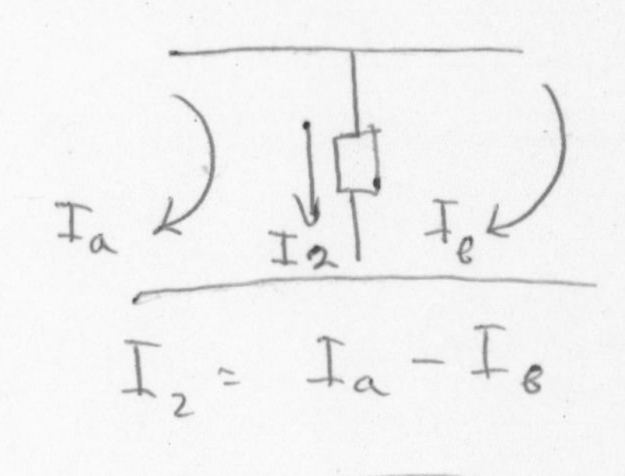

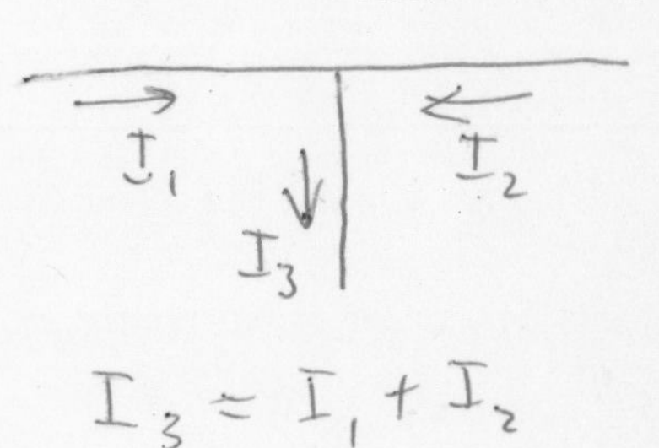

FUNDAMENTALS OF CIRCUIT THEORY

FUNDAMENTALS OF

This book is part of the Allyn and Bacon

Electrical Engineering Series

under the consulting editorship of Norman Balabanian

CIRCUIT THEORY

N O R M A N   B A L A B A N I A N

Electrical Engineering Department
Syracuse University

A L L Y N   A N D   B A C O N ,   I N C . ,   1 9 6 1

PRINTED IN THE UNITED STATES OF AMERICA.
LIBRARY OF CONGRESS CATALOG CARD NUMBER: 61-13555

To the fervent hope that through the scientific spirit of free inquiry man can rise from a life of fear and ignorance to one in which reason, tempered by human compassion, will prevail in all his affairs.

PREFACE

Within the last decade, stimulated by an ever-expanding technology, great upheavals have taken place in the undergraduate electrical engineering curriculum. Many engineering teachers and institutions have been restudying the subject matter and organization of the courses considered to be basic to the curriculum. As a result, in the last few years a large number of introductory level books have appeared in the areas of both circuits and electromagnetic fields. The present book is my attempt at giving an account of the concepts and techniques of circuit theory. In this preface I wish to describe some of my thoughts underlying the organization and emphasis of the book.

In the past, the subject of linear circuit theory has suffered from schizophrenia. On the one hand, dc circuits, ac circuits, and transients have been treated in separate blocks with very little interaction among the blocks. On the other hand, "active" networks have been isolated from "passive" networks to the extent that some students have not even been able to recognize that the analysis of these networks is susceptible to common techniques.

My own view is that, in seeking to explain and predict electrical phenomena in the physical world of circuits, we follow what is often described as "the scientific method." Over the years, the result of observations and experiments in the physical world has been a set of laws (Ohm's law, Kirchhoff's laws, Faraday's law, etc.) which are enunciated in terms of a hypothetical model of that part of the physical world which is under discussion. The model consists of a number of elements. These hypothetical elements do not actually exist in the physical world, although the behavior of *some* devices under *some* conditions of operation may approximate that of one or another of the elements. However, our theory is built on the assumption that the electrical behavior of a physical device can be approximated by a suitable interconnection of elements of the model.

We may find in some cases that the simple model adequate to describe the behavior of some circuits is not sufficient to describe the behavior of others. Hence, following the scientific method, the model must be modified or augmented by the introduction of additional elements. This is the case when coupled coils, or transistors or vacuum tubes are studied. The same approach can be taken in introducing the subject of distributed circuits, although this subject is not treated in the present book.

In analyzing a physical problem, the first task of the engineer is to arrive at a hypothetical model consistent with the conditions of the problem. Once the model is at hand, analytical techniques (I include graphical techniques in this category) are employed to solve the mathematical problem. The solution obtained is then interpreted in the light of the physical situation. All of these specific parts of the overall problem are important and, at some time or other, anyone aspiring to become an engineer should acquire the knowledge and facility for carrying them out. However, it is possible, and desirable, to separate these diverse aspects of the problem into specific units and to study each unit separately. Thus, the study of analytical techniques for the solution of circuit problems can be carried out without reference to the origin of the particular model under investigation.

This is not to say that, when an engineer is presented with a problem, he solves the problem in isolated parts. Certainly the solution of an engineering problem requires the application of analytical techniques, it requires intimate knowledge of physical principles, and it requires what may be described as engineering judgment—and these ingredients are intermingled. Yet in the *learning*, as distinct from the *practice*, of engineering, it is desirable to dissect the overall problem and to study the component parts individually. This is not much different from what a doctor does, for example. It is certainly important for a doctor in his practice to have a knowledge of psychology and human behavior; but he does not learn these while studying surgical techniques. In this book, then, we are concerned primarily with the network model.

In a rapidly moving era of technological advance, a textbook in a technical scientific subject should do much more than present the facts of the subject under study, even in a classical subject such as circuit theory. Whatever the technical content of the book may be, the most beneficial legacy which can be left to the student is a deep appreciation of the open-minded, critical spirit of scientific inquiry, an intimate acquaintance with patterns of discovery in science and the attitudes which lead to successful investigation and solution of problems.

The deductive process—deriving results from given laws—is usually well illustrated in textbooks. In this book I have in addition tried to emphasize the inductive process, which involves the development of a general law, or a generally valid procedure, from the examination of a few specific cases. Sometimes, the study of one or more specific cases leads to a conjecture about a generally valid result which can then be derived by deduction from previously established results. This was the pattern in Chapter 7 for example, in showing that the complex locus of a network function is a circle when a single network element varies.

The generalizations arrived at by induction are always candidates for modification if upon examination they are found to be invalid under conditions not visualized when the generalization was originally made. This concept is illustrated in several places. Thus, after examining a few simple cases it is found that the forced response of a network has the same waveform as the excitation in each case. The generalization is then made by induction that this will be true for all excitations; the method of undetermined coefficients is based on this assumption. However, it is later found that the generalization is not valid for all waveforms—triangular waves, for example. Nevertheless, the generalization can be modified when the Fourier theorem is introduced, since the waveform is then expressed as a superposition of elementary waveforms for each of which the assumption is valid.

When a subject, such as circuit theory, reaches maturity, there is a tendency for textbooks to acquire some of the characteristics of an encyclopedia by treating each subject exhaustively. This robs the student of all the joys of discovery; he is given the complete story and all he is asked to do is to learn it. I have tried to avoid such a pitfall, and have left for the student the pleasure of developing many well-known results, rather than cataloguing for him the sum total of our knowledge in a given area. I have attempted to leave the student with the impression that circuit theory still has plenty of vitality, that it can still challenge his inventiveness and that it is not a "finished" subject.

In organizing the material I have often sacrificed systematic presentation for the purpose of treating a topic when the need for it arises naturally in a particular development, or when its relevance is a product of that development. Thus, complex numbers are treated, not on an *ad hoc* basis in anticipation of discussing the sinusoidal steady state solution—in which I, as authority, know that complex numbers are useful—but as a consequence of the demonstrated need for complex numbers arising from a study of the natural response. As another example, the reciprocity theorem is not introduced simply as another network theorem when these are being discussed, but arises naturally from the investigation of the transfer functions of two-ports. Again, the Laplace transforms of a derivative and an integral are not discussed because I happen to know this knowledge will be useful later: they are brought up as a consequence of a demonstrated need for that information in solving problems.

In conclusion, I would like to acknowledge my indebtedness to the College of Engineering of Syracuse University for the opportunities made available in the evolution of this book from notes. I have benefited also from the counsel of many of my colleagues in the Electrical Engineering Department.

NORMAN BALABANIAN

CONTENTS

THE FUNDAMENTAL LAWS

1.1 INTRODUCTION

1 AN electrical engineer is concerned with *energy* and *information*—their *generation*, their *conversion* from one form to another, their *transmission* from one place to another, and their subsequent *utilization*. You can give countless examples of these four tasks from your own experience. Electrical power is converted from water power, organic fuels, or nuclear fuel. It is then transmitted to homes or to industry, where it is utilized to provide light and heat, and to operate hundreds of different electrical devices, large and small. The information which we receive in the form of a television picture or a radio program is generated in a studio somewhere, is converted to a form suitable for transmission, is transmitted, and is finally reconverted to the desired form by the utilization devices which we call television and radio receivers. These are but two of the most familiar examples.

In performing these four operations, or functions—generation, conversion, transmission, and utilization of energy and information—the electrical engineer must deal with many aspects of the physical world. These operations are performed by interconnecting many physical devices and pieces of equipment. The electrical engineer, then, must be able to describe the electrical behavior of an interconnection of devices. The variables in terms of which this behavior is described are most often voltage and current, but also may be electric charge, magnetic flux, or electrical and magnetic field intensities. The description of the behavior of a device involves a statement about the relationship among the variables.

To give an example from another field, the behavior of a body of matter subjected to an external force is described in terms of the force and the velocity (or acceleration), and we say that the force is equal to the mass of the body times the acceleration, which is the time rate of change of the velocity.

In describing the electrical behavior of devices, we would be hopelessly lost if we had to discover the relationships among the variables separately for every conceivable device. Instead, as in all of science, we try to establish a model; not a physical model but an abstract model, a conceptual scheme, a theory. If we are successful, the behavior of the model will be very similar to the behavior of the actual device or the interconnections of devices.

The way in which we develop this model, of course, is through observation and experimentation in the physical world. We make many repeated observations of the same or of similar phenomena, and we note the trends that our results follow. We try to vary one of the factors that might affect the result, holding other factors constant, and we note how the results change.

A classical example of this procedure is Galileo's experiment to determine how long it takes a body to fall from a given height. He dropped bodies of various sizes, shapes, and weights, and measured the time it took for each one to fall a given distance. He found that, for the most part, all the bodies took the same length of time, except the lightest ones such as pieces of paper or feathers, which took longer. Even these light bodies, however, tended to fall faster if the air through which they were falling was not turbulent or windy. On the basis of these experiments, Galileo contended that all bodies, even the lightest ones, would take the same length of time to fall a given distance, provided they were in a vacuum, so that the air resistance to the fall was removed.

This conclusion appears perfectly reasonable if it is assumed that the acceleration of gravity g is a constant. With such an assumption, if a body is dropped with no initial velocity, the distance it falls in a time t is $x = gt^2/2$. This expression is independent of the weight, size, and shape of the body. If the distance of fall is fixed, then the time it takes will be constant, independent of the properties of the body.

On this basis we can postulate, or hypothesize, that the acceleration of gravity is constant. We have just seen that, on the basis of this postulate, we can predict something which is pretty closely confirmed by experimentation. Armed with this information, we can now go on to predict other results concerning motions of bodies under the influence of gravity. Perhaps, as we go along, we will need to make additional hypotheses. If, at any stage of the development, a result predicted from our assumptions fails to agree with our observations (assuming that the error made in the measurement is less than the observed discrepancy), then we must re-examine our postulates and modify them, in order that the observed results can again be predicted from the new postulates.

What we have described is the scientific method. The first step is to make observations, to experiment. The next step is to postulate a model of the physical world (that portion of it which is under observation) and to generalize the results of the observations to this model. This is done by stating a relationship among the variables used to describe the phenomenon under consideration. This relationship comes to be known as a law. Newton's law relating force and acceleration, and Ohm's law (which we shall discuss below) relating voltage and current, are examples.

On the basis of the postulated laws, we next try to compute other relationships among the variables in the model. In other words, we *derive additional results* based on the assumed laws. Finally, we *test the validity* of these results by making more observations. If these observations confirm our predictions, we gain confidence in our original "laws" and in our model. On the other hand, if the observations fail to confirm our predictions, we seek modifications of the model.

The conservation laws of science are perhaps the most fundamental and universal postulates in physical science. The law or principle of *conservation of energy* states that *the total energy in a closed system must remain constant.* Men have tried unsucessfully to find situations in which these principles are violated, but they have not succeeded. Many "perpetual-motion" machines have been invented, but nature has always foiled the inventor. Any additional postulates which we may introduce in developing a theory of electric networks must be consistent with the conservation laws.

In addition to the conservation laws, any theory we develop must be consistent with the *atomic theory of matter.* This theory postulates the existence of atomic charge of two kinds—positive and negative. Charge is associated with the elementary particles known as electrons and protons. *All electrical effects are caused by electric charges, their spatial distribution, and their motion.*

Charge in motion constitutes an electric current. This is a qualitative statement; let us give it some quantitative content. For simplicity, assume that a large number of particles, carrying the same kind of charge, are moving at the same speed in the same direction. Consider a surface perpendicular to the direction of flow. During any interval of time, a certain number of these particles will pass through the surface. We define the *current* through the surface as the amount of charge (number of particles times the charge of each one) passing the surface per unit time.

Although this simplified picture is not the most general way in which charge can flow, it is approximated in some very important situations. Even if all the charges are not moving at the same velocity, a satisfactory description is often obtained if we determine the average velocity and assume that all the charges are moving at the average velocity. This might be done, for example, in an electron beam. In a metallic conductor the moving charges are continually colliding with the fixed nuclei, but there is an average *drift* velocity at which each of the charges can be assumed to be moving. For our purposes, this simplified description of electric current is adequate. As you go on later to the study of electric fields and electronic devices, you will get a more complete understanding of the concept of current.

The theory which will be developed in this book will serve to explain electrical phenomena which take place at the terminals of electrical devices. It will not explain such phenomena as the propagation of electromagnetic waves in the atmosphere or in waveguides. It will not explain the microscopic phenomena associated with electric charges which are either stationary or in motion. These phenomena are explained by a much more extensive conceptual scheme called *field theory.*

In developing network theory, we shall have occasion to mention some concepts, such as potential difference and magnetic flux, which are in the proper domain of field theory. We shall not digress for the purpose of developing these concepts, together with others on which they are based, since to do so would take us far afield. You will be introduced to these concepts in your later study of field theory, at which time you should make an effort to determine the relationships between the two theories, their regions of validity, and their limitations.

1.2 KIRCHHOFF'S LAWS

Let us now proceed to the task of developing a model which will explain and predict the electrical behavior of an interconnection of electrical devices. The foundation of electrical network theory was laid by Kirchhoff in the 1840's when, after repeated observations, he announced the laws which today bear his name. Let us here conceptually repeat Kirchhoff's experiments.

Before we proceed to this task, we should define some terms which we shall need. A *circuit* is an interconnection of electrical devices—actual physical devices in the real world. A *branch* is a component part of a circuit which is characterized by two terminals to which connections can be made. A *junction* is formed where two or more branches are connected. In the physical world the branches of a circuit are interconnected by conductors, usually (but not always) in the form of wires.

Let us now consider an electrical circuit and look at one of its junctions. Such a junction is shown in Fig. 1-1. We are not concerned

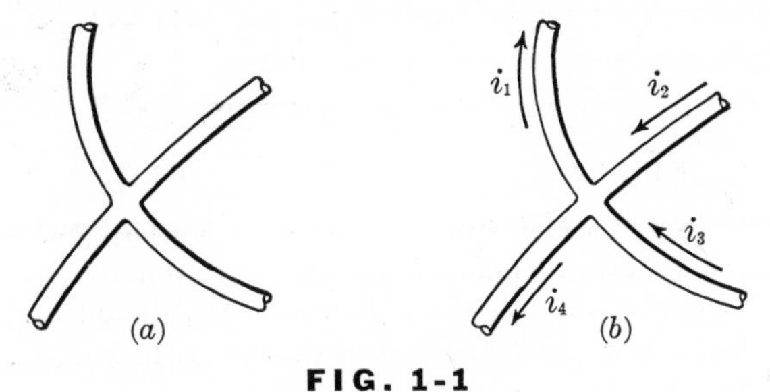

FIG. 1-1

Current junction

with the devices that may be connected to the other ends of the wires, as long as there is current flowing in them.

Now, by some means, let us measure the currents in the branches at each instant of time. At any given time some of the currents may be flowing toward the junction, while some may be flowing away. In order to describe the current at any time, then, we should tell both how much current is flowing and in what direction. This is most easily done by choosing a *reference* for the current in each wire. We can agree to say that the current is *positive* when it is flowing away from the junction, and *negative* when it is toward the junction. We can also choose just the opposite convention if we wish. More generally, we can choose one of the directions as the reference in some of the conductors, and the opposite direction in others. As a matter of fact, since the wires go from one junction to another, once we choose a reference for one of the currents—say, toward one of the junctions—this reference will automatically be away from the other junction.

How we choose the reference for current is unimportant. What is important is to recognize that a reference *must* be chosen. The current reference is indicated by drawing an arrow beside the wire. One possible choice for the references is shown in Fig. 1-1(*b*). What does the arrow, say on i_1, indicate about the actual direction of current flow? The answer is, *nothing.* It just means that we have agreed to call the current positive when it is flowing in the arrow direction, and otherwise, negative. For example, Fig. 1-2 shows the variation of current i_1 with time.

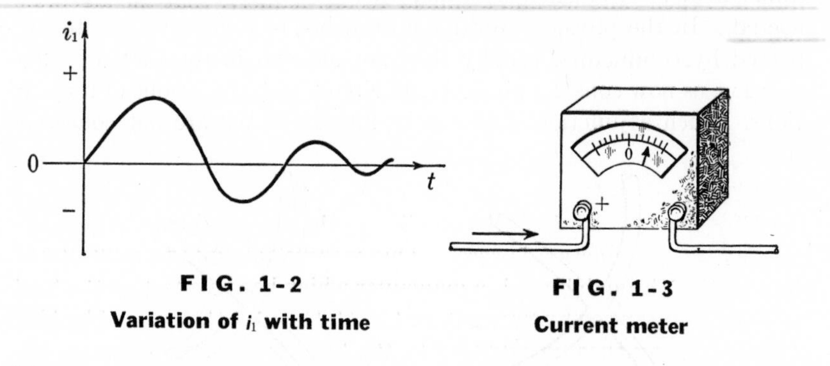

FIG. 1-2
Variation of i_1 with time

FIG. 1-3
Current meter

Whenever the curve is above the horizontal axis, the current is flowing away from the junction in Fig. 1-1, and whenever the curve is below the horizontal axis, it is flowing toward the junction.

If we wish to measure the current, we should have a meter that tells us both the amount of the current and its direction. Let us think of a meter with a needle that can swing in either direction about its equilibrium point, which corresponds to zero current. A sketch is shown in Fig. 1-3. The meter can be connected in two ways in the branch. When it is connected one way, the needle will swing in one direction for a given current. If the connections of the meter are reversed, the needle will swing in the other direction for the same current. Here, again, we have a choice of how to connect the meter. How we do it is immaterial, as long as we agree on the correspondence between the direction of the current and the direction of swing of the needle. This agreement is made by placing some mark on the terminals of the meter, say a + on one of the terminals, which indicates the direction of swing of the needle when the current enters the + terminal at a given time. Let us agree to connect the meter so that the needle swings to the right when the current enters the + terminal. The current reference for a given branch is then toward the + terminal of the meter when it is connected in this manner.

Having disposed of these important considerations, let us return to our experiment of measuring the currents at the junctions. Suppose the meters are connected according to the references shown in Fig. 1-1(*b*). We will find, as did Kirchhoff, that

$$i_1 - i_2 - i_3 + i_4 \doteq 0 \tag{1-1}$$

at all instants of time, within the limits of accuracy of the meters. If we repeat this experiment at other junctions with different numbers of branches, each time we find a similar result, namely, that the algebraic sum of the currents is approximately zero.

Based on this observation, we now make the postulate that such a result is true at all junctions of all electrical circuits. This result is called *Kirchhoff's current law.* It states that *the algebraic sum of all currents leaving a junction is zero at each instant of time, for all junctions of an electric circuit.*

Of course, we might just as well say, the algebraic sum of the currents *entering* a junction is zero. This merely amounts to multiplying the equation through by −1, a maneuver which has no effect on the final result, as you can see by trying it on Eq. (1-1). Another way of looking at this is to say that, in addition to the branch-current references, the junction itself has a reference or an orientation. In our statement of Kirchoff's law, we have assumed that "away from the junction" is the junction reference. If the branch-current reference coincides with the junction reference, the symbol for the corresponding current is preceded by a + sign; otherwise, a − sign. This scheme was used in writing Eq. (1-1). Note that there is nothing basic about this junction reference; there is no rule to be memorized here.

An important property of the flow of charge can be demonstrated by means of Kirchhoff's current law. Let us integrate Eq. (1-1) over a time interval t_1 to t_2. The result will be

$$\left(\int_{t_1}^{t_2} i_1\,dt + \int_{t_1}^{t_2} i_4\,dt\right) - \left(\int_{t_1}^{t_2} i_2\,dt + \int_{t_1}^{t_2} i_3\,dt\right) = 0$$

From the definition of current, each of the integrals represents the charge transported through the corresponding branch toward or away from the junction (depending on the current references) in the time interval $t_2 - t_1$. Hence the equation states that the net charge transported toward or away from the junction over any interval of time is zero. That is, *no charge can accumulate at a junction, or be removed therefrom.* This conclusion is valid no matter how long the time interval is, even if it is infinitesimally small. The flow of charge at a junction must be continuous.

Sometimes this principle is referred to as the *conservation of charge.* It might be considered as the starting point from which Kirchhoff's current law itself follows as a logical consequence and, instead of choosing Kirchhoff's current law as a basic postulate, we can choose the law of conservation of charge. Then Kirchhoff's current law can be derived as a consequence.

There is a second relationship which bears Kirchhoff's name. In an electrical circuit the branches are connected and form closed paths. One such closed path is shown in Fig. 1-4. This may be the only one in

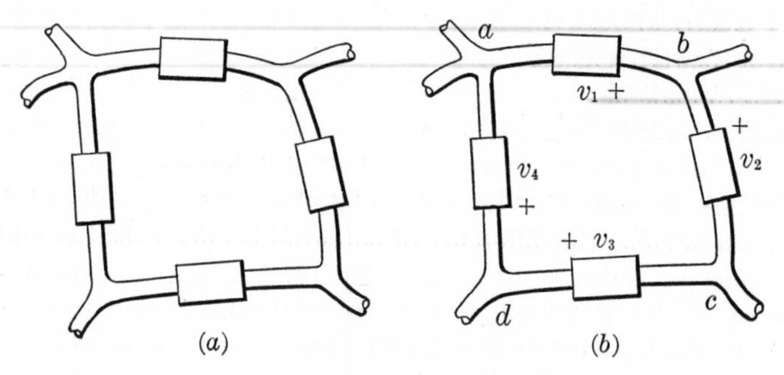

FIG. 1-4
Closed path

the circuit, or there may be others. At the present we are not interested in what the branches consist of, so we just show them as rectangles (sometimes referred to as "black boxes"). What we plan to do is to measure the voltage of the branches at each instant of time.

At this point we should clarify the terminology used in connection with the concept of voltage. A number of different terms, such as potential difference, electromotive force (emf), and electromotance are used to describe concepts related to that of voltage. The electrostatic *potential difference* between two points is defined as the work done by a unit positive "test" charge in going from the first point to the other. The reason why work must be done is because other charges exert a force on the test charge. This force is a consequence simply of the presence and distribution of the charges. If the charges which are exerting a force on the test charge are in motion, they will exert on the test charge an additional force which is not accounted for by the potential difference. The concept of *electromotive force*, or *electromotance*, incorporates this additional component of force on the test charge.

In circuit theory we regard an electrical device as a branch which has an external pair of terminals. The only electrical effects we observe take place at the terminals of the branch. We do not observe the components of force on electric charges owing to the positions and motions of other individual charges; we observe simply the integrated effects of these forces. These integrated effects are described in terms of the *voltage* of the branch. The voltage is, thus, similar to a potential difference as far as its effect at the branch terminals are concerned, except that it includes the integrated effect of forces due to the motions of charges. Some authors go to the extent of using the two terms, voltage and potential difference, synonomously. This is valid as long as we do not inquire into the electrical behavior *within* a device. In this book we shall avoid the use of the term electromotive force. We shall define the voltage at a pair of terminals as the total work done against electrical forces by a unit positive charge in going from one terminal to the other. We shall treat it as being synonymous with potential difference.

Let us now return to the contemplated experiment in Fig. 1-4. At any instant of time, the difference of potential between the two ends of a branch may be either positive or negative. In order to describe the voltage of a branch, we must tell both how much the voltage is, and which end of the branch is at a higher potential. This is most conveniently done by choosing a reference for the voltage of each branch, just as we did in the case of current. We can agree to say the voltage of a branch, say the one between junctions a and b in Fig. 1-4, is positive when the terminal at a is at a higher potential than the terminal at b. Or we can choose the opposite reference. The actual choice of reference is immaterial; but it is important to recognize that a reference *must* be chosen.

Any symbol can be used to show the voltage reference. Sometimes an arrow is used, but this can be confused with the reference for current. In this book we shall use a + sign at one end of the branch to designate the reference, as shown in Fig. 1-5(a).

A system of double subscripts is also used for the designation of voltage (as well as current). If the terminals of a branch are labeled as in Fig. 1-5(b), then the symbol v_{ab} is positive; terminal a is at a higher potential than terminal b. Clearly, $v_{ab} = -v_{ba}$. Similarly, the symbol i_{ab} indicates that the current-reference arrow is directed from a to b. Also, $i_{ba} = -i_{ab}$.

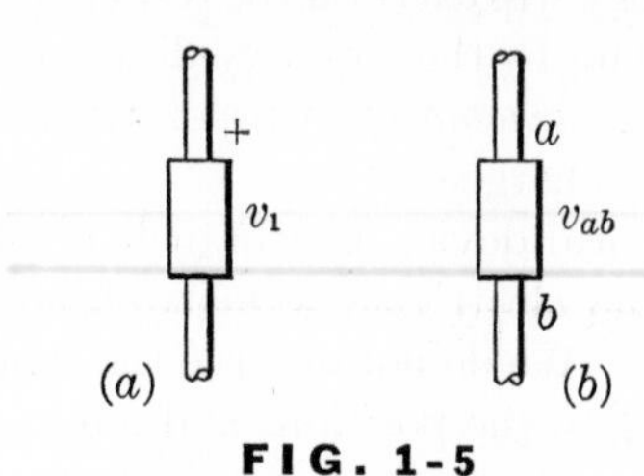

FIG. 1-5
Voltage reference

As in the case of the current, the voltage reference tells nothing about the sign, or *polarity*, of the actual voltage. It just means that we have agreed to call the voltage positive when the + terminal is at a higher potential than the other one.

To measure the voltage we again need a meter whose needle can swing in both directions. Again the meter can be connected in two ways; which way we connect it is immaterial, as long as we agree on the correspondence between the polarity of the voltage and the direction of swing of the needle. Again we place + on one terminal of the meter, which indicates the direction of swing of the needle when the terminal of the branch at a higher potential is connected to this terminal of the meter. Let us agree to connect the meter so that the needle swings upscale when the higher-potential terminal of the branch at any given time is connected to the + terminal of the meter. Thus the + which indicates the voltage reference and the + on the meter fall at the same end of the branch.

Now let us return to the experiment and measure the voltages of the branches that form the closed path in Fig. 1-4, with the meters connected according to the references shown. We will find, as did Kirchhoff, that at each instant of time

$$-v_1 + v_2 - v_3 + v_4 \doteq 0 \tag{1-2}$$

within the limits of accuracy of our meters. If we repeat this experiment on other closed paths, or *loops*, with different numbers of branches, we will find a similar result—that the algebraic sum of the voltages is zero. Based on these observations, we now make the postulate that this is a general result which applies for all closed loops in all electric circuits.

This result is called *Kirchhoff's voltage law*. It states that *the algebraic sum of all branch voltages around a closed loop is zero at all instants of time, for all loops in an electric circuit.*

In summing the voltages around the loop, we have two possible orientations to follow—clockwise or counterclockwise. Thus the loop also has a reference, in addition to the voltages. This is analogous to the junction reference in the case of the current law. In writing Eq. (1-2), we implicitly took the loop reference to be clockwise. There is nothing basically important about this; if we had chosen the opposite reference, the equation would have been multiplied through by -1.

Kirchhoff's two laws are the fundamental postulates which form the cornerstones of network theory. It is possible to adopt a point of view different from the one we have adopted and to say that Kirchhoff's laws are experimental or empirical laws. But we cannot possibly test the laws at all junctions and all closed loops in the world. Nevertheless, all the measurements we make at junctions or closed loops always lead to the same results, within experimental accuracy. Thus, our measurements suggest universal relationships; we then assume that these relaships, or laws, are universally valid, and we pursue their consequences.

Notice that these laws are not dependent on the constituents of the branches which form the loops or the junctions. They depend only on the existence of the loops and junctions. This is a property of the geometry, or *topology*, of the network. Many important results about networks can be derived merely from topological considerations, without consideration of the types of branches involved. We shall discuss some topological concepts in the next chapter.

1.3 NETWORK PARAMETERS

To continue with the development of a model of electric circuits, we must now consider the detailed relationships among the variables that describe the behavior of the circuit. Let us consider a branch of a circuit as illustrated in Fig. 1-6. The effect of the rest of the circuit to which the branch is connected is to cause a current to flow in the branch, a voltage to appear at the terminals, and some charge to accumulate on the terminals. These variables are not all independent, however. Furthermore, the current, or the voltage, or the charge will depend on the properties of the branch. Let us enumerate the quantities on which the voltage, say, can

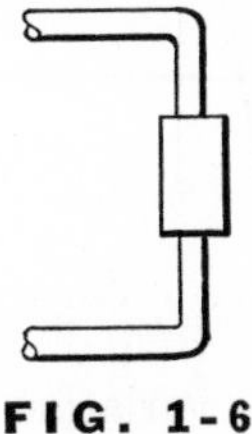

FIG. 1-6

Circuit branch

be expected to depend. First of all, the voltage will depend on the current and the charge. We might also expect it to depend on whether the current is changing, or how fast it is changing. Then we would also expect it to depend on properties of the material of the branch, and on its geometrical shape, or dimensions. It can conceivably depend also on the temperature. We would not, of course, expect the voltage to depend explicitly on the geographic location of the circuit, or atmospheric pressure, or the phase of the moon, or other such variables!

Resistance Parameter

Let us set up an experiment to measure the voltage of a branch as we vary the current, holding constant all other variables which might affect the voltage. The current is to be varied slowly from one steady value to another, so that the rate of change of current will not be a factor. Depending on what constitutes the branch, many different types of relationships between the voltage and current will be obtained. Two possible curves, which correspond to some common electrical devices, are shown in Fig. 1-7. The curve in Fig. 1-7(a) is typical of a device called

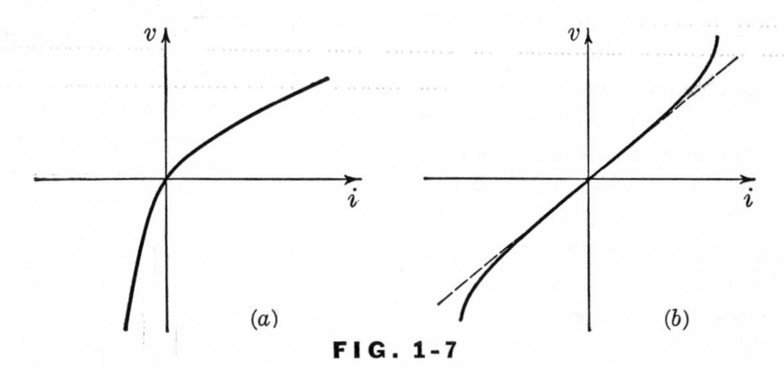

FIG. 1-7

Possible v-i relationships of a branch

a *diode.* The variation of voltage with current is seen to be quite different for negative values from what it is for positive values. Because of this lack of symmetry of the voltage-current (v-i) relationship, such a branch is called a *unilateral branch.*

The curve in Fig. 1-7(b) does not have this one-sided behavior. It is only natural to call the corresponding branch a *bilateral branch.* In the vicinity of the origin in Fig. 1-7(b), the curve is almost a straight line. It begins to deviate farther and farther from a straight line as the magnitude of the current becomes greater. This departure of the curve from linearity depends on the branch component.

The first extensive measurements of electricity were made by Georg Simon Ohm early in the nineteenth century. His experiments involved finding the relationship between voltage and current in lengths of conducting materials of different geometrical shapes, using a galvanic cell as a source of current. The results he obtained were similar to the curve in Fig. 1-7(b). The range of voltage that Ohm had available, however, was limited by the voltage obtainable from a galvanic cell, so that his measurements fell near the origin of Fig. 1-7(b). On this basis he stated the famous law named after him, asserting the proportionality of voltage and current. For metallic conductors this proportionality remains valid up to very high values of current. For other materials, however, the nonlinear nature of the curve becomes evident at relatively low values of current.

Having made these observations, we are now ready to start building our model. We postulate the existence of a hypothetical branch whose v-i relationship is a straight line for all values of current. For this branch we can write

$$\boxed{v = Ri} \tag{1-3}$$

where R is the slope of the straight line. It is a constant which we call the *resistance parameter*, or simply the *resistance*. The hypothetical branch itself is an *element* of the model. It is called a *resistor*. Note the distinction between a resistor, which is a branch, and the mathematical quantity R, the resistance. Dimensionally, R is the ratio of voltage to current; in the meter-kilogram-second system, its unit is the *ohm*. The expression in Eq. (1-3) is *Ohm's law*. This expression can also be written

$$i = \frac{1}{R}v = Gv \tag{1-4}$$

where $G = 1/R$ is called the *conductance*.

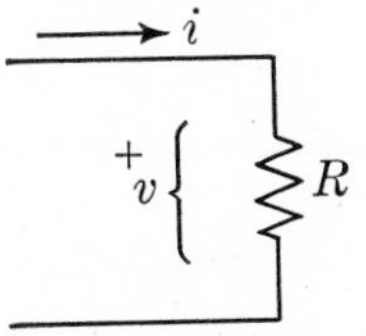

FIG. 1-8
Symbol for *R*

The graphical symbol representing a resistor is shown in Fig. 1-8. Recall that v and i are algebraic quantities whose references can be chosen arbitrarily. If we want R to be a positive constant (which we do), then Eq. (1-3) will be valid only for a particular set of references. These are shown in Fig. 1-8. If either one of these references is reversed, then the sign in Eq. (1-3) must be changed to $v = -Ri$.

It is possible to take another viewpoint in examining the v-i relationship of a circuit branch. Either of the curves in Fig. 1-7 can be expressed analytically as $v = f(i)$, where the function f depends on the curve. In the case of a straight line, we have written this as $v = Ri$, where R is a constant independent of i. However, R does depend on all the other quantities, such as temperature, type of material, etc., which were held constant. The variation of R with any one of these other variables can be obtained by holding all other variables constant except this one, and determining experimentally the ratio of v to i for different values of this one variable. We can then plot R against this variable. The kind of curve we get will depend on the kind of branch under consideration.

Suppose we wish to find the dependence of R on the properties of the material of which the branch is constructed. The experimental procedure would be to take pieces of different materials having the same shape and to measure R, holding temperature and other variables constant. If we do this, we shall find a wide range of values among the selected materials. The property of a material which expresses how well it conducts current is given the name *conductivity*. It is measured in units of *mhos per meter*. (A mho is ohm spelled backward.) If a material has high conductivity, it conducts current well. The reciprocal of conductivity is called the *resistivity*. Its unit, being the reciprocal of that of conductivity, is the *ohm-meter*. When the resistivity is small, the material is a good conductor. The best conductors are metals, and, of these, silver, copper, and aluminum are the best, in that order.

Let us also say a word about the variation of R with temperature. We find, by performing experiments on specimens of material with a given shape, that the resistance of metals increases with temperature along a straight line in the temperature range approximately 200° K above and below room temperature. This is no longer true at higher and at lower temperatures. The resistance of other materials, such as carbon, also varies with temperature, but in the opposite direction, being lower at room temperature than at higher temperature.

Another experimental observation concerns the variation of the resistance of a conductor with the rate of change of current. When the current is varying at a fast rate (as, for example, when the current varies as $\sin \omega t$ with a large value of ω), then the current is not uniformly distributed through the conductor. It tends to be concentrated near the surface with very little current flowing near the center of the conductor.

This effect is so pronounced that the center part of the conductor can be removed without greatly changing the current-carrying ability of a given conductor. We refer to it as the *skin effect.*

Having made these observations about the physical world, let us turn back to Eq. (1-3) in which R is a constant. We have seen that, in the actual physical world, there is no electrical branch whose ratio of voltage to current is really constant but depends on several variables. We postulate the existence of a hypothetical element, however, which is characterized by a parameter which we call the resistance. This element constitutes one of the building blocks of the model which we are constructing.

Inductance Parameter

To continue with our development of a model, we turn again to observations in the physical world. In our preceding observations, the time variation of current was not a factor. The measurements we made involved only steady values of current. Let us now consider making observations of the relationship between the voltage and current of a branch when these are varying with time. The results will again depend on the particular branch under observation. If the branch is again a length of conducting material, nothing new will be observed (except for the previously mentioned skin effect). For two different types of branch, however, radically different behavior is observed.

Let us consider a branch consisting of a coil of wire wound on a core. The coil may consist of one or more turns, and the core may be solid or it may be simply air. Observations of the v-i relationships of such a branch were first made almost simultaneously, approximately five years following the publication of Ohm's law, by Michael Faraday and by Joseph Henry. If the voltage across the branch is increased suddenly, we find that the current does not follow this change instantaneously, but builds up gradually. Furthermore, if the voltage is suddenly reduced, the current again decreases slowly. This property is similar to the property of inertia of a moving body. The rate at which the current follows a sudden change in the voltage is found to depend on the geometry (size and shape) of the coil and the material of the core.

Faraday advanced the idea of *magnetic flux* in an effort to explain his observations. It is difficult for us to give a definition of magnetic flux which is both accurate and satisfactory, without invoking additional field concepts. It is enough here to say that whenever a current exists, a magnetic field exists in the space surrounding the current. In the case of a permanent magnet, the field is produced by the electronic spin inside the magnetic material. One aspect of the magnetic field is the magnetic flux, which is given the symbol ϕ and is measured in *webers*. An elementary way of visualizing the flux is to think of lines which fill up all of space and whose direction at a point represents the direction of the magnetic field at the point. The density of the lines passing through any area is a measure of the strength of the magnetic field there. The total number of lines passing through an area is the flux through that area. The magnetic lines of flux are closed upon themselves. This fact is readily observed by noting the distribution of iron filings in the vicinity of a magnet.

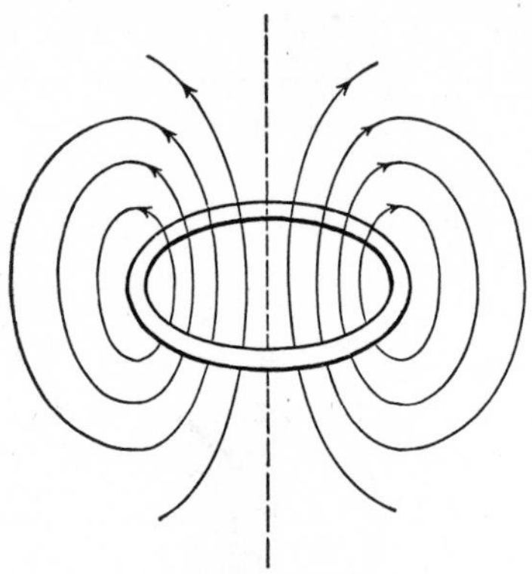

FIG. 1-9

Magnetic flux

Figure 1-9 shows a single turn of wire which is carrying a current. (The source of current is not shown.) Some of the flux lines are shown in the figure. If we think of the flux lines as piercing the plane of the turn of wire with a certain density, then the total flux is the number of lines which pass through, or link, the turn of wire. We refer to this total flux as the *flux linkage*, and we give it the symbol λ. If a coil of wire has N turns and the same flux ϕ links each turn, then the flux linkage will be $N\phi$. The flux linkage has the same unit as flux, although we often refer to its unit as a *weber-turn*.

Let us consider the effect of a changing magnetic flux on the current flowing in a coil or in a closed conducting path. One way of obtaining a changing flux is illustrated in Fig. 1-10. A closed path is formed by sliding a conductor on a pair of conducting rails. A voltage v is applied across the other end of the rails through a battery and potentiometer arrangement. A uniform magnetic field directed into the paper is present. (The source of this field is unimportant for the present discussion; assume it is caused by a large permanent magnet.) As the slide-wire moves to the right, the area in the plane of the closed path is increasing, and so is the flux linking the path. The rate of increase of flux depends on the velocity of motion. We observe that the current decreases during the motion. This is partly to be expected, since the length of rail, and so the resistance, included in the path is increasing. The observed reduction of current, however, is much greater than that which would be caused by this increase in resistance.

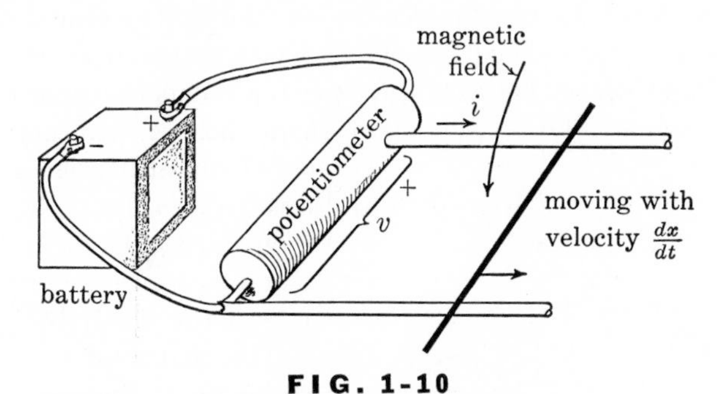

FIG. 1-10

An experimental observation

By means of the potentiometer, it is possible for us to increase the voltage v during the motion. In this way we can maintain the current constant. We find that the increase in voltage needed to maintain a constant current (above the value needed to account for the increased resistance) is almost exactly equal to the rate of change of the flux linkage.

In this experiment the change in flux linkage is caused by a motion which changes the dimensions of the closed path. It is possible for the changing flux linkage to be caused by other means also. For example, if a bar magnet is rapidly inserted through or withdrawn from a coil of wire, the flux linkage is changed by changing the strength of the magnetic field, rather than by changing the area. In this case a change in the current is observed (using either a galvanometer or an oscilloscope). The more rapidly the flux linkage is changed, the greater the current. If we use coils with different-sized wires made with different materials, so that the resistance is different, we find that the current is different for the same rate of change of flux linkage. This fact leads us to suspect that it is not the current which is fundamentally related to the changing magnetic flux.

In his experiments Faraday concluded that a voltage is *induced* by the changing magnetic flux, its value being very closely the same as the time rate of change of the flux linkage. The differences in the currents which are observed with the different coils are then easily explained in terms of the resistances.

From such observations Faraday stated the following law which we now refer to as *Faraday's law*.

$$v = \frac{d\lambda}{dt} \tag{1-5}$$

This law is one of the basic postulates on which all of electrical science is founded. You may have previously seen such an expression written with a negative sign. The appropriate sign will depend on the references chosen for voltage and flux linkage.† Consider the situation shown in Fig. 1-11. Faraday's law will be written with a positive sign if the voltage and flux references are chosen as illustrated.

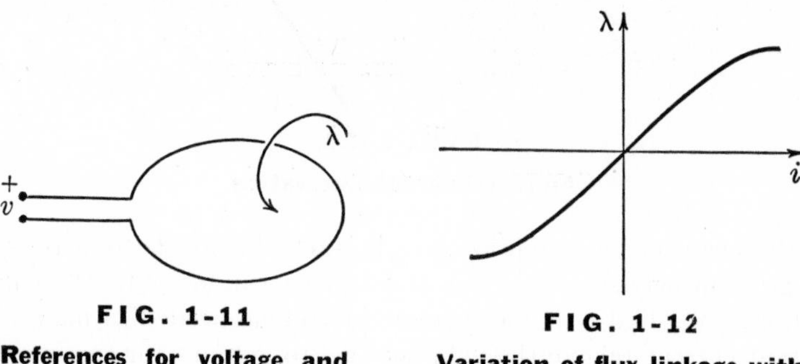

FIG. 1-11

References for voltage and flux linkage

FIG. 1-12

Variation of flux linkage with current

We still do not have a relationship between the voltage and current of a circuit branch which consists of a coil of wire. We noted previously, however, that a magnetic field will exist whenever there is current. We can, therefore, expect that the flux linkage of a coil is dependent on all the currents flowing in neighboring circuits. In the special case in which the flux linkage is produced only by the current in the coil itself, we can write

$$\lambda = f(i) \tag{1-6}$$

where i is the current and f is a function describing the variation of flux linkage with current. This function can be expected to depend on the geometry of the core and the magnetic properties of the core material.

In a given case, if we were to plot the flux linkage as a function of the current, a curve such as the one in Fig. 1-12 might be obtained.‡

† The law is usually stated in terms of the electromotive force. As previously noted, this concept is not needed in circuit theory; consequently, we shall avoid introducing it.

‡ You will probably question how the flux linkage is measured. It is not our purpose here to give a complete account of all the empirical evidence we have on which our model is based (we did not describe Ohm's equipment either). What we wish to emphasize is the method of arriving at a model. The complications introduced by the phenomenon of hysteresis are also neglected.

The function $f(i)$ in Eq. (1-6) is an analytical representation of the curve. Near the origin the curve is approximately a straight line and hence can be represented by

$$\lambda = Li \tag{1-7}$$

where L is the slope of the line, a constant.

Just as we discussed in the case of the resistance element, it is possible to write the functional relationship between λ and i as $\lambda = Li$, even if the curve is not a straight line. In this case L will be not a constant but a function of current, and should be written $L(i)$ to indicate this functional dependence.

Let us now substitute Eq. (1-7), with L assumed to be a function of i, into Faraday's law. The result will be

$$v = \frac{d}{dt}(Li) = \frac{d}{di}(Li)\frac{di}{dt} = \left(L + i\frac{dL}{di}\right)\frac{di}{dt} \tag{1-8}$$

To obtain the final expression, we used the formulas for differentiation of an implicit function, and differentiation of a product. Of course, if L is a constant (independent of i), then the second term on the right side will vanish.

We now postulate the existence of a hypothetical circuit branch whose v-i relationship is given by

$$v = L\frac{di}{dt} \tag{1-9}$$

in which L is a positive constant called the *inductance* (or self-inductance). The hypothetical branch is an *element* of the model and is called an *inductor*. L has the dimensions of a voltage divided by a time derivative of current. From Eq. (1-7), however, it also has the dimensions of flux linkage per unit current. Its unit is the *henry*, named in honor of Joseph Henry. The symbol which represents an inductor is shown in Fig. 1-13. This is suggestive of a coil of wire. In order for Eq. (1-9) to be valid with a positive sign, the references for voltage and current must be chosen as shown in Fig. 1-13. If either one of the references is reversed, the sign of the equation must be changed.

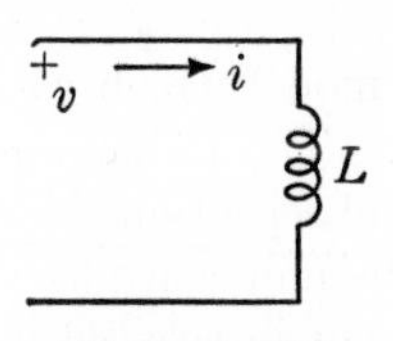

FIG. 1-13
Symbol for *L*

Equation (1-9) gives the voltage in terms of the current. If we want to solve for the current in terms of the voltage, we should integrate

Eq. (1-9). In most problems we are interested in the behavior of a circuit after some particular instant of time. Let us choose this time as the origin, $t = 0$, and integrate Eq. (1-9) from zero to some arbitrary time t.

$$\int di = \frac{1}{L}\int_0^t v\,dt \tag{1-10}$$

A brief comment on this expression is in order. In the first place, the variable of integration of the right-hand integral is t, while the upper limit is also t. In any definite integral, the variable of integration is a dummy variable; it takes on all values from the lower limit up to the upper limit. Hence, to avoid confusion, let us change the dummy variable of integration to something else, say x. As for the left-hand integral, its lower limit should be the value of i at $t = 0$, and its upper limit the value of i at t. These will be labeled $i(0)$ and $i(t)$, respectively. With these comments, Eq. (1-10) becomes

$$i(t) = \frac{1}{L}\int_0^t v(x)\,dx + i(0) \tag{1-11}$$

We refer to $i(0)$ as the *initial value* of the current. It is sometimes convenient to deal with the reciprocal of inductance since $1/L$ appears in this expression. Just as a symbol is defined for the reciprocal of resistance, so also a symbol is sometimes used for the reciprocal of inductance; it is simply an inverted ell, ⅂. We shall not use this symbol very often.

Capacitance Parameter

The model which we have been developing has now grown to two elements. The inductance element was postulated as an idealization based on observations involving a coil of wire.

Let us now consider a branch which consists of a conductor which has been cut in two, as shown in Fig. 1-14. The effect can be emphasized by supplying to the ends so formed a set of conducting plates. We assume that there is a current in the branch and a voltage across it. The current is, by definition, the charge that flows past a given cross section of the conductor per unit time; that is,

$$i = \frac{dq}{dt} \tag{1-12}$$

where q is the charge flowing across any cross section of the conductor.

Compare this expression with Eq. (1-5), which is Faraday's law; the forms are the same. Charge and current are related in the same way as are flux linkage and voltage. But, whereas Eq. (1-12) is obtained simply by the definition of current, Faraday's law is arrived at after a considerable amount of difficulty. The reason for this is that electric charge and current appear to us as less abstract than magnetic flux and potential difference, since the latter are "field" concepts. Nevertheless, the analogy between Eqs. (1-5) and (1-12) should be noted.

The assertion that current is flowing in a branch which is cut in two may be perplexing. How can the charge jump across the gap? Actually, charge is transferred in and out of the terminals, but none is conducted across the gap. The continuity of current is maintained by the *displacement current*, a concept which was advanced by James Clerk Maxwell. As far as circuit theory is concerned, only the terminals of a branch are available for external observation. What transpires within a branch is of no concern. At the terminals, we do observe a current.

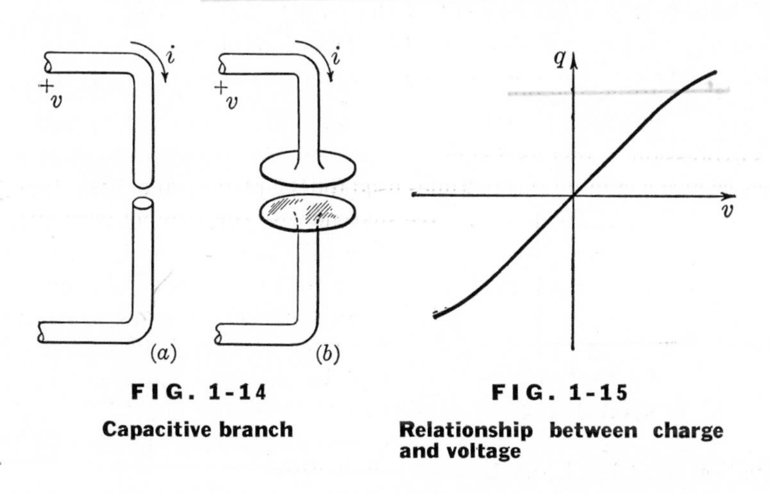

FIG. 1-14

Capacitive branch

FIG. 1-15

Relationship between charge and voltage

The question now arises as to how the voltage in Fig. 1-14 is related to the charge q. To answer this question, we must fall back on observations. It is reasonable to expect that the amount of charge will depend on the geometry (size and shape of the plates, and their distance apart) as well as the material between the plates. If we make a plot of the charge as a function of voltage with a wide variety of geometrical shapes and materials used for the plates, we find curves such as the one in Fig. 1-15. This curve is very similar to the one in Fig. 1-12 relating flux linkage and current, the main difference being that the present one does not seem to be as nonlinear; that is, the q-v curve is almost a straight line. An analytical expression for the relationship can be written as

$$q = Cv \tag{1-13}$$

If the curve is a straight line, then C is the constant slope of the line. In the more general case, C is a function of v. In either case, Eq. (1-13) can be inserted into Eq. (1-12) to give

$$i = \frac{d}{dt}(Cv) = \frac{d}{dv}(Cv)\frac{dv}{dt} = \left(C + v\frac{dC}{dv}\right)\frac{dv}{dt} \tag{1-14}$$

In case C is a constant, the second term on the right will vanish. Based on these observations, we now postulate the existence of a hypothetical circuit branch whose v-i relationship is given by

$$i = C\frac{dv}{dt} \tag{1-15}$$

in which C is a positive constant called the *capacitance*. The hypothetical branch is another element of the model and is called a *capacitor*. From Eq. (1-13), capacitance has the dimensions of charge per unit voltage. Its unit is the *farad*, named in honor of Faraday. Since a farad is an inconveniently large unit, a smaller fraction, microfarad (μf), is usually used.†

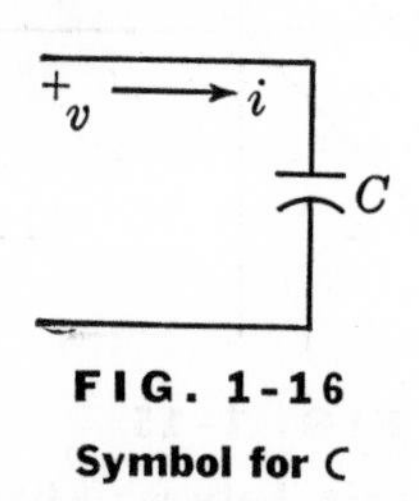

FIG. 1-16
Symbol for C

The symbol a for capacitor is shown in Fig. 1-16. This is suggestive of a pair of metal plates. In order for Eq. (1-15) to be valid with a positive sign, it is necessary that the references for voltage and current be chosen as indicated in the figure. If either one of the references is reversed, the sign of the equation must be changed.

It is often convenient to have the voltage across a capacitor expressed in terms of the current. In this case we must integrate Eq. (1-15), just as we did in the case of the inductor. Again we integrate between the limits 0 and t; the result will be

$$v = \frac{1}{C}\int_0^t i(x)\,dx + v(0) \tag{1-16}$$

where $v(0)$ is the initial value of the voltage. Note again that the

† To appreciate how large a unit a farad is, the area of the plates of a parallel plate capacitor with an air dielectric and a spacing of 1 millimeter must be of the order of 100 sq km for a capacitance of 1 farad.

dummy variable of integration is changed to x. The reciprocal of capacitance is given the symbol D; its unit is the *daraf* (farad spelled backward).

Sources

Up to this point our model contains three elements. We refer to them as *passive* elements, since, if they were left alone and no external voltages or currents were supplied, they would remain dormant and would exhibit no electrical effects. Since we know that currents can be made to flow in circuits, our model is still incomplete; we must still provide branches which are sources of electrical current or voltage.

In the physical world, many devices can be observed which seem to generate electrical energy. Some of these devices may be familiar to you, others still strange. Of course, the energy is not really generated; it is converted from some other form. There are the rotating electrical generator, the work horse which supplies electrical energy "in bulk"; the storage battery which converts chemical energy; the photoelectric cell, or electric eye, which converts light energy to electric current; the thermocouple which converts a difference in temperature to an electric voltage; and so on.

A common characteristic of these devices is the fact that a voltage or current which depends on some nonelectrical quantities is supplied at their terminals. We also observe that the voltage or current depends, to a greater or lesser degree, on the circuit which is connected at the terminals of the device. You have probably made at least one such observation if you have noticed the dimming of the electric lights when an additional electrical appliance (such as a refrigerator) is suddenly connected to the power line. The extent of this dependence on the external circuit varies from one physical device to another. Ideally, there should be no influence from the external circuit.

Based on these observations (and more detailed ones into which we shall not delve here), we postulate the existence of two hypothetical branches. One of these we call a *voltage source*, or *voltage generator*. It is characterized by a function of time $v_g(t)$ which specifies the voltage *wave shape*—its variation with time. This function of time is independent of the current flowing in the branch; that is, no matter what other branches are connected at the terminals, causing varying amounts of current, the voltage will remain $v_g(t)$. A voltage source is symbolized by a circle with the symbol $v_g(t)$ written alongside and with the reference clearly shown, as indicated in Fig. 1-17(a).

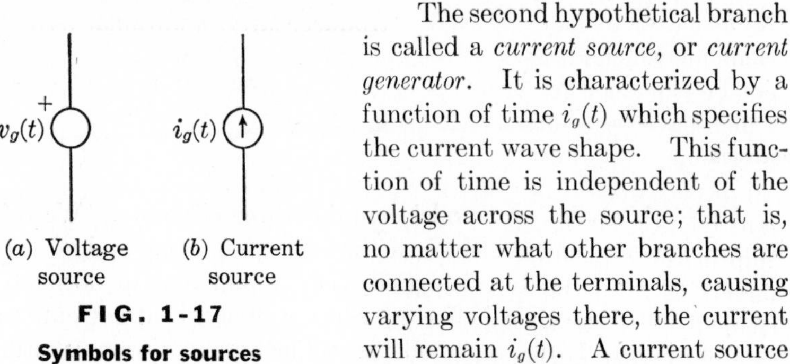

(a) Voltage source (b) Current source

FIG. 1-17

Symbols for sources

The second hypothetical branch is called a *current source*, or *current generator*. It is characterized by a function of time $i_g(t)$ which specifies the current wave shape. This function of time is independent of the voltage across the source; that is, no matter what other branches are connected at the terminals, causing varying voltages there, the current will remain $i_g(t)$. A current source is symbolized by a circle with $i_g(t)$ written alongside. The reference for the current is indicated by an arrow either alongside or inside the circle, as shown in Fig. 1-17(b). In contrast with the R, L, and C passive elements, the sources are referred to as *active* elements.

TABLE 1-1

Element	Parameter	Symbol	Equation
Resistor	Resistance R	$+\ v \rightarrow i$	$v = Ri$
	Conductance G		$i = Gv$
Inductor	Inductance L	$+\ v \rightarrow i$	$v = L\dfrac{di}{dt}$
	Inverse inductance Γ		$i = \dfrac{1}{L}\displaystyle\int_0^t v(x)\,dx + i(0)$
Capacitor	Capacitance C	$+\ v \rightarrow i$	$i = C\dfrac{dv}{dt}$
	Inverse capacitance D		$v = \dfrac{1}{C}\displaystyle\int_0^t i(x)\,dx + v(0)$
Current source		i_g	$i = i_g$
Voltage source		$+\ v_g$	$v = v_g$

The elements which we have now postulated are not the only conceivable ones which would be useful in our model of electric circuits. As a matter of fact, we shall introduce, later, additional hypothetical elements which are needed to explain observable phenomena which cannot be explained in terms of a model consisting only of those elements which have been introduced up to this point.

Some important general characteristics of our model are that the v-i relationship of the resistor is a straight line; so is the flux-linkage–current relationship of the inductor, as well as the charge-voltage relationship of the capacitor. Furthermore, the straight line extends in both directions from the origin. We say that a circuit consisting of such elements is a *linear, bilateral* circuit. Since each element is localized at a fixed place in the circuit, whereas the observed effect for which the element accounts is distributed, we say the circuit is *lumped.*

Let us collect in Table 1-1, for easy reference, the symbols and v-i relationships representing the elements which we have defined.

1.4 PHYSICAL COMPONENTS

There are no devices in the physical world which behave exactly like the hypothetical elements we have introduced. One of the greatest difficulties is the fact that all physical devices behave in a nonlinear manner, if the range of operation of the variables is extended far enough. Hence, if we use the model to represent a given physical situation, we must make sure that the conditions under which the model is valid are not violently violated.

Assuming that the linearity assumption is valid, it is still true that physical devices do not behave exactly like any one single element of the model. If our model is to be successful in explaining and predicting observable phenomena, however, we should be able to represent any electrical device in terms of combinations of the elements in the model.

As an example, let us find a model which will represent the device known as an electric battery. Consider the situation shown in Fig. 1-18(a). A rheostat is connected to the terminals of a battery in order to permit variation of the current, and measurements of the voltage and current are made. These data are presented as a plot of voltage against current in Fig. 1-18(b). The curve is approximately a straight line,

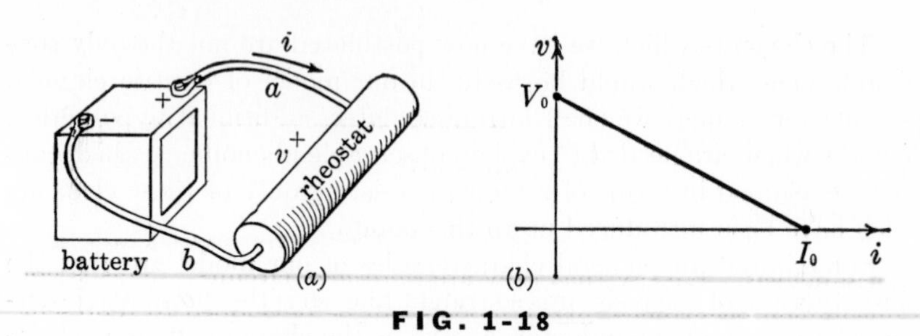

FIG. 1-18

Developing a model for a battery

with intercepts on both the voltage and current axes. The equation of a straight line with intercepts V_0 and I_0 is

$$v = V_0 - R_0 i \tag{1-17}$$

where R_0 is written for V_0/I_0.

The objective is to find a combination of elements of our model such that the relationship between the voltage and current at the terminals will be the same as that of the battery given by Eq. (1-17). Such a combination of elements is shown in Fig. 1-19(*a*). That Eq. (1-17) describes the v-i relationship at the terminals can be demonstrated by applying Kirchhoff's voltage law and Ohm's law. Figure 1-19(*b*) shows the more standard symbol for a battery. It consists of two parallel lines, one short and one long. The long line is at the terminal with positive polarity.

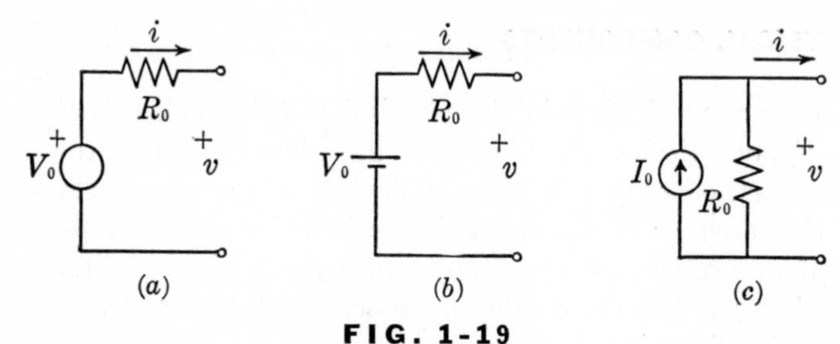

FIG. 1-19

Model of a battery

Figure 1-18(*b*) is plotted with voltage as ordinate and current as abscissa. In the same vein, Eq. (1-17) is written as if current were the independent variable. We can just as well interchange the axes in Fig. 1-18(*b*), however, or solve Eq. (1-17) for i. The result of this last step will be

$$i = \frac{V_0}{R_0} - \frac{1}{R_0} v = I_0 - \frac{1}{R_0} v \tag{1-18}$$

This expression suggests the circuit model shown in Fig. 1-19(*c*). You can verify that the v-i relationship is given by Eq. (1-18) by applying Ohm's law and Kirchhoff's current law at the upper terminal.

We have now obtained two circuit models which represent the same physical device. We say that they are *equivalent* circuits. This is not an uncommon occurrence; many models can be developed to represent the same physical situation.

The first task in seeking to explain and predict observable effects in the physical world is to establish a model; up to this point, it is this task with which we have concerned ourselves. Having established a model, we can then solve problems by *analyzing* the model. This will be our objective throughout most of the subsequent work. However, an engineer must *synthesize* systems and *design* physical equipment, to perform particular jobs. Starting from a given requirement, he first designs a model (paper model) of the equipment. Then he must find physical components which *realize* the ideal elements in the model. This task is difficult and can be done only approximately.

In designing electric circuits to be used in the actual world, we are very much interested in having physical components whose behaviors approach those of the ideal elements. For example, a piece of copper wire has a v-i relationship which quite closely approximates that of a resistor. The value of the resistance of a short length of wire, however, will be quite small. For a larger value of resistance, we need to have a longer length of wire. Of course it would be impractical to have long lengths of wire, as such, so the wire must be compressed into some compact form, for example, a coil.

A physical circuit component which is designed so that the relationship between its voltage and current is approximately the same as that of the ideal resistor element is called a *resistor*. It is not necessary that resistors be constructed with metallic conductors; many resistors are made of carbon.

Similarly, we give the name *inductor* to a physical component which is designed so that the relationship between its voltage and current is approximately the same as that of an ideal inductor. A physical inductor usually consists of a coil of wire, and is sometimes called a *choke*. Finally, *capacitor* is the name given to a physical component whose v-i relationship is approximately the same as that of the ideal capacitor element.

These names are somewhat unfortunate, since the same name is used for the ideal element in the model as for its physical embodiment. Whenever there may be any possibility of confusion, we shall precede the name with the words *physical* or *ideal* to distinguish the meaning. Thus, a physical inductor might take the form of a coil of wire, whereas an ideal inductor is a hypothetical element defined in terms of an equation representing the v-i relationship at its (paper) terminals. When no adjective precedes the name, we shall mean the ideal element.

It must be clearly understood that the symbols given in Table 1-1 do not represent physical components. A physical component may sometimes be approximated by one of the ideal elements, but, more often, a combination of ideal elements are needed to represent a physical component with any degree of accuracy. As an example, the electrical behavior of a coil of wire may, under appropriate conditions, be approximated by that of an inductor. The coil, however, is made of a metallic conductor and, hence, will have resistance as well. Thus a model for a coil of wire may take the form shown in Fig. 1-20(*b*), but even this may

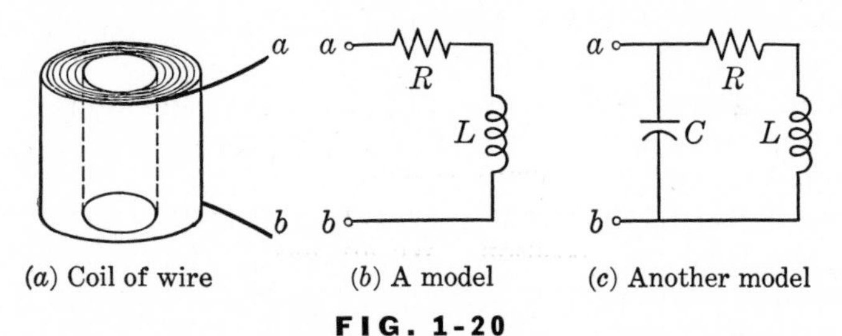

(*a*) Coil of wire (*b*) A model (*c*) Another model

FIG. 1-20

Model of a coil of wire

not be adequate in some cases. Since the turns which make up the coil are conductors in close proximity, there will be some capacitance between the turns; hence the model shown in Fig. 1-20(*c*) may be more adequate in some cases.

We shall not here concern ourselves with the task of designing a physical inductor, resistor, or capacitor to yield a specified value of inductance, resistance, or capacitance, respectively. For this purpose it would be necessary to call upon concepts of field theory for which we are not at present equipped. These gaps in your total education in electrical engineering are being temporarily left unfilled. Subsequently, you should always make an effort to integrate your knowledge in one area with that from another area.

1.5 ENERGY AND POWER

Our interest in electrical circuits lies in the fact that they can transmit energy and information from one place to another. Up until now, we have dealt almost exclusively with voltage and current as variables. The question arises as to how these quantities are related to energy, or power, and information. To answer this, as far as information is concerned, is not easy, because "information" is such a vague concept. The concept, however, can be made quantitative, and the "information" aspect of a voltage or current is a matter of great concern to an electrical engineer. We shall not reach a level of competence which will permit us to discuss such topics in this book. In this section we shall concern ourselves with energy—how it is associated with and modified by the elements in an electrical network.

The concept of *energy* is also a very difficult one to define in its broad generality. We say that energy is the ability to do work, but this definition does not attach any quantitative meaning to energy. Perhaps our earliest contact with energy is with mechanical energy. Here we find two forms: *kinetic energy* associated with motion, and *potential energy* associated with position. We are also familiar with the idea that heat is a form of energy. We shall find the counterparts of these forms in electrical energy. The unit of work or energy in the mks system is the *newton-meter* or *joule*.

Power is defined as the rate of doing work, or the rate of change of energy. If we let w represent energy and p power, then we can write

$$p = \frac{dw}{dt} \qquad (a)$$

$$w = \int p\,dt \qquad (b) \qquad (1\text{-}19)$$

Consider a circuit branch as shown in Fig. 1-21. By definition, the voltage across the terminals is the work done by a unit positive charge in going from the positive terminal to the other; that is, it is the energy given up by the charge. (Or, equivalently, it is the work done *on* a unit-

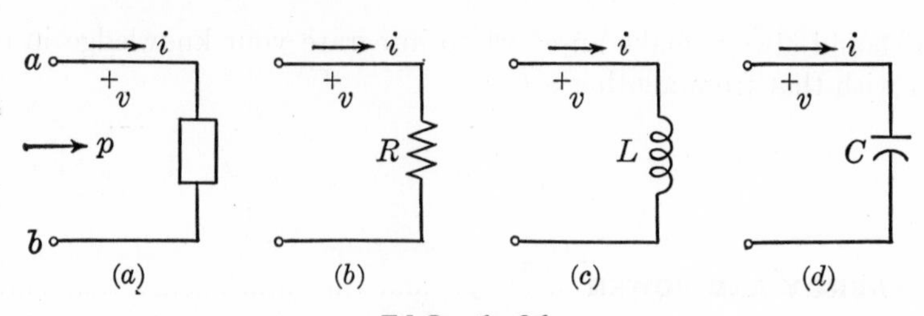

FIG. 1-21
Electrical power

positive charge in carrying it from the lower-potential terminal to the higher-potential terminal.) Suppose a charge q moves into terminal a and an equal charge leaves terminal b.

The energy given up by the transfer of charge will be

$$w = qv \tag{1-20}$$

To determine the rate at which this work is done, let us assume that the charge in question is an incremental one Δq, and that its transfer from a to b takes place in Δt second. The incremental energy given up by the charge will then be Δw. Hence we can write

$$\Delta w = v\,\Delta q \tag{1-21}$$

If we divide both sides by the time increment Δt, and then take the limit as $\Delta t \to 0$, we will get

$$p = \frac{dw}{dt} = v\frac{dq}{dt} = vi \tag{1-22}$$

This is a fundamental relationship relating the voltage and current of a branch to the power entering the branch. Note that this expression is valid for the voltage and current references shown in Fig. 1-21; that is, power is an algebraic quantity and, hence, has a reference. The reference for power is related to those of voltage and current as shown in Fig. 1-21. If either the voltage or the current reference is reversed, the power reference will be reversed. The choice of references means that the power will actually flow toward the branch if the voltage and current are positive, with the indicated references.

From the definition in Eq. (1-19), power is dimensionally joules per second, or *watts*. But, from Eq. (1-22), electrical power has the dimensions volt-ampere. Thus a *volt-ampere* is dimensionally the same as a watt.

Let us now turn back to Fig. 1-21 and assume that the branch in question consists of either a resistor, an inductor, or a capacitor. Consider first the resistor in Fig. 1-21(*b*). If we write the v-i relationship (Ohm's law) and multiply by the current, we will get

$$v = Ri \quad (a)$$
$$p = iv = Ri^2 \quad (b) \qquad (1\text{-}23)$$

The last expression gives the rate at which energy is flowing into the resistor from the external circuit. To find the total energy that enters the resistor between two instants of time t_1 and t_2, we should integrate this equation. The result will be

$$w_R = \int_{t_1}^{t_2} Ri^2\, dt \qquad (1\text{-}24)$$

The question naturally arises as to the disposition of this energy: What happens to it? Can we get it back from the resistor? The answer is supplied by noting that, even though the current may be positive or negative at different times, its square is always positive. Hence w_R can never be negative. It cannot even be zero unless there is no current. Saying that the energy which enters the resistor can never be negative is another way of saying that electrical energy can never leave the resistor.

This answer is corroborated by observations in the physical world. Whenever current passes through a conducting medium, there is an attendant appearance of heat. (The electric toaster operates on this principle.) This conversion of the electrical energy into heat is an irreversible process; we say that the energy is *dissipated* in the resistor.† In the physical world this conversion of energy into heat is caused by the collisions of the charge carriers in the conductor, the electrons, with other charge carriers or with atomic nuclei. Quantitative measurements of the heat produced by the passage of an electric current through a conductor were first made by James Joule in the 1840's. In fact, Eq. (1-24) is often referred to as *Joule's law.*

Since current and voltage are proportional in a resistor, both the energy dissipated and the power, the rate at which it is dissipated, can be written in terms of voltage. Thus

$$p = Ri^2 = R\left(\frac{v}{R}\right)^2 = Gv^2 \quad (a)$$
$$w_R = \int_{t_1}^{t_2} Gv^2\, dt \quad (b) \qquad (1\text{-}25)$$

† Of course, the heat might be used to generate steam, which can drive a turbine, which drives an electric generator, thus yielding electrical energy again.

In a specific situation these expressions may be more convenient to use than Eqs. (1-23) and (1-24).

Next let us consider the inductor shown in Fig. 1-21(c). Again we write the v-i relationship and then multiply by i. The result will be

$$v = L \frac{di}{dt} \qquad (a)$$

$$p = iv = Li \frac{di}{dt} \qquad (b) \qquad (1\text{-}26)$$

The last expression gives the power flowing into the inductor. To find the energy which enters the inductor between times t_1 and t_2, we integrate. The result will be

$$\int_{t_1}^{t_2} Li \frac{di}{dt}\, dt = \int_{i_1}^{i_2} Li\, di = \frac{1}{2} L(i_2^2 - i_1^2) \qquad (1\text{-}27)$$

where i_1 and i_2 are the currents at times t_1 and t_2, respectively. We see that the right-hand side is the difference between two positive numbers; hence it may be positive or negative, depending on the relative magnitudes of i_1 and i_2. This is equivalent to saying that the net energy over a period of time may enter the inductor or it may leave it. Any energy which may be supplied to the inductor is available to be returned. We say that the energy is *stored* in the inductor. (In terms of field theory, it is stored in the magnetic field.) The form of the result in Eq. (1-27) suggests that the energy stored at any instant of time is

$$w_L = \frac{1}{2} Li^2 \qquad (1\text{-}28)$$

Equation (1-27) then gives the difference in the stored energy at the two times, t_1 and t_2.†

Finally, let us consider the capacitor in Fig. 1-21(d). After writing the v-i relationship, let us multiply by the voltage. The result will be

$$i = C \frac{dv}{dt} \qquad (a)$$

$$p = vi = Cv \frac{dv}{dt} \qquad (b) \qquad (1\text{-}29)$$

† Actually, any arbitrary constant can be added to $Li^2/2$, and its derivative will still be $Li\, di/dt$. Also, the difference in stored energy will still be given by Eq. (1-27). Since the zero reference level of energy is arbitrary (just as that of temperature is), we choose to say that the energy is zero when i is zero. This makes the arbitrary constant zero.

The last expression gives the power flowing into the capacitor. It is quite similar to Eq. [1-26(b)] except that C and v replace L and i, respectively. To find the energy which enters the capacitor between times t_1 and t_2, we again integrate and get

$$\int_{t_1}^{t_2} Cv \frac{dv}{dt} = \int_{v_1}^{v_2} Cv\,dv = \frac{1}{2} C(v_2^2 - v_1^2) \tag{1-30}$$

where v_1 and v_2 are the voltages at times t_1 and t_2, respectively. Again, the right-hand side is the difference between two positive numbers. This also implies that the energy may flow away from the capacitor just as well as toward it. Any energy which is supplied to the capacitor is stored and is available to be returned. (In terms of field theory, it is stored in the electric field.) At any instant of time the energy stored in the capacitor can be written†

$$w_C = \frac{1}{2} Cv^2 \tag{1-31}$$

To summarize, we have now found that one of the three passive elements in our model is an energy-dissipating element, and the other two are energy-storing elements. Let us collect here the most important relationships we have developed.

$$p_R = Ri^2 = Gv^2 \qquad (a)$$

$$w_L = \frac{1}{2} Li^2 = \frac{1}{2} i\lambda = \frac{1}{2L} \lambda^2 \qquad (b) \qquad (1\text{-}32)$$

$$w_C = \frac{1}{2} Cv^2 = \frac{1}{2} qv = \frac{1}{2C} q^2 \qquad (c)$$

In the last two equations we have used the relationships $\lambda = Li$ and $q = Cv$.

Note that in the expressions for energy stored or dissipated, no restriction is placed on the form of variation of the voltage or current with time. They may have any waveshapes, and the expressions will still apply at each instant of time.

The energy stored in a capacitor depends only on the positions of charges, the fact that positive and negative charges exist separated from each other. (In a physical device the charges appear on opposite plates with a gap between them.) This energy, then, is analogous to potential energy. On the other hand, the energy stored in an inductor depends on the existence of current, which is a motion of charge. In this respect it is analogous to kinetic energy.

† See the preceding footnote.

Let us now question how the active elements—voltage and current sources—behave with respect to power and energy. Consider the voltage source shown in Fig. 1-22(a). With the references for voltage and current chosen as shown, the reference for power is away from the source. Its value will be $p = iv_g$. Since either v_g or i may be negative as well as positive, the power may actually flow in either direction. Also, by definition, v_g does not depend on the value of i. Hence, any amount of power can be supplied by a voltage source and any amount of power can be accepted by it. The same thing is true about the energy. A voltage source is a source of any amount of energy; it is also a sink of any amount of energy. (A sink is a consumer as opposed to a generator or producer.) The same arguments can be applied to a current source, and with the same conclusions.

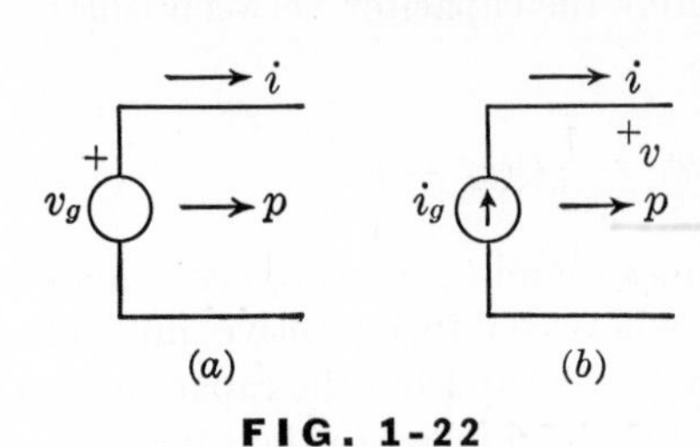

FIG. 1-22

Power supplied by sources

These conclusions may appear troublesome if you think that a voltage source or current source represents a physical source. We know that physical sources cannot deliver infinite energy. For example, the chemicals in a battery will eventually be used up after a finite amount of energy has been delivered. Even the nuclear fuel supplying nuclear energy has a finite lifetime. Our model, however, takes no account of these limitations. If, in a given problem, we find that infinite energy is involved, we will know that the model we are working with does not accurately represent the physical situation in that case.

1.6 SIGNALS

Up to this point we have concentrated on the elements that make up our model of electrical circuits. Eventually, we shall be interested in the electrical behavior of various interconnections of these elements. By "electrical behavior" we mean the way in which the electrical variables (usually voltage and current) vary. We can look upon these variables as *signals*. These signals appear at various places in a circuit;

they are modified by the circuit elements as they are transmitted from one portion of the circuit to another.

One method of classifying signals is by their *waveform* or waveshape. The waveform describes the variation of the signal with time, its shape. Signals that appear naturally in the physical world have relatively complicated waveforms. As an example, Fig. 1-23 shows the waveform of a

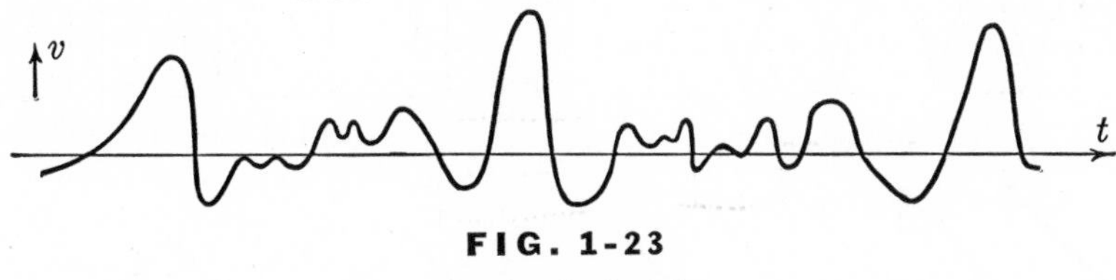

FIG. 1-23

A speech signal

speech signal. (This is the variation of a voltage which is proportional to the sound pressure in front of the mouth of a person who is speaking.) This particular signal has a certain amount of regularity about it. Other signals, such as the noise generated in a physical resistor due to the Brownian motion of the charge carriers, are completely irregular and are called *random signals.* It will obviously be impractical, if not impossible, to obtain a useful analysis of a circuit when the signal waveshapes are so complicated.†

Just as it is possible to make models to represent complicated interconnections of electrical devices, however, so also is it possible to construct models of signals. The elementary signals in the model may not be exactly what is found in the physical world, but it is hoped that combinations of these elementary signals can represent any arbitrary waveshape. Then, once it is found how a given circuit acts on a simple signal, the action of the circuit on a more complicated signal will be known, because the complicated signal can be decomposed into simpler ones. (As we shall discuss more fully later, this procedure is valid only for linear circuits.)

There is a wide variety of simple waveforms that can be used as elementary signals. They can be classified in a general way as *periodic* or *aperiodic.* Some examples of periodic waveshapes are shown in Fig. 1-24 and of aperiodic waveshapes in Fig. 1-25.

† In the case of random signals, it is possible to carry out an analysis based on the very randomness itself. The properties of such signals can be described in terms of probability theory. In this book we shall not discuss this type of analysis.

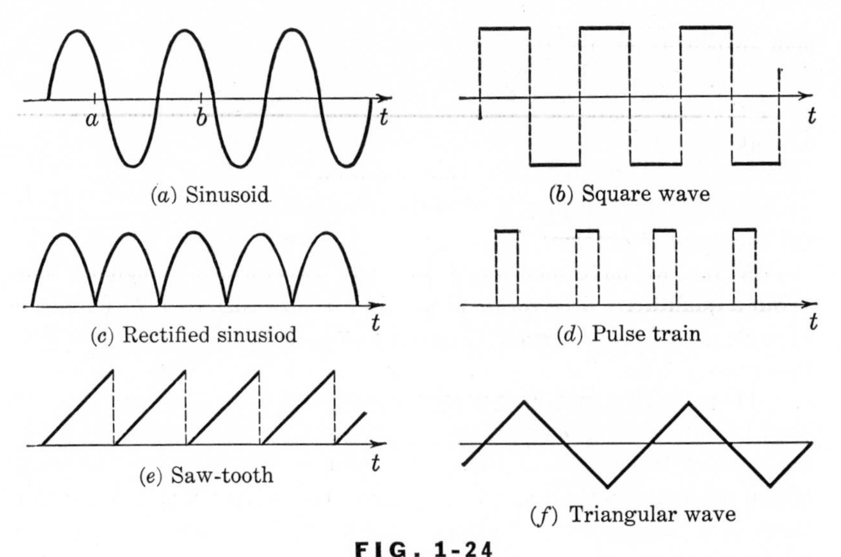

(*a*) Sinusoid (*b*) Square wave (*c*) Rectified sinusiod (*d*) Pulse train (*e*) Saw-tooth (*f*) Triangular wave

FIG. 1-24

Periodic waveforms

A periodic function is one whose form repeats itself over and over again in intervals of time, called the *period*; that is, if $f(t)$ is the periodic function and its period is T, then

$$f(t + T) = f(t) \tag{1-33}$$

Whatever value the function has at time t, it will have the same value T units of time later. The interval from a to b in Fig. 1-24(a) is one period. The sequence of values of the function over the interval of one period is called a *cycle*. Thus, periodic functions have no beginning or end. They started in the dim past of antiquity (minus infinity) and

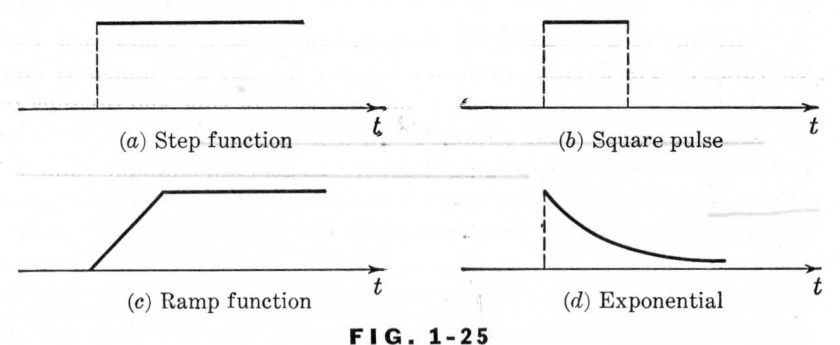

(*a*) Step function (*b*) Square pulse (*c*) Ramp function (*d*) Exponential

FIG. 1-25

Aperiodic waveforms

will outlast doomsday (plus infinity). Very often, however, we are interested in signals which have started at a finite point in time, and thereafter take the shape of a periodic function. We should not make the mistake of calling these periodic functions. The effects of such signals on a circuit a long time after they have been switched on, however, is practically indistinguishable from the effects of a truly periodic function (of the same waveform).

So far, we have mentioned only the waveform of a signal. This is but a qualitative description of it. Let us now turn to a consideration of some quantitative aspects of signals. We shall consider only periodic functions.

Probably the first quantitative feature that comes to mind is the *peak* value—the *amplitude* at the highest point. This is certainly a useful number to describe a function. Such a number alone, however, would not permit the distinction between a function which is small over most of its period but shoots up to a high peak momentarily, and one which remains at the same peak value over a long period of time. Two such functions are shown in Fig. 1-26.

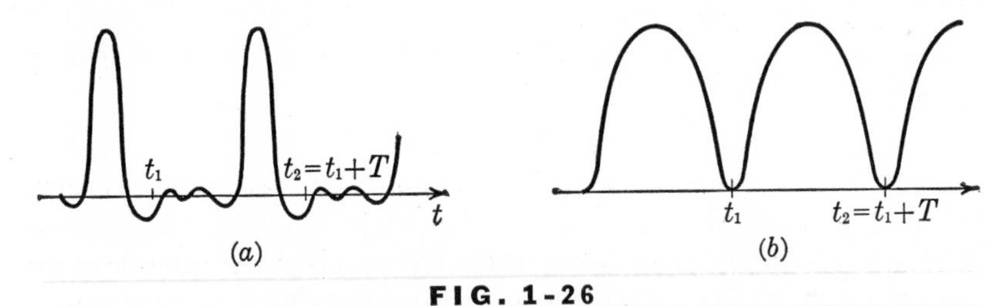

FIG. 1-26

Two functions with the same peak

One way to distinguish between such functions is to find their *average* values. The average value of a function of time is obtained by finding the area under the curve over some interval of time and dividing by the interval. The value obtained will depend on the length of the interval over which the averaging is done. In general, a unique value will be obtained only if we agree to average over one or more complete cycles of a periodic function. Thus, in terms of the signals in Fig. 1-26, we find the area from t_1 to t_2 and divide by $t_2 - t_1 = T$. Although both functions have the same peak value, their *full-cycle* average will be different.

Another quantitative measure of a signal may be obtained in the following way. For purposes of discussion, let us assume that the signal is a current. If a current signal i goes through a resistor R for an interval of time, some energy will be dissipated as heat. A quantitative measure of the signal is obtained if we compare this heat with the heat dissipated when a standard current signal goes through the same resistor. Again, to obtain a unique value, the time interval is chosen as one period. As a standard signal we choose a constant. Let this constant current be designated I. Using Eq. (1-24), we find the amount of energy dissipated as heat over one cycle, from time t_1 to $t_1 + T$, where T is the period, to be

$$w_R = \int_{t_1}^{t_1+T} RI^2\,dt = RI^2T \tag{1-34}$$

Notice that the starting time t_1 does not enter into the result at all. It doesn't matter where we start, as long as we cover one period. Hence it is often convenient to take the starting point to be $t = 0$.

For the current signal i, the energy dissipated in the same interval is

$$w_R = \int_{t_1}^{t_1+T} Ri^2\,dt \tag{1-35}$$

Since these two amounts of energy are to be the same, we equate them and get

$$TI^2 = \int_{t_1}^{t_1+T} i^2\,dt \tag{1-36}$$

and finally,

$$I = \sqrt{\frac{1}{T}\int_{t_1}^{t_1+T} i^2\,dt} \tag{1-37}$$

Remember that I is a constant current which gives the same heating effect as the current i. For this reason it is called the *effective* value of i. Note from Eq. (1-37) that to find the effective value of a signal we first square it, then find the average value of the result (this involves integrating and dividing by the period), and finally take the square root. Because of these operations, the effective value is also called the *root-mean-square* value (abbreviated rms).

Although we developed the concept of the rms value in terms of the dissipation of heat, this fact is not apparent when we look at Eq. (1-37). In fact, the rms concept arises in other ways also, without considering heat dissipation. Indeed, the variable, or signal, need not be a current or voltage; it may be any periodic function $f(t)$. The rms value of $f(t)$ is then defined as

$$F_{\text{rms}} = \sqrt{\frac{1}{T}\int_0^T f^2(t)\,dt} \tag{1-38}$$

(The starting point has been chosen as the origin of time, since it has no influence on the result.)

The introduction of the rms value of a periodic signal permits us to express not only the average power dissipated in a resistor when subjected to a periodic signal, but also the average energy stored in an inductor or in a capacitor. Thus, to find the average energy stored in an inductor, we should integrate Eq. (1-28) over one cycle and divide by the period. Correspondingly, for a capacitor the same operation should be applied to Eq. (1-31). The results will be

$$W_L = \frac{1}{2} L \frac{1}{T} \int_0^T i^2\, dt = \frac{1}{2} L I^2_{\text{rms}} \qquad (a)$$

$$W_C = \frac{1}{2} C \frac{1}{T} \int_0^T v^2\, dt = \frac{1}{2} C V^2_{\text{rms}} \qquad (b) \qquad (1\text{-}39)$$

Capital W's have been used to represent the average values.

The Sinusoidal Function

The sinusoidal function of time is perhaps the most important periodic function in engineering. Let us now discuss in some detail the quantitative features of a sinusoidal function. For purposes of discussion, assume that the signal is a voltage. A sinusoidal function is written

$$v(t) = V_m \cos(\omega t + \theta) \qquad (1\text{-}40)$$

A plot of this function is shown in Fig. 1-27. Two scales are shown for the abscissa; one for t and one for ωt. As time goes on, the function alternately takes on positive and negative values; hence it is referred to

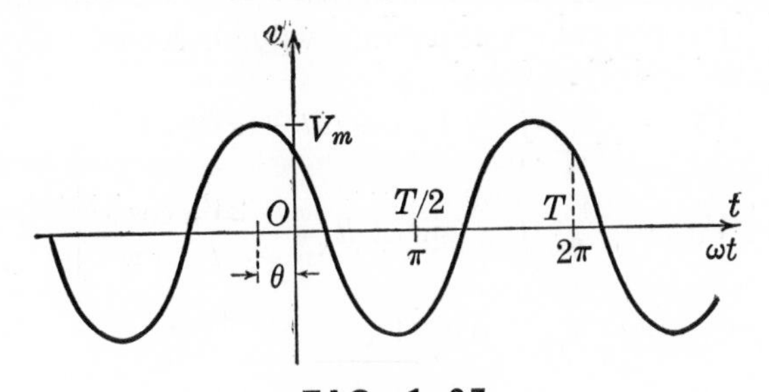

FIG. 1-27
Sinuloidal function

as an *alternating* function. To determine the number of cycles that the function goes through per unit time, note that the change in ωt over one period is 2π. Hence the time elapsed during one cycle, which is the period, is $2\pi/\omega = T$. The number of cycles per unit time, which is called the *frequency* f, is simply the reciprocal of this value, or

$$\boxed{f = \frac{\omega}{2\pi} = \frac{1}{T}} \tag{1-41}$$

ω is referred to as the *angular frequency*. (It is also called the *radian frequency* and the *angular velocity*.) The peak or amplitude is V_m. The symbol θ represents the entire argument of the cosine function at $t = 0$, and so it is called the *initial angle*, or simply the *angle*. It is measured from the positive peak; that is, $\theta = 0$ means that the vertical axis has been chosen to go through the positive peak. In Fig. 1-27 θ is a positive number.

To clarify the terminology, note that the sine function $V_m \sin \omega t$ and the cosine function $V_m \cos \omega t$ are special cases of the sinusoid, as written in Eq. (1-40), with the special values $\theta = -\pi/2$ and $\theta = 0$, respectively. (Verify this statement.) The most general sinusoid can be written either as a sine or as a cosine, but with arguments that differ by $\pi/2$ radians. This follows from the trigonometric identities

$$\cos x = \sin (x + \pi/2)$$

$$\sin x = \cos (x - \pi/2)$$

Let us now calculate for the sinusoid the quantitative values discussed in the preceding part of this section. Since the area under the negative half of the cycle is equal to the area under the positive half, the full-cycle average value is zero. Another useful measure is obtained, however, by defining the *half-cycle average* value. This is defined as the average value of the positive half cycle.† The positive half cycle will be covered if we choose the time origin so that $\theta = -\pi/2$ and integrate from 0 to $T/2$. Thus the half-cycle average value becomes

$$V_{\text{avg}} = \frac{1}{T/2} \int_0^{T/2} V_m \cos (\omega t - \pi/2)\, dt$$

$$V_{\text{avg}} = \frac{2V_m}{T} \int_0^{T/2} \sin \omega t\, dt = -\frac{2V_m}{T} \frac{\cos \omega t}{\omega}\bigg|_0^{T/2}$$

$$= \frac{2}{\pi} V_m \doteq 0.637 V_m \tag{1-42}$$

† One reason why this measure is useful is that the readings of some instruments are proportional to the rectified average, which is the half-cycle average for a sinusoid.

Let us next calculate the rms value of a sinusoid. For convenience, we shall this time choose the origin of time to make $\theta = 0$. Then

$$V_{\text{rms}} = \sqrt{\frac{1}{T}\int_0^T V_m^2 \cos^2 \omega t \, dt} = \sqrt{\frac{V_m^2}{2T}\int_0^T (1 + \cos 2\omega t) \, dt}$$

$$= \sqrt{\frac{V_m^2}{2T}\left(t + \frac{\sin 2\omega t}{2\omega}\right)\Bigg|_0^T} = \frac{V_m}{\sqrt{2}} \doteq 0.707 V_m \tag{1-43}$$

In the second step we used the trigonometric identity $2 \cos^2 x = 1 + \cos 2x$. We shall have many occasions to use this result throughout the subsequent work.

The sinusoid is important not only in its own right as a signal but also because almost any other signal can be represented as a combination of sinusoids. We shall not discuss this aspect of signal theory here, and you are not yet expected to understand the full implication of the previous statement. We shall be content with showing one simple example whereby a relatively complicated signal waveform can be resolved into a sum of sinusoids.

Consider the waveform shown in Fig. 1-28. This is a so-called

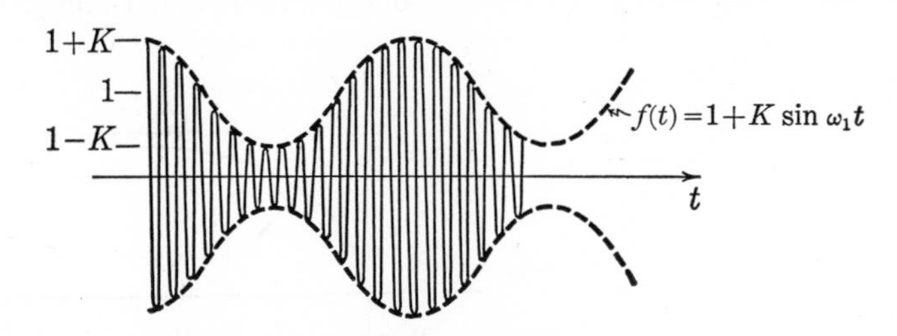

FIG. 1-28

Amplitude modulated wave

amplitude-modulated wave. A general expression for such a wave can be written

$$g(t) = f(t) \sin \omega_0 t \tag{1-44}$$

where $f(t)$ is any function of time multiplying the sinusoid $\sin \omega_0 t$, which is called the *carrier*. The function $f(t)$ plays the role of the amplitude of $\sin \omega_0 t$. In the case of Fig. 1-28, the function $f(t)$ is

$$f(t) = 1 + K \sin \omega_1 t \tag{1-45}$$

so that the entire waveform is given by

$$g(t) = (1 + K \sin \omega_1 t) \sin \omega_0 t$$

$$= \sin \omega_0 t + K \sin \omega_1 t \sin \omega_0 t \tag{1-46}$$

The second term of this expression can be modified by using the trigonometric identity

$$2 \sin x \sin y = \cos (x - y) - \cos (x + y) \tag{1-47}$$

The final result will be

$$g(t) = \sin \omega_0 t + \frac{K}{2} \cos (\omega_0 - \omega_1)t - \frac{K}{2} \cos (\omega_0 + \omega_1)t \tag{1-48}$$

We see that the relatively complicated waveform shown in Fig. 1-28 can be expressed as a combination of sinusoids of different frequencies.

Some Elementary Aperiodic Signals

Let us now turn our attention to some simple aperiodic signals. Some aperiodic waveforms were shown in Fig. 1-25. You might be concerned with these choices, since signals containing discontinuities or consisting of segments of straight lines do not occur in nature. Consider, however, the waveform given by the solid curve in Fig. 1-29(*a*). This

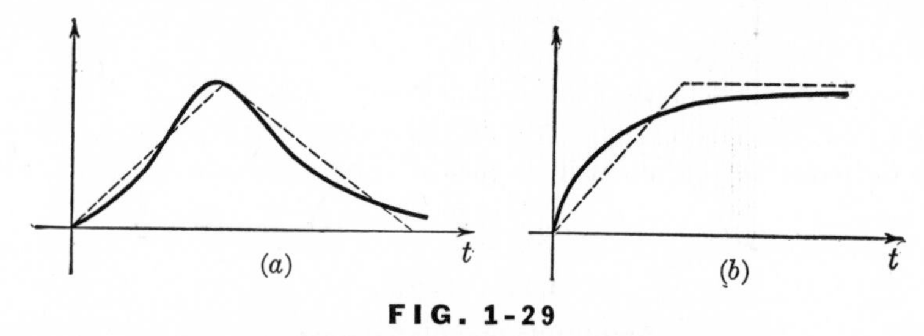

FIG. 1-29
Approximation of aperiodic physical signals

is a smooth curve which might represent a voltage or current in a physical circuit. The triangle shown by the dashed curve is an approximation of the solid curve. We would not expect the electrical behavior of a circuit in response to the triangle to differ greatly from its behavior in response to the solid curve. Our reason for dealing with the triangle rather than the solid curve is the greater ease of handling it mathematically. The same type of discussion applies to the curves in Fig. 1-29(*b*).

Consider the waveforms shown in Fig. 1-30(*a*). The one on top is called a *ramp function.* The curve rises linearly from zero for a time interval Δt, then remains constant. The slope of the curve is zero everywhere except in the interval Δt, where it is a constant. Thus the pulse

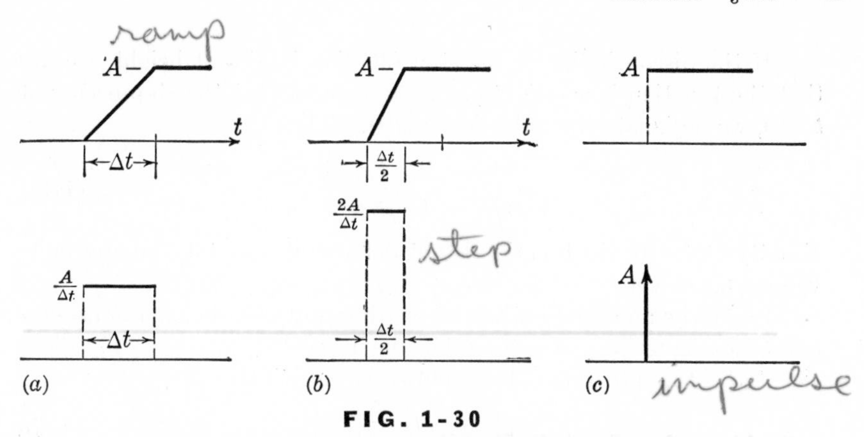

FIG. 1-30

Ramp function and pulse, step function and impulse

waveform shown below the ramp is the derivative of the ramp. Conversely, the ramp is the integral of the pulse. This derivative-integral relationship is easily visualized from the graphs. At the ends of the interval Δt, the derivative is not defined. The area under the pulse is the height A of the ramp.

Now suppose that the interval Δt is made smaller, while the height A of the ramp is held fixed. This means that the slope of the ramp will increase. Thus the pulse will get narrower and taller, as shown in Fig. 1-30(b), but the area under it will stay constant at the value A. If we keep on reducing the interval Δt until it reaches zero in the limit, the ramp will take the form of a discontinuous jump of height A, as shown in Fig. 1-30(c). This waveform is called a *step function.* The pulse will get narrower and taller until, in the limit, it attains an infinite height for a zero time interval. This waveform is called an *impulse.* At each stage, as Δt is reduced, the area under the pulse remains constant. This is assumed to be true even in the limit as $\Delta t \to 0$. The area A is called the *strength* of the impulse. An impulse is represented graphically by a vertical arrow, as shown in Fig. 1-29(c), the strength of the impulse being written alongside the arrow. For each value of Δt, the pulse is the derivative of the ramp and the ramp is the integral of the pulse, except at the ends of the interval where the derivative is not defined.

Note that in the limit, as $\Delta t \to 0$, the end points of the interval merge. Thus, strictly speaking, the impulse and the step do not have the derivative-integral relationship of the ramp and the pulse. It is conceptually useful, however, to interpret the derivative of a function at a discontinuity as an impulse.

If the discontinuity of the step function is of unit height, we say that the function is a *unit step*. If we assume that the step occurs at $t = 0$, we designate the unit step as $u(t)$. Thus

$$\begin{aligned} u(t) &= 0; \quad t < 0 \\ u(t) &= 1; \quad t > 0 \end{aligned} \tag{1-49}$$

Exactly at $t = 0$, the function is not defined; it may take on any finite value whatsoever.

Similarly, an impulse whose strength is unity is called a *unit impulse* and is designated $\delta(t)$, assuming that the impulse occurs at $t = 0$. Since the step is the integral of the impulse, we can write

$$\int_{-\infty}^{t} \delta(x)\, dx = u(t) \tag{1-50}$$

Thus, if the upper limit of the integral is negative, the integrand is zero over the entire range of integration, and so the integral will be zero. This agrees with the definition of $u(t)$ for negative values of t. The only contribution to the integral comes from the point $t = 0$, so that, if the upper limit of the integral is positive, the integral will be the strength of the impulse, which is unity.

Just as an arbitrary function can be represented as a combination of sinusoids of different frequencies, so also it can be represented by a combination of step functions. Figure 1-31 shows a smooth curve which

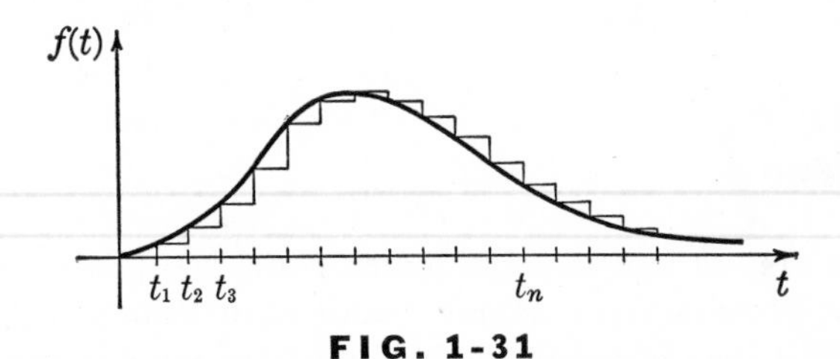

FIG. 1-31

Approximation by step functions

is approximated by the staircase function shown. This staircase is a sum of step functions, each one displaced from the preceding one. The approximation can be improved by taking smaller intervals between steps.

In order to write an analytical expression for the staircase function, we must have some way of writing a step function which occurs at a time other than $t = 0$. Note that $u(t)$ denotes the function which is zero whenever its argument t is negative, and 1 whenever its argument is positive. If we wish to designate a step which occurs at, say, $t = 5$, we should write u with an argument which is negative for $t < 5$ and positive for $t > 5$. This argument is, obviously, $t - 5$.

In general, then, the step function which occurs at $t = t_1$ can be written as $u(t - t_1)$.

$$\begin{aligned} u(t - t_1) &= 0 \quad t < t_1 \\ u(t - t_1) &= 1 \quad t > t_1 \end{aligned} \tag{1-51}$$

This notation agrees with Eq. (1-49) for $t_1 = 0$.

The function in Fig. 1-31 can now be approximated by a sum of displaced steps as follows.

$$f(t) = K_1 u(t - t_1) + K_2 u(t - t_2) + \cdots + K_n u(t - t_n) + \cdots \tag{1-52}$$

where the K's are the jumps at the discontinuities. If the intervals between steps are allowed to approach zero, the summation will become an integral and the approximate equality will become exact.

We shall not pursue this topic any further here. It is not our purpose to be exhaustive, but to give an indication of how the elementary signals we have introduced can be used to represent any arbitrary signal. These ideas are the basis for powerful methods of analysis which we shall discuss later.

1.7 SUMMARY

In this chapter we laid the groundwork for a study of network theory. We reviewed and examined some of the experimental evidence concerning the laws of nature as they relate to electric circuits. This evidence has been accumulated over the last $1\frac{1}{2}$ to 2 centuries.

Based on the available evidence, we postulated a model consisting of a number of elements: three passive elements and two active ones. The active elements account for the behavior of devices which generate (or convert) electrical energy. The passive elements account for the appearance of heat and the storage of energy. It will be our major purpose throughout the remainder of the book to study interconnections of these elements and to predict their behavior based on deductions arising logically from the postulated laws. Many of these predictions can then be subjected to verification by observations in the physical world.

Not all observations in our domain of the physical world can be explained by means of our present model; hence we shall find it necessary, later on, to introduce additional elements.

The model we have introduced can be described by certain adjectives. It is, first of all, a linear model, the relationships between appropriate variables being straight lines. Second, it is bilateral, since the effects are the same for one direction as for the other. Third, we can call it a lumped model. By this we mean that the effects which are to be accounted for by the elements are assumed to be localized within the terminals of the elements. For example, in the physical world the magnetic field resulting from a current flowing in a closed circuit permeates all of space. We assume, however, that its effects are localized, or lumped, in one part of the circuit, and we idealize this effect in terms of the inductance. Finally, it is time invariant; that is, the parameters do not change with time. An example of a parameter varying with time is a model of a physical capacitor whose plates are in motion relative to each other.

There are many physical situations in which one or more of these adjectives are not valid. In some cases, such as transmission lines, the lumped restriction must be removed, but satisfactory results can still be obtained by retaining the linearity. In contrast to the term "lumped," we refer to such networks as being *distributed*. Nevertheless, the linear, bilateral, lumped, time-invariant theory which we shall develop is a prerequisite for the more complicated theories needed to study distributed or nonlinear systems.

In this chapter we also dealt with some of the qualitative and quantitative aspects of signals. We discussed the characteristics of a sinusoid, a unit step, and a unit impulse. We indicated that complicated signal waveforms can be considered to be composed of these elementary signals. Hence, when we are determining the behavior of electric circuits in the presence of the elementary signals, we shall, at the same time, be determining their behavior in the presence of more complicated waveforms.

Problems

1-1 The functions in Fig. P1-1 are the branch voltages of an inductor or capacitor. Sketch the corresponding branch currents and the stored energy as a function of time.

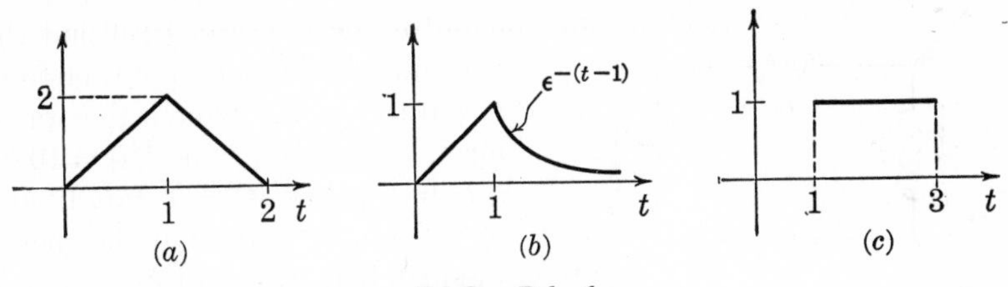

FIG. P1-1

1-2 Let the functions in Fig. P1-1 represent the branch currents of an inductor or capacitor. Sketch the corresponding branch voltages and stored energy.

1-3 Repeat Prob. 1-1 and 1-2 for the functions given in Figs. 1-24 and 1-25 in the text.

1-4 Write Kirchhoff's-law equations at all the junctions and all the closed loops of the circuits shown in Fig. P1-4. Direct particular attention to the references.

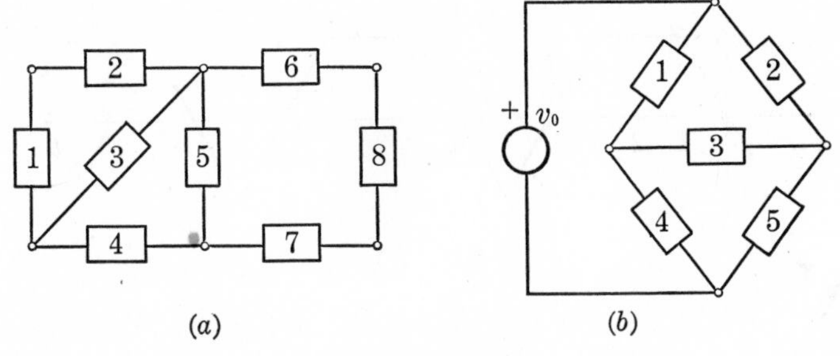

FIG. P1-4

1-5 (a) In Fig. P1-5 find the voltage v_4 as a function of time from $t = 0$ to $t = 2\pi$. Sketch the waveform. Assume $i_1 = 10 \sin t$ and $v_2 = 5t$, $R_1 = R_2 = 2$ ohms, $C = 1$ farad, and $L = 2$ henrys. (b) Find the energy stored in L and C at each instant of time.

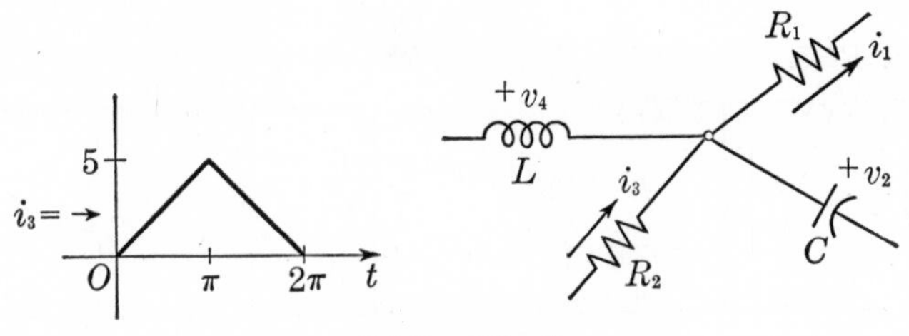

FIG. P1-5

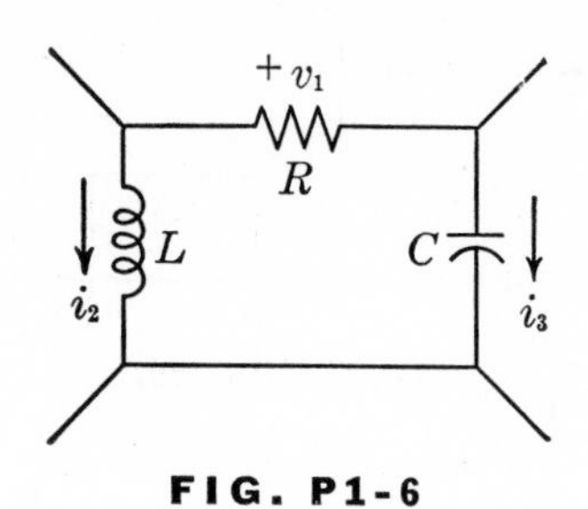

FIG. P1-6

1-6 (*a*) In Fig. P1-6 find the current i_3 as a function of time from $t = 0$ to $t = 2$. Sketch the waveform. Assume $v_1 = 5\epsilon^{-t}$, $i_2 = 10 \sin t$, $L = 1$ henry, $C = 1$ farad, and $R = 2$ ohms. (*b*) Find the energy dissipated in R over this period of time.

1-7 Find the average and rms values of the waveforms shown in Fig. P1-7.

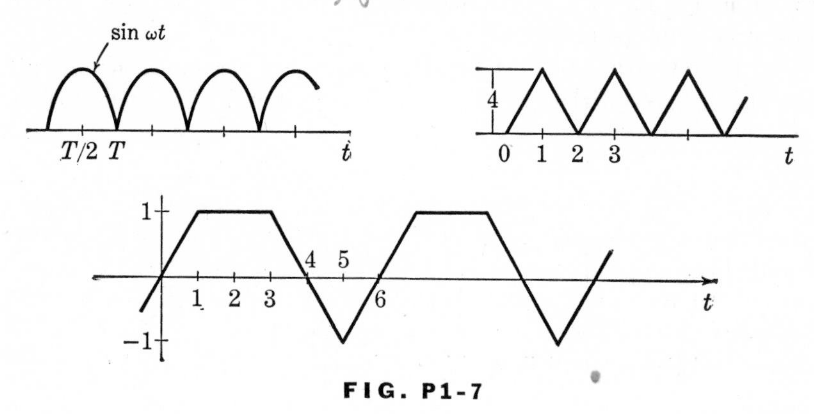

FIG. P1-7

1-8 (*a*) Let each of the waveforms of Fig. P1-7 be the voltage across a resistor. Find the average power dissipated. (*b*) Suppose each of these waveforms is the voltage across an inductor or the current in a capacitor. Determine the average value of the stored energy.

1-9 (*a*) In Fig. P1-9 it is required to find the voltage v as a function of time using fundamentals laws. The known quantities are $v_g = 53 \cos 40t$ and $v_1 = 32 \cos 40t + 24 \sin 40t$. (*b*) Find also the value of R_1.

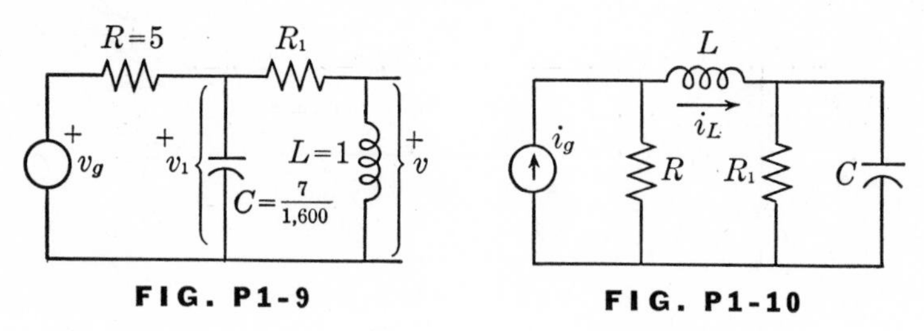

FIG. P1-9 FIG. P1-10

1-10 Using the fundamental laws, find the current in the capacitor of Fig. P1-10 in terms of the source current i_g and the current i_L in the inductor, which is assumed known.

1-11 Let $v_1 = K_1 \cos(\omega t + \alpha)$ and $v_2 = K_2 \cos(\omega t + \beta)$ be two sinusoids of the same frequency ω. Prove that their sum $v_1 + v_2$ is also a sinusoid of the same frequency.

1-12 The *external characteristic* (the terminal v-i relationship) of a d-c self-excited shunt generator is shown in Fig. P1-12. Find a linear model of this generator which is valid for currents $0 < i < 40$. Specify numerical values.

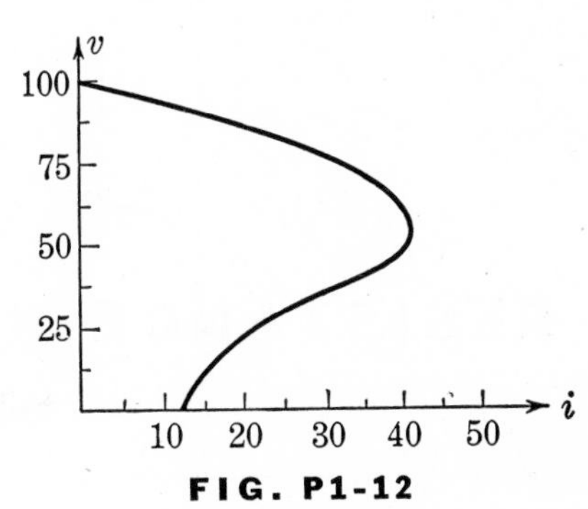

FIG. P1-12

1-13 In each of the diagrams in Fig. P1-13, the designated voltage or current is known and is specified for $t > 0$. By applying the fundamental laws, calculate the required waveform of the sources. All initial capacitor voltages and initial inductor currents are assumed to be zero. Calculate also the instantaneous values of the energy stored in each inductor and the power delivered by each source.

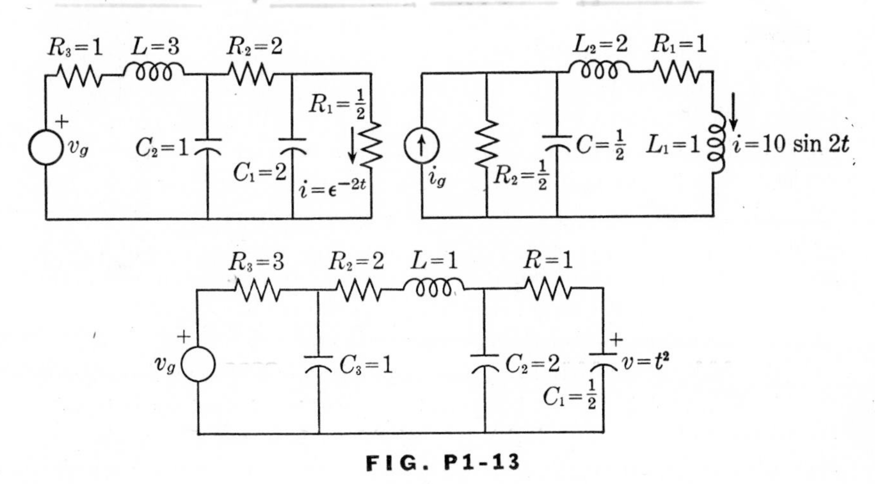

FIG. P1-13

RESISTANCE NETWORKS

2.1 ELECTRICAL NETWORKS

2

IN Chap. 1 we developed a model for electric circuits. This model, at present, contains five elements: resistance, capacitance, inductance, voltage source, and current source. We hope that the electrical behavior of interconnections of these elements will approximate the behavior of physical devices to any desired degree. Many of the problems which we are called upon to solve involve finding the voltage or current in a physical device when one or more sources of energy are connected to the device. The first task in solving this is to replace the physical devices by their models. Once we have the model, then we apply the laws of equilibrium, which are Kirchhoff's laws. Application of Kirchhoff's laws will lead to a set of equations involving the branch currents and voltages. These must then be solved subject to the constraints imposed by the v-i relationships of the elements.

The task of approximating the physical circuit with a model is a very important part of this procedure. It has no bearing, however, on the method of procedure for obtaining a solution once the model has been obtained. In this book we shall be concerned mainly with the model, and most of the time the problems will be formulated in terms of models. From time to time, however, we shall formulate some problems in terms of physical equipment, just to remind you that this aspect of the complete problem should not be forgotten.

In the last chapter we defined an electrical circuit as an interconnection of physical electrical devices. Since we now have a model, and since we shall be interconnecting elements of the model, we should have a name for such an interconnection. Let us use the word *network* to designate an interconnection of hypothetical branches of the model. Thus the words "circuit" and "network" refer to similar things; circuit applies to physical devices, whereas network applies to models. We shall not remain completely consistent in this terminology, however, since we shall often use the word "circuit" also when referring to the model. Nevertheless, when there is any reason to distinguish between the two, we shall be careful to do so.

The aspect of electrical engineering with which we shall be dealing is variously called *circuit theory*, *circuit analysis*, *network theory*, or *network analysis*. The main objective is to be able to calculate the current or voltage signal (this is called the *output* or *response*) in a network branch when sources of one kind or another are connected to the network. (These are called *inputs* or *excitations*.)

The converse of this process is called *synthesis;* that is, the desired form of a current of voltage output signal is given when a voltage or current input signal is specified. It then remains to find a network which will perform in the desired manner. This task is much more difficult than the task of analysis. We shall here be concerned only with the task of network analysis.

An electrical network has two distinct aspects. The elements which form the network are connected at their terminals. The junctions formed at the common terminals of two or more elements are the nodes of the network. One or more closed loops are formed by the branches. Thus, no matter what the constituents of a branch may be, there is a definite geometrical aspect to a network. There is a considerable amount of knowledge in a branch of mathematics called *topology* that deals with these geometrical properties of a network.

The second aspect deals with the modification of signals by the network. According to Ohm's law, the only way in which a resistor modifies a signal is to multiply it by a constant. On the other hand, inductors and capacitors perform the operations of differentiation and integration. Thus, when a signal is transmitted through a network consisting of R's, L's, and C's, it is modified by different combinations of differentiation, integration, and multiplication by a constant. If you will now pause to think of other operations which might conceivably be performed on signals—such as addition, changing of the average value (this is called *clamping*), limiting of the amplitude (this is called *clipping*), delaying the occurrence of the signal, modulation, etc.—you will appreciate that our model, so far, is quite limited in the ways that it can modify a signal. We shall later introduce additional elements in our model which can perform some additional operations besides the ones we now have available. As long as we restrict ourselves to a linear model, however, some desirable operations will always remain beyond our reach. In order to bring these operations within our domain, we must resort to nonlinear devices.

The line of thought followed in the previous paragraph leads to another consideration. When we started our study of electrical circuits, we made observations in the physical world and postulated a model which would serve to explain the observed phenomena. We now see a whole new vista opening before us. When confronted with the task of synthesizing a network to perform a particular function, we need not limit our thinking to those operations which can be performed on signals by presently available devices. In our design we might ask for a device which is to perform some operation which heretofore has not been performed. This, in turn, might spur investigation and eventually invention of one or more devices to perform the desired operation. This is an exciting thought. Instead of developing a model to account for the behavior of physical devices, we first postulate a model of a desired physical device; then we seek to invent a device whose behavior will be approximately the same as that of the model.

We shall not reach a level of maturity in this book which will permit us to explore the ideas which have just been presented. Nevertheless, these possibilities should be kept in mind as one of the goals toward which you should strive in your professional education.

Let us now return to our main stream of thought. In seeking to explain the electrical behavior of a network, it is clear that the geometry alone is not sufficient. Certainly, the network performance will depend on the detailed properties of each branch. On the other hand, certain procedures in solving network problems are, to a large extent, independent of the types of branches in the network. Hence, initially, we shall deal with networks containing only resistors and sources, in order to focus attention on these concepts without becoming burdened with the complexities introduced by the mathematical operations performed by inductors and capacitors. After we have gained some familiarity with methods of solution, we shall turn to networks containing inductors and capacitors as well.

2.2 VOLTAGE AND CURRENT DIVISION

Before we come to the end of this chapter, we shall have discussed general methods of analysis which can be applied to an electrical network of arbitrary structural complexity. Before discussing such involved methods, however, we shall devote some time to a few relatively simple procedures which can be applied in a number of special configurations. These configurations arise so often that they are quite important.

Consider the situation portrayed in Fig. 2-1(a). A portion of a

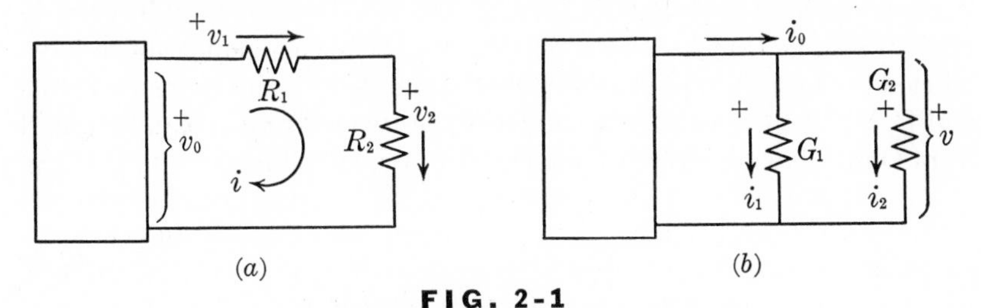

FIG. 2-1

Voltage divider and current divider

network consists of the series connection of two branches, the voltage across the combined branch being v_0. Two branches are said to be *in series* if they are connected end to end in such a way that a current in one branch will also flow in the other. The part of the network shown by the rectangle in Fig. 2-1 can be anything, as long as the voltage at the terminals is v_0. For simplicity, you can think of it as a voltage source. With the current references shown in the figure, we see that the branch currents are the same. Hence, using Ohm's law and then Kirchhoff's law, we get

$$i = \frac{v_1}{R_1} = \frac{v_2}{R_2} \qquad (a)$$

$$v_0 = v_1 + v_2 = i(R_1 + R_2) \qquad (b) \qquad (2\text{-}1)$$

When the last equation is solved for i and the result is substituted into the previous two, we get

$$v_1 = \frac{R_1}{R_1 + R_2} v_0 \qquad (a)$$

$$v_2 = \frac{R_2}{R_1 + R_2} v_0 \qquad (b) \qquad (2\text{-}2)$$

$$\frac{v_1}{v_2} = \frac{R_1}{R_2} \qquad (c)$$

We say that the voltage v_0 "divides" between R_1 and R_2 in the ratio of the two resistances, as shown in the last equation. The network is called a *voltage divider*. The fraction of the total voltage which appears across R_2 is in the ratio of R_2 to the total series resistance $R_1 + R_2$, and similarly for the voltage across R_1.

Another network similar to the voltage divider is shown in Fig. 2-1(*b*). Here, two resistances are connected in parallel, the current into the combined branch being i_0. Two branches are said to be *in parallel* if their ends are connected in pairs between two terminals so that any voltage across one branch will also be across the other. Again, the network within the rectangle in Fig. 2-1(*b*) can be anything. For simplicity, assume that it is a current source. (For convenience, we have labeled the branches in terms of the conductances G_1 and G_2 instead of R_1 and R_2.) The common voltage across the two resistances can be found from Ohm's law and Kirchhoff's current law, as follows:

$$v = \frac{i_1}{G_1} = \frac{i_2}{G_2} \qquad (a)$$
$$i_0 = i_1 + i_2 = (G_1 + G_2)v \qquad (b) \qquad (2\text{-}3)$$

Solving the last equation for v and the preceding one for i_1 and i_2 gives

$$i_1 = \frac{G_1}{G_1 + G_2} i_0 \qquad (a)$$
$$i_2 = \frac{G_2}{G_1 + G_2} i_0 \qquad (b) \qquad (2\text{-}4)$$
$$\frac{i_1}{i_2} = \frac{G_1}{G_2} \qquad (c)$$

Thus the current i_0 "divides" between G_1 and G_2 in the ratio of the two conductances. The network is called a *current divider*. The fraction of the total current which goes through G_1 is in the ratio of G_1 to the total parallel conductance $G_1 + G_2$. Similarly for the current through G_2.

2.3 EQUIVALENT SOURCES

In introducing voltage and current sources in Chap. 1, we emphasized that these are hypothetical idealizations. No physical device behaves exactly like a voltage source or a current source. The terminal behavior of many sources, however, is approximately the same as that of a voltage source in series with a resistor, or a current source in parallel with a resistor. An illustration of this fact was given in Chap. 1 by examining the behavior of a storage battery. Let us now examine this problem with a view toward determining the generality of these two methods of representation.

Consider the network shown in Fig. 2-2(*a*) which shows a voltage

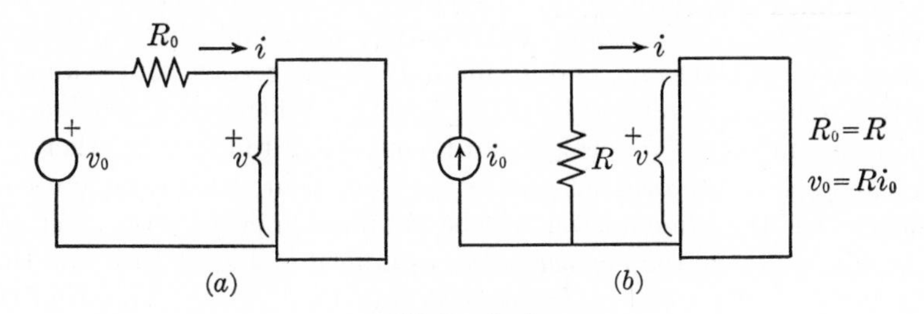

FIG. 2-2

Source equivalents

source v_0 in series with a resistance R_0 connected to an arbitrary network which we can consider to be a *load*. Similarly, the network in Fig. 2-2(*b*) shows a current source i_0 in parallel with a resistance R connected to the same load. In Chap. 1 we discovered that two such networks could be made equivalent by proper choice of the parameters, for the special case of the battery. We are now considering a general source having any waveform whatsoever.

Using Kirchhoff's voltage law and Ohm's law, the v-i relationship for the first figure can be written

$$v = v_0 - R_0 i \tag{2-5}$$

while, from Kirchhoff's current law and Ohm's law, that of the second figure is

$$\begin{aligned} i &= i_0 - \frac{v}{R} && (a) \\ v &= Ri_0 - Ri && (b) \end{aligned} \qquad (2\text{-}6)$$

If the two networks are to be equivalent, then the terminal voltage in Eq. (2-5) should be the same as that in Eq. [2-6(*b*)], assuming identical loads. Thus

$$v_0 - R_0 i = Ri_0 - Ri \qquad (a)$$

or

$$(v_0 - Ri_0) + i(R - R_0) = 0 \qquad (b) \qquad (2\text{-}7)$$

Now this equation should be true for all possible loads, which means for all values of i; that is, the equivalence of the two networks should not depend on what is connected at the terminals. As i varies, the left side of the last equation will vary, and hence will not be zero, unless

$$\begin{aligned} R &= R_0 \qquad (a) \\ v_0 &= R_0 i_0 \qquad (b) \end{aligned} \tag{2-8}$$

As far as the behavior at the terminals is concerned, the two networks in Fig. 2-2 are completely equivalent if, and only if, the two resistances are the same, and if, and only if, the voltage source and the current source are related by Eq. [2-8(*b*)]. Whenever a problem involves one of these two forms and it is convenient to have the other one, we can replace the original one by its equivalent, with complete confidence that the same answer will be obtained in either case. We say that Fig. 2-2(*b*) is the *current-source equivalent* of Fig. 2-2(*a*), and that Fig. 2-2(*a*) is the *voltage-source equivalent* of Fig. 2-2(*b*). Note that the equivalences refer to the voltage source in series with the resistor, and the current source in parallel with the resistor, not to the sources alone.

Although we are considering only resistive networks in this chapter, this is a natural place to discuss equivalent sources when the resistors in Fig. 2-2 are replaced by inductors or by capacitors. We shall, therefore, digress momentarily in order to consider these two situations.

Figure 2-3 shows a voltage source v_0 in series with an inductor, and a current source i_0 in parallel with another inductor. We wish to determine the conditions under which these two configurations are equivalent at their terminals. For the first network, Kirchhoff's voltage law and the v-i relationship of the inductor lead to

$$v = v_0 - L_0 \frac{di}{dt} \tag{2-9}$$

For the second network, let us write Kirchhoff's current law.

$$i_L = i_0 - i \tag{2-10}$$

If we differentiate this equation and multiply by L, the result will be

$$L \frac{di_L}{dt} = L \frac{di_0}{dt} - L \frac{di}{dt} = v \tag{2-11}$$

The last step follows from the v-i relationship of the inductor. By the same argument as before, we set the two voltages equal and get

$$\left(v_0 - L\frac{di_0}{dt}\right) + (L - L_0)\frac{di}{dt} = 0 \tag{2-12}$$

This result should be true, no matter what the network connected at the terminals; that is, for all values of di/dt. This is possible if, and only if,

$$L = L_0 \qquad (a)$$

$$v_0 = L\frac{di_0}{dt} \qquad (b) \qquad (2\text{-}13)$$

That is, the two networks in Fig. 2-3 will be equivalent at the terminals, if the two inductors are the same and if the two sources are related by Eq. [2-13(b)].

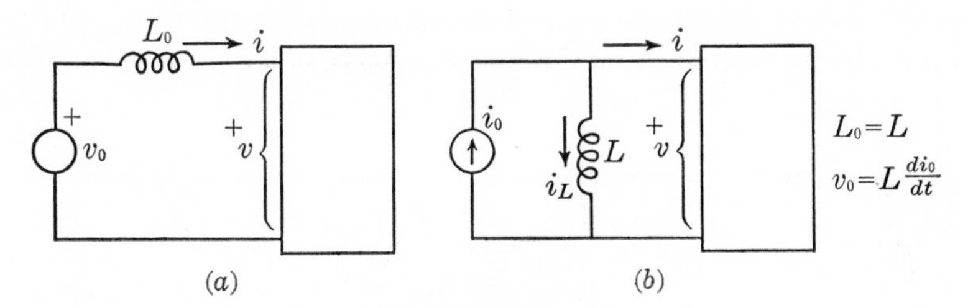

FIG. 2-3

Source equivalents

Finally, let us turn to Fig. 2-4 in which a voltage source is in series with a capacitor and a current source is in parallel with another capacitor. By repeating the arguments leading to Eqs. (2-13) and (2-8), you

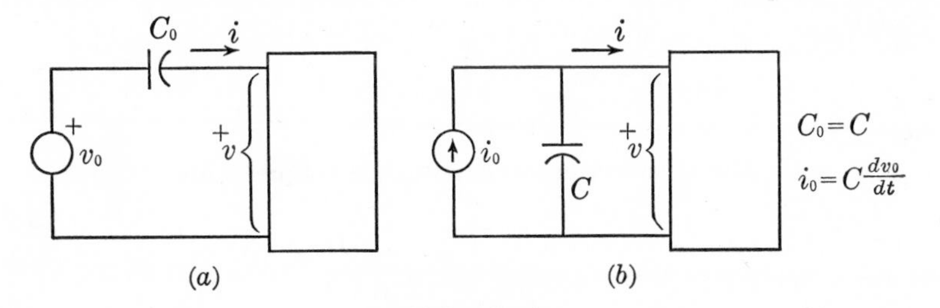

FIG. 2-4

Source equivalents

will find that equivalence of the two networks at the terminals requires that

$$C_0 = C \quad (a)$$

$$i_0 = C\frac{dv_0}{dt} \quad (b) \qquad (2\text{-}14)$$

The details will be left for you to carry out.

At this point a note of warning is in order. We have been discussing equivalence of two networks. The equivalence applies only to the v-i relationship at the terminals. No claim is made regarding voltages and currents anywhere within the equivalent networks. For example, the current in the resistor of Fig. 2-2(a) is not the same as that in the resistor of Fig. 2-2(b).

Let us now use these ideas of source equivalents to find a solution for one of the variables in a network. Figure 2-5(a) shows a fairly

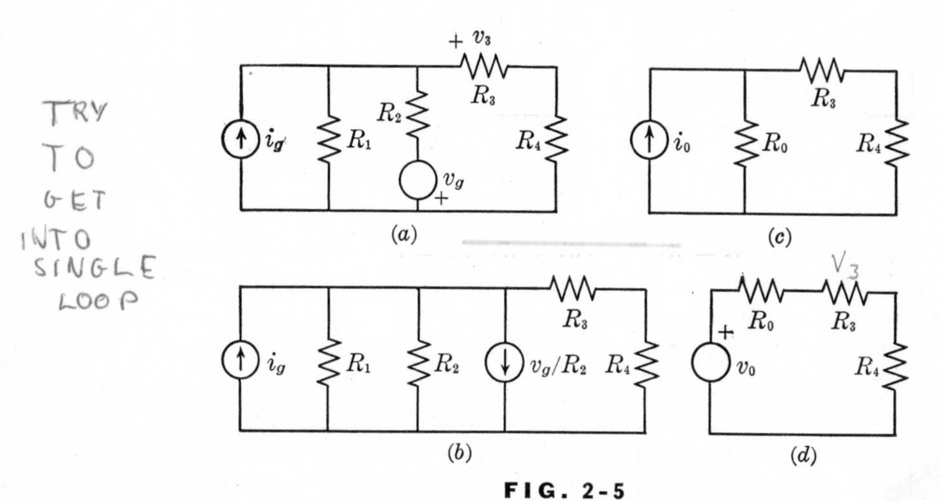

FIG. 2-5

Use of source equivalents in solving problems

extensive network which includes both a voltage source and a current source. Let us find the voltage v_3 of the branch R_3. The voltage source v_g in series with R_2 may be replaced by a current source v_g/R_2 in parallel with R_2. The resulting network will take the form shown in Fig. 2-5(b). Note carefully the reference of the current source v_g/R_2 relative to the reference of v_g.

We should now inquire into the terminal behavior of the two current sources in parallel. If the combination is considered as a single branch, the total current in the branch will be the algebraic sum of the two source currents. Hence the combination is equivalent to a single current source having a current which is the algebraic sum of the two individual source currents; that is,

$$i_0 = i_g - \frac{v_g}{R_2} \tag{2-15}$$

Similarly, the parallel combination of the two resistances R_1 and R_2 is equivalent to a single resistance whose value is

$$R_0 = \frac{R_1R_2}{R_1 + R_2} \tag{2-16}$$

(This is left for you to prove as an exercise.) When the branches of Fig. 2-5(*b*) are combined in this way, the result is Fig. 2-5(*c*). Now the current source i_0 in parallel with R_0 can be converted into an equivalent voltage source in series with R_0, as shown in Fig. 2-5(*d*). The value of v_0 will be

$$v_0 = R_0i_0 = \frac{R_1R_2}{R_1 + R_2}\left(i_g - \frac{v_g}{R_2}\right) \tag{2-17}$$

The resulting network has the form of a voltage divider, if we consider R_0 to be combined with R_4. Hence the voltage v_3, in which we are interested, is found by using the voltage-divider relationship. The result is

$$v_3 = \frac{R_3}{R_0 + R_3 + R_4} v_0 = \frac{R_1R_3(R_2i_g - v_g)}{(R_1 + R_2)(R_3 + R_4) + R_1R_2} \tag{2-18}$$

The final form is obtained by substituting for R_0 and v_0 from Eqs. (2-16) and (2-17).

In the technique just used, the solution proceeds step by step in gradually converting parts of the network to equivalent forms. The network structure is gradually simplified until there remains a very simple structure containing the branch in which our interest is centered. This structure reduction is achieved by using (1) source equivalents, (2) series combination of branches, and (3) parallel combination of branches, over and over again.

You should note, however, that this method does not give a complete solution for all the variables in the network. When sources are converted to their equivalents, only the terminal behavior is conserved; the details of currents and voltages within the terminals are lost. Once part of the solution has been obtained by this procedure, however, we can go back to the original network and solve for the remaining unknowns. The job is now simpler because some of the previously unknown variables are now known.

2.4 SOURCE TRANSFORMATIONS

We have just seen how a network solution is simplified by conversion of a voltage source in series with a resistor to a current source in parallel with a resistor, and vice versa. This simplification, however, is not available to us if a voltage source appears in a network without having a branch in series with it, or if a current source appears without having a branch in parallel.

To meet these difficulties, consider the portion of a network shown in Fig. 2-6(a). The voltage source does not have a single branch in

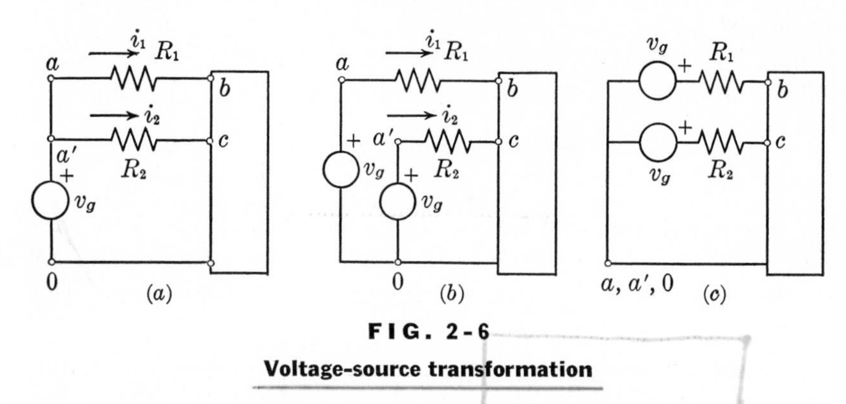

FIG. 2-6

Voltage-source transformation

series with it. But suppose we take the node marked (a, a'), which is a single node, and split it in two; then connect a voltage source v_g from each of these to point 0, as shown in Figure 2-6(b). Note in this diagram that points a and a' have the same voltage, just as they do in the first diagram.

To see what effect this maneuver will have on the branch v-i relationships, let us write the equations for voltages v_{b0} and v_{c0} in the original network. Using Kirchhoff's voltage law, we will get

$$v_{b0} = v_{ba} + v_g = v_g - R_1 i_1 \qquad (a)$$
$$v_{c0} = v_{ca'} + v_g = v_g - R_2 i_2 \qquad (b) \qquad \text{(2-19)}$$

But these are exactly what we get if we write the branch equations for Fig. 2-6(b). That is, the modification of the network has left these particular branch equations unaltered. All other branch v-i relationships are, of course, unchanged. Hence, these two networks are equivalent.

The result of this step is a network with two voltage sources, each of which is in series with a resistor. We can now utilize the procedure of replacing these combinations with their current-source equivalents and continue with the solution.

Another way of viewing the result we have achieved is shown in Fig. 2-6(*c*). Instead of thinking of separating the node (a-a') into two parts, assume that the source v_g has moved through the node into each of the branches connected there. Node (a-a') is now short-circuited to node 0 and thereby disappears. The result shown in Fig. 2-6(*c*), of course, has the same topology, or geometry, as Fig. 2-6(*b*).

Although this discussion was carried on in terms of two branches connected to the voltage source, the arguments that were used are valid no matter how many branches are connected there. This result is therefore a general theorem, known as the voltage-shift theorem, which we can state as follows:

A voltage source can be moved through one of its terminals into each of the branches connected there, leaving its original position short-circuited, without affecting the voltages and currents anywhere else in the network.

Note, however, that the identity of the original voltage source is lost. Hence, if we want to find the current through the source, we shall not be able to do so from the modified network. In such a case, after the remaining unknowns are found in the modified network, we can go back to the original network. The only remaining unknown there is the current through the voltage source, which can now be found. Thus, in terms of Fig. 2-6(*a*), once v_{b0} and v_{c0} are found, we can find i_1 and i_2 from Eq. (2-19). The current through v_g is then $i_1 + i_2$ (with the appropriate reference).

As a simple example, let us consider the special case in which a voltage source appears directly across a branch, as illustrated in Fig. 2-7(*a*). With the voltage-shift theorem, v_g is moved into branches R_1 and R_2, leaving its original position shorted. This is illustrated in Fig. 2-7(*b*). As far as the network to the right of points a, b is concerned, the branch containing v_g and R_1 is shorted, and it might as well be removed. We say this branch is not coupled to the rest of the network. This leaves the network shown in Fig. 2-7(*c*). Comparison with the original network shows that branch R_1, which appears across the voltage source, has no influence on the behavior of the rest of the network. Such a branch can simply be neglected.

The current in the original voltage source, however, does depend on R_1. If it is desired to find this current, we return to the original network after finding the current in R_2 from Fig. 2-7(*c*). The current in v_g is now obtained by Kirchhoff's current law.

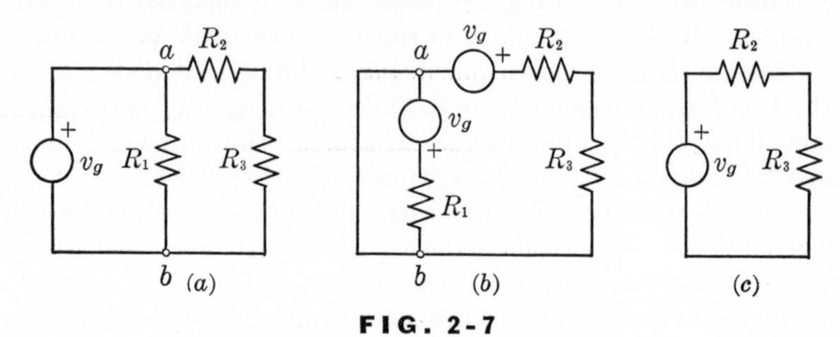

FIG. 2-7

Illustration of the voltage shift

What we have done with a voltage source can be duplicated, in an analogous manner, with a current source. Thus, consider the portion of a network shown in Fig. 2-8(*a*). A closed path is formed by a current source i_g and two resistors R_1 and R_2. Suppose we make two closed paths out of this one, placing a current source i_g in each path, as shown in Fig. 2-8(*b*). This step is analogous to splitting the node in Fig. 2-6.

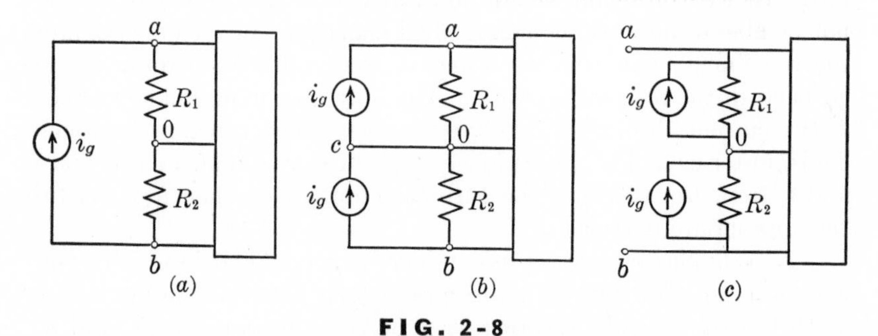

FIG. 2-8

Current-source transformation

Note that no current will flow in the short circuit between points *c* and 0. (Show this by applying Kirchhoff's current law at node *a*.) The current balance at nodes *a*, *b*, and 0 is not changed by this modification; hence the new network will be equivalent to the original one. Now, however, we have two current sources, and each one is in parallel with a resistor. These combinations can be transformed to their voltage-source equivalents and the solution of the problem continued.

Another way of looking at these results is shown in Fig. 2-8(*c*). Instead of forming two closed paths, as in Fig. 2-8(*b*), assume that the current source has moved through the loop until it is across each of the other branches in the loop. This leaves points *a* and *b* open-circuited, thus destroying the original loop formed by i_g, R_1, and R_2. Of course this network is the same as the one in Fig. 2-8(*b*), only slightly redrawn.

Our discussion has been carried out in terms of only two branches on the closed path, besides the current source. All the arguments are still valid, however, no matter how many branches there are on the loop. Hence we can state this result as a general current-shift theorem, as follows:

A current source can be moved through any loop which it forms with other branches and placed across each of these branches, leaving its original position open-circuited, without affecting the voltages and currents anywhere else in the network.

Again note that the identity of the original current source is lost. If we wish to find the voltage across this current source, we must return to the original network after a solution of the modified network is obtained.

As an illustration, let us find the output voltage v in Fig. 2-9. The steps are illustrated in the figure. All numerical values are those of

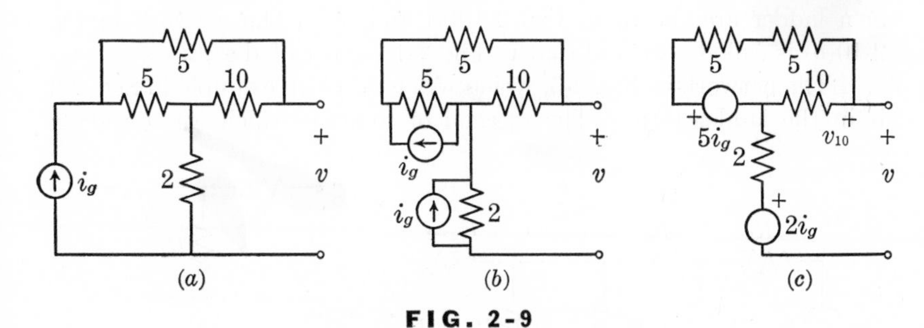

FIG. 2-9

Illustration of current shift

resistance in ohms. In part (*b*) we apply the current-shift theorem. Then we replace the current sources in parallel with the resistors with their voltage-source equivalents to obtain part (*c*). From this figure, Kirchhoff's voltage law gives

$$v = v_{10} + 2i_g \tag{2-20}$$

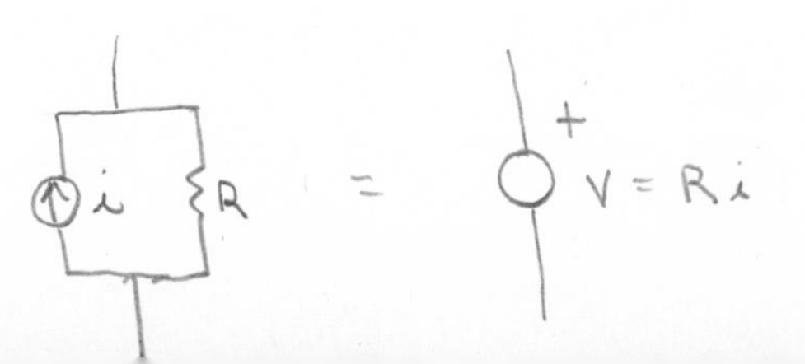

where v_{10} is the voltage across the 10-ohm resistor. Note that there is no current in the 2-ohm resistor. Also, from the figure, note that v_{10} can be obtained from the voltage-divider law applied to the closed loop. It is

$$v_{10} = \frac{10}{10 + 5 + 5} 5i_g = \frac{5}{2} i_g \tag{2-21}$$

When this is substituted into Eq. (2-20), the result becomes

$$v = \tfrac{5}{2} i_g + 2i_g = \tfrac{9}{2} i_g \tag{2-22}$$

The voltage-shift and current-shift theorems permit modification of any network so that all voltage sources appear in series with a branch and all current sources in parallel with a branch, thus permitting the use of source equivalents in all cases.

2.5 LADDER NETWORKS

One of the most common and important structures to be found among electric networks is the so-called ladder network, shown in Fig. 2-10(a). The name derives from the form of the network. The odd-numbered branches are referred to as the *series branches*, and the even-numbered branches are called the *shunt branches*. Two special cases of a ladder are shown in Fig. 2-10(b) and (c). The network in Fig. 2-10(b) is called a *tee* and that in Fig. 2-10(c) is called a *pi*.

The network in Fig. 2-5, which was used as an example in Sec. 2-3, is in the ladder form. The method of solution which we considered

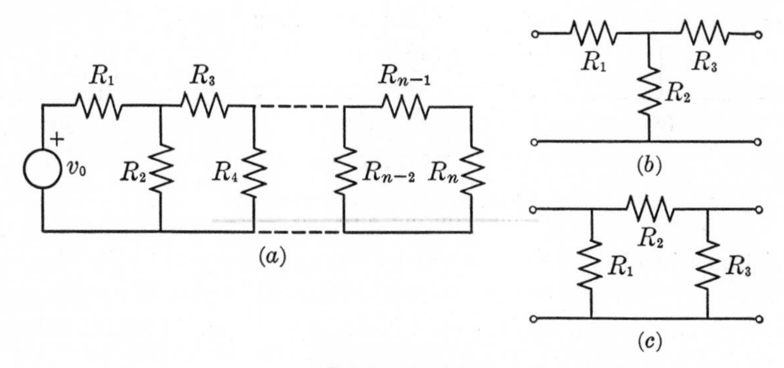

FIG. 2-10

A ladder network and two special ones, a tee and a pi

there, starting from the left-hand end (input) and alternately changing from a voltage-source equivalent to a current-source equivalent, and back again, can be used for any ladder network. That procedure, however, leads to a solution for the voltage or current at the right-hand end (output) only. Nevertheless, after the output voltage or current is found, we can then backtrack and "unwind" all the source changes one by one, solving for each branch voltage and current as we proceed toward the front of the ladder. As a matter of fact, there is no need to *calculate* the output voltage and then work back; we can *assume* an output voltage and work toward the input, eventually finding what the source voltage or current must be to produce the assumed value.

To illustrate this procedure, consider the ladder shown in Fig. 2-11.

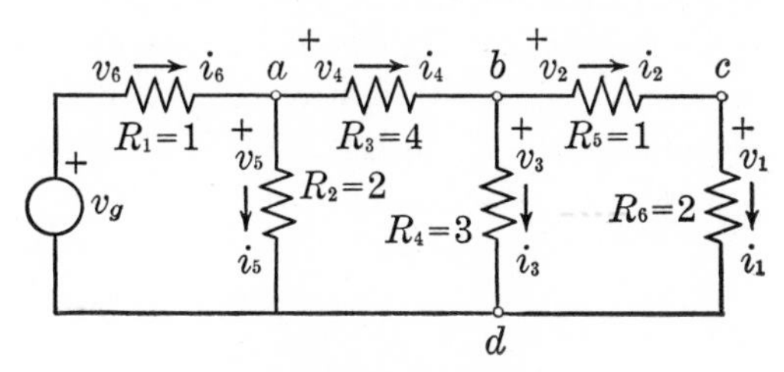

FIG. 2-11
Numerical example

The branch resistances have been given numerical values in order to avoid algebraic complexity which will detract from the desired concentration on the method of solution.

We start from the output end (the right-hand end) by calling the output voltage v_1. (Keep referring to the diagram and arrange not to be distracted until you have finished this paragraph.) Then $i_1 = v_1/2$ by Ohm's law. Since i_2 is the same as i_1, from Kirchhoff's current law applied to node c, we find v_2 to be $v_2 = v_1/2$. Now v_3 is the sum of v_2 and v_1 by the voltage law, and so, $v_3 = 3v_1/2$. For i_3 we then get $i_3 = v_3/3 = v_1/2$ by Ohm's law. At node b we find that i_4 is the sum of i_2 and i_3; so $i_4 = i_2 + i_3 = v_1/2 + v_1/2 = v_1$. From i_4 we find $v_4 = 4i_4 = 4v_1$. Now, by the voltage law, $v_5 = v_4 + v_3 = 4v_1 + 3v_1/2 = 11v_1/2$; hence $i_5 = v_5/2 = 11v_1/4$. From the current law at node a, we get $i_6 = i_4 + i_5 = v_1 + 11v_1/4 = 15v_1/4$. Then $v_6 = i_6 = 15v_1/4$. Finally, from the voltage law, $v_g = v_6 + v_5 = 15v_1/4 + 11v_1/2 = 37v_1/4$.

Thus, to get an output voltage v_1, the source voltage must be $v_g = 37v_1/4$. This means that for a source voltage v_g, the output voltage will be $v_1 = 4v_g/37$. Since all the remaining branch voltages and currents are expressed in terms of v_1 somewhere in the last paragraph, they can now be expressed in terms of v_g. The solution is thus complete.

We now have two methods of solution for a ladder network. In the first method we start from the input end and alternately change from voltage-source equivalent to current-source equivalent, and vice versa, until the output is reached. This procedure is most useful when only the output voltage or current is desired, because, when the process is complete, the structure of the network has been destroyed.

In the second method we assume a voltage at the output and work toward the input, employing Ohm's law, or the voltage law, or the current law at each step, eventually expressing the output voltage and all intermediate voltages and currents in terms of the input voltage or current. The structure of the network is not changed.

In case the desired quantity is the input current in Fig. 2-11, it is possible to proceed in still another way. This involves alternately combining series and parallel branches, working from the output end toward the input end. The following calculations illustrate this process.

R_5 and R_6 in series: $R_5 + R_6 = 3$

R_4 in parallel with this result: $\dfrac{3R_4}{3 + R_4} = \dfrac{3}{2}$

R_3 in series with this result: $R_3 + \frac{3}{2} = \frac{11}{2}$

R_2 in parallel with this result: $\dfrac{\frac{11}{2}R_2}{\frac{11}{2} + R_2} = \dfrac{22}{15}$

R_1 in series with this result: $R_1 + \frac{22}{15} = \frac{37}{15}$

Thus the network, looking toward the right from the terminals of the source, is equivalent to a resistance of $\frac{37}{15}$. Hence

$$i_6 = \frac{15v_g}{37} \tag{2-23}$$

To see whether this agrees with our previous result, note that in our previous calculation we had found $i_6 = 15v_1/4$ and $v_1 = 4v_g/37$. Hence, the present result agrees with the previous one.

2.6 METHOD OF BRANCH VARIABLES

Up to this point we have considered some simple methods of solution which can be applied to networks having certain simple configurations. We are now ready to consider some general methods of solution whose validity is not restricted to special configurations.

Consider the situation portrayed in Fig. 2-12(*a*). A battery charger is to charge a storage battery; a variable resistor is also placed in the

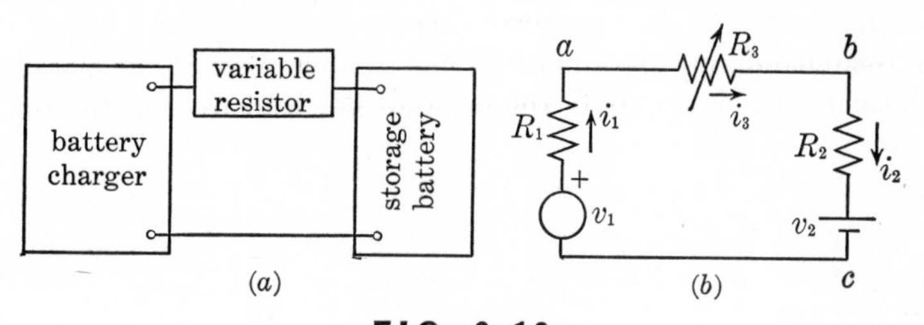

FIG. 2-12

Charging a battery

circuit, in order to control the amount of current. It is desired to find the current supplied by the charger.

The first task is to replace the physical circuit with a model. We shall assume that the battery charger, as well as the battery, can be replaced by a series connection of a voltage source and a resistance, as shown in Fig. 2-12(*b*). Also, it is assumed that the physical resistor can be replaced by an ideal resistor having a resistance R_3. The arrow on the symbol for the resistor indicates that it is variable. With the model at hand, we can now proceed to solve the problem.

Let us first consider the topological properties of the network. As we develop the theory of network analysis, we shall discuss several topological concepts. At the present, however, the only topological properties we have mentioned are the concepts of branches, nodes, and loops. In Fig. 2-12(*b*) let us count the voltage source v_1 in series with R_1 as a single branch, and let us do the same for v_2 in series with R_2. (This step will be clarified shortly.) Thus the network has three branches. As for nodes, there are three of them in the figure, labeled *a*, *b*, and *c*. Finally, there is a single closed loop.

For each of the branches there are two variables—the voltage and the current; hence there are a total of six unknowns. In order to solve for six unknowns, we must have at least six equations all of which are *independent*. We shall illustrate what we mean by independent equations shortly. We can write three equations expressing the voltage-current relationship of the three branches as follows:

Branch a-c: $\quad v_{ac} = v_1 - R_1 i_1 \quad (a)$

Branch b-c: $\quad v_{bc} = v_2 + R_2 i_2 \quad (b) \quad$ (2-24)

Branch a-b: $\quad v_{ab} = R_3 i_3 \quad (c)$

In these expressions v_1 and v_2 as well as R_1, R_2, and R_3 are assumed known.

Now there are three equations left. Since there is one closed loop in the network, Kirchhoff's voltage law will lead to one equation, as follows:

$$v_{ab} + v_{bc} + v_{ca} = 0 \tag{2-25}$$

We need two more equations. We can get three equations by writing Kirchhoff's current law at each of the three nodes. The three equations are:

Node a: $\quad -i_1 + i_3 = 0 \quad (a)$

Node b: $\quad i_2 - i_3 = 0 \quad (b) \quad$ (2-26)

Node c: $\quad i_1 - i_2 = 0 \quad (c)$

For the simple case at hand, it is not necessary to be so formal and write these three equations. From a quick glance at the network, we can see that the same current flows in all branches, so that $i_1 = i_2 = i_3$, which is the same information we get from the equations. Let us, however, keep the three equations for emphasis, to remind us, in more complicated cases, that what we are doing is applying Kirchhoff's current law at each of the nodes.

By examining these equations, we notice that each one of them can be obtained by adding the other two. A set of quantities (in the present case, equations) are said to be *independent* if no one of them can be obtained as a linear combination of others; that is, no one of them can be obtained by multiplying the others by appropriate constants and adding. If it is possible to obtain one of the quantities in terms of the others by a linear combination, then the set is said to be *dependent.* The three equations we have are, therefore, dependent.

This result can also be obtained by examining the network. If Kirchhoff's current law is satisfied at any two of the three nodes, it will automatically be satisfied at the third one. Hence we really have only two independent equations arising from Kirchhoff's current law in this network—just the number we need. Which two of the three we choose to keep is immaterial. For purposes of illustration, let us retain the equations at nodes a and b.

1 Branch equations
2 Loop eqs.
3 Node eqs.

We now have six equations in six unknowns, and the job is to solve these equations. To do this, we can proceed in several ways. As one possibility, let us substitute the three v-i relationships from Eq. (2-24) into Eq. (2-25). The result will be

$$R_3 i_3 + (v_2 + R_2 i_2) - (v_1 - R_1 i_1) = 0 \tag{2-27}$$

This eliminates all the branch-voltage variables, leaving just the three currents. We can now use the Kirchhoff current-law equations from Eq. (2-26) to eliminate two of the branch currents, leaving just the one we want. (In the present case this step is almost trivial, since $i_1 = i_2 = i_3$.) Finally, we get

$$i_2 = \frac{v_1 - v_2}{R_1 + R_2 + R_3} \tag{2-28}$$

This is the desired result. Knowing i_2, we can now solve for i_1 and i_3 from Eq. (2-26); then we can solve for the branch voltages from Eq. (2-24), thus finding all the unknowns in the network.

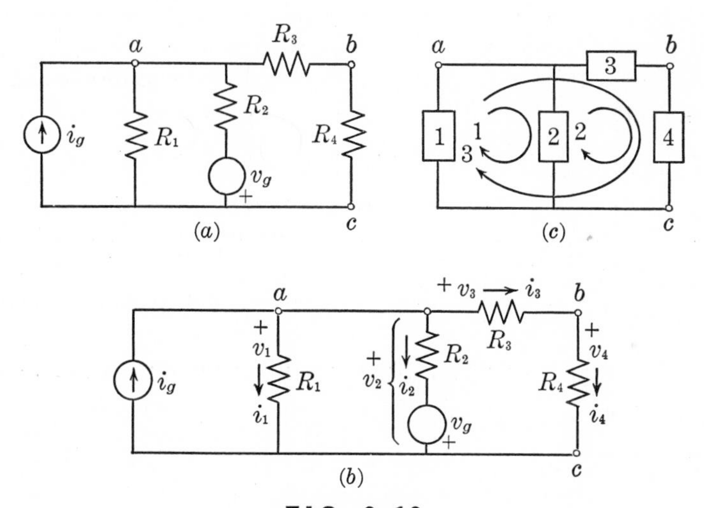

FIG. 2-13

Network example

Let us now consider another example. The network of Fig. 2-5 is redrawn in Fig. 2-13. Let us choose the current source in parallel with resistance R_1 as a single branch, and do the same for the voltage source in series with R_2. Thus the network has four branches, and their v-i relationships are as follows:

number of branches = number of passive elements (ie) R, C, L

$$
\begin{aligned}
v_1 &= R_1 i_1 && (a) \\
v_2 &= R_2 i_2 - v_g && (b) \\
v_3 &= R_3 i_3 && (c) \\
v_4 &= R_4 i_4 && (d)
\end{aligned}
\qquad (2\text{-}29)
$$

There are three nodes in the network, labeled a, b, and c. Again we note that, if Kirchhoff's current law is satisfied at two of the nodes, it will automatically be satisfied at the other one also. Hence let us choose nodes a and b at which to write Kirchhoff's current-law equations. The result will be

$$
\begin{aligned}
&\text{Node } a\text{:} && i_1 - i_g + i_2 + i_3 = 0 \\
&\text{Node } b\text{:} && -i_3 + i_4 = 0
\end{aligned}
\qquad (2\text{-}30)
$$

Finally, there remains Kirchhoff's voltage law. In Fig. 2-13(c) we have drawn the outline of the network using rectangles to represent the branches. This avoids confusion when our interest is centered only on the geometry. We see that there are a total of three closed loops in the network, as indicated by the arrows labeled 1, 2, and 3. When we apply Kirchhoff's voltage law at these loops, we get

$$
\begin{aligned}
&\text{Loop 1:} && -v_1 + v_2 = 0 && (a) \\
&\text{Loop 2:} && -v_2 + v_3 + v_4 = 0 && (b) \\
&\text{Loop 3:} && -v_1 + v_3 + v_4 = 0 && (c)
\end{aligned}
\qquad (2\text{-}31)
$$

A brief examination shows that these equations are not independent; the last one, for example, is the sum of the first two. Hence only two of the equations in Eq. (2-31) need be used; which two are chosen is immaterial.

We now have eight equations in the eight unknowns. Four of these are the v-i relationships of the branches; two of them express the current law; the remaining two express the voltage law. To carry through the solution, let us now substitute the branch v-i relationships from Eq. (2-29) into the current-law equations in Eq. (2-30). The result will be

$$
\begin{aligned}
\frac{v_1}{R_1} - i_g + \frac{v_2}{R_2} + \frac{v_g}{R_2} + \frac{v_3}{R_3} &= 0 && (a) \\
-\frac{v_3}{R_3} + \frac{v_4}{R_4} &= 0 && (b)
\end{aligned}
\qquad (2\text{-}32)
$$

These are two equations in the four unknown branch voltages. We still have the voltage-law equations. Let us choose Eqs. [2-31(a)] and [2-31(b)] as the two independent Kirchhoff voltage-law equations, thus adding two more equations in the same four variables to the ones in Eq. (2-32). We can eliminate v_1 by substituting Eq. [2-31(a)] into Eq. [2-32(a)]. Similarly, we can eliminate v_4 by substituting Eq. [2-32(b)] into Eq. [2-31(b)]. The results will be

$$\frac{v_2}{R_1} + \frac{v_2}{R_2} + \frac{v_3}{R_3} = i_g - \frac{v_g}{R_2} \qquad (a)$$

$$-v_2 + v_3 + \frac{R_4}{R_3} v_3 = 0 \qquad (b) \tag{2-33}$$

Now we have only two equations in two unknowns. We can solve the last equation for v_2 and substitute into the previous one, leaving only one equation with v_3 as an unknown. Finally, we solve that one for v_3. The result will be

$$\frac{R_3 + R_4}{R_3}\left(\frac{1}{R_1} + \frac{1}{R_2}\right) v_3 + \frac{v_3}{R_3} = i_g - \frac{v_g}{R_2} \tag{2-34}$$

or

$$v_3 = \frac{R_1 R_3 (R_2 i_g - v_g)}{(R_1 + R_2)(R_3 + R_4) + R_1 R_2} \tag{2-35}$$

This expression can now be substituted into the preceding equations and all the unknowns determined. You should carry out the details which have been omitted.

2.7 THE NUMBER OF NETWORK EQUATIONS

Let us now look back over the two examples that we considered in order to determine if there is any general information that can be gleaned. In the first place, we notice that the sets of equations which are obtained from an application of each of Kirchhoff's two laws are not independent. In each of the examples we found that the number of independent current-law equations is one less than the number of nodes. This result is actually a general one and applies universally to all networks, as we shall discover shortly.

Let us designate by N_v the number of nodes in a network. (The subscript v stands for vertex.) It is easy to see that, of the N_v Kirchhoff current-law equations in a network, at least one is superfluous. Consider, for example, the general network shown in Fig. 2-14, in which one of the nodes a is isolated, and write current-law equations at all the nodes except a; then add them. Each branch connects just two nodes; hence each branch current will appear in exactly two of the equations, except for those branches which are connected to node a. In the internal branches the current references will be toward one node and away from the other. Hence, when the equations are added, these internal-branch currents will cancel; only the branches which connect node a to the other nodes will remain. Thus Kirchhoff's current law at node a is satisfied automatically if it is satisfied at all the other nodes.

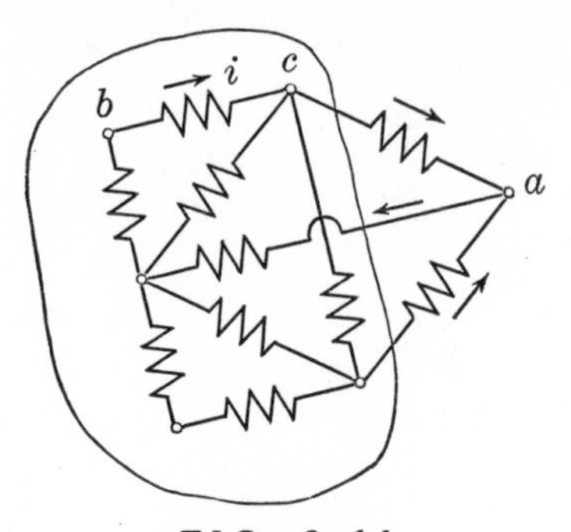

FIG. 2-14
Arbitrary Network

It might appear that this discussion proves that the number of independent Kirchhoff's current-law equations is exactly one less than the number of nodes, $N_v - 1$. This is not so. We have proved that one of the N_v equations is superfluous, but we have not eliminated the possibility that additional ones may be superfluous also. It may seem intuitively true, and for any specific case we find exactly $N_v - 1$ equations *are* independent, but this does not constitute a proof.

One avenue open to us is to accept this result as a hypothesis based on the available evidence. This is a legitimate procedure. It appears, however, that nature operates under the law of parsimony. If the number of hypotheses on which a theory is based becomes very large, there will be a strong suspicion that the theory does not correspond to reality, and that a different theory will explain the observed phenomena just as well with fewer hypotheses.

Actually in the next section we shall complete the proof that *the number of independent Kirchhoff current-law equations is one less than the number of nodes, $N_v - 1$.* Some of you, however, may wish to skip that section. Hence we shall accept this result as proved.

As for the Kirchhoff voltage-law equations, we saw in the second example of the last section that, of the three possible equations, only two were independent. From this single result it is not easy to see what the generalization should be. In order to get a clue, let us see how many voltage-law equations are needed in order to obtain a complete solution of a network problem.

To start with, let N_b be the number of branches in a network. We must first inquire into the number of unknowns. Any branch consisting of an R, L, or C element will have two unknowns, the voltage and the current. If we consider a voltage source alone, only the current is unknown. Similarly, a current source alone has only one unknown—the voltage. Suppose, however, that we consider a voltage source in series with a resistor as a single branch. For this combined branch there will be only two unknowns; there is but a single current, and, once the voltage across the entire branch is known, that across the resistor alone will be known also. Likewise, a branch consisting of a current source in parallel with a resistor will contain only two unknowns.

In case a voltage source is not initially in series with a single element, the voltage-shift theorem leads to a modification which permits the voltage source to acquire an element in series. Likewise, if a current source is not initially in parallel with a single element, the current-shift theorem permits each current source to acquire a parallel resistor. With these considerations, we conclude that voltage sources and current sources can be disregarded in counting the number of unknowns. Since a voltage source in series with a resistor does not influence the number of variables, we can assume the voltage source to be a short circuit for the purpose of counting unknowns. Similarly, we can assume that current sources are open circuits for the purpose of counting unknowns. (This discussion should clarify the reason for the choice of branches in the illustrative examples of the last section.)

With a network containing N_b branches, there will be $2N_b$ unknowns. To solve for $2N_b$ unknowns, we must have $2N_b$ independent equations. Of course these unknowns are related by N_b v-i relationships (Ohm's law); this leaves N_b more which are still required. But we have found that $N_v - 1$ independent equations are provided by Kirchhoff's current law. Hence we need exactly $N_b - (N_v - 1)$ independent equations to be provided by Kirchhoff's voltage law.

Let us check the examples in the last section to see how many such equations we had. In the first example there were three branches and three nodes, giving $N_b - (N_v - 1) = 1$; and there was only one voltage-law equation. In the second example we had $N_b = 4$, $N_v = 3$, and $N_b - (N_v - 1) = 2$, which was exactly the number of independent voltage-law equations. In the examples, at least, the number of independent voltage-law equations was just the number needed according to our reasoning.

In the next section we shall prove that, in a network having N_b branches and N_v nodes, *the number of independent Kirchhoff voltage-law equations is* $N_m = N_b - (N_v - 1)$. From here on, we shall assume that this result is proved.

We are now reassured that in each problem there are exactly the right number of independent equations to determine all the unknowns—$2N_b$ of them. "Solving a problem" then consists of finding a simultaneous solution of all these equations. This is no mean task, since $2N_b$ is a rather large number of simultaneous equations to solve. We have already indicated one method of solution in the examples. This consists of systematically eliminating variables until a single equation in only one unknown is obtained. This is not always the easiest method to use, nor is it the most general. In the pages to come, we shall develop additional methods for performing this task.

Choosing an appropriate set of nodes for applying Kirchhoff's current law is relatively easy; all we need to do is to omit one node—any one—and write equations at all the other nodes. The situation is not so simple when it comes to choosing an appropriate set of loops for applying Kirchhoff's voltage law. In writing the equations, we certainly should make sure that each branch appears at least once in an equation; otherwise, we cannot account for its influence on the solution. Although this condition is necessary, it is not sufficient to guarantee that the correct number of equations has been chosen.

There is a rather large and important class of networks for which it is easy to choose an appropriate set of loops for the application of Kirchhoff's voltage law. Figure 2-15(a) shows such a network. It

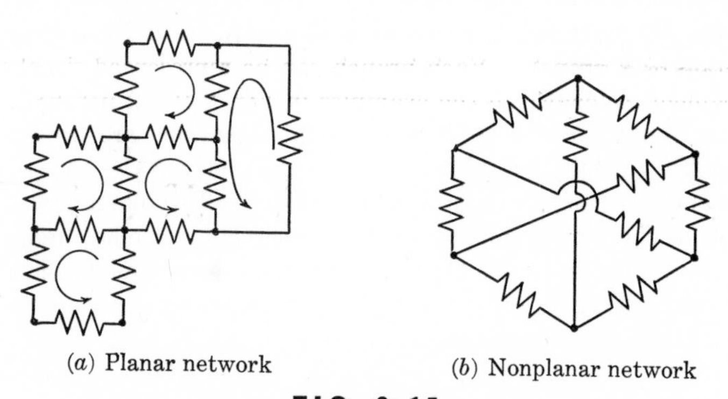

(a) Planar network (b) Nonplanar network

FIG. 2-15

Planar and nonplanar networks

can be drawn on a plane without having any branches cross each other; it is called a *planar* network. In contrast to this is a *nonplanar* network, shown in Fig. 2-15(b). No matter how this network is redrawn on the paper, at least two of the branches will cross each other.

For a planar network a simple and appropriate set of loops consists of the small "meshes" or "windows," as indicated by the arrows in Fig. 2-15(a). (For simplicity, it is more convenient to choose all the loop orientations similarly—all clockwise or all counterclockwise—although this is not necessary.) This choice of loops will always lead to an independent set of Kirchhoff voltage-law equations. In the case of nonplanar networks, it is not so easy to choose an appropriate set of closed loops; but, even in this case, we know that the correct number of equations is $N_b - (N_v - 1)$, and we know that each branch should be included in at least one equation.

2.8 NETWORK TOPOLOGY

Network topology is a generic name given to the geometric properties of a network that depend on the existence of a network as an interconnection of branches. In order to gain some reliable knowledge concerning the number of independent equations in a network of arbitrary structural complexity, we shall turn to a brief study of network topology.

Since we are concerned only with the properties of figures that are made up of branches connected at nodes, it is not necessary to show the details of a branch. Each branch can be represented simply by a line segment, as shown in the examples in Fig. 2-16. Such figures are

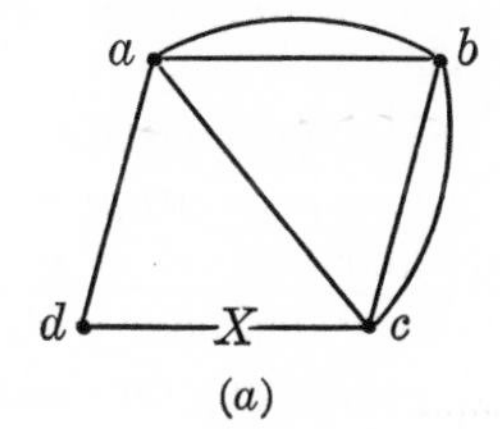

(a)

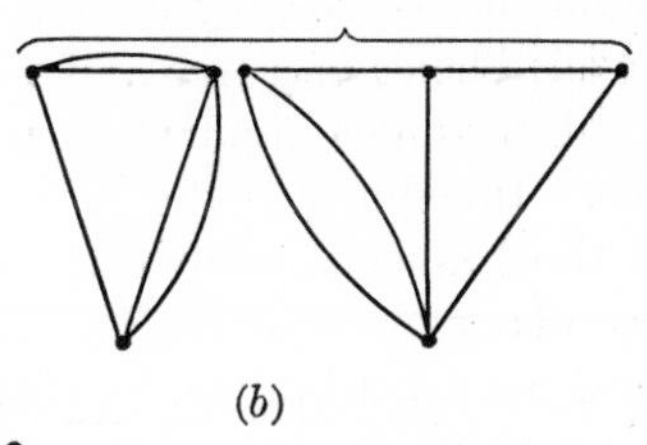

(b)

FIG. 2-16

Network graphs

called network *graphs*. A graph is made up of a number of nodes which are interconnected by branches. If there is a path which can be traced from any node to every other node, the graph is said to be connected. [See Fig. 2-16(a).] Otherwise, the graph will consist of several parts; the network in Fig. 2-16(b) is in two parts. We shall initially deal with connected networks only.

In the graph of Fig. 2-16(a), the branch between nodes c and d, marked with an X, can be removed, and the graph will still remain connected. Some closed loops formed with that branch, however, will have been destroyed. On the other hand, if all the branches are removed, the graph will become completely disconnected. It is clear that there must be a state in which the graph remains connected but all the closed loops are destroyed.

We shall define a *tree* of the graph as a set of branches which connect all the nodes without forming a closed path. It should be clear that a given network can have many trees. For example, Fig. 2-17

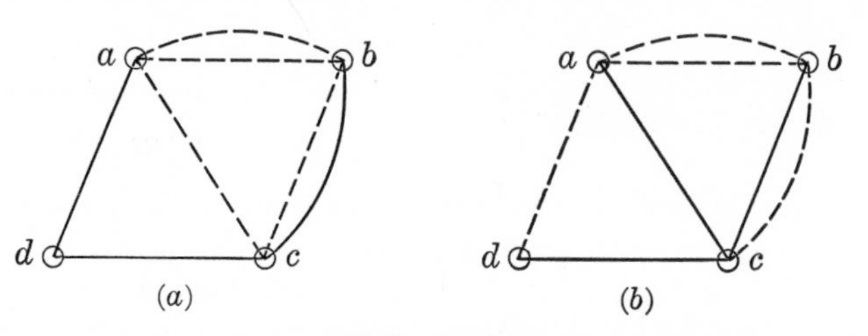

FIG. 2-17

Trees (solid lines) and links (dashed lines)

shows two possible trees (solid lines) of the graph of Fig. 2-16(a). Each branch of a tree is, appropriately enough, called a *tree branch*, or a *twig*. For a given tree, the branches which are not on the tree are called *links* (dashed lines in the figures). Note that a given branch of a network cannot be irrevocably designated as a twig or a link; it may be a twig for one choice of a tree and a link for a different choice.

Let us now inquire into the number of branches which constitute a tree. Imagine that the N_v nodes of a graph are placed in position with all the branches removed. Start with any node and connect it to a second one by one of the branches. Now connect the second node with a third one by a branch. Continue in this fashion, connecting one node at a time until all the nodes are exhausted. The result is a tree. Each successive branch connects one additional node, except for the first branch which connects two nodes. Hence there will be one more node than branches on the tree, which is equivalent to saying that there will be one less branch than nodes; that is, *a tree consists of* $N_v - 1$ *branches*. Since the total number of branches in a network is N_b, *the number of links in a network is* $N_b - (N_v - 1)$.

So far we have discussed Kirchhoff's current law in terms of a summation of currents at a node. Suppose we now consider a group of nodes together, as shown within the circle in Fig. 2-18(*a*). The branches connected to the nodes of this group fall in two sets: those that run between nodes *within* the group, and those that run between one node of the group and a node external to the group. The branches marked with an X in the figure constitute this second set. If we write Kirchhoff's current law at each node in the group, and add the equa-

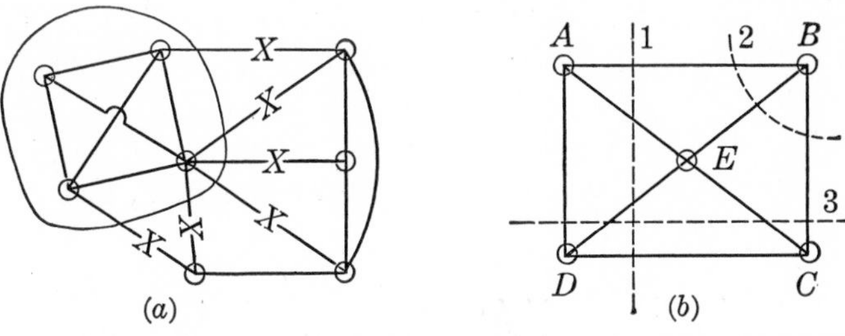

FIG. 2-18
Cut sets

tions, the internal branch currents will cancel each other, just as we discussed with regard to Fig. 2-14. Only the X-marked branch currents will remain. Thus, if we were to conceive of the group of nodes together as constituting one giant node, Kirchhoff's current law would be satisfied at this giant node. This is true for any group of nodes taken together.

Observe that the X-marked branches in Fig. 2-18 have the distinctive property that, if they were removed from the network, the network would become disconnected, or cut, into two parts. We make this property the basis for a definition, as follows:

A cut set is a set of the smallest number of branches of a network which, when cut, will divide the network into two parts. Thus the X-marked branches in Fig. 2-18 constitute a cut set. Note that it is possible to cut some additional branches in Fig. 2-18 (find them) without increasing the number of separate parts, but if any one of the X-marked branches is not cut, the network will remain connected. This is the reason for specifying the number of branches in the cut set to be the minimum number. Figure 2-18(*b*) shows some of the cut sets for a given network. Each of the dashed lines divides the network into two parts, and the branches crossing a dashed line constitute a cut set.

A cut set divides the network into two groups of nodes, or two giant nodes, and Kirchhoff's current law will be satisfied at each of these giant nodes, as we have seen. The Kirchhoff current-law equation, written at either of the giant nodes into which the network is divided by a cut set, is called a *cut-set equation.* (One equation will be just the negative of the other.) Each cut-set equation is a linear combination (simply a sum) of current-law equations written at nodes. Hence, if we determine the number of independent cut-set equations, we shall have determined also the number of independent Kirchhoff current-law equations.

Let us proceed by considering a tree; it has $N_v - 1$ branches. If any one of these branches is cut, the tree will be separated into two parts; for example, a tree of a network is shown by solid lines in Fig. 2-19. Cutting branch *A-B* will divide the nodes, and hence the tree, into two groups—nodes *A* and *D* in one group and nodes *B*, *C*, and *E* in the other.

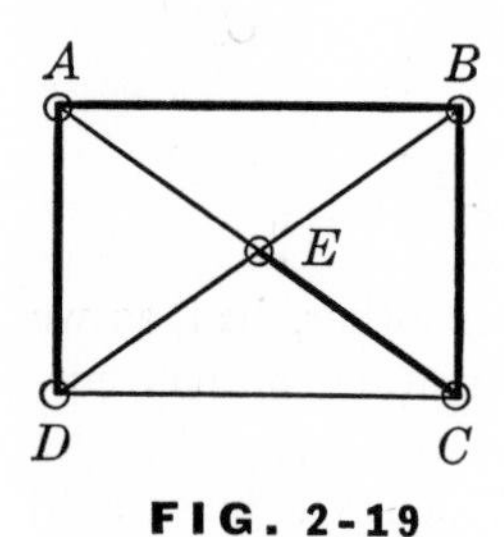

FIG. 2-19

Definition of fundamental cut sets.

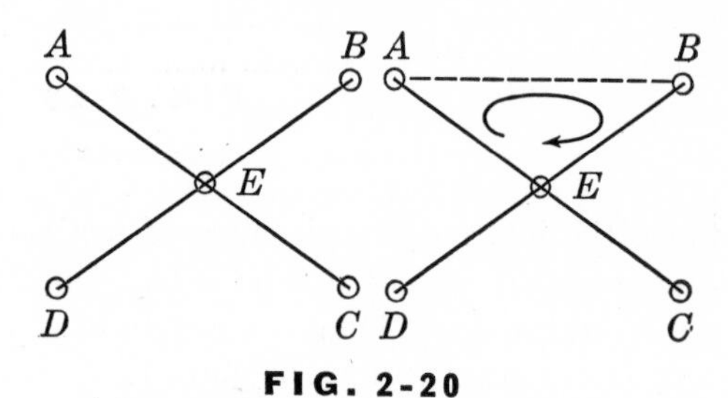

FIG. 2-20

Definition of fundamental loops

Now consider the links that go from one part of the network to the other. These links, together with the tree branch which was cut, constitute a cut set. Thus, links *A-E*, *D-E*, and *D-C*, together with twig *A-B* in Fig. 2-19, constitute a cut set. [This is cut set 1 in Fig. 2-18(*b*).] Each tree branch determines such a cut set. Since there are $N_v - 1$ tree branches, there will be $N_v - 1$ such cut sets, each containing only one tree branch. We call these cut sets which have only one twig the *fundamental cut sets.* The cut-set equations of these cut sets will be independent, since each one will contain a tree-branch current not appearing in any other equation, which means that this current cannot be obtained by linear combinations of the other equations. Thus we have found that at least $N_v - 1$ cut-set equations are independent. We still do not know, however, whether other cut-set equations, when added to the fundamental cut-set equations, are also independent. We shall now show that there are no other independent cut-set equations.

First of all, note that any cut set must contain at least one twig; it cannot consist of links only, because, even if all the links are cut, a tree will still remain, and, by definition, a tree connects all the nodes. Thus any cut set which is not a fundamental cut set must contain more than one twig. In Fig. 2-18(*b*) cut set 1 contains one twig, cut set 2 contains two twigs, and cut set 3 contains three twigs, with the tree chosen as in Fig. 2-19.

Let us contemplate writing a cut-set equation for a cut set having two or more tree branches. From the fundamental cut-set equations, each twig current can be expressed in terms of link currents. Using these expressions, we can eliminate the twig currents in the cut-set equation in question, leaving only link currents. Furthermore, the coefficient of each link current in the resulting expression must become zero, so that the resulting equation must reduce to an identity $0 = 0$. If this were not true, it would mean the existence of a current equation containing link currents only. This would permit one link current, say i_1, to be expressed as a linear combination of other link currents. But if all link currents except i_1 were now reduced to zero by cutting the links, this expression would force the vanishing of i_1 as well. But this is not possible, since, even if all links except one are cut, this branch, together with tree branches, will form a closed path whose current will in no way be affected by the vanishing of all other link currents. (Draw a network and remove all links but one to demonstrate this fact to yourself.)

The conclusion is that *any cut-set equation can be obtained as a linear combination of the fundamental cut-set equations*; no other cut-set equations are independent. This completes the proof that, in a network having N_v nodes, the number of independent current-law equations is $N_v - 1$.

Note that each twig current can be expressed in terms of link currents, as is evident from the fundamental cut-set equations. Thus, once all the $N_b - (N_v - 1)$ link currents are determined by some process, all the branch currents in the network will become known.

Let us now turn our attention to Kirchhoff's voltage law. We have already established that the number of links in a network is $N_b - (N_v - 1)$. Now consider a network with only the tree branches in place, as shown in Fig. 2-20(*a*). The network is the same as the one in Fig. 2-19. Now add one link, as shown by the dotted line in Fig. 2-20(*b*). This branch, together with tree branches, will form a closed path or loop. If the links are added to the tree one at a time, each link will form such a loop. Since there are $N_b - (N_v - 1)$ links, there will be $N_b - (N_v - 1)$ such loops, each one containing only one link. We call these loops which have only one link the *fundamental loops.* They are also sometimes called the *fundamental tie sets,* a tie set being a set of branches comprising a closed path.

The Kirchhoff voltage-law equations written around these fundamental loops will be independent, since each equation contains a branch voltage (that of the link) not contained in any of the other equations. Thus there are at least $N_b - (N_v - 1)$ Kirchhoff voltage-law equations.

We must now show that all other Kirchhoff voltage-law equations are dependent on the equations around the fundamental loops, thus showing that there are no other independent ones. Note that any closed loop must contain at least one link; it cannot consist of twigs only, because, by the definition of a tree, the twigs alone form no closed paths. Thus any loop which is not a fundamental loop must contain two or more links.

Let us write a voltage-law equation around a loop containing two or more links, say loop 1. The link voltages can be expressed in terms of twig voltages from the voltage equations for the fundamental loops. If we now eliminate the link voltages from the voltage equation of loop 1, leaving only twig voltages, the result must be an identity $0 = 0$; otherwise, a voltage equation would exist containing twig voltages only. This would permit one twig voltage, say v_1, to be expressed in terms of other twig voltages. If all tree-branch voltages except v_1 were now reduced to zero by short-circuiting the tree branches, the expression relating v_1 to twig voltages would force the vanishing of v_1 also. But this is not possible, since short-circuiting all tree branches but one will still not cause this one to be short-circuited. (Draw a tree and demonstrate this to yourself.)

The conclusion is that *the voltage equation around any loop can be obtained as a linear combination of voltage equations around the fundamental loops;* no other voltage equations are independent. This completes the proof that, in a network of N_b branches and N_v nodes, the number of independent voltage-law equations is $N_m = N_b - (N_v - 1)$.

Note that each link voltage can be expressed in terms of twig voltages, as is evident from the voltage equations of the fundamental loops, each of which contains one link voltage only. Thus, once all the $N_v - 1$ twig voltages are determined by some process, all the branch voltages of the network will become known.

The topological concepts which have been introduced in this section have proved fruitful in determining the number of independent equations obtained from Kirchhoff's two laws. These concepts find application in other topics in network theory as well. You are urged to consult additional references on the subject.† We shall carry out our subsequent work more or less independently of the discussion in this section. Where appropriate, however, we shall allude to topological concepts in parallel discussions of the topics under consideration.

2.9 LOOP EQUATIONS

In the preceding sections of this chapter, we have found that the branch voltages and currents of a network must satisfy three sets of equations: (1) $N_v - 1$ Kirchhoff current-law equations; (2) $N_b - (N_v - 1)$ Kirchhoff voltage-law equations; and (3) N_b branch v-i relationships. These three sets together contain $2N_b$ equations which must all be employed in a complete solution of any network problem.

In Sec. 2-6 we discussed a method of solution using the branch variables which required the simultaneous solution of $2N_b$ equations. We shall now consider a method which avoids the simultaneous solution of $2N_b$ equations. The method applies to any network of arbitrary structure. We shall introduce the method by means of an example, but we shall point out its general applicability at various points as we go along.

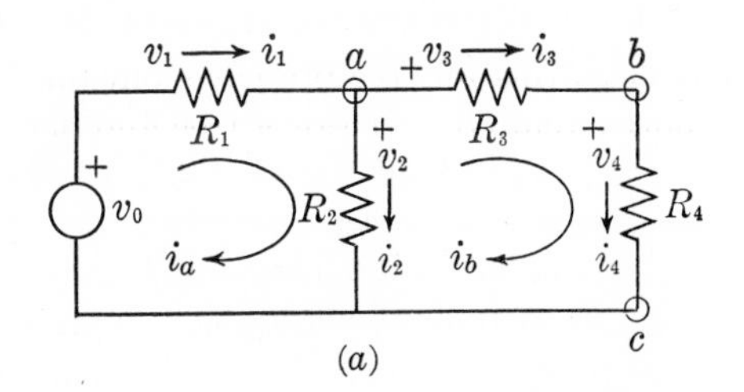

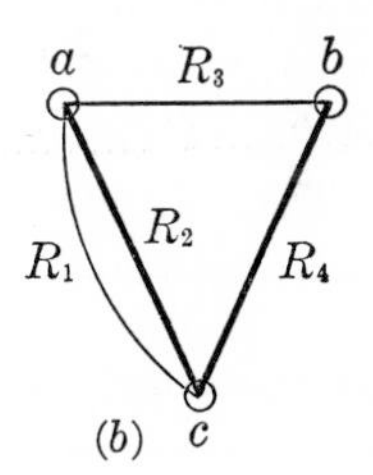

FIG. 2-21

Example for loop equations

Consider the network shown in Fig. 2-21. It has four branches and three nodes. There will be $N_v - 1 = 2$ independent Kirchhoff

† E. A. Guillemin, *Introductory Circuit Theory*, John Wiley & Sons, Inc., New York, 1955.

current-law equations. Let us write current equations at nodes a and b.

$$-i_1 + i_2 + i_3 = 0 \qquad (a)$$
$$-i_3 + i_4 = 0 \qquad (b) \qquad (2\text{-}36)$$

There are only two equations but four unknowns. Such a set of equations cannot be solved completely. Suppose i_1 and i_3, however, are known (or can be found from other information). Then the other two currents can be found from Eq. (2-36).

$$i_2 = i_1 - i_3 \qquad (a)$$
$$i_4 = i_3 \qquad (b) \qquad (2\text{-}37)$$

Now look at Fig. 2-21 and assume that there are two circulating currents around the closed loops, indicated by i_a and i_b. These fictitious circulating currents are called *loop currents*. We see that i_a, being the only circulating current going through R_1, is identical with i_1. Similarly, i_b is identical with i_3. Thus, if we were able to find i_a and i_b by other methods, then the remaining branch currents would follow from Eq. (2-37), and the problem would be solved.

Note that the number of loop currents needed, in terms of which all the other currents are expressed, is two in this example. Since $N_b = 4$ and $N_v = 3$, this number is also equal to $N_b - (N_v - 1)$, which is the number of independent Kirchhoff voltage-law equations. This is not an accident; it is a general result which is universally valid.

[To see its universality, refer to Sec. 2-8. Once the link currents are known, all other branch currents will become known. The loop currents are link currents for a particular choice of tree, and there are $N_b - (N_v - 1)$ of them. The graph of the network in the illustrative example is shown in Fig. 2-21(b). Choosing R_2 and R_4 as a tree (dark lines) leaves R_1 and R_3 as links. Thus, knowing i_1 and i_3 will determine i_2 and i_4 as well.]

We can make use of this universal fact in the following way. First we write an appropriate number of Kirchhoff voltage-law equations, namely $N_b - (N_v - 1)$. (The unknowns here are the N_b branch voltages.) Then we substitute the branch v-i relationships into these equations; the unknowns now become the N_b branch currents. Finally, we express the branch currents in terms of the loop currents.

Let us now follow this procedure in the illustrative example. In Fig. 2-21 there are three possible closed paths (the two meshes and the outside contour) around which Kirchhoff's voltage law can be applied, of which only two are independent. In the present case any two of these three will lead to independent equations, and which two we choose is immaterial. Let us choose the two meshes. The equations will be

$$\begin{aligned} v_1 + v_2 &= v_0 && (a) \\ -v_2 + v_3 + v_4 &= 0 && (b) \end{aligned} \tag{2-38}$$

The next step is to substitute the branch v-i relationships (Ohm's law) into these equations. The result will be

$$\begin{aligned} R_1 i_1 + R_2 i_2 &= v_0 && (a) \\ -R_2 i_2 + R_3 i_3 + R_4 i_4 &= 0 && (b) \end{aligned} \tag{2-39}$$

Finally, using Eq. (2-37) and remembering that the loop currents are identical with i_1 and i_3, we get

$$\begin{aligned} R_1 i_a + R_2(i_a - i_b) &= v_0 && (a) \\ -R_2(i_a - i_b) + R_3 i_b + R_4 i_b &= 0 && (b) \end{aligned} \tag{2-40}$$

It remains now to solve these two equations in two unknowns. One method is the process of elimination. We can solve the second equation for i_a, for example, and substitute into the first, leaving only i_b as unknown there. This equation is then solved for i_b.

There is also a more formal method of solving simultaneous algebraic equations which involves determinants. In preparation for this method of solution, let us rearrange the equations by grouping all the terms involving the same loop current, as follows:

$$\begin{aligned} (R_1 + R_2)i_a - R_2 i_b &= v_0 && (a) \\ -R_2 i_a + (R_2 + R_3 + R_4)i_b &= 0 && (b) \end{aligned} \tag{2-41}$$

These are a set of simultaneous linear algebraic equations which have the following form:

$$\begin{aligned} a_{11}x_1 + a_{12}x_2 &= y_1 && (a) \\ a_{21}x_1 + a_{22}x_2 &= y_2 && (b) \end{aligned} \tag{2-42}$$

where the a coefficients are known quantities. This set of equations can be solved by multiplying the first equation by a_{22}, the second by a_{12}, and subtracting. Alternatively, the first equation can be multiplied by a_{21}, the second by a_{11}, and the first subtracted from the second. The results will be

$$\begin{aligned} (a_{11}a_{22} - a_{12}a_{21})x_1 &= a_{22}y_1 - a_{12}y_2 && (a) \\ (a_{11}a_{22} - a_{12}a_{21})x_2 &= -a_{21}y_1 + a_{11}y_2 && (b) \end{aligned} \tag{2-43}$$

When solved for x_1 and x_2, these become

$$x_1 = \frac{a_{22}y_1 - a_{12}y_2}{a_{11}a_{22} - a_{12}a_{21}} \quad (a)$$

$$x_2 = \frac{-a_{21}y_1 + a_{11}y_2}{a_{11}a_{22} - a_{12}a_{21}} \quad (b) \qquad (2\text{-}44)$$

These expressions can be put into a neat, compact form in terms of determinants. The *determinant* of the set of equations is defined as

$$\Delta = \begin{vmatrix} a_{11} & a_{12} \\ a_{21} & a_{22} \end{vmatrix} = a_{11}a_{22} - a_{12}a_{21} \qquad (2\text{-}45)$$

The minor of a determinant is again a determinant with one row and one column removed; it is denoted by the symbol M, with two subscripts representing the row and column removed. Thus M_{12} is the determinant obtained by removing row 1 and column 2 from Δ. A *cofactor* is defined to be the same as a minor except for sign. The cofactor is denoted by the symbol Δ with two subscripts identical with those of the corresponding minor. Thus

$$\Delta_{ij} = (-1)^{i+j}M_{ij} \qquad (2\text{-}46)$$

When the sum of the subscripts is even, the cofactor has the same sign as the minor; when the sum of the subscripts is odd, the cofactor has the sign opposite from that of the minor.

In terms of these definitions, Eq. (2-44) can be written as follows:

$$x_1 = \frac{\Delta_{11}y_1 + \Delta_{21}y_2}{\Delta} \quad (a)$$

$$x_2 = \frac{\Delta_{12}y_1 + \Delta_{22}y_2}{\Delta} \quad (b) \qquad (2\text{-}47)$$

This form of the solution is often called Cramer's rule. Although we have shown it only for the case of two simultaneous equations, it can be extended to a set of more equations. For a set of n equations, the solution can be written formally as follows:

$$x_1 = \frac{\Delta_{11}y_1 + \Delta_{21}y_2 + \Delta_{31}y_3 + \cdots + \Delta_{n1}y_n}{\Delta}$$

$$x_2 = \frac{\Delta_{12}y_1 + \Delta_{22}y_2 + \Delta_{32}y_3 + \cdots + \Delta_{n2}y_n}{\Delta} \qquad (2\text{-}48)$$

$$\cdots\cdots\cdots\cdots\cdots\cdots\cdots\cdots\cdots\cdots$$

$$x_n = \frac{\Delta_{1n}y_1 + \Delta_{2n}y_2 + \Delta_{3n}y_3 + \cdots + \Delta_{nn}y_n}{\Delta}$$

Let us now return to the loop equations in Eq. (2-41) and use the ideas which we have just been discussing. The determinant will be given by

$$\Delta = \begin{vmatrix} R_1 + R_2 & -R_2 \\ -R_2 & R_2 + R_3 + R_4 \end{vmatrix}$$
$$= (R_1 + R_2)(R_2 + R_3 + R_4) - R_2^2 \quad (2\text{-}49)$$

and the solution for the two loop currents will become

$$i_a = \frac{(R_2 + R_3 + R_4)v_0}{\Delta} \quad (a)$$
$$i_b = \frac{R_2 v_0}{\Delta} \quad (b) \qquad (2\text{-}50)$$

From the loop currents, we now find the branch currents. Actually, i_1 is identical with i_a; i_3 and i_4 are identical with i_b. That leaves i_2, which can be found from Eq. (2-37). The result is

$$i_2 = i_a - i_b = \frac{R_3 + R_4}{\Delta} v_0 \quad (2\text{-}51)$$

This completes the solution, since the branch voltages can now be easily obtained from the branch currents.

A word of caution should be injected here. Although the solution in terms of determinants has been written in a nice, elegant form, this form of solution is not necessarily the best to use for numerical computations. In numerical work perhaps it is easiest to use the successive-elimination method for solving a set of simultaneous equations. When the objective is not the numerical solution of a problem but the exploration of general solutions and properties of solutions, however, then the compact form in terms of determinants leads to insights which a long algebraic expression will tend to mask.

Let us now pause and reflect on the procedure we have outlined. The first step is to choose a set of $N_b - (N_v - 1)$ loops around which to write Kirchhoff voltage-law equations. The loops we choose must be appropriate in that the resulting equations must be independent. [One such set of loops is the set of fundamental loops defined in Sec. 2-8. For a given tree, the fundamental loops contain one link and any number of tree branches. Since there are $N_b - (N_v - 1)$ links, the fundamental loops will give us the correct number of equations and they will all be independent.]

Next we choose a set of loop currents, also $N_b - (N_v - 1)$ in number. *The loops defined by the loop currents need not be the same as the loops chosen for writing the voltage-law equations.* There is no reason for not choosing them in this fashion, however, and there is a good reason for so choosing them, which we shall discuss shortly. Hence we shall always do so, unless explicitly stated otherwise.

Into the Kirchhoff voltage-law equations we now substitute the branch v-i relationships. Since all the branch currents can be expressed rather simply in terms of the loop currents, we can perform this step mentally while substituting the branch v-i relationships. The job is now complete. All that remains is to solve the resulting set of simultaneous equations, which are called the *loop equations.* After a moderate amount of experience, the loop equations in the form of Eq. (2-41) can be written down directly from the network diagram. Our present objective, however, is not to acquire such facility but to obtain a thorough understanding of the ingredients of the process.

One other point should be noted here. *The loop equations do not depend on the branch-voltage and branch-current references.* This statement can be verified by considering the references of any of the branches in Fig. 2-21, say branch R_3. Suppose we reverse the reference for v_3. Then the sign of v_3 in Eq. (2-38) will be changed. The sign of the v-i relationship of the branch will also change, however, becoming $v_3 = -R_3 i_3$. Hence the sign of the corresponding term in Eq. (2-39) will remain unchanged. A similar argument applies if the branch-current reference is reversed. Hence there is no reason to show branch references when writing loop equations. On the other hand, the signs in the loop equations will depend on the loop-current references, and these must be shown.

This completes our discussion of the method of solution which we call *loop analysis.* It is a powerful method of analysis and we shall extend its usefulness in later chapters, emphasizing in somewhat greater detail the formal aspects. For the present the fundamentals of the method are the important things.

2.10 NODE EQUATIONS

Let us now duplicate the loop analysis discussed in the last section, but with the roles of the voltage and current laws interchanged. As a starting point, consider again the network in Fig. 2-21. There are two independent Kirchhoff voltage-law equations, one possible set of which is given in Eq. (2-38). There are four unknowns here, two more than the number of equations. If we assume that an appropriate pair of these, say v_2 and v_4, are known (or can be found by other means), then the others can be found in terms of these. From Eq. (2-38) we get

$$\begin{aligned} v_1 &= v_0 - v_2 && (a) \\ v_3 &= v_2 - v_4 && (b) \end{aligned} \tag{2-52}$$

Note from the network diagram that v_2 and v_4 are the voltages of nodes a and b, respectively, relative to that of node c.

In general, for an arbitrary network, any node of a network can be chosen as a *datum* (or a *reference*) node. We shall refer to the voltages of the other nodes relative to the datum node as the *node-to-datum*, or simply as the *node* voltages. Since every branch of a network lies between two nodes, every branch voltage can be expressed as the difference between two node voltages. (For some branches, one of these two nodes will be the datum node, so the branch voltage will be identical with the node voltage.) In any network there are exactly $N_v - 1$ node voltages.

For the present example $N_v - 1$ is 2. With node c chosen as a datum, v_{ac} and v_{bc} constitute the node voltages. But these are identical with v_2 and v_4, respectively. Thus Eq. (2-52) expresses the branch voltages v_1 and v_3 in terms of the node voltages.

Now let us turn to the independent Kirchhoff current-law equations, of which there are two in this example (or $N_v - 1$ in general). They are given in Eq. (2-36). Into these we substitute the v-i relationships of the branches, getting

$$\begin{aligned} -G_1v_1 + G_2v_2 + G_3v_3 &= 0 && (a) \\ -G_3v_3 + G_4v_4 &= 0 && (b) \end{aligned} \tag{2-53}$$

(Note that, for convenience, we have used conductances instead of the corresponding resistances.) These two equations contain four branch-voltage unknowns. Replacing the branch voltages by their node-voltage equivalents from Eq. (2-52), however, reduces the number of unknowns to two. Thus

$$\begin{aligned} -G_1(v_0 - v_2) + G_2v_2 + G_3(v_2 - v_4) &= 0 && (a) \\ -G_3(v_2 - v_4) + G_4v_4 &= 0 && (b) \end{aligned} \tag{2-54}$$

There are now two equations in the two node-voltage unknowns. We call them the *node equations*. The terms can be rearranged by grouping all the terms involving the same variable. The result will be

$$\begin{aligned} (G_1 + G_2 + G_3)v_2 - G_3v_4 &= G_1v_0 \qquad (a) \\ -G_3v_2 + (G_3 + G_4)v_4 &= 0 \qquad (b) \end{aligned} \tag{2-55}$$

The solution of this set of equations can now be obtained as before. If we call the determinant of the set Δ, then

$$\Delta = \begin{vmatrix} (G_1 + G_2 + G_3) & -G_3 \\ -G_3 & (G_3 + G_4) \end{vmatrix} = (G_1 + G_2 + G_3)(G_3 + G_4) - G_3^2 \tag{2-56}$$

The solution for the node voltages can now be written formally as

$$\begin{aligned} v_2 &= \frac{G_3 + G_4}{\Delta} G_1v_0 \qquad (a) \\ v_4 &= \frac{G_3G_1v_0}{\Delta} \qquad (b) \end{aligned} \tag{2-57}$$

The remaining two branch voltages are found, from Eq. (2-52), to be

$$\begin{aligned} v_1 &= \frac{G_2G_3 + G_2G_4 + G_3G_4}{\Delta} v_0 \qquad (a) \\ v_3 &= \frac{G_1G_4}{\Delta} v_0 \qquad (b) \end{aligned} \tag{2-58}$$

This completes the solution.

Let us now turn to a consideration of the generality of this procedure. In any network there are $N_v - 1$ independent current-law equations in the N_b branch unknowns. When we substitute the v-i relationships into these equations, the number of unknowns stays the same, but the unknowns are now branch voltages. When we express each branch voltage as the difference between two node voltages and substitute into the equations, the result will be a set of $N_v - 1$ equations in $N_v - 1$ node-voltage unknowns. The only remaining question is whether or not Kirchhoff's voltage law will be satisfied around all closed loops. The answer to this question is in the affirmative. Expressing the branch voltages in terms of node voltages is equivalent to Kirchhoff's voltage law.

[To prove the statement in the preceding paragraph, refer to Sec. 2-8. Each node voltage is the voltage from a node to the datum node. Since all nodes lie on a tree, a path consisting of twigs only can be found from any node to the datum node. Hence any node voltage is a sum of twig voltages. Suppose that voltage-law equations are written around all the fundamental loops of the network. Each of these loops contains one link and several twigs. We now express the branch voltages in terms of node voltages. The result must be the identity $0 = 0$. This is true because each node voltage is a sum of twig voltages. If the result were not an identity, this would mean that there exists a linear combination of twig voltages which vanishes, so that one twig voltage could be expressed as a linear combination of others. But this is impossible, since forcing all twig voltages but one to vanish would then cause this one also to vanish, whereas we know that short-circuiting all twigs but one will still not short-circuit this one. Hence, expressing the branch voltages as a difference of two node voltages automatically satisfies Kirchhoff's voltage law around all closed paths.]

Let us now make some comments about the *node analysis* which we have just discussed. The procedure seems to be quite analogous to loop analysis, except that voltage and current variables are interchanged. Here, we start from the current-law equations, into which we substitute the branch v-i relationships. When we next express the branch voltages in terms of the node voltages, the result becomes the node equations. In fact, after some practice, the last two steps can be performed mentally while writing the current-law equations, without the need for actually writing them down separately.

Again, as for the loop equations, the branch references have no effect on the node equations. Reversing a branch voltage or current reference will cause two sign changes with no net effect. The node-voltage references, however, do affect the result. On the other hand, the references can always be chosen at the nondatum nodes; so let us make the convention that this will always be done, except if otherwise stated.

In writing node equations, two specific choices of nodes are made: (1) A datum node for defining node equations is chosen. (2) One node is chosen to be omitted when writing current-law equations. It is not necessary that these two nodes be the same. On the other hand, there is no reason not to make this choice, and we shall always do so. Later, we shall even see an advantage in this choice.

Finally, note that no difficulty was encountered from the presence of a voltage source when writing node equations. Node analysis can be applied whether there are current or voltage sources, or both, present in the network. The same is true for a loop analysis. After you have had some practice and start writing loop or node equations directly from the network diagram, however, you may find it more convenient to have only voltage sources when writing loop equations, and only current sources when writing node equations. To satisfy this desire, it is possible to convert from one type of source to its equivalent by means of the source transformations.

Since both the method of loop equations and the method of node equations involve a similar sequence of steps, the question will arise as to which one to choose in a given problem. The answer to this question is not clarified by recourse to our example, since, in both cases, there were the same number of simultaneous equations. This is *not* a general result. In a given network we have seen that there are $N_b - (N_v - 1)$ loop equations and $N_v - 1$ node equations. In our example these two numbers happened to be the same. More generally, they will be different. Hence one or the other method may lead to a fewer number of equations to solve simultaneously.

2.11 ILLUSTRATIVE EXAMPLE

In this section we shall illustrate, by means of a numerical example, some of the general methods of solution that were discussed in the last several sections. A network with a relatively complicated structure is shown in Fig. 2-22. The numerical values given are those of resistance.

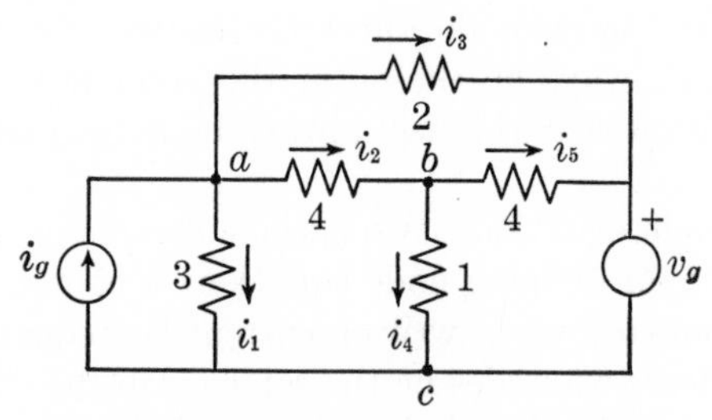

FIG. 2-22
Numerical example

Let us first analyze the network in terms of branch variables. The branch-current references are shown on the diagram; each voltage plus sign is assumed to be at the tail of the corresponding current arrow, and so they are not shown explicitly. In order to count branches and nodes, let us assume that the voltage source is shorted while the current source is opened. Thus $N_b = 5$, $N_v = 3$. There will be $N_v - 1 = 2$ independent current equations, and $N_b - (N_v - 1) = 3$ independent voltage equations.

Let us choose the windows for writing voltage equations. The result will be

$$\begin{aligned} -v_1 + v_2 + v_4 &= 0 && (a) \\ -v_4 + v_5 + v_g &= 0 && (b) \qquad (2\text{-}59) \\ -v_2 + v_3 - v_5 &= 0 && (c) \end{aligned}$$

As for current equations, let us omit node c and write equations at nodes a and b. The result will be

$$\begin{aligned} i_1 + i_2 + i_3 &= i_g && (a) \\ -i_2 + i_4 + i_5 &= 0 && (b) \end{aligned} \qquad (2\text{-}60)$$

We now have a total of five equations in the 10 branch currents and voltages. Substituting the branch v-i relationships will reduce the unknowns to five. There is some flexibility in this step, because it is possible to eliminate either the current or the voltage of each branch. If all the currents are eliminated, there will remain five equations in the branch voltages. Similarly, if all the voltages are eliminated, the branch currents will remain as variables. For purposes of illustration, let us eliminate i_3 from Eq. (2-60) and all voltages except v_3 from Eq. (2-59) by substituting the v-i relationships. The result will be

$$\begin{aligned} -3i_1 + 4i_2 + i_4 &= 0 && (a) \\ -i_4 + 4i_5 &= -v_g && (b) \\ -4i_2 + v_3 - 4i_5 &= 0 && (c) \qquad (2\text{-}61) \\ i_1 + i_2 + \tfrac{1}{2}v_3 &= i_g && (d) \\ -i_2 + i_4 + i_5 &= 0 && (e) \end{aligned}$$

This is a set of five equations in five unknowns. We shall leave to you the task of finding the solution. Observe that the problem is one of solving simultaneously a set of five algebraic equations.

Next let us attempt the solution by the method of loop equations. The first step is to choose a set of loop currents. In this example we need $N_b - (N_v - 1) = 3$ loop currents. Let us choose the meshes for defining loop currents, as shown in Fig. 2-23(a). The branch references that were chosen before are also shown. The basic ingredients of the loop method of solution are still the current-law and voltage-law equations, as well as Ohm's law.

The first step is to write three voltage-law equations, and this step

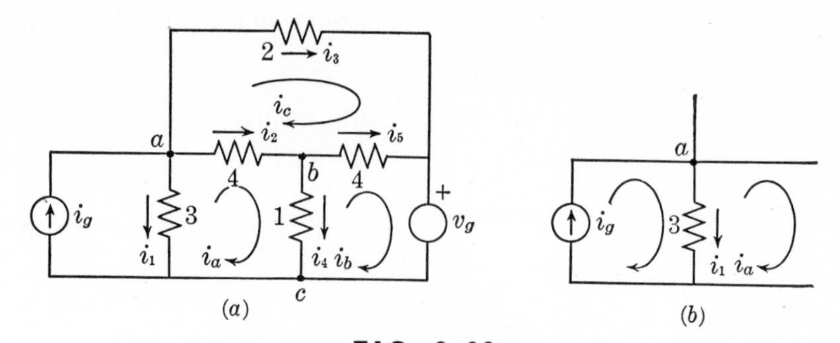

FIG. 2-23

Choice of loops in numerical example

is already performed in Eq. (2-59). Next we substitute the v-i relationships in terms of branch currents. The result will be

$$-3i_1 + 4i_2 + i_4 = 0 \qquad (a)$$
$$-i_4 + 4i_5 = -v_g \qquad (b) \qquad (2\text{-}62)$$
$$-4i_2 + 2i_3 - 4i_5 = 0 \qquad (c)$$

The next step is to express the branch currents in terms of the loop currents. By examining the network diagram, we get

$$i_1 = i_g - i_a \qquad (a)$$
$$i_2 = i_a - i_c \qquad (b)$$
$$i_3 = i_c \qquad (c) \qquad (2\text{-}63)$$
$$i_4 = i_a - i_b \qquad (d)$$
$$i_5 = i_b - i_c \qquad (e)$$

Note that when these expressions are substituted into the current-law equations in Eq. (2-60), the result is an identity $0 = 0$, which means that the current law is automatically satisfied when branch currents are expressed in terms of loop currents.

The expression for branch current i_1 may need some comment. One point of view which can be taken in arriving at this expression is to write the current law at node a and then replace all branch currents except i_1 by their equivalents in terms of the loop currents. (Do this.) Another point of view is the following: Consider a pseudo loop current in the loop formed by the current source and its parallel branch, as shown in Fig. 2-23(b). This loop current is identical with i_g; hence the expression for i_1 as the difference between i_g and i_a follows immediately.

Finally, we substitute Eq. (2-63) into Eq. (2-62), getting

$$\begin{aligned} -3(i_g - i_a) + 4(i_a - i_c) + (i_a - i_b) &= 0 && (a) \\ -(i_a - i_b) + 4(i_b - i_c) &= -v_g && (b) \\ -4(i_a - i_c) + 2i_c - 4(i_b - i_c) &= 0 && (c) \end{aligned} \qquad (2\text{-}64)$$

which, upon rearrangement, become

$$\begin{aligned} 8i_a - i_b - 4i_c &= 3i_g && (a) \\ -i_a + 5i_b - 4i_c &= -v_g && (b) \\ -4i_a - 4i_b + 10i_c &= 0 && (c) \end{aligned} \qquad (2\text{-}65)$$

Here we have three equations in the three loop currents. Again, the task of carrying out the algebraic details of solution will be left to you. Note that a set of three equations is normally easier to solve than is a set of five.

After the fundamental concepts involved in the method of loop equations are understood, it is not necessary to go through each of the steps explicitly. The first step is still a choice of loop currents. The next step is to write voltage-law equations around these same loops in the reference direction of the loop current. As we come to each branch, we assume that the branch-current reference coincides with that of the loop current under consideration, and that the voltage plus is at the tail of the current reference, so that the v-i relationship of the branch carries a plus sign. These choices are possible because the references can be chosen arbitrarily. By examining the network diagram, we note the dependence of the branch current on the loop currents. Thus Eq. (2-63) need never be written explicitly. Mentally, we say: "The voltage of this branch is R times the branch current, which is (say) this loop current minus that one." Thus, instead of writing the voltage equations in terms of branch voltages, then substituting the v-i relationships, then replacing branch currents by loop currents, we mentally perform all these steps at once and write down the final result.

Suppose we follow this scheme in writing the loop equations for Fig. 2-23. (Ignore the branch references shown there.) The result will be

$$\begin{aligned} &\text{Loop } a: & 4(i_a - i_c) + (i_a - i_b) + 3(i_a - i_g) &= 0 && (a) \\ &\text{Loop } b: & 4(i_b - i_c) + v_g + (i_b - i_a) &= 0 && (b) \\ &\text{Loop } c: & 2i_c + 4(i_c - i_b) + 4(i_c - i_a) &= 0 && (c) \end{aligned} \qquad (2\text{-}66)$$

When these equations are rearranged, the result is again Eq. (2-65).

Finally, let us solve the same problem by the method of node equations. The first step is to choose a datum node. This choice is guided by convenience. Usually, it is best to choose the datum node as the one which is connected to most other nodes. In this way the node voltages will coincide with many branch voltages. Let us suppose that node c is chosen as the datum. The next step is to write current-law equations at the other nodes. This has already been done in Eq. (2-60). We now substitute the v-i relationships into these equations. The result will be

$$\frac{v_1}{3} + \frac{v_2}{4} + \frac{v_3}{2} = i_g \qquad (a)$$
$$-\frac{v_2}{4} + v_4 + \frac{v_5}{4} = 0 \qquad (b) \qquad (2\text{-}67)$$

Next we express the branch voltages in terms of the node voltages. By examining the network diagram, we find

$$v_1 = v_a \qquad (a)$$
$$v_2 = v_a - v_b \qquad (b)$$
$$v_3 = v_a - v_g \qquad (c) \qquad (2\text{-}68)$$
$$v_4 = v_b \qquad (d)$$
$$v_5 = v_b - v_g \qquad (e)$$

Note that when these expressions are substituted into the voltage-law equations in Eq. (2-59), the result is an identity $0 = 0$, which means that the voltage law is automatically satisfied when branch voltages are expressed in terms of node voltages.

Finally, we substitute these expressions into Eq. (2-67), getting

$$\frac{v_a}{3} + \frac{v_a - v_b}{4} + \frac{v_a - v_g}{2} = i_g \qquad (a)$$
$$-\frac{(v_a - v_b)}{4} + v_b + \frac{v_b - v_g}{4} = 0 \qquad (b) \qquad (2\text{-}69)$$

which, upon rearrangement, become

$$\frac{13}{12} v_a - \frac{1}{4} v_b = i_g + \frac{v_g}{2} \qquad (a)$$
$$-\frac{1}{4} v_a + \frac{3}{4} v_b = \frac{v_g}{4} \qquad (b) \qquad (2\text{-}70)$$

Here we have two equations in the two node voltages. Purely on the basis of the number of equations that must be solved simultaneously, it appears that the method of node equations is the simplest method of solution of the present problem. The remainder of the solution is left to you.

As in the case of loop equations, it is not necessary to write all the steps explicitly. After a datum node is chosen, the next step is to write current-law equations at the other nodes. We assume that all node-voltage references are at the nondatum nodes. As we come to each branch, we assume that the branch-current reference is away from the node in question, and that the branch-voltage reference plus is at the node in question, so that the v-i relationship carries a plus sign. By observing the diagram, we note the dependence of the branch voltage on the node voltages. Mentally, we say: "The current of this branch is the conductance G times the branch voltage, which is this node voltage minus that one." Thus all the intermediate steps are performed mentally, and the current-law equations are written down in terms of node voltages in one step.

Let us follow this scheme in writing the node equations for Fig. 2-22. (Ignore the branch references shown there and assume that the node-voltage references are at nodes a and b.) The result will be

$$\begin{aligned} &\text{Node } a\text{:} \quad -i_g + \frac{v_a}{3} + \frac{1}{4}(v_a - v_b) + \frac{1}{2}(v_a - v_g) = 0 \quad (a) \\ &\text{Node } b\text{:} \quad \tfrac{1}{4}(v_b - v_a) + v_b + \tfrac{1}{4}(v_b - v_g) = 0 \quad (b) \end{aligned} \qquad (2\text{-}71)$$

Upon collecting terms and rearranging, these equations become Eq. (2-70).

2.12 DUALITY

Next we shall discuss a phenomenon some of whose aspects may have intruded themselves upon your consciousness in the development of the fundamental laws in the last chapter, and in the discussion of methods of analysis in this chapter. The laws and the procedures under discussion, the ideas and the physical quantities, seem to occur in pairs. First there are Kirchhoff's two laws which refer to current and voltage, respectively. Then there is Ohm's law, which can be written as $v = Ri$ or $i = Gv$. The v-i relationships of an inductor and a capacitor have the same form, except that current and voltage appear to be interchanged. (Glance back at Table 1-1.) We have a certain number of independent current equations and a certain number of independent voltage equations. We have cut sets and we have tie sets; we have trees and we have links. There are voltage dividers and current dividers. There are the voltage-shift theorem and the current-shift theorem.

We refer to these similarities, analogies, and parallelisms between certain quantities and laws—this "twosome" property—by the general term duality. This is a very vague term, and to be useful we must specify more precisely what we mean by duality. We shall say that two things are dual if they play similar roles in the formulations of the laws and procedures of network theory which come under the general heading of loop analysis and node analysis. This definition may still not be extremely lucid, but we shall clarify it further as we proceed.

Some of the aspects of duality are topological, whereas others are physical. Kirchhoff's voltage law refers to a loop and Kirchhoff's current law refers to a node. Thus a loop and a node are topological duals. Voltage and current are, themselves, dual physical quantities. Similarly, a cut set (a particular set of branches which connects two groups of nodes) is the topological dual of a tie set (a set of branches which lie on a loop). A twig is the topological dual of a link.

On the other hand, note the v-i relationships of the network elements.

$$v = Ri; \qquad i = Gv \qquad (a)$$

$$v = L\frac{di}{dt}; \qquad i = C\frac{dv}{dt} \qquad (b) \qquad (2\text{-}72)$$

$$v = \frac{1}{C}\int_0^t i\,dx + v(0); \quad i = \frac{1}{L}\int_0^t v\,dx + i(0) \qquad (c)$$

The right-hand column can be obtained from the left-hand column by interchanging v and i, R and G, and L and C. In the second line, for example, L on the left plays the same role as C on the right. Hence inductance is the dual of capacitance (and vice versa), and conductance is the dual of resistance. These are not topological duals. Table 2-1 gives a partial tabulation of duals. Throughout the remainder of the text, as we introduce additional topics, we shall further extend this table.

We say that *two branches are dual if the expression for the voltage in terms of the current of the first branch has the same form as the expression for the current in terms of the voltage of the second branch.*

We say that *two networks, N_1 and N_2, are dual networks if: (1) the current equations of N_1 are the same as the voltage equations of N_2, provided*

TABLE 2-1

Dual Quantities and Laws

Voltage	Current
Kirchhoff's voltage law	Kirchhoff's current law
Loop	Node
Number of loops	Number of nodes
Node voltage	Loop current
Resistance	Conductance
Inductance	Capacitance
Twig	Link
Tie set	Cut set
Series	Parallel
Short circuit	Open circuit
Loop equations	Node equations

current is replaced by voltage; (2) the voltage equations of N_1 are the same as the current equations of N_2, but with voltage replaced by current; and (3) the branches of N_1 are the duals of the corresponding branches of N_2.

Clearly, two dual networks must have the same number of branches; $N_{b1} = N_{b2} = N_b$. The first two conditions of duality require that the number of independent current equations of N_1 be the same as the number of independent voltage equations of N_2; that is,

$$N_b - (N_{v_1} - 1) = N_{v_2} - 1 \tag{2-73}$$

When two networks are dual, the loop equations of one network will be identical with the node equations of the second, and vice versa, but with an interchange of the dual quantities. This idea reduces by a factor of 2 the totality of all networks which are candidates for analysis, since, in a solution obtained for one network, it is only necessary to interchange dual quantities in order to have a solution for its dual network.

To see how the concept of duality can be of use in analysis, let us return to a consideration of the current divider in Fig. 2-1(*b*) and the expressions for the division of current given in Eq. (2-4). The current divider, which consists of a parallel connection of two branches to which a total current i_0 is supplied, is the dual of the voltage divider, which consists of the series connection of two branches to which a total voltage v_0 is supplied. Hence we should expect that the equations expressing the division of current can be obtained from the corresponding equations expressing the division of voltage of a voltage divider, with all quantities replaced by their duals. This is indeed the case, as you can verify by comparing Eqs. (2-2) and (2-4) and noting the duality between voltage and current, and between resistance and conductance. On this basis, having obtained Eq. (2-2), we could have immediately written Eq. (2-4) without going through any further analysis. In later work we shall use this concept of duality to obtain results concerning one network when the corresponding results concerning the dual network are at hand.

2.13 SUMMARY

This chapter was concerned with the problem of finding a voltage or current signal at any point in a resistance network in response to one or more voltage or current signals at other points. The fundamental laws on which the solution is based are Kirchhoff's two laws and the v-i relationship of the branches. (In the present case this relationship is Ohm's law, since the branches are resistors.)

Application of Kirchhoff's current law leads to $N_v - 1$ independent equations in a network having N_v nodes and N_b branches. Similarly, the voltage law leads to $N_m = N_b - (N_v - 1)$ independent equations. In counting branches and nodes, voltage sources are assumed to be short circuits and current sources open circuits.

We discussed three general methods of solution. The first of these utilizes $2N_b$ equations in mixed voltage and current variables. The second method utilizes a set of fictitious loop currents as variables and requires the simultaneous solution of N_m equations. In the third method one node is chosen as a datum. The variables are then the other node voltages relative to this datum. The number of equations required to be solved simultaneously is $N_v - 1$.

In addition to these general methods, we discussed some special methods of solution which apply to networks with particular structures. Particularly useful are the source transformations: the voltage-shift theorem, the current-shift theorem, and the equivalence between a voltage source in series with a resistor and a current source in parallel with it. Consecutive application of these transformations often leads to a solution with much less computation than the other methods.

A brief discussion of network topology was also given as a basis for proving the number of independent equations.

Problems

(All numerical-element values are those of resistance in ohms.)

2-1 For the networks shown in Fig. P2-1: (*a*) Find the number of branches and the number of nodes. (*b*) From these, find the number of independent loops. (*c*) Determine the networks which are

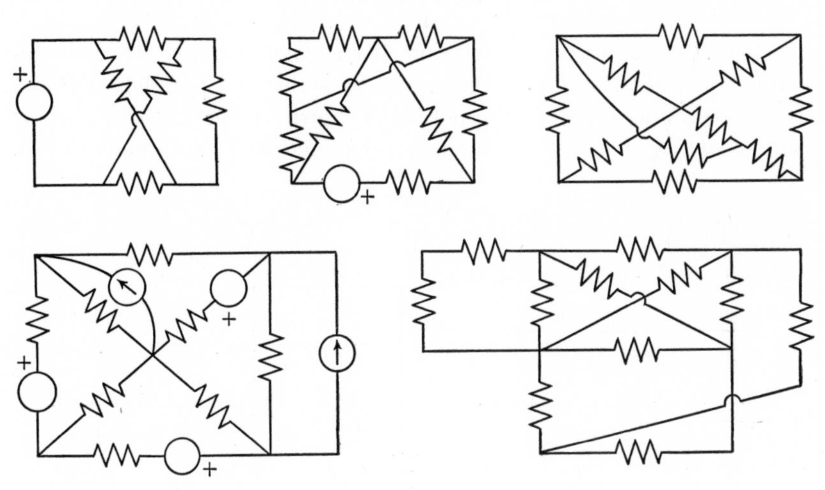

FIG. P2-1

planar and which nonplanar. (*d*) Redraw those which are planar so that none of the branches cross each other.

2-2 Write the current-law equations at each of the nodes in the network shown in Fig. P2-2. (Be careful of references.) Show how any one of them can be obtained from the others, indicating that they are not all independent.

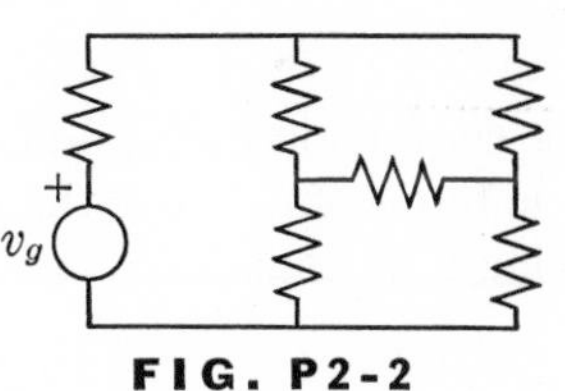

FIG. P2-2

2-3 Write the Kirchhoff voltage-law equations around all the closed loops in Fig. P2-2. (How many are there?) Show that the equations around those loops which are not meshes can be obtained from those which are.

2-4 Use the method discussed in Sec. 2-5 to solve for the branch variables in the ladder networks shown in Fig. P2-4. Check your solution for the input current for (*a*) and the input voltage for (*b*) by finding the equivalent resistance of the network at the source terminals.

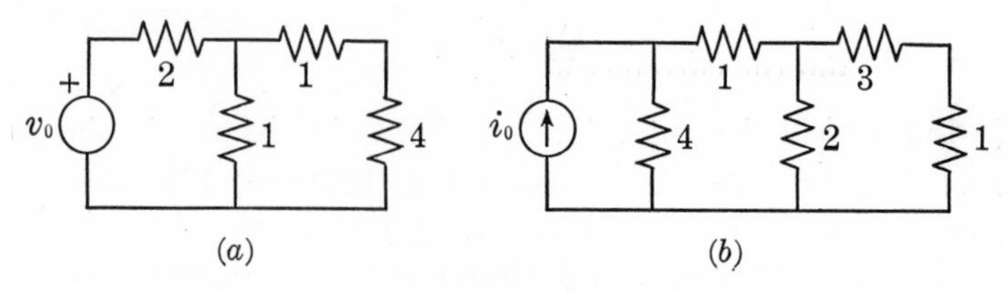

FIG. P2-4

2-5 By successive conversion from current-source to voltage-source equivalent, and vice versa, and by series and parallel combination of resistances, reduce the networks shown in Fig. P2-5 to a single voltage source in series with a single resistance at the terminals shown.

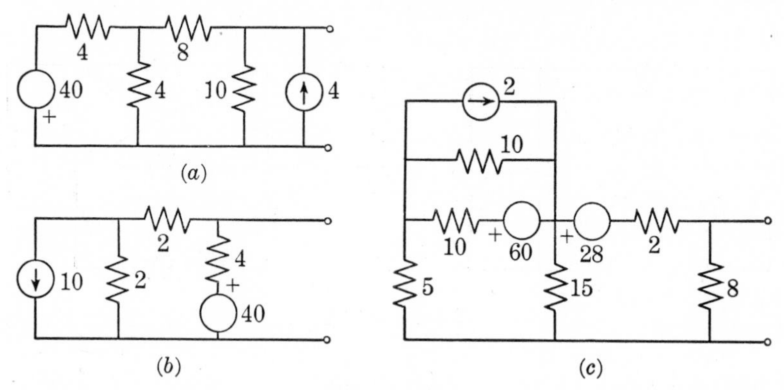

FIG. P2-5

2-6 In the networks of Fig. P2-6, use the voltage-shift theorem, source conversions, and branch combinations to find the output voltage v in terms of v_g. Use the voltage-shift theorem at node a and repeat by using it at node b.

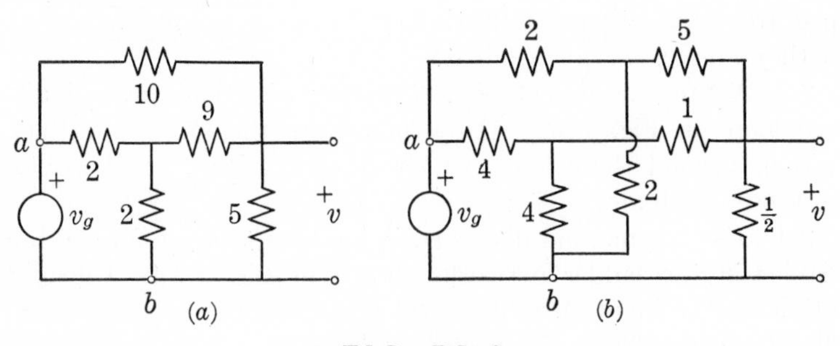

FIG. P2-6

2-7 (*a*) Without writing loop equations or node equations, find the voltage v and the current i in Fig. P2-7, in terms of v_g. (*b*) Repeat, using loop equations. (*c*) Repeat, using node equations. (*d*) Compute the power delivered by the source and the power dissipated in the 4-ohm resistor.

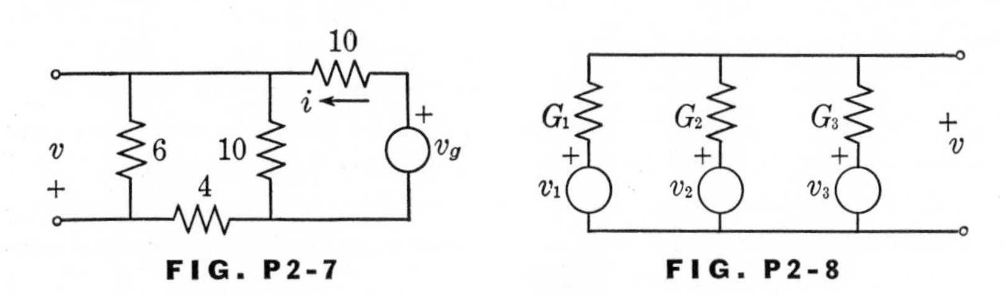

FIG. P2-7 FIG. P2-8

2-8 For the network shown in Fig. P2-8, prove that the voltage v is given by $v = \dfrac{G_1 v_1 + G_2 v_2 + G_3 v_3}{G_1 + G_2 + G_3}$.

2-9 (*a*) Using the theorem in Prob. 2-8, find the current i in Fig. P2-9. Check your answer by solving the problem another way. (*b*) Insert a 1-ohm resistance in parallel with the 5/2-ohm resistance and again find i.

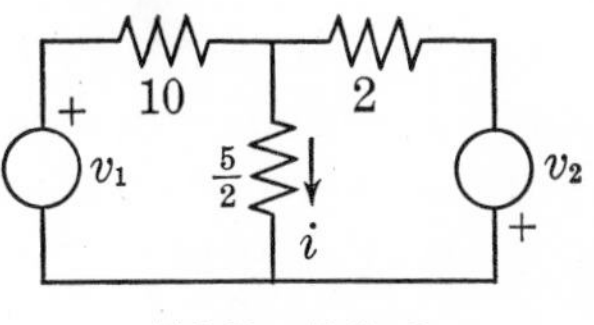

FIG. P2-9

2-10 Consider the equivalent sources shown in Fig. P2-10. Compute the power supplied by each of the two sources and note that they are not equal. Convince yourself why this is not unreasonable.

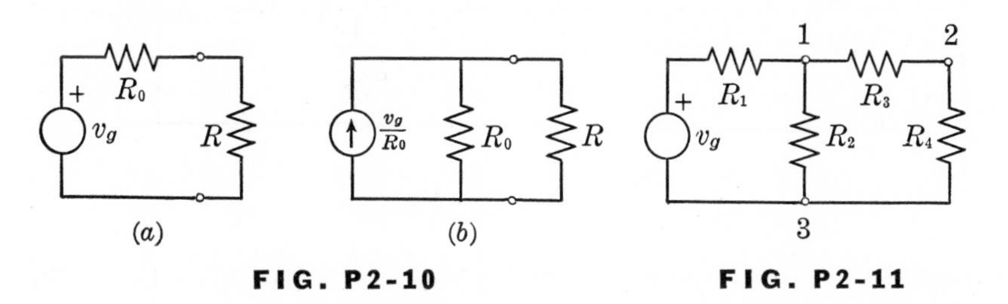

FIG. P2-10 FIG. P2-11

2-11 (*a*) In the network of Fig. P2-11, define loop currents in terms of the "windows" and write loop equations using these same loops for summing voltages. (*b*) Now define the same loop currents to write a set of loop equations, but this time use one of the windows and the outer contour for summing voltages. (*c*) As a third alternative, define the loop currents in terms of one of the windows and the outer contour, and sum voltages around the two windows. Note that, even though the equations will not be the same, the determinants of the three systems of equations will be identical.

2-12 (*a*) In the same network let node 3 be the datum node. Write node equations in the usual way by omitting node 3 when writing Kirchhoff's current law. (*b*) Use the same datum node to write a set of node equations, but this time use node 3 and one other node in summing currents. (*c*) Finally choose node 2 as a datum node and sum currents at nodes 1 and 2. Compare the three determinants.

2-13 (*a*) For the networks shown in Fig. P2-13, write independent current-law and voltage-law equations, and the branch v-i relationships. Solve these equations by successive elimination. (*b*) Solve for all branch variables by the method of loop equations. (*c*) Repeat, using node equations. (*d*) Repeat, using source equivalences, voltage- and current-divider concepts, and series and parallel combinations of branches. (*e*) In the topological graph of these networks, show as many trees as you can. (*f*) For one of these trees, show the fundamental cut sets and write the current equations for these cut sets. (*g*) For the same tree, show the fundamental loops and write the voltage equations for these loops.

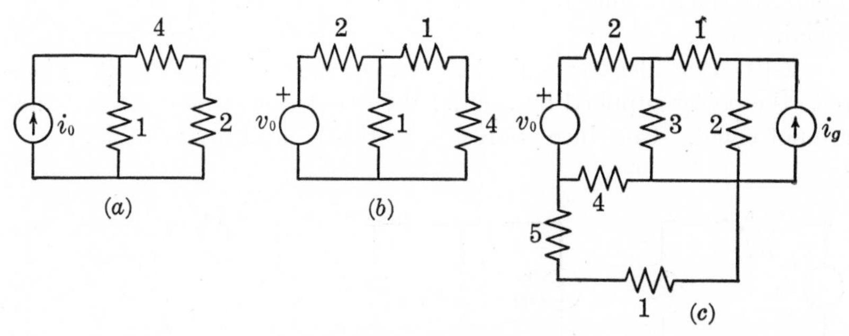

FIG. P2-13

2-14 (*a*) Show as many trees as you can for the topological graph of the network in Fig. P2-2. (There are a total of 17 trees.) (*b*) For a given tree, show the fundamental cut sets and the fundamental loops.

2-15 Repeat the previous problem for the networks in Figs. P2-5 and P2-6.

2-16 Figure P2-16 shows a model of a three-wire power-distribution system. Find the values of the source voltages which will produce the load powers at the voltages given in the figure. Find also the power lost in the transmission system. Calculate the power delivered by each source and verify that the total power delivered is equal to the power lost plus the load power.

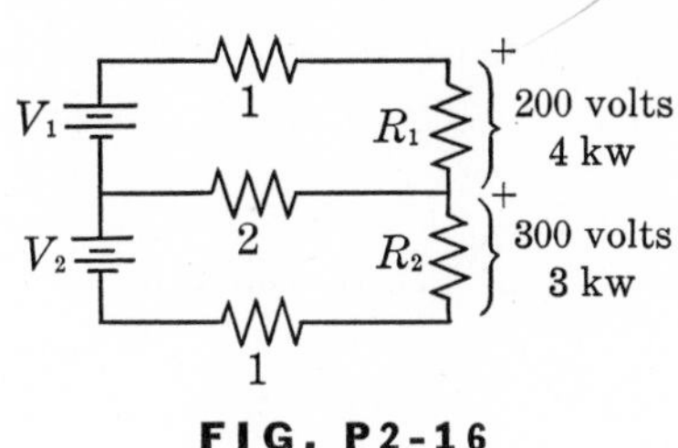

FIG. P2-16

2-17 In Fig. P2-16 let $R_1 = 10$, $R_2 = 15$, and $V_1 = V_2 = 250$. Calculate the power delivered to each load: (*a*) Do this first by loop equations. (*b*) Repeat by first shifting V_2 through its upper node and using source conversions to find the power in R_1, then shifting V_1 down through its lower node and using source conversions to find the power in R_2.

2-18 Figure P2-18 shows a suggested arrangement for an instrument to be used as a combination voltmeter and ammeter. The meter *m* has a resistance R_m of 100 ohms and a maximum deflection when the current in it is 1 ma. When connected as an ammeter (terminals *A*-*O*), the arrangement should measure currents up to 1 amp. At this maximum current the voltage at terminals *A*-*O* should be $\frac{1}{2}$ volt. When connected as a voltmeter (terminals *V*-*O*), it should measure voltages up to 10 volts. Find the values of R_1, R_2, and R_3 to meet these requirements.

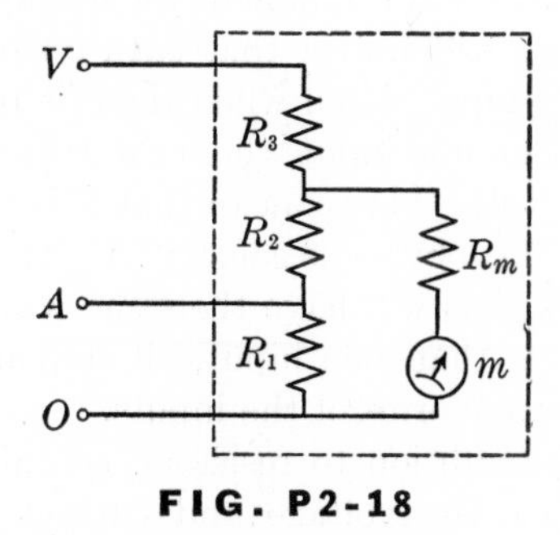

FIG. P2-18

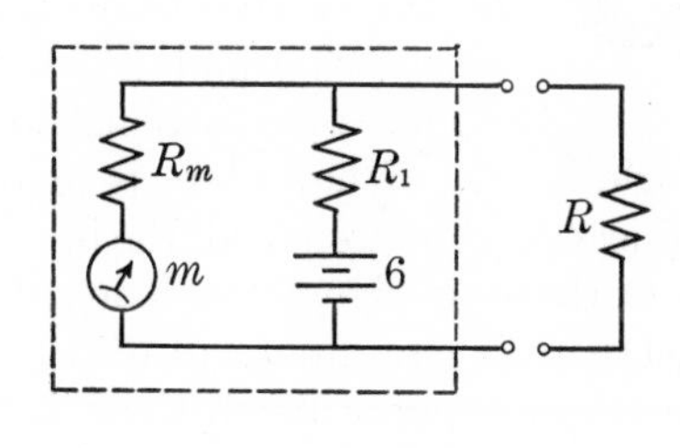

FIG. P2-19

2-19 Figure P2-19 shows an arrangement which is to be used as an ohmmeter. The meter *m* is the same as in Prob. 2-18. Find the value of R_1 which will cause maximum deflection of the meter when no resistor is connected at the terminals. Find the value of *R* for which the deflection will be half scale, assuming a linear relationship between needle deflection and meter current.

COMPLETE RESPONSE OF SIMPLE NETWORKS

3

CHAPTER 2 was devoted to the problem of calculating output signals in response to known input signals in networks containing resistors only. If you will glance back at any of the solutions which were obtained, for example Eqs. (2-35) and (2-50), you will notice that each output signal is proportional to the input signal when there is but a single input. When there is more than one input, the output signal is a linear combination of all the inputs. (By this we mean that it is the sum of the input signals, each one multiplied by some constant.) Thus, when there is one input, all response signals will have the same waveform as the input. When there is more than one input, all response waveforms will be combinations of the waveforms of the inputs.

Now consider a network which, in addition to resistors, contains inductors or capacitors, or both. Since the voltage and current in these elements are not simply proportional, but one is the derivative of the other, we would not expect the waveform of an output signal to be the same as the waveform of an input signal.

We are now ready to take up the study of arbitrary networks made up of any combination of all the elements in our model. We shall challenge only one complexity at a time, however, and start by considering networks containing but a single inductor or single capacitor. Since both inductors and capacitors are energy-storage elements, we refer to such networks as *single-energy* networks.

3.1 FREE RESPONSE OF RESISTANCE-CAPACITANCE NETWORK

As a starting point, we shall discuss the behavior of a network which is not subjected to an external excitation but reacts to an internal condition. Consider the network shown in Fig. 3-1(a). A double-

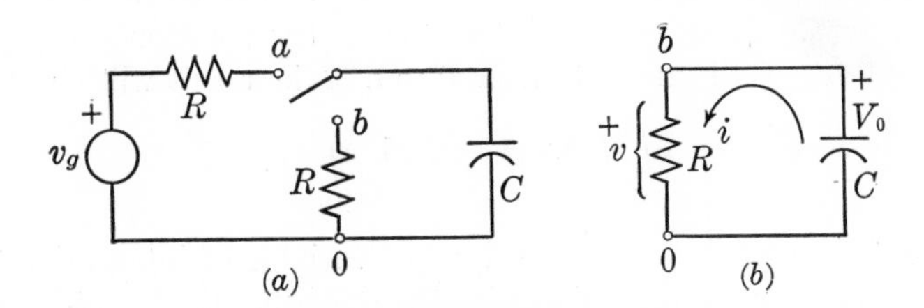

FIG. 3-1
Discharging a capacitor

throw switch can be moved to either position a or position b. Suppose that the switch has been in position a for a while, when, at a time which we designate $t = 0$, the switch is thrown to position b. After the switching, the network takes the form of Fig. 3-1(b). Just before the switching, suppose that the voltage across the capacitor is V_0 with the reference shown in Fig. 3-1(b). It is required to find the current in the resistor subsequent to the closing of the switch. Since the voltage across a resistor is proportional to the current, finding the voltage will be equivalent to finding the current.

Let us solve the problem by the use of node equations. There will be only one node equation, which is obtained by writing Kirchhoff's current law at, say, node b and then substituting the branch v-i relationship. The result will be

$$i_C + i_R = 0; \quad C\frac{dv}{dt} + \frac{v}{R} = 0$$

or

$$\frac{dv}{dt} + \frac{1}{RC}v = 0 \tag{3-1}$$

Here we meet something new. Whereas in resistance networks all the equations that arise are algebraic equations, we now have an equation which contains a derivative, as well as the algebraic variable itself. Such an equation is called a *differential equation*. In order to find the voltage v, we shall have to solve this equation. By *solving* a differential equation, we mean finding a function of the dependent variable which, when substituted into the equation, will cause it to be *satisfied*. When we say that an equation is satisfied by a function, we mean that insertion of the function into the equation will lead to an identity, such as $0 = 0$. A large part of network analysis is devoted to the solution of such differential equations.

Let us now consider solving the equation at hand. One method of solution is to integrate Eq. (3-1) directly. Let us do so after first dividing by v. Since we are interested in the solution for any time after $t = 0$, we shall integrate from 0 to t. The result will be

$$\int_0^t \frac{1}{v}\frac{dv}{dx}\,dx = -\frac{1}{RC}\int_0^t dx \tag{3-2}$$

Note that we have changed the dummy variable of integration from t to x in order to avoid confusion with the upper limit. On the left-hand side the integration with respect to x can be changed to an integration with respect to v, since $(dv/dx)dx = dv$. The limits must also be changed appropriately. The upper limit is $v(t)$ which is the voltage corresponding to any time t.

As for the lower limit, it will be the value of v at $t = 0$. Here we meet a slight problem. Just before the switching operation, the capacitor voltage is V_0. The differential equation applies for the time following the switching, however, and the lower limit of the integral must be the value of the voltage immediately after the switching. The distinction is made by designating these two values before and after switching as $v(0-)$ and $v(0+)$, respectively. The question becomes: "Can we say anything about the zero-plus value of the capacitor voltage knowing the zero-minus value?"

The answer to the question is given by the law of conservation of charge, which was discussed in Chap. 1. As applied to a network, this principle states that electric charge cannot be transferred into or out of a junction instantaneously. In terms of Fig. 3-1(b), the charge at node b or node 0 cannot be changed instantaneously. Since the capacitor voltage is proportional to its charge, it follows that the voltage cannot change instantaneously either. Whatever the voltage was just prior to the switching, that is what it must be just afterward. The lower limit of the integral must be V_0.

In the present simple case we can also reason as follows: Suppose the voltage and charge on the capacitor do change instantaneously at some instant of time; that is, the voltage and charge will contain a step discontinuity. The current, being the derivative of the charge, will then become infinite at the discontinuity; it will contain an impulse. Since the resistor voltage is Ri, this will require the voltage to become infinite also. But the resistor voltage is the same as the capacitor voltage, which is not infinite. Hence, a discontinuous change in the capacitor voltage is not possible.

After changing the integration variable and changing the limits, Eq. (3-2) becomes

$$\int_{V_0}^{v} \frac{du}{u} = \ln u \Big|_{V_0}^{v} = -\frac{t}{RC} \tag{3-3}$$

Again we have changed the symbol for the dummy variable of integration. (In the future, we shall not continually call attention to this step whenever it is performed.) After the limits are inserted, this expression becomes

$$\ln \frac{v}{V_0} = -\frac{t}{RC} \tag{3-4}$$

Since the inverse of a logarithm is an exponential, we get, finally,

$$v = V_0 \epsilon^{-t/RC} \tag{3-5}$$

A plot of Eq. (3-5) is shown in Fig. 3-2. The voltage of the capacitor is seen to start at V_0 and to decay gradually as time goes on.

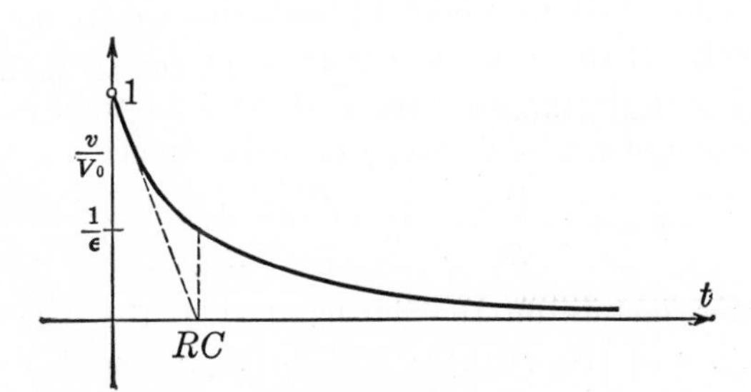

FIG. 3-2
Exponential decay

The exponential function does not become zero until t approaches infinity. It becomes so small after a relatively short time, however, that it can no longer be measured; for practical purposes, we can then assume that it is zero.

The rate at which the curve falls toward zero is determined by the product of R and C which appears in the exponent. By differentiating

Eq. (3-5), we find that $-1/RC$ is the slope of the exponential evaluated at $t = 0$. That is to say, if the curve were to fall with the same slope it has at $t = 0$, it would reach zero at $t = RC$. This is illustrated by the dashed line in Fig. 3-1. The RC product is referred to as the *time constant* of the network. (Check to see that its dimensions are indeed those of time.) When $t = RC$ (one time constant), Eq. (3-5) shows that the voltage has reached $1/\epsilon \doteq 37$ per cent of its initial value; when $t = 3$ time constants, the voltage will have reached $1/\epsilon^3 \doteq 5$ per cent of its initial value.

The exponential function (with a modification of the exponent which we shall later discuss) is perhaps the most important single function which you will meet in engineering, and you should become intimately acquainted with it in all its aspects.

Let us turn back to Eq. (3-1) and examine it in some detail with a view toward finding an alternate method of solution which might be useful in other cases besides the one under discussion. The equation states that the rate of change of the function v which we are seeking is proportional to the function itself.† Now the only mathematical function in our experience which has the property that its derivative has again the same form as the function itself is the exponential. So let's take a guess that the solution might be an exponential, and tentatively write

$$v = K\epsilon^{st} \tag{3-6}$$

where K and s are constants about whose values we are not committing ourselves as yet. If this is to be a solution of Eq. (3-1), we should be able to insert it into the equation and find out if the equation is satisfied. Upon substitution of Eq. (3-6) into Eq. (3-1), we get

$$Ks\epsilon^{st} + \frac{K\epsilon^{st}}{RC} = K\epsilon^{st}\left(s + \frac{1}{RC}\right) = 0 \tag{3-7}$$

There are three factors on the left-hand side: the constant K, the exponential ϵ^{st}, and the quantity $s + 1/RC$ in the parentheses. K cannot be zero, otherwise the voltage will be identically zero, according to Eq. (3-6).‡ The exponential function is not zero for any finite value

† This type of situation is encountered over and over again in many branches of engineering. To name a few, it arises in the radioactive decay of matter and in the rate of recombination of ions in a solution. If you obtain a fundamental grasp of the problem and its solution here, you will, at the same time, obtain some insight into other problems whose formulation leads to a similar equation.

‡ This is actually a trivial possibility. If there were no initial charge on the capacitor when the switch was closed, nothing would happen in the network, and the voltage would indeed remain zero. But this eventuality is of no interest to us.

of the exponent. This leaves $s + 1/RC$, which will be zero if, and only if,

$$s = -1/RC \tag{3-8}$$

With this value of s, our assumed solution in Eq. (3-6) will satisfy the equation, and it will do so no matter what the value of K may be. To determine the value of K for our problem, we can note that Eq. (3-6) should be a solution which is valid at all times; in particular, at time $t = 0$ just after the switching. By the law of conservation of charge, we know that the capacitor voltage at this time is V_0. Putting $t = 0$ and $v = V_0$ in Eq. (3-6) leads to $K = V_0$. With these values of K and s, the tentatively assumed solution becomes identical with the one we found before by another method. The advantage of the present method is that it can be applied in other, more difficult, cases when direct integration is not possible, as we shall soon demonstrate. Since $i = v/R$, the current in the resistor will have the same waveform as the voltage but will differ by a constant multiplier.

Note that there is no external source in this network. The signal which arises is a consequence of the energy originally stored in the capacitor. Hence we refer to the response that we get as the *natural response*, or the *free* or *unforced response*. The natural behavior of the network is an exponential decay. We will have much more to say about this as we go along.

Let us make a computation to determine the eventual disposition, after the switching, of the initial energy stored in the capacitor. Expressions for the energy stored in a capacitor and that dissipated in a resistor from the time of switching up to time t are

$$w_C(t) = \tfrac{1}{2}Cv^2 \qquad (a)$$

$$w_R(t) = R\int_0^t i(x)\,dx \qquad (b) \tag{3-9}$$

These were first discussed in Sec. 1-5. Note that these are instantaneous expressions which are valid at any time t. Since the initial capacitor voltage is V_0, the initial energy stored in the capacitor is $w_C(0) = CV_0^2/2$. The energy dissipated in the resistor from the time of switching up to time t can be calculated by inserting Eq. (3-5) into Eq. [3-9 (*b*)]. The result will be

$$\begin{aligned} w_R(t) &= R\int_0^t \left(\frac{V_0}{R}\,\epsilon^{-x/RC}\right)^2 dx = \frac{V_0^2}{R}\int_0^t \epsilon^{-2x/RC}\,dx \\ &= \frac{V_0^2}{R}\left(\frac{\epsilon^{-2x/RC}}{-2/RC}\right)\Bigg|_0^t = \frac{1}{2}CV_0^2 - \frac{1}{2}CV_0^2\epsilon^{-2t/RC} \end{aligned} \tag{3-10}$$

But since $v = V_0\epsilon^{-t/RC}$, the last term on the right is seen to be $Cv^2/2$, which is the energy stored in the capacitor at any instant. Thus

$$w_R(t) = \tfrac{1}{2}CV_0^2 - w_C(t) \qquad (a)$$

or

$$w_C(t) = w_C(0) - w_R(t) \qquad (b) \qquad (3\text{-}11)$$

In words, the last equation states that the energy stored in the capacitor at any time t is equal to the initial energy stored there diminished by the energy dissipated in the resistor. It is certainly heartening to find that the law of conservation of energy is upheld by our solution.

3.2 DRIVEN RESISTANCE-CAPACITANCE NETWORK

Let us now turn back to the network in Fig. 3-1(a) and assume that this time the switch is thrown from position b to a. Again we choose the time of switching as $t = 0$. The resulting network is shown in Fig. 3-3(a). It is required to find the voltage across the capacitor

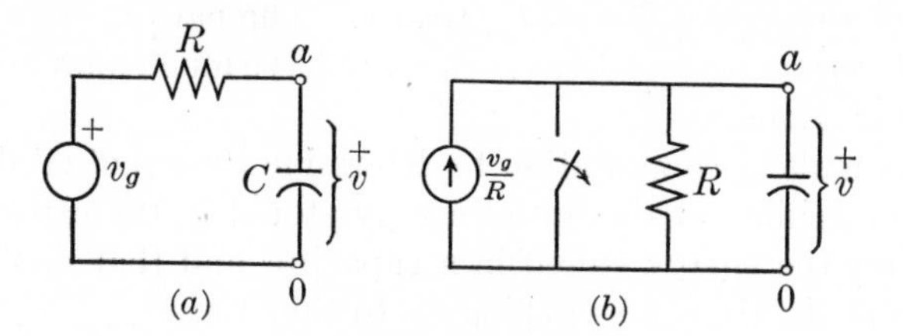

FIG. 3-3

Charging a capacitor

for all time following the switching. Assume that, prior to the switching, the capacitor voltage is again V_0. Since a voltage is required, let us again write a node equation at node a, choosing node O as datum. For $t > 0$ the result will be

$$C\frac{dv}{dt} + \frac{(v - v_g)}{R} = 0$$

or

$$\frac{dv}{dt} + \frac{1}{RC}v = \frac{1}{RC}v_g(t) \qquad (3\text{-}12)$$

(We have explicitly written the functional dependence of v_g on t as $v_g(t)$ to emphasize that the source voltage is a function of time.)

Note that this same result can be obtained by making a source conversion which leads to the network in Fig. 3-3(*b*). In this network the switch has been closed until $t = 0$, when it is opened. If we again choose node O as datum and write a node equation at node a, then Eq. (3-12) will again follow.

Compare this equation with Eq. (3-1). The left-hand sides are identical, but the right-hand side is no longer zero. They are distinguished from each other by labeling the previous equation (with zero on the right) *homogeneous*, whereas the present equation is labeled *nonhomogeneous*. In contrast with the previous equation, we say the present one has a *forcing*, or *driving*, function. The solution of the equation will certainly depend on the waveform of the forcing function.

Let us initially assume that $v_g(t)$ is a constant dc voltage; $v_g(t) = V_1$. Then Eq. (3-12) can be rewritten as

$$\frac{1}{v - V_1}\frac{dv}{dt} = -\frac{1}{RC} \tag{3-13}$$

This equation can now be integrated between the limits 0 and t. The result will be

$$\int_0^t \frac{1}{v - V_1}\frac{dv}{dx}\,dx = -\int_0^t \frac{dx}{RC}$$

or

$$\int_{V_0}^{v} \frac{du}{u - V_1} = -\frac{t}{RC} \tag{3-14}$$

Again we have used the principle of conservation of charge in setting the lower limit at V_0. The left-hand side will again lead to a logarithmic function, and the final result will be

$$\ln\frac{v - V_1}{V_0 - V_1} = -\frac{1}{RC}t$$

or

$$v = V_1 - (V_1 - V_0)\epsilon^{-t/RC} \tag{3-15}$$

Recall that V_1 and V_0 are simply constants. Hence this solution consists of a constant plus an exponential, the exponential being the same as the one in the preceding problem, except for its multiplier. Note that if the source voltage is zero, the solution will become identical with that of the previous problem. This is certainly reasonable, since the network then becomes identical with the previous network.

A plot of Eq. (3-15) is shown in Fig. 3-4 for the case in which the initial voltage V_0 of the capacitor is less than the source voltage. The two dashed curves represent the two individual terms in Eq. (3-15). The difference of the two dashed curves gives the solid curve, which is the total voltage. In case $V_0 > V_1$ then the solid curve will approach V_1 asymptotically from above rather than below.

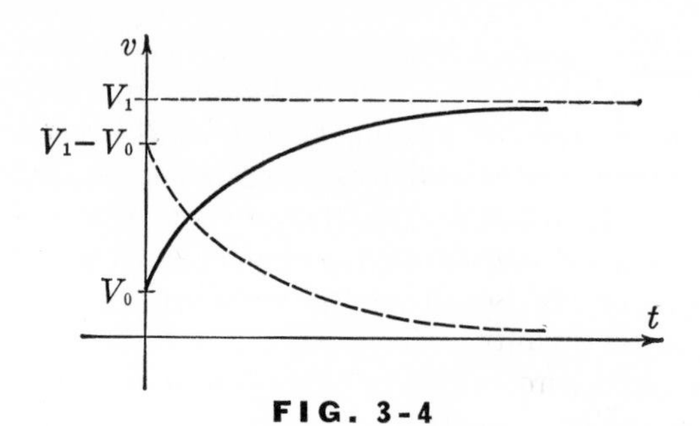

FIG. 3-4

Charging a capacitor

Note that the exponential eventually dies out and passes away. For this reason, we say that this part of the solution is *transient*. It is identical in waveform with the natural solution which we found before. In contrast to this, the constant term endures for all t, just as long as the source remains connected. This part of the solution is forced by the driving function or excitation, so we refer to it as the forced response. In the present case the forced response approaches a steady condition for large values of t. This leads to the term *steady state* for this part of the total solution. Let us designate the forced response as v_f and the natural response as v_n. Then the total response can be written

$$v = v_f + v_n \tag{3-16}$$

When the excitation is applied to the network, it seems to demand that the response take on the waveform of the excitation. The network, however, has its own natural manner in which it would like to behave. As a matter of fact, for an interval of time it tries to behave in its natural manner, and a transient period ensues during which a combination of the forced and natural behaviors takes place. But, eventually, the forcing function wins out, and the natural response of the network dies down.

Since Eq. (3-15) is a complete solution of the original differential equation, it will satisfy the equation if we substitute it there. Let us suppose that only the forced solution $v_f = V_1$ is substituted into Eq. (3-12) (taking $v_g = V_1$, of course). We will find that the equation is satisfied. (Do it.) With this in mind, let us substitute the total solution written as the sum of forced and natural response in the form of Eq. (3-16), into Eq. (3-12), still taking $v_g = V_1$. The result will be

$$\frac{d}{dt}(v_f + v_n) + \frac{1}{RC}(v_f + v_n) = V_1/RC$$

$$\left(\frac{dv_f}{dt} + \frac{1}{RC}v_f\right) + \left(\frac{dv_n}{dt} + \frac{1}{RC}v_n\right) = V_1/RC \qquad (3\text{-}17)$$

The second form is obtained by collecting terms after noting that the derivative of a sum is the sum of the derivatives. (Differentiation and addition are commutative.) Since, as we observed above, the forced response alone satisfied the original differential equation, the quantity within the first set of parentheses on the left side will equal V_1/RC. Hence the quantity within the second set of parentheses on the left must equal zero. That is, the natural response satisfies the homogeneous equation

$$\frac{dv_n}{dt} + \frac{1}{RC}v_n = 0 \qquad (3\text{-}18)$$

Let us now turn back to Eq. (3-12) and observe what difficulties will be encountered in its solution if $v_g(t)$ is anything but a constant. In such a case it will be impossible for us to integrate directly, as we did in Eq. (3-14). Hence we must seek an alternate method of solution. The considerations leading to Eqs. (3-17) and (3-18) provide us with such a method; that is, instead of finding the complete solution at one stroke, we can find the natural part and the forced part separately. The two ideas that permit such an attack are, first, that the natural solution satisfies a homogeneous equation such as Eq. (3-18), in which the driving function is set equal to zero; and, second, that the forced solution has the same waveform as the forcing function. Of course, we have not proved this last statement; we have simply observed it to be true in one simple case. Suppose we tentatively accept its general validity, however, and proceed to obtain solutions accordingly. After we obtain a function which we claim is a solution, it is always possible to determine whether or not it really is by substituting it into the original equation. If the equation is satisfied, the claim is validated; otherwise not.

Suppose we illustrate this approach using the case we have just treated, with $v_g(t) = V_1$. The forced response is to have the same waveform as the driving function; hence it should be a constant, say A. Its value is still unknown to us, so we substitute $v_f = A$ in the original equation. The equation should be satisfied, since this is the forced solution. From Eq. (3-12) we get

$$\frac{1}{RC} A = \frac{1}{RC} V_1$$

$$A = V_1 \tag{3-19}$$

since the derivative of a constant is zero. This result agrees with our previous result.

To find the natural solution, we set the excitation in Eq. (3-12) equal to zero. In the present case we have already found the waveform to be $K\epsilon^{st}$ with $s = -1/RC$ in Eq. (3-6). Hence the complete solution is

$$v = V_1 + K\epsilon^{-t/RC} \tag{3-20}$$

There remains now the problem of determining the unknown constant K.

To determine K, we must have one additional piece of information. In the present case we have such a piece of information in our knowledge of the value of the capacitor voltage V_0 immediately *before* the close of the switch at $t = 0$. By the law of conservation of charge, this is the capacitor voltage just *after* the switching also. Thus, setting $t = 0$ and $v = V_0$ in Eq. (3-20) leads to

$$V_0 = V_1 + K\epsilon^0$$

or

$$K = V_0 - V_1 \tag{3-21}$$

With this value of K, Eq. (3-20) becomes identical with Eq. (3-15), and our solution is complete.

In order to gain familiarity with this method of solution, let us use it to solve the same problem with which we have been dealing but with a different excitation function. Let us take $v_g(t)$ to be the ramp function

$$\boxed{v_g(t) = t \quad t > 0} \tag{3-22}$$

With this as the driving function, Eq. (3-12) becomes

$$\frac{dv}{dt} + \frac{1}{RC} v = \frac{t}{RC} \tag{3-23}$$

Our plan is to find the forced response and the natural response separately. We have already found the latter to have the form of an exponential. It remains to find the forced response. According to the foregoing discussion, we expect the forced response to have the same waveform as the excitation. Now the excitation is a linear function of t, a straight line. We would therefore expect the forced response to be a straight line also. Of course, we do not know whether it will have the same slope as the excitation, or the same intercept. Hence, to cover all contingencies, we write

$$v_f = At + B \tag{3-24}$$

in which we do not commit ourselves as to the constants A and B. If this is to be the forced response, it should satisfy Eq. (3-23) identically. Substituting Eq. (3-24) into Eq. (3-23) and performing the indicated differentiation leads to

$$\left(A + \frac{B}{RC}\right) + \frac{A}{RC}t \equiv \frac{t}{RC} \tag{3-25}$$

Note that this is to be an identity; that is, it is to be true for all values of t, not just one value or some values. The only way this is possible is for coefficients of like powers of t to be equal on both sides of the equation. (A constant can be considered as a coefficient of t^0.) Thus

$$\frac{A}{RC} = \frac{1}{RC}; \quad A + \frac{B}{RC} = 0$$

or

$$A = 1 \qquad (a)$$
$$B = -RC \qquad (b) \tag{3-26}$$

With these values of A and B, the forced response becomes

$$v_f = t - RC \tag{3-27}$$

The method of finding the forced response which we have been describing is called the *method of undetermined coefficients*, for an obvious reason.

The complete response is obtained by adding the natural response $K\epsilon^{-t/RC}$ to Eq. (3-27), where K is still unknown. Thus

$$v = t - RC + K\epsilon^{-t/RC} \tag{3-28}$$

Again, using the initial value of the capacitor voltage, we can determine K. We set $t = 0$ and $v = V_0$ in Eq. (3-28) and find $K = V_0 + RC$. The total response then becomes

$$v = t - RC + (V_0 + RC)\epsilon^{-t/RC} \tag{3-29}$$

Note that the known initial value of the voltage refers to the total voltage of the capacitor. Hence, when using it to determine the unknown constant in the natural response, it is necessary to use the expression for the total response. This is frequently a source of student error.

As a final illustration of this method of solution let us consider the same problem but with a sinusoidal excitation function. Let

$$v_g(t) = V_m \cos \omega t \tag{3-30}$$

With this function as a source voltage, Eq. (3-12) becomes

$$\frac{dv}{dt} + \frac{1}{RC} v = \frac{V_m}{RC} \cos \omega t \tag{3-31}$$

Again we assume that the forced response has the same waveform as the excitation; that is, a sinusoid. The amplitude of the sinusoid and the initial angle, however, are unknown; they are probably not the same as the amplitude and initial angle of the excitation. Hence let us write the forced solution as

$$v_f = K_1 \cos (\omega t - \phi) = A \cos \omega t + B \sin \omega t \tag{3-32}$$

The second form on the right follows by expanding the first form, using the following trigonometric identity:

$$\cos (x - y) = \cos x \cos y + \sin x \sin y \tag{3-33}$$

A and B are two constants whose values we do not yet know but expect to find by substituting the assumed solution into Eq. (3-31). If we do this, we shall get

$$(-A\omega \sin \omega t + B\omega \cos \omega t) + \frac{1}{RC} (A \cos \omega t + B \sin \omega t) = \frac{V_m}{RC} \cos \omega t$$

or

$$\left(B\omega + \frac{A}{RC} - \frac{V_m}{RC}\right) \cos \omega t + \left(\frac{1}{RC} B - \omega A\right) \sin \omega t = 0 \tag{3-34}$$

In the last equation we collected together all the terms involving $\cos \omega t$ and all those involving $\sin \omega t$. This equation is to be valid for all values of t. That is, its truth should not depend on t; it should be an identity. But it is impossible for the sum of a sine function and a cosine function to be zero for all time except if the multipliers of each one are individually zero. Hence we get

$$\begin{aligned} \frac{1}{RC} A + \omega B &= \frac{V_m}{RC} & (a) \\ -\omega A + \frac{1}{RC} B &= 0 & (b) \end{aligned} \tag{3-35}$$

Here we have a pair of simultaneous algebraic equations, the unknowns being A and B. We can solve these equations by solving for A or B in the last equation and substituting into the first. Alternatively, we can use determinants and Cramer's rule. Using the latter procedure, we get

$$A = \frac{V_m}{R^2C^2\Delta} = \frac{V_m}{1 + (\omega CR)^2} \qquad (a)$$
$$B = \frac{\omega V_m}{RC\Delta} = \frac{\omega CRV_m}{1 + (\omega CR)^2} \qquad (b) \qquad (3\text{-}36)$$

where

$$\Delta = \frac{1}{(RC)^2} + \omega^2 \qquad (3\text{-}37)$$

is the determinant of the system.

With these values of A and B, the forced-response voltage in Eq. (3-32) becomes

$$v_f = \frac{V_m}{1 + (\omega CR)^2} \cos \omega t + \frac{\omega CRV_m}{1 + (\omega CR)^2} \sin \omega t \qquad (3\text{-}38)$$

We can put this expression in the form of single sinusoidal function by using the identity in Eq. (3-33) in reverse order. If we let x in that equation stand for ωt and y for ϕ, then we see that $\tan \phi$ will be given by the ratio of the coefficients of $\sin \omega t$ and $\cos \omega t$ in Eq. (3-38); that is,

$$\tan \phi = \frac{\dfrac{\omega CRV_m}{1 + (\omega CR)^2}}{\dfrac{V_m}{1 + (\omega CR)^2}} = \omega CR \qquad (3\text{-}39)$$

From this we can find $\cos \phi$ and $\sin \phi$ with the help of the diagram shown in Fig. 3-5. We form a right triangle with one side labeled ωCR and the other side 1. The hypotenuse is then computed as the square root of the sum of the squares of the sides. From the diagram, then,

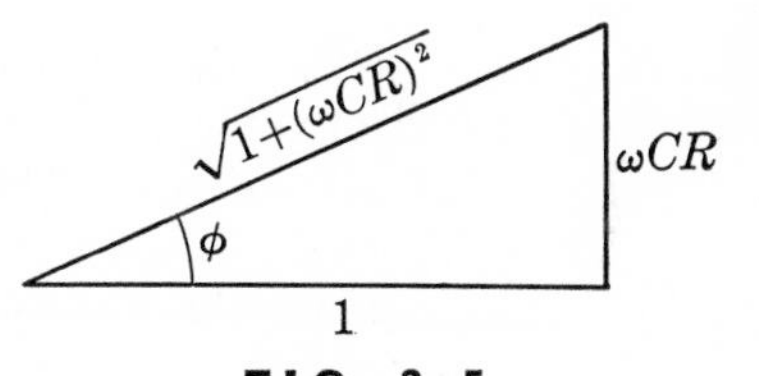

FIG. 3-5
Graphical aid

$$\cos \phi = \frac{1}{\sqrt{1 + (\omega CR)^2}} \qquad (a)$$
$$\sin \phi = \frac{\omega CR}{\sqrt{1 + (\omega CR)^2}} \qquad (b) \qquad (3\text{-}40)$$

Using these expressions in Eq. (3-38), we can now write the forced response as

$$v_f = \frac{V_m}{\sqrt{1 + (\omega CR)^2}} \cos(\omega t - \phi) \qquad (a)$$

$$\phi = \tan^{-1}(\omega CR) \qquad (b) \tag{3-41}$$

This completes the calculation of the forced response. We can now write the complete response as the sum of the forced response and the natural response as follows:

$$\begin{aligned} v &= \frac{V_m}{\sqrt{1 + (\omega RC)^2}} \cos(\omega t - \phi) + K\epsilon^{-t/RC} \\ &= \frac{V_m}{1 + (\omega RC)^2} (\cos \omega t + \omega CR \sin \omega t) + K\epsilon^{-t/RC} \end{aligned} \tag{3-42}$$

There still remains the task of computing the constant K which is as yet unknown. As in the previous case, it can be determined by utilizing the knowledge of the initial capacitor voltage. If we let $v = V_0$ and $t = 0$ in the last equation, we shall get

$$V_0 = \frac{V_m}{1 + (\omega RC)^2} + K \tag{3-43}$$

which determines K. The final solution now becomes

$$V = \frac{V_m}{\sqrt{1 + (\omega RC)^2}} \cos(\omega t - \phi) + \left[V_0 - \frac{V_m}{1 + (\omega CR)^2} \right] \epsilon^{-t/RC} \tag{3-44}$$

That this is really the solution can be checked by inserting it into the original equation, and by noting that it gives the correct known value at $t = 0$.

In order to get a graphical picture of the result, let us assume the following numerical values:

$$\omega CR = 1$$

$$\phi = \tan^{-1}(1) = \pi/4$$

$$V_0 = \tfrac{3}{2} V_m$$

With these values, Eq. (3-44) becomes

$$\begin{aligned} \frac{v}{V_m} &= \frac{1}{\sqrt{2}} \cos(\omega t - \pi/4) + \epsilon^{-t/RC} \\ &= \frac{1}{\sqrt{2}} \cos(\omega t - \pi/4) + \epsilon^{-\omega t} \end{aligned} \tag{3-45}$$

The exponent in the last line is obtained by the process of multiplying and dividing the exponent in the previous line by ω, and then noting that $\omega CR = 1$. This expression is plotted in Fig. 3-6. The dashed curve is the source voltage given for comparison purposes. Notice that after about one cycle of the sinusoidal voltage, the total response differs from the forced response by an imperceptible amount.

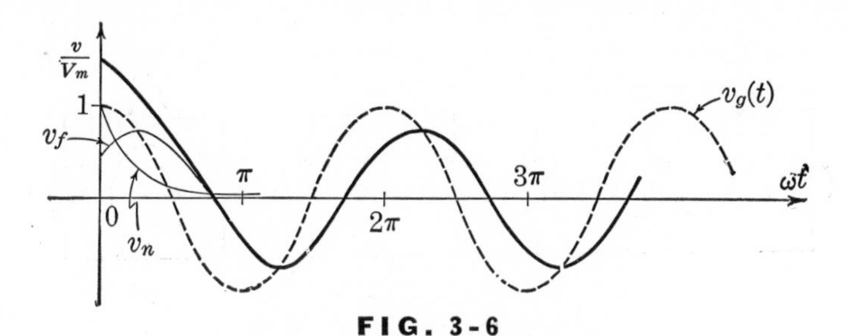

FIG. 3-6

Solution for capacitor voltage

Since the forced response is periodic and continues indefinitely, it is again called the steady state. In fact, the term "steady state" originated in connection with the forced response of electrical networks to sinusiodal excitations.

We could now go on and find the response for other excitation functions in the same manner. But, no matter what the excitation, the natural response is always an exponential with the same exponent, whose value depends only on the network parameters R and C. Its multiplying constant, however, will depend on the value of the excitation at the instant of switching, and on the initial value of the capacitor voltage, as we have seen in the preceding examples.

Note that the total response of a network to an excitation must satisfy two things: In the first place, it must be a solution of the differential equation which results from an application of the fundamental laws. Second, it must fit the state of the network at the initial instant; it must satisfy the initial conditions. The constant K which appears in the natural component of the response provides the flexibility required in order that the solution will fit any arbitrarily prescribed initial capacitor voltage.

3.3 SIMPLE RESISTANCE-INDUCTANCE NETWORK

Let us now consider a network which contains an inductor and a resistor. Such a network is shown in Fig. 3-7(a). The switch has been in position a for some time when it is suddenly switched to position b at a time which we take as $t = 0$. At this time there is a current of value I_0 in the inductor. After the switching, the network has the form shown in Fig. 3-7(b). It is required to find the current in the inductor for all time following the switching.

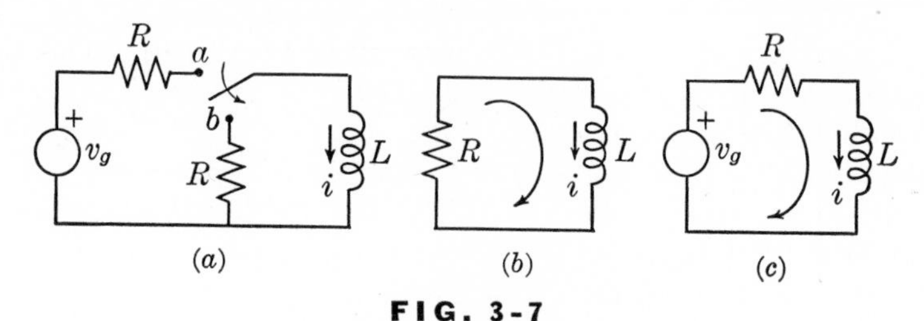

FIG. 3-7

Current in an inductance

The simplest procedure here is to write a loop equation. Choosing the loop current to be the same as the inductor current, we get

$$L\frac{di}{dt} + Ri = 0$$

or

$$\frac{di}{dt} + \frac{R}{L}i = 0 \tag{3-46}$$

Again we have a differential equation, this time with the current as a variable.

In fact, the network of Fig. 3-7(b) is the dual of that of Fig. 3-1(b). Hence the loop equation which we have here should be identical with Eq. (3-1), which was the node equation of the R-C network, except for an interchange of the dual quantities i for v, L for C, and G for R. Actually, the resistor in Fig. 3-7 has not been labeled with the conductance G but with the resistance R, which is $1/G$. This is the reason why R/L appears in Eq. (3-46) rather than $1/GL$.

The initial current I_0 plays the role of the initial voltage V_0. Hence the solution of this equation can be written down immediately from Eq. (3-5) with a suitable change of symbols. It is

$$i = I_0\epsilon^{-Rt/L} \tag{3-47}$$

You can verify that this is the solution by inserting it into the differential equation, and by noting that the correct initial current is obtained.

A plot of the solution will again take the form shown in Fig. 3-2.

The ratio of L to R is called the *time constant*. (Again, check the dimensions.) We see that the current dies out with time at a rate determined by the time constant. Since the response is a result of the natural behavior of the network, it is again called the natural response.

Let us now turn back to the network in Fig. 3-7(a). This time, assume that the switch has been in position b for some time and is suddenly switched to position a. Let us assume that, prior to the switching, the current in the inductor is not zero but some value I_0. The network takes the form shown in Fig. 3-7(c). Again there is a single loop, but now the network is driven by a voltage source. Hence the loop equation becomes

$$L\frac{di}{dt} + Ri = v_g$$

or

$$\frac{di}{dt} + \frac{R}{L}i = \frac{1}{L}v_g \tag{3-48}$$

This is to be compared with Eq. (3-12). The equations are identical in form, except that the variable is now current instead of voltage. The time constant L/R takes the place of the time constant RC. The duality will be complete if, in Eq. (3-12), we recognize that v_g/R represents the current source of Fig. 3-3(b) and can be replaced by i_g. The two equations will then become identical except for an interchange of dual quantities.

Again we would expect the total solution to depend on the exact form of the driving function v_g. The natural response will always be an exponential such as the one in Eq. (3-47); only the multiplier will be different in different cases. Hence the total response can be written

$$i(t) = i_f(t) + K\epsilon^{-Rt/L} \tag{3-49}$$

where $i_f(t)$ is the forced response. Its form will depend on the excitation. Let us temporarily assume that i_f has been determined. The only remaining unknown is then the constant K. To determine the value of K, we need to know one condition, one value of the current at a known time. We do know the value of the current immediately before the switching. The equation is valid, however, only after the switching. How is the current after the switching related to the current before the switching?

A similar question with regard to a capacitor voltage was answered by appealing to the law of conservation of charge. To answer the present question, we must again have recourse to observations in the physical world. It was mentioned in Chap. 1 that inductance in an electrical system is analogous to mass in a mechanical system. Now the distinctive property of mass is inertia. The state of motion of a massive body cannot be changed suddenly. As a matter of fact, a fundamental postulate of mechanics, called the *law of conservation of momentum*, states that the total momentum of a system of particles cannot be changed instantly (except by an infinite force).

We also noted in Chap. 1 that flux linkage is analogous to mechanical momentum. Thus it is reasonable to assume that a relationship similar to the law of conservation of momentum will apply to flux linkage. Such an assumption appears reasonable from observations that can be made showing that the current in a coil of wire does not instantaneously follow any rapid changes in the voltage across it; instead, it changes gradually.

On this basis we postulate that *the sum of all the flux linkages around any closed path will be continuous and cannot be changed instantly.* We refer to this postulate as the *principle of conservation of flux linkage.*

In case there is only a single inductor on a closed loop (together with other elements), the conservation of flux linkage reduces to the statement that *the current in an inductor must be continuous and cannot be changed instantly.* This follows because the total flux linkage in such a case will be simply Li. Note the analogy between the conservation of flux linkage and the conservation of charge.

Let us now return to the problem at hand. We can now state that the current just after the switching must be equal to the current just before the switching. Since this value is known to be I_0, we can determine the constant K in Eq. (3-49). Setting $t = 0$ and $i = I_0$, we will get

$$I_0 = i_f(0) + K$$

and, finally,

$$i = i_f(t) + [I_0 - i_f(0)]\epsilon^{-Rt/L} \tag{3-50}$$

This constitutes the complete response of the network.

There now remains the task of finding the forced response. We have already discussed one procedure, the method of undetermined coefficients. We assume the forced-response waveform to be the same as that of the excitation, but with parameters (such as amplitude and angle in the case of sinusoidal excitation) still undetermined. These parameters are determined by substituting the assumed forced response into the differential equation, and arguing that the resulting expression must be valid for all time.

Let us illustrate this method once again, and this time take the excitation to be a parabolic function, $v_g(t) = t^2$ for $t > 0$. We assume that the forced response is also a parabola, but possibly with a different vertex and different intercepts; that is, for the response function we assume the general form of a parabola.

$$i_f(t) = at^2 + bt + c \tag{3-51}$$

in which a, b, and c are still unknown. This is to be substituted into Eq. (3-48), in which we also insert $v_g = t^2$. The result will be

$$(2at + b) + \frac{R}{L}(at^2 + bt + c) = \frac{1}{L}t^2 \tag{3-52}$$

Let us now collect all terms involving the same powers of t. We shall get

$$\left(\frac{R}{L}a - \frac{1}{L}\right)t^2 + \left(2a + \frac{R}{L}b\right)t + \left(b + \frac{R}{L}c\right) \equiv 0 \tag{3-53}$$

This expression states that the function of time on the left (in this case simply a polynomial) must be identically zero; that is, it must be zero for all values of time, if the assumed solution is really a solution. Now we know that a quadratic polynomial, which the left-hand side is, can go to zero for just two values of t. The only way the quadratic can be identically zero is for all the coefficients to be identically zero. Hence we must have

$$\frac{R}{L}a - \frac{1}{L} = 0 \qquad (a)$$

$$2a + \frac{R}{L}b = 0 \qquad (b) \qquad (3\text{-}54)$$

$$b + \frac{R}{L}c = 0 \qquad (c)$$

From the first of these we find a; with this value inserted into the second equation, we find b; and, finally, with this value of b inserted into the third equation, we find c. (Go through the details.) The forced solution now takes the form

$$i_f(t) = \frac{1}{R}\left(t^2 - \frac{2L}{R}t + \frac{2L^2}{R^2}\right) \tag{3-55}$$

The complete response is now obtained by inserting this expression into Eq. (3-50).

Note that the forced response is in fact a parabola, as dictated by the driving function. Since there is nothing steady about its behavior as time goes on, however, we do not refer to it as the steady-state response.

3.4 PAUSE AND CONSOLIDATION

Let us now pause briefly and try to obtain a perspective view of what we have done so far in this chapter, to see if there are any universal aspects in the method of solution we have employed which can be used for other networks than the ones we have studied.

We considered two very simple networks—one consisting of a single capacitor and resistor, the other of a single inductor and resistor (not counting sources). In each case, application of the network equations (Kirchhoff's laws and the v-i relationships) resulted in a differential equation. Differential equations are classified according to the derivatives present. The *order* of the equation refers to the highest derivative which appears in it. Thus, the equations which we have met in this chapter are of the *first order*, since only the first derivative appears. Equations are also classified by degree. In a differential equation the dependent variable or its derivatives may be raised to any power. There may also be products of derivatives of different order. The *degree* of a differential equation is the highest power to which the dependent variable or any of its derivatives are raised (or the sum of the powers of any cross products). If all derivatives are raised to the first power, the equation is of the first degree, or *linear*. The equations with which we shall deal in this book are all linear with coefficients which are constant, such as R/L or RC.

In case a network contains only one energy-storing element but has an otherwise extensive structure containing resistors only, we can still handle the situation by the methods we have discussed.

As an example, consider the network shown in Fig. 3-8(a). We can proceed in one of two ways. One possibility is to write the appropriate number of network equations, choosing loops or nodes in such a way that the capacitance C appears in only one of the equations. Hence only this one will be a differential equation, all the others being algebraic. We can then eliminate variables in all the algebraic equations, leaving just the one first-order differential equation. The solution now proceeds in the manner that we have already discussed.

The second alternative is to transform the network by making source transformations, and series or parallel combinations of resistors, until only one loop or one pair of nodes remains. Such a transformation is shown in Fig. 3-8(b). (Go through the details of obtaining this.) If a loop equation is written for this network, again a first-order differential equation will result.

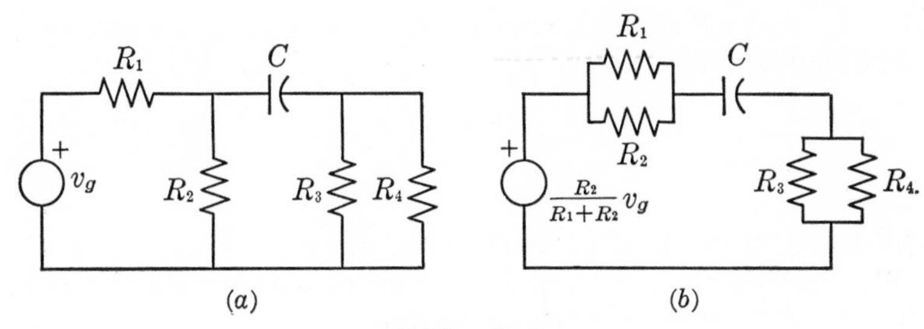

FIG. 3-8
Illustrative example

Thus we are now able to find the complete solution of a network problem for a network of any structural complexity as long as only one energy-storing element is present. We found that the solution of the first-order differential equation that results from such a network consists of two parts. The natural component of the solution is characterized by an exponential with a negative exponent; as time goes on, this term approaches zero. For this reason we refer to this part of the solution as the transient. The other part of the solution has the same waveform as the excitation or driving function; we refer to it as the forced response. When the driving function has a steady form, such as a periodic function or a constant, then the forced response will also have a steady form; so we call it the steady state.

Now suppose that a network contains more than one energy-storing element. It is not hard to appreciate that each inductor and each capacitor will lead to a derivative. Instead of having a set of algebraic equations to solve, we shall have a set of simultaneous differential equations. The solution will now take on a higher degree of complexity.

Let us here review the steps that are involved in solving a differential equation by the method we have discussed. There are three distinct steps: (1) finding the forced response; (2) finding the general form of the natural response in which there appear one or more constants, whose values must be determined so as to permit the solution to be adjusted to accommodate any given initial conditions; and (3) finding these constants by applying the laws of conservation of charge and of flux linkage.

In the examples which we discussed, the driving functions were simple functions of t, such as $\cos \omega t$, or t, or simply a constant. Suppose the driving function consists of the sum of several terms. The question is: Must we find the forced response in one step, or is it possible to find the forced response for each term in the driving function, and then add? To answer this, let us write a first-order linear equation in general form, as follows:

$$\frac{dx}{dt} + ax = y \qquad (a)$$
$$y = y_1 + y_2 \qquad (b) \qquad (3\text{-}56)$$

x and y may represent either voltages or currents; y is the excitation, and x is the response. We assume that the excitation is the sum of two terms y_1 and y_2. Let x_1 be the forced response of the equation when y_1 alone is the excitation, and let x_2 be the forced response when y_2 alone is the excitation; that is, x_1 and x_2 satisfy the equations

$$\frac{dx_1}{dt} + ax_1 = y_1 \qquad (a)$$
$$\frac{dx_2}{dt} + ax_2 = y_2 \qquad (b) \qquad (3\text{-}57)$$

If we now add these equations and note that differentiation and addition are commutative, we will get

$$\frac{d}{dt}(x_1 + x_2) + a(x_1 + x_2) = y_1 + y_2 = y \qquad (3\text{-}58)$$

From this result we see that the forced solution under the combined excitation $(y_1 + y_2)$ is truly the sum of the forced solutions under each of the excitations acting alone. This is a very important result. It is a direct consequence of the linearity of the differential equation. (Demonstrate that the same conclusion is not true for, say, $dx/dt + ax^2 = y_1 + y_2$.) We refer to this result as the *principle of superposition*, which can be stated as follows: *The forced response of a linear network to a sum of different excitations is the same as the sum of the forced responses to these same excitations when they are applied individually, all others being removed.*

Although we have demonstrated the validity of the superposition principle only for a first-order linear equation, you can readily appreciate its validity for higher orders as well, since addition and differentiation of any order are commutative. It is, then, a fundamental property of linear equations.

3.5 SECOND-ORDER NETWORKS

We shall now turn our attention to networks which contain more than one energy-storage device, with a view toward learning more about the natural behavior of electrical networks.

As an initial example, consider the series connection of R, L, and C shown in Fig. 3-9. We can assume that the capacitor has an initial voltage V_0 just before a switch is closed establishing the loop. It is required to find the capacitor voltage v or the current i as a function of time.

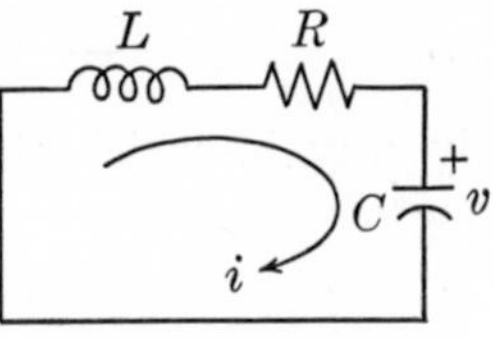

FIG. 3-9

Discharge of capacitor through series resistance and inductance

Let us apply Kirchhoff's voltage law around the loop and, at the same time, insert the v-i relationships of R and L. The result will be

$$L\frac{di}{dt} + Ri + v = 0 \tag{3-59}$$

There are two unknowns in this equation, v and i. Two alternative approaches are available to us at this point. We can insert for v in this equation the v-i relationship for the capacitor, leaving an equation with i as the unknown, or we can substitute $i = C\dfrac{dv}{dt}$ and leave v as the unknown. If we choose this second alternative, we will get

$$LC\frac{d^2v}{dt^2} + RC\frac{dv}{dt} + v = 0$$

or

$$\frac{d^2v}{dt^2} + \frac{R}{L}\frac{dv}{dt} + \frac{1}{LC}v = 0 \tag{3-60}$$

We again have a linear differential equation and, since the highest derivative is the second, the equation is of the second order. To find the solution of this equation, we again resort to an intelligent, reasoned guess. In the case of the first-order equation we found the solution to be an exponential. Furthermore, the present equation requires that a linear combination of v and its first two derivatives be zero. This condition demands that v and its derivatives have the same waveform. Now we know this to be true for an exponential. On the basis of these arguments, let us again try an exponential as a tentative solution. Assume

$$v = K\epsilon^{st} \tag{3-61}$$

in which K and s are again unknown. Substituting this into the differential equation, we get

$$K\left(s^2 + \frac{R}{L}s + \frac{1}{LC}\right)\epsilon^{st} = 0 \tag{3-62}$$

The only way in which this equation can be satisfied (besides the trivial one $K = 0$) is

$$s^2 + \frac{R}{L}s + \frac{1}{LC} = 0 \tag{3-63}$$

We now have a quadratic equation to solve. No matter what the roots might be, we know that there will be two of them, which we can label s_1 and s_2. If we use either of these roots in the assumed solution given in Eq. (3-61), the equation will be satisfied. As a matter of fact, if we take the sum of two exponentials, one with each of these roots in the exponent, this will also be a solution. Thus

$$v = K_1\epsilon^{s_1 t} + K_2\epsilon^{s_2 t} \tag{3-64}$$

will be the general solution of the equation. [Go through the details of inserting this expression into Eq. (3-60) and show that the latter is satisfied.] The constants K_1 and K_2 are still to be determined from the initial conditions, but let us not concern ourselves with the algebraic details here. Instead, let us see what we can learn about the natural behavior of the network from a detailed consideration of Eqs. (3-63) and (3-64).

Equation (3-63) is a quadratic which has as coefficients certain combinations of the network parameters. It is thus characteristic of the network. In fact, we refer to it as the *characteristic equation*. Its roots are the *characteristic values* which play a vital role in network behavior. We will eventually give these roots a more descriptive name.

The two roots of the quadratic in Eq. (3-63) can easily be found from the quadratic formula as

$$\begin{aligned} s_1 &= -\frac{R}{2L} + \sqrt{\left(\frac{R}{2L}\right)^2 - \frac{1}{LC}} && (a) \\ s_2 &= -\frac{R}{2L} - \sqrt{\left(\frac{R}{2L}\right)^2 - \frac{1}{LC}} && (b) \end{aligned} \tag{3-65}$$

The character of the roots is dependent in part upon the discriminant (the quantity under the radical). The roots will be real and distinct if the discriminant is positive. In case the discriminant is zero, the two real roots will be identical. On the other hand, if the discriminant is negative, the two roots will be complex. In such a case, remembering that these roots are to go into the exponents in Eq. (3-64), we shall have the task of interpreting the meaning of an exponential with a complex exponent.

It is assumed that you are familiar with the elements of the algebra of complex numbers. We shall here digress temporarily, however, for the purpose of summarizing those aspects which we shall make use of in the sequel. For a more thorough treatment, consult a mathematics text. Those of you who do not need this summary can turn directly to Sec. 3-7.

3.6 COMPLEX NUMBERS

A complex number is a particular combination of two real numbers, an ordered pair. When we say the pair of real numbers is ordered, we mean the sequence of writing the numbers, their order, is important. Such a number is written (a, b); a comma separates the two members of the pair. A complex number has two parts each of which is a real number. The terminology in common use is to call the first of the real numbers the *real part* and the second one the *imaginary part*. These names are unfortunate, since there is nothing more "imaginary" about an imaginary number than there is about a real number. However, we shall continue to use these names. A real number is a special case of a complex number which has zero imaginary part; that is, complex numbers of the form $(a, 0)$ are real. When the real part of a complex number is zero, the complex number is said to be an imaginary number; imaginary numbers have the form $(0, b)$.

We must now define elementary operations for complex numbers. We define the following:

Equality. Two complex numbers $A = (a_1, a_2)$ and $B = (b_1, b_2)$ are equal if, and only if, $a_1 = b_1$ and $a_2 = b_2$; that is, the two real parts as well as the two imaginary parts must be equal.

Addition. The sum of two complex numbers $A = (a_1, a_2)$ and $B = (b_1, b_2)$ is defined as a complex number $C = (c_1, c_2)$ such that

$$C = A + B = (a_1 + b_1, a_2 + b_2)$$

that is, the sum has a real part which is the sum of the real parts of the two numbers, and an imaginary part which is the sum of the imaginary parts.

Subtraction. The difference of two complex numbers A and B is defined as $A - B = A + (-B)$; that is, change the sign of the subtrahend and add.

Multiplication. The product of two complex numbers $A = (a_1, a_2)$ and $B = (b_1, b_2)$ is defined as

$$C = AB = (a_1b_1 - a_2b_2, a_2b_1 + a_1b_2)$$

The question might arise as to the reasons for choosing this particular definition for multiplication and not one of a hundred other possible combinations of a_1, a_2, b_1, and b_2. Of course, any definition which is chosen must be consistent with the definition of multiplication of real numbers, since real numbers are special kinds of complex numbers. The only other criterion is that the definition should be useful. We cannot now see that the adopted definition is useful, but we can certainly check it to see that it reduces to the definition of multiplication of two real numbers when the imaginary parts a_2 and b_2 are zero.

Using the definitions of addition and multiplication, we can now show that any complex number can be written in a particular form as follows:

$$\begin{aligned}(a_1, a_2) &= (a_1, 0) + (0, a_2) \\ &= (a_1, 0) + (0, 1)(a_2, 0)\end{aligned} \tag{3-66}$$

The complex number has been written as the sum of two complex numbers. The first of these has a zero imaginary part, so it is a real number which can be written simply as a_1. The second of these is the product of two other complex numbers, one of which is again simply a real number a_2. The second number of the product is a special imaginary number. It arises so often that we give it a special symbol. Thus

$$(0, 1) = j \tag{3-67}$$

With this definition, the complex number in Eq. (3-66) can now be written

$$(a_1, a_2) = a_1 + ja_2$$

The right-hand side is the usual form in which complex numbers are written, and we shall use this form most of the time from now on. The symbol i is used by mathematicians instead of j, but we shall use j in order to avoid confusion with the symbol for current.

Again using the definition of multiplication, let us find the product of j by itself.

$$j^2 = (0, 1)(0, 1) = (-1, 0) = -1 \tag{3-68}$$

Thus the square of j is the real number -1. From this, it follows that j is the square root of -1.

$$j = \sqrt{-1} \tag{3-69}$$

With the use of the symbol j we can now show that multiplication can be performed in a more convenient way than is indicated by the original definition. The usual rules of algebra will be obeyed. Thus, using the rules of algebra, we get

$$\begin{array}{r} a_1 + ja_2 \\ b_1 + jb_2 \\ \hline a_1b_1 + ja_2b_1 + ja_1b_2 + j^2a_2b_2 \\ = (a_1b_1 - a_2b_2) + j(a_2b_1 + a_1b_2) \end{array}$$

which, of course, agrees with the original definition.

Division. The quotient of two complex numbers A and B is defined as the complex number C which, when multiplied by B, gives A. This definition is consistent with the definition of division of real numbers. Thus

$$C = \frac{A}{B} \quad \text{means} \quad BC = A$$

If we write $A = a_1 + ja_2$, $B = b_1 + jb_2$, and $C = c_1 + jc_2$, then $BC = A$ becomes

$$(b_1 + jb_2)(c_1 + jc_2) = a_1 + ja_2$$

from which, by the definitions of equality and multiplication, we get

$$b_1c_1 - b_2c_2 = a_1$$

$$b_2c_1 + b_1c_2 = a_2$$

Remembering that the unknowns here are c_1 and c_2, we can solve this pair of equations (by means of determinants, for example) to find c_1 and c_2. (Go through the details.) Then, finally, we get

$$C = \frac{A}{B} = c_1 + jc_2 = \frac{a_1b_1 + a_2b_2}{b_1^2 + b_2^2} + j\,\frac{a_2b_1 - a_1b_2}{b_1^2 + b_2^2} \tag{3-70}$$

Conjugate. The conjugate of a complex number $A = a_1 + ja_2$ is defined as the complex number $A^* = a_1 - ja_2$; that is, the conjugate of a number has the same real part but an imaginary part which is opposite in sign. It is designated by an asterisk as shown.

With the concept of conjugate, we can develop a process whereby the quotient of two complex numbers can be taken directly without the necessity for remembering Eq. (3-70). After writing $a_1 + ja_2$ over $b_1 + jb_2$, let us multiply both numerator and denominator by the conjugate of the denominator. Thus

$$\frac{a_1 + ja_2}{b_1 + jb_2} = \frac{(a_1 + ja_2)(b_1 - jb_2)}{(b_1 + jb_2)(b_1 - jb_2)} = \frac{a_1b_1 + a_2b_2}{b_1^2 + b_2^2} - j\frac{a_2b_1 - a_1b_2}{b_1^2 + b_2^2} \tag{3-71}$$

After multiplying, we find that the denominator becomes a real number. The result agrees with the definition of division. The process of multiplying numerator and denominator by the conjugate of the denominator is called *rationalization.*

Having defined the basic algebraic operations, let us now consider a geometrical representation of complex numbers. We know that the set of all real numbers can be put in a one-to-one correspondence with points on a straight line. Any real number can be represented by the distance of a point on the line from another point which is called the origin. A complex number consists of a pair of two real numbers, so one line alone will not be sufficient to represent it. Instead of a line, we can use a plane to represent a complex number.

Consider Fig. 3-10 which shows a pair of axes at right angles to each other. Each part (real and imaginary) of a complex number can be

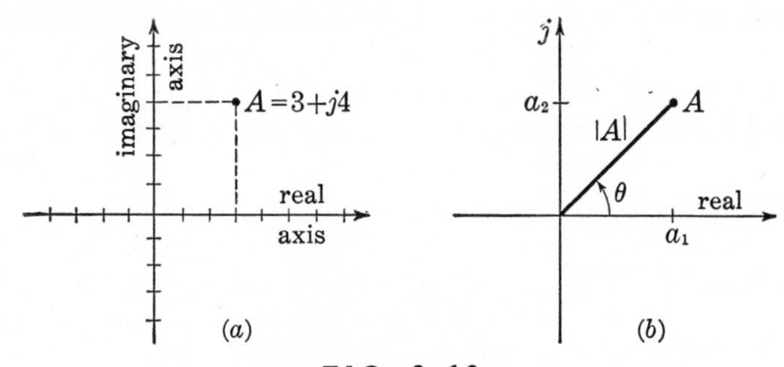

FIG. 3-10
The complex plane

put in a one-to-one correspondence with points on one of these lines. The horizontal axis is called the *real* axis while the vertical one is called the *imaginary* axis or the j axis. Any point in the plane can be described in terms of its horizontal and vertical coordinates. Hence any point in the plane can represent a complex number.

As an example, consider the complex number $A = 3 + j4$. This is represented in the figure by the point marked A having a horizontal coordinate (real part) of 3 and a vertical coordinate (imaginary part) of 4.

Now a point in a plane, besides being describable in rectangular coordinates, can also be described in polar coordinates. The polar coordinates of a point are (1) the distance of the point from the origin and (2) the angle which the line joining the origin to the point makes with the horizontal axis. This is illustrated in Fig. 3-10(*b*).

Thus a complex number A can be described in terms of the quantities $|A|$, its *magnitude*, and θ, its *angle*, in addition to its real part a_1, and its imaginary part a_2. From the geometry of the figure, it is clear that

$$\begin{aligned} a_1 &= |A| \cos \theta && (a) \\ a_2 &= |A| \sin \theta && (b) \end{aligned} \qquad (3\text{-}72)$$

Hence a complex number can now be written in the so-called trigonometric form as

$$A = |A| \cos \theta + j|A| \sin \theta \qquad (3\text{-}73)$$

To describe the complex number, it is enough to state its magnitude and its angle. This fact leads to a symbolic way of writing the complex number as follows

$$A = |A| \underline{/\theta} \qquad (3\text{-}74)$$

(We read this as "magnitude A at an angle θ.") Be careful to regard this notation as simply symbolic and not a form on which mathematical operations are to be performed.

Multiplication and division of complex numbers are greatly simplified in terms of the magnitude and angle. Let us write two complex numbers A and B in trigonometric form.

$$A = |A| (\cos \alpha + j \sin \alpha)$$

$$B = |B| (\cos \beta + j \sin \beta)$$

Their product C will become

$$\begin{aligned} C &= AB \\ &= |A|\,|B|\,[(\cos \alpha \cos \beta - \sin \alpha \sin \beta) + j\,(\sin \alpha \cos \beta + \cos \alpha \sin \beta)] \\ &= |A|\,|B|\,[\cos (\alpha + \beta) + j \sin (\alpha + \beta)] \end{aligned} \qquad (3\text{-}75)$$

The last step was obtained by using the trigonometric identities involving the sine and cosine of a sum of angles. The final result is in the form

$$C = |C| (\cos \theta + j \sin \theta) \qquad (3\text{-}76)$$

where

$$\begin{aligned} |C| &= |A|\,|B| && (a) \\ \theta &= \alpha + \beta && (b) \end{aligned} \qquad (3\text{-}77)$$

That is, the product of two complex numbers is a complex number whose magnitude is the product of magnitudes of the two numbers, and whose angle is the sum of angles of the two numbers. For computational purposes, this is the form in which multiplication is most easily carried out.

This same result gives us a method for finding the quotient of two numbers. If it is required to find the quotient B/C, by definition this is the complex number A such that $AB = C$. Equation (3-77) gives the relationships among the magnitudes and angles. Hence, symbolically, we can write

$$B = |B| \underline{/\beta}$$

$$C = |C| \underline{/\theta}$$

$$A = \frac{B}{C} = \frac{|B|}{|C|} \underline{/\beta - \theta} \tag{3-78}$$

That is, the quotient of two complex numbers is a number whose magnitude is the quotient of the magnitudes of the two, and whose angle is the difference of the angles. In numerical computations this expression for division is easy to carry out.

Let us illustrate these concepts by means of an example. Let it be required to find the complex number

$$E = \frac{AB}{CD}$$

where

$$A = 2 + j1 \doteq 2.23 \underline{/26.5^\circ}$$

$$B = 3 - j4 \doteq 5 \underline{/-53.1^\circ}$$

$$C = 6 + j1 \doteq 6.1 \underline{/9.5^\circ}$$

$$D = 4 + j5 \doteq 6.4 \underline{/51.4^\circ}$$

The magnitude of E is equal to the product of 2.23 and 5 divided by the product of 6.1 and 6.4. The angle of E is the sum of 26.5 deg and -53.1 deg minus the sum of 9.5 deg and 51.4 deg. Symbolically, the result will be

$$E = \frac{(2.23)(5) \underline{/26.5^\circ - 53.1^\circ}}{(6.1)(6.4) \underline{/9.5^\circ + 51.4^\circ}} = 0.285 \underline{/-87.5^\circ}$$

Let us now see how the conjugate of a complex number can be written in terms of the magnitude and angle. Figure 3-11 shows a number $A = a_1 + ja_2$ located in the complex plane. It has an angle θ. The conjugate of A has the same real part but a negative imaginary part. Hence it has a position in the plane as shown in the figure. It is clear that A^* has the same magnitude as A but an angle which is the negative of the angle of A. Thus, in the symbolic notation of Eq. (3-74), A^* can be written

$$A^* = |A| \underline{/-\theta} \tag{3-79}$$

Suppose we take the product of A by its conjugate. Symbolically, we shall get

$$AA^* = |A| \underline{/\theta} |A| \underline{/-\theta} = |A|^2 \tag{3-80}$$

That is, a complex number multiplied by its conjugate is a real number equal to the square of the magnitude of the complex number.

The preceding discussion indicates that a complex number has many properties in common with a two-dimensional vector. Thus a complex number can be regarded as a directed line from the origin to a point in the plane. The magnitude of the complex number is the length of the vector, and the angle specifies the direction of the vector. Furthermore, if the vector from the origin is translated parallel to itself anywhere in the plane, it still represents the complex number, since its length and angle with the horizontal axis do not change.

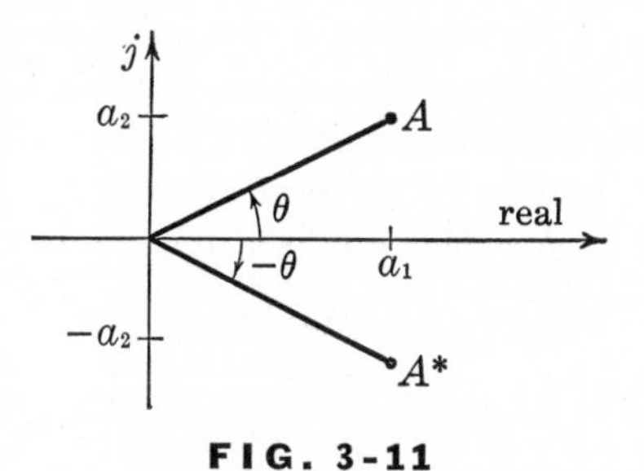

FIG. 3-11

A number and its conjugate in the complex plane

The operations of addition and subtraction can be performed in a graphical manner very easily in the complex plane. As an example, consider Fig. 3-12(a). The parallelogram law for adding vectors is

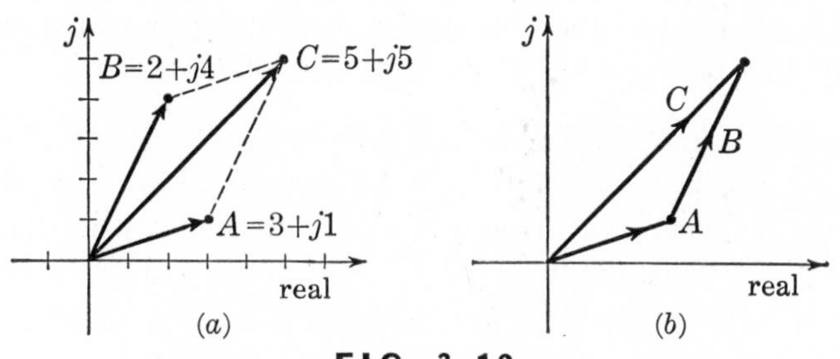

FIG. 3-12

Graphical addition of complex numbers

valid, as the figure illustrates. Perusal of Figure 3-12(b) shows that the polygon rule for addition of vectors is also valid for complex numbers; that is, the tail of each vector in a sum is joined to the head of the preceding one. The sum is then the vector joining the tail of the first vector to the head of the last one.

So far, we have been discussing the algebra of complex numbers. If a complex quantity takes on a sequence of values, we can think of it as a *variable.* If the values which one complex variable takes on are dependent on the values of a second complex variable, the first one is said to be a *function* of the second. Just as we have functions of real variables, we also have functions of complex variables.

The simplest functions to think of are powers. Thus let us designate a general complex variable as $z = x + jy$. Then by z^2 we mean the variable z multiplied once by itself. Higher powers of z mean that z is multiplied by itself more than once.

A *polynomial* is a function of the form

$$f(z) = z^n + a_1 z^{n-1} + \cdots + a_{n-1} z + a_n \tag{3-81}$$

A *rational function* is a ratio of two polynomials. There is no difficulty in defining powers of z, polynomials, and rational functions, but we do encounter difficulty in attempting to define transcendental functions of complex variables, such as trigonometric functions and exponentials.

For example, the tangent of a real number is defined in trigonometry as the ratio of two sides of a right triangle, one of the angles of which is the given number. Now what could we possibly mean by taking the tangent of a complex number? There is no such thing as a complex angle in the physical world.

For our present purposes we are interested only in the exponential function. Recall that the exponential function of a real variable is represented by means of an infinite series, as follows:

$$\epsilon^x = 1 + x + \frac{x^2}{2!} + \frac{x^3}{3!} + \cdots \tag{3-82}$$

In the case of a complex variable $z = x + jy$, we simply make the definition that the exponential function of z, written ϵ^z, is given by the same series, with z replacing x. Thus by definition,

$$\epsilon^z = 1 + z + \frac{z^2}{2!} + \frac{z^3}{3!} + \cdots \tag{3-83}$$

In the special, but important, case in which the real part of z is zero, so that $z = jy$ is purely imaginary, the expression reduces to

$$\epsilon^{jy} = 1 + jy + \frac{(jy)^2}{2!} + \frac{(jy)^3}{3!} + \cdots \tag{3-84}$$

Recalling from Eq. (3-68) that $j^2 = -1$, we see that all terms involving even powers will be real, and all terms involving odd powers will be imaginary. If we collect all the real terms and all the imaginary terms, the last equation becomes

$$\epsilon^{jy} = 1 - \frac{y^2}{2!} + \frac{y^4}{4!} - \cdots + j\left(y - \frac{y^3}{3!} + \frac{y^5}{5!} - \cdots\right) \tag{3-85}$$

Now trigonometric functions of real variables are also represented by means of infinite series. The series for $\sin y$ and $\cos y$ are as follows:

$$\cos y = 1 - \frac{y^2}{2!} + \frac{y^4}{4!} - \cdots \qquad (a)$$

$$\sin y = y - \frac{y^3}{3!} + \frac{y^5}{5!} - \cdots \qquad (b) \tag{3-86}$$

If we compare these expressions with the preceding one, we come to the important conclusion that

$$\epsilon^{jy} = \cos y + j \sin y \tag{3-87}$$

This result is known as *Euler's theorem;* it is of inestimable value in network theory.

With Euler's theorem it is possible to express sinusoids in terms of exponentials. Thus, in Eq. (3-87), let y be replaced by $-y$. Then

$$\epsilon^{-jy} = \cos(-y) + j \sin(-y) = \cos y - j \sin y \tag{3-88}$$

The last form follows from the fact that the cosine is an even function [$\cos(-y) = \cos y$] and the sine is an odd function [$\sin(-y) = -\sin y$]. Now, by adding and subtracting the last two equations, we get

$$\cos y = \tfrac{1}{2}(\epsilon^{jy} + \epsilon^{-jy}) \qquad (a)$$

$$\sin y = \frac{1}{2j}(\epsilon^{jy} - \epsilon^{-jy}) \qquad (b) \tag{3-89}$$

These expressions, together with Euler's theorem, arise very often in our subsequent work.

An alternative method of expressing the sine and cosine in terms of an exponential follows by noting from Euler's theorem that ϵ^{jy} is a complex quantity whose real part is a cosine and whose imaginary part is a sine. Hence we can write

$$\cos y = Re(\epsilon^{jy}) \qquad (a)$$
$$\sin y = Im(\epsilon^{jy}) \qquad (b) \qquad (3\text{-}90)$$

The notation $Re(A)$ is read "real part of A" and $Im(A)$ is read "imaginary part of A."

The law of exponents is valid for complex exponentials, just as it is for real exponentials. This can be demonstrated by means of the infinite-series definition, but we shall not take the trouble to do so here. That is, if z_1 and z_2 are two complex numbers, then

$$\epsilon^{z_1+z_2} = \epsilon^{z_1}\epsilon^{z_2} \qquad (3\text{-}91)$$

In particular,

$$\epsilon^{z} = \epsilon^{x+jy} = \epsilon^{x}\epsilon^{jy} = \epsilon^{x}(\cos y + j \sin y) \qquad (3\text{-}92)$$

The last step follows from Euler's theorem.

Let us now compare Euler's theorem with the trigonometric form for expressing complex numbers given in Eq. (3-73). From the comparison it is clear that a complex number can be written in still another form as

$$A = |A|\epsilon^{j\theta} \qquad (3\text{-}93)$$

This form also serves to put into evidence the magnitude and angle, just as the symbolic form given in Eq. (3-74) does.

3.7 SECOND-ORDER NETWORKS, CONTINUED

After this digression into a discussion of complex numbers, let us now return to the study of the natural behavior of the second-order network which we started in Sec. 3-5. There we found the natural response of the network shown in Fig. 3-9 to be

$$v = K_1\epsilon^{s_1t} + K_2\epsilon^{s_2t} \qquad (3\text{-}94)$$

where s_1 and s_2 are the roots of the characteristic equation

$$s^2 + \frac{R}{L}s + \frac{1}{LC} = 0 \qquad (a)$$
$$s_1 = -\frac{R}{2L} + \sqrt{\left(\frac{R}{2L}\right)^2 - \frac{1}{LC}} \qquad (b) \qquad (3\text{-}95)$$
$$s_2 = -\frac{R}{2L} - \sqrt{\left(\frac{R}{2L}\right)^2 - \frac{1}{LC}} \qquad (c)$$

We are now in a position to interpret the nature of these roots for all values of the discriminant. Before we consider the detailed behavior of the response, however, let us turn our attention to the constants K_1 and K_2 appearing in Eq. (3-94). As far as the solution of the differential equation is concerned, the values of these constants are arbitrary; that is, no matter what their values may be, the equation will be satisfied. In a given problem, however, not only must the differential equation be satisfied, but also the conditions imposed by the state of the network at the moment of switching must be satisfied. In the present case, whatever the eventual value of the capacitor voltage may be, the value at $t = 0$ must be that value which is dictated by the initial voltage.

But this is only one condition; yet there are two constants whose values can be adjusted. Is there any other condition imposed by the initial state of the network which will help determine the constants? A glance at the network will assure us that there is: The current in the inductor must be continuous, as required by the continuity of flux linkages. This will lead to one other condition, which makes just the right number to determine the two constants. An expression for the current is obtained from the v-i relationship of the capacitor. Thus

$$i = C \frac{dv}{dt} = C(K_1 s_1 \epsilon^{s_1 t} + K_2 s_2 \epsilon^{s_2 t}) \tag{3-96}$$

We are now ready to find the two constants K_1 and K_2. In Eq. (3-94) we substitute the values $t = 0$, $v = V_0$; in Eq. (3-96) we substitute the values $t = 0$, $i = 0$. The results will be

$$\begin{aligned} K_1 + K_2 &= V_0 && (a) \\ K_1 s_1 + K_2 s_2 &= 0 && (b) \end{aligned} \tag{3-97}$$

Here we have two equations in the two unknowns K_1 and K_2. The solution of these two equations leads to

$$\begin{aligned} K_1 &= \frac{s_2}{s_2 - s_1} V_0 && (a) \\ K_2 &= \frac{-s_1}{s_2 - s_1} V_0 && (b) \end{aligned} \tag{3-98}$$

(Go through the details of the solution.) In this way the constants of integration are determined. Note that in the present problem the natural response is also the total response, since there is no forced component. This justifies using what appears to be only the natural response when applying the initial conditions.

Let us now turn back to the characteristic values in Eq. (3-95). If the discriminant is positive, no difficulty is encountered. Each of the numbers s_1 and s_2 is real and negative. The constants K_1 and K_2 will also be real numbers, as seen from Eq. (3-98). The response then consists of a pair of damped exponentials. This is then a direct extension of the situation in a first-order network.

If the discriminant is negative, however, s_1 and s_2 contain the square root of a negative number. We are now in a position to deal with such a situation. For the sake of simplicity, let us define the following notation:

$$\frac{1}{LC} = \omega_0^2 \qquad (a)$$

$$\frac{R}{2L} = \alpha \qquad (b) \qquad (3\text{-}99)$$

$$\omega_d^2 = \omega_0^2 - \alpha^2 \qquad (c)$$

Then we can write

$$\sqrt{\left(\frac{R}{2L}\right)^2 - \frac{1}{LC}} = \sqrt{-\left[\frac{1}{LC} - \left(\frac{R}{2L}\right)^2\right]} = j\sqrt{\omega_0^2 - \alpha^2} = j\omega_d \qquad (3\text{-}100)$$

Hence Eq. (3-95) becomes

$$s_{1,2} = -\alpha \pm j\sqrt{\omega_0^2 - \alpha^2} = -\alpha \pm j\omega_d \qquad (3\text{-}101)$$

These values for s_1 and s_2 are to be substituted into the exponents in Eq. (3-94). In the present case the characteristic values are complex; in fact, they form a complex conjugate pair with a real part which is negative. The corresponding values of K_1 and K_2 are found from Eq. (3-98) to be

$$K_1 = \frac{V_0}{2}\left(1 - j\frac{\alpha}{\sqrt{\omega_0^2 - \alpha^2}}\right) = \frac{V_0}{2}\left(1 - j\frac{\alpha}{\omega_d}\right) \qquad (a)$$

$$K_2 = \frac{V_0}{2}\left(1 + j\frac{\alpha}{\sqrt{\omega_0^2 - \alpha^2}}\right) = \frac{V_0}{2}\left(1 + j\frac{\alpha}{\omega_d}\right) = K_1^* \qquad (b) \qquad (3\text{-}102)$$

We see that K_1 and K_2 are also complex; they form a pair of complex conjugates.

In the special case in which $\alpha = 0$, ω_0 and ω_d are identical. This case corresponds to a network in which $R = 0$, as shown in Fig. 3-13. For this case the characteristic values become

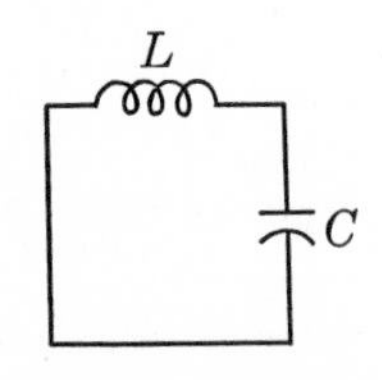

FIG. 3-13
Lossless network

$$s_{1,2} = \pm j\omega_0 \tag{3-103}$$

Thus the characteristic values in this case are purely imaginary. When these values are substituted into the expression for the voltage given in Eq. (3-94) we get

$$\begin{aligned} v &= K_1\epsilon^{j\omega_0 t} + K_2\epsilon^{-j\omega_0 t} \\ &= K_1 (\cos \omega_0 t + j \sin \omega_0 t) + K_2 (\cos \omega_0 t - j \sin \omega_0 t) \\ &= A \cos \omega_0 t + B \sin \omega_0 t \end{aligned} \tag{3-104}$$

The last step is obtained by collecting the terms involving the cosine and the sine. A and B are two constants which are combinations of K_1 and K_2; they are

$$\begin{aligned} A &= K_1 + K_2 = V_0 && (a) \\ B &= j(K_1 - K_2) = \frac{\alpha}{\omega_d} V_0 && (b) \end{aligned} \tag{3-105}$$

The right-hand sides are obtained by using Eq. (3-102). In the present case ($\alpha = 0$) Eq. (3-102) shows that K_1 and K_2 are real and equal, so that B in Eq. (3-104) vanishes.

We see that, when there is no resistance in the network of Fig. 3-9, the natural behavior of the response is a free, undamped oscillation $v = V_0 \cos \omega_0 t$ with an angular frequency ω_0. What could be more natural than to refer to ω_0 as the *natural frequency?*

Let us now turn to the general case with $\alpha \neq 0$. The characteristic values are now complex and are given by Eq. (3-101). Substituting these values into Eq. (3-94), we get

$$\begin{aligned} v &= K_1\epsilon^{(-\alpha+j\omega_d)t} + K_2\epsilon^{(-\alpha-j\omega_d)t} = \epsilon^{-\alpha t}(K_1\epsilon^{j\omega_d t} + K_2\epsilon^{-j\omega_d t}) \\ &= \epsilon^{-\alpha t}(A \cos \omega_d t + B \sin \omega_d t) = V_m\epsilon^{-\alpha t} \cos (\omega_d t - \theta) \end{aligned} \tag{3-106}$$

where

$$\begin{aligned} V_m &= \sqrt{A^2 + B^2} = \sqrt{V_0^2 + \left(\frac{\alpha V_0}{\omega_d}\right)^2} = \frac{\omega_0}{\omega_d} V_0 && (a) \\ \tan \theta &= \frac{B}{A} = \frac{\alpha}{\omega_d} && (b) \end{aligned} \tag{3-107}$$

The right-hand sides are obtained by inserting Eq. (3-105) for A and B and using $\alpha^2 + \omega_d^2 = \omega_0^2$. To get the last line of Eq. (3-106) we used Euler's theorem and again collected the cosine terms and the sine terms. The constants A and B are still given by Eq. (3-105). The present result is to be compared with Eq. (3-104). We still have a sinusoidal oscillation, but this time it is multiplied by a damped exponential, which causes the response to die down eventually. Another difference is that the frequency is now smaller than in the previous case (for the same values of L and C). Since the oscillation is damped, we refer to ω_d as the *damped natural frequency.* (In contrast to this, ω_0 is sometimes referred to as the *undamped natural frequency.*)

The current flowing in the network can be obtained from the v-i relationship of the capacitor. Using the left side of the last line of Eq. (3-106), and inserting the values of A and B from Eq. (3-105), we get

$$i = C\frac{dv}{dt} = -\omega_0 C V_m \epsilon^{-\alpha t} \sin \omega_d t \tag{3-108}$$

3.8 THE NATURAL FREQUENCIES OF ELECTRICAL NETWORKS

Let us now glance back over the preceding discussion and pick out the salient features when separated from the algebraic details. The natural response of the series RLC network consists of a sum of two exponential functions of time. Depending on the relative values of the network parameters, the exponents may be either real or complex. When the exponents are real, the response is the sum of two decaying exponentials (nonoscillatory case). When the exponents are complex, they are a pair of conjugates, and the exponentials can be written in terms of a sinusoidal function multiplied by a real decaying exponential. The response is then basically oscillatory but decays exponentially. In the special case in which $R = 0$, there is no damping and the response is purely oscillatory. (In this case would it not be senseless to refer to the natural response as "transient"?)

Sketches of the response voltage and current waveforms are shown in Fig. 3-14. The damped oscillations and the nonoscillatory case are both shown. In the oscillatory case the exponential $\epsilon^{-\alpha t}$ forms an "envelope" beyond which the oscillations do not extend.

Suppose that the resistance in a series RLC network is varied from zero up to a high value. The response at first will be oscillatory, the angular frequency being ω_0. As R is increased, the frequency of oscillation ω_d will change, as observed from Eq. (3-100). As the resistance goes through the critical value for which $\alpha = \omega_0$, the response will change to a nonoscillatory character. But, instead of changing our terminology, we can still speak of an "oscillation" and a "frequency," even though these terms with their usual meaning are no longer descriptive of the response behavior.

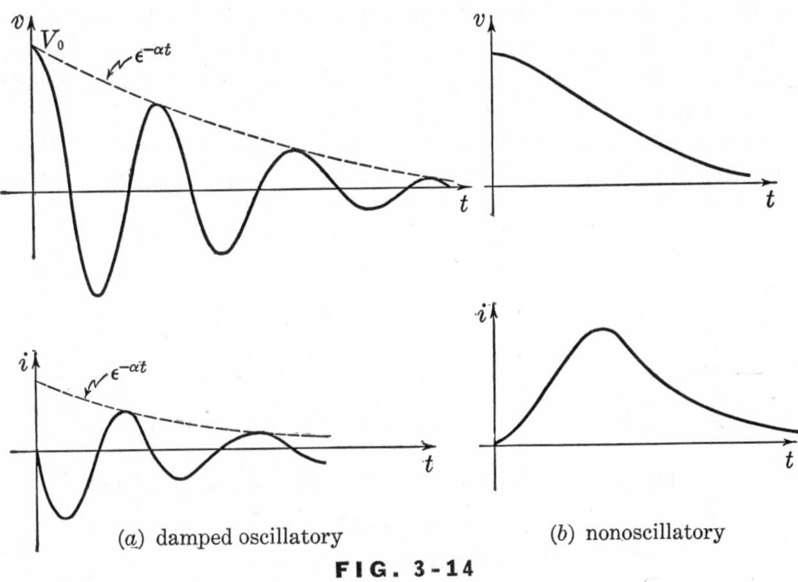

FIG. 3-14

Natural response

Nevertheless, if we do think of any value whatsoever of the quantity s as being a "frequency," a door will be opened before us presenting a whole new vista of exciting possibilities. This may not be apparent to you immediately, but, as we proceed, the implications of this process of thought and the power of the techniques which can be brought to bear on network problems with this point of view will be amply demonstrated.

Let us therefore call the characteristic values which are obtained as the roots of the characteristic equation of a network the *natural frequencies*. Since these values in general are complex, let us further call them the *complex natural frequencies*. It may seem strange, if not downright fantastic, to think of such a thing as a "complex frequency." What can this possibly mean? What it means is that we are giving the name "complex natural frequency" to the roots of a certain polynomial which is characteristic of a given network; and that is all right now.

The variable in terms of which this polynomial is written is s. Hence, we refer to s as the *complex frequency variable*. In the case of the series RLC network, we found two specific values of this variable, which we labeled s_1 and s_2, and these we called the natural frequencies of the network. Earlier, we found that networks with one energy-storing element are characterized by a linear equation which, of course, has but a single root. The natural frequency in this case is a real number. But real numbers are special cases of complex numbers. Hence, here also, we have a complex frequency; it so happens that the imaginary part is zero.

These ideas can be further clarified by means of diagrams. Consider the complex plane shown in Fig. 3-15. Any point in the plane

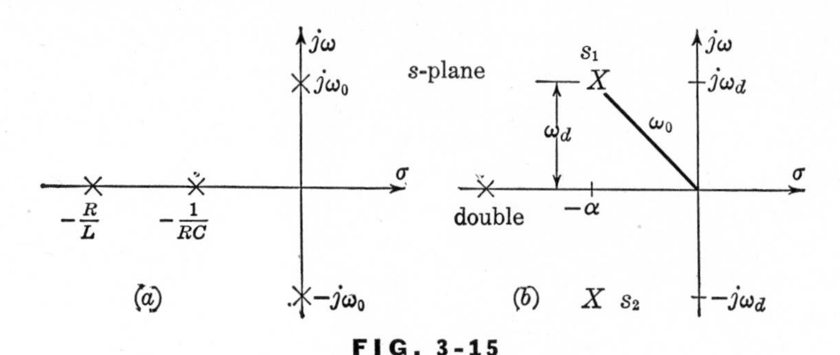

FIG. 3-15
The complex-frequency plane

represents a value of s. Hence we call it the s plane or the *complex-frequency* plane. Since s is complex, we can write it in the form

$$s = \sigma + j\omega \qquad (a)$$
$$\sigma = Re(s) \qquad (b) \quad (3\text{-}109)$$
$$\omega = Im(s) \qquad (c)$$

The real part of s is given the symbol σ, and the imaginary part of s is given the symbol ω. The axes in the figure are labeled in terms of these quantities. All real numbers are represented by points on the σ axis; all imaginary numbers are represented by points on the imaginary axis. For example, in the case of the first-order RC and RL networks, we found the natural frequencies to be $s = -1/RC$ and $s = -R/L$. These are illustrated by the X's on the negative real axis in the figure. Similarly, in the undamped oscillatory case the two natural frequencies are at $s = \pm j\omega_0$. These points are shown by X's on the imaginary axis.

In the damped oscillatory case the natural frequencies have both real and imaginary parts. The two complex conjugate points s_1 and s_2 are marked by X's in Fig. 3-15(b). Note that the real part $-\alpha$ shows how fast the response dies out, whereas the imaginary part ω_d gives the angular frequency of oscillation.

Representation of the natural frequencies in the complex plane is a clear, compact way of describing the natural behavior of a network.

We shall make extensive use of this method of representation, both for computational purposes and as an aid in developing our ideas.

In discussing the natural frequencies of the series *RLC* network, there is one special situation which we did not consider—the case in which two characteristic values are equal. This is the boundary beween the oscillatory and the nonoscillatory case. In this case the analytical form of the solution will no longer be given by Eq. (3-94). Nevertheless, the qualitative behavior of the solution will be much the same as that of the nonoscillatory case. We shall not look into the details of this special case here, since it would not add significantly to our understanding of the natural behavior of networks. The double natural frequencies are shown in Fig. 3-15(*b*) by the double ×.

Although we have carried on the discussion in this section in terms of the series *RLC* network, the basic nature of the results will remain unaltered if we deal instead with any network that leads to a second-order differential equation. In particular, the dual of the series *RLC* network, which is a parallel *RLC* network, will lead to such an equation. It should be clear from the v-i relationships that each inductor and each capacitor in a network will contribute a derivative to the differential equation satisfied by the response.† Thus the same type of natural response will result from many specific networks whose response function satisfies a second-order differential equation.

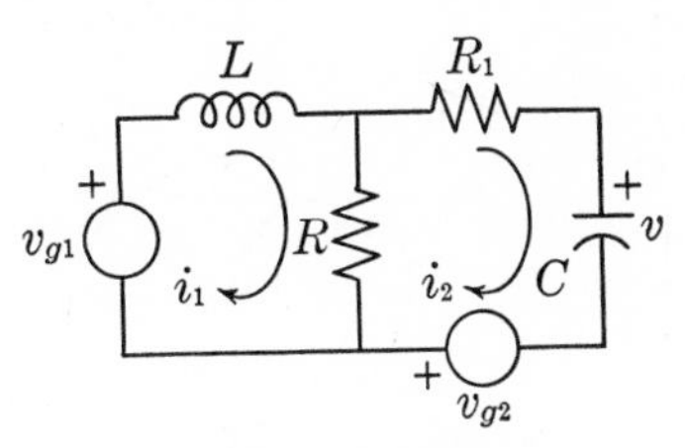

FIG. 3-16
Illustrative example

Let us illustrate the discussion of the last paragraph by means of the network shown in Fig. 3-16. Choosing loop currents as shown, the loop equations can be written

$$L\frac{di_1}{dt} + Ri_1 - Ri_2 = v_{g1} \qquad (a)$$

$$(R_1 + R)i_2 - Ri_1 + v = v_{g2} \qquad (b) \qquad (3\text{-}110)$$

Here there are three unknowns: i_1, i_2, and v. Let us replace i_2 by $C\,dv/dt$ and solve the second equation for i_1. The result will be

† This remark should be qualified in certain rather trivial situations, such as, for example, two inductors in series. Other similar situations are structurally more complicated than this but may be recognized in terms of certain topological concepts. In such cases the order of the differential equation will be less than the number indicated by counting all the inductors and capacitors.

$$L\frac{di_1}{dt} + Ri_1 - RC\frac{dv}{dt} = v_{g1} \qquad (a)$$

$$i_1 = \frac{1}{R}\left[(R_1 + R)C\frac{dv}{dt} + v - v_{g2}\right] \qquad (b) \qquad \text{(3-111)}$$

If we now substitute i_1 from the second equation into the first, the final result will be

$$\frac{LC}{R}(R_1 + R)\frac{d^2v}{dt^2} + \left(\frac{L}{R} + R_1C\right)\frac{dv}{dt} + v = v_{g1} + v_{g2} + \frac{L}{R}\frac{dv_{g2}}{dt} \qquad \text{(3-112)}$$

This is a second-order differential equation, as anticipated.

Let us now make some brief comments on what to expect about the natural behavior of networks containing an arbitrary number of energy-storing elements. We shall not go through a detailed development, since we merely wish to get a qualitative understanding at the present time.

When we write the network equations, whether these are loop equations, node equations, or a mixed set, each of these will be a differential equation. If we now eliminate all the variables except one in these equations, we shall have a single differential equation; we would anticipate its order to be the sum of all capacitors and inductors in the network, since each capacitor and each inductor contributes a derivative (limited by the comments in the footnote on page 145). In fact, this equation will have the form

$$\frac{d^nr}{dt^n} + a_{n-1}\frac{d^{n-1}r}{dt^{n-1}} + \cdots + a_1\frac{dr}{dt} + a_0 = f(e_1, e_2 \cdots) \qquad \text{(3-113)}$$

We have used the symbol r to stand for "response," since this may be either a voltage or a current. Similarly, we have used the symbol e to stand for "excitation," since both voltage sources and current sources might be involved. Generally, there will be more than one excitation, and these have been labeled e_1, e_2, etc. The right-hand side includes all the excitations and possibly derivatives of some of them. Equation (3-112) is of this form, for example.

If we go through the process of assuming an exponential ϵ^{st} for the natural response and substituting into the homogeneous equation, a polynomial of degree n will result as the characteristic equation, where n is the order of the differential equation. This equation will have n roots, all of which will be natural frequencies of the network. The natural response will then be of the form

$$r_n(t) = K_1\epsilon^{s_1t} + K_2\epsilon^{s_2t} + K_3\epsilon^{s_3t} + \cdots K_n\epsilon^{s_nt} \qquad \text{(3-114)}$$

The s exponents are the natural frequencies; they may be either real or complex. Suppose that one of the natural frequencies is positive. The corresponding exponential will increase indefinitely with time, and this will be true no matter what the excitations may be. The power dissipated in a resistor, being proportional to the square of the response, will also increase indefinitely. Such a state of affairs is impossible, since it demands a continually increasing amount of energy. The same conclusion will be true if one of the complex natural frequencies has a real part which is positive. Hence we conclude that, for a passive network, the natural frequencies are either real and negative, or complex with real parts that are negative. As a limiting case, the real part might be zero, corresponding to a sustained, undamped, oscillation which occurs when resistance is absent. Of course, in an actual physical circuit (as opposed to a network model), resistance will always be present, however slight, and this case cannot occur.

We shall have much more to say about natural frequencies and the natural behavior of networks in later chapters.

3.9 SUMMARY

In this chapter we took the second step in solving for any voltage and current in an arbitrary electrical network. The first step involved finding methods of solution in resistive networks; this was undertaken in the previous chapter. In the second step, energy-storing elements were introduced into the network; at first only one, then two and more.

We found, in each case, that application of the fundamental laws leads to one or more differential equations to be solved. These equations all fall in a class which can be characterized by the words *ordinary* (as opposed to *partial* in which partial derivatives appear); *linear* (because all variables and their derivatives appear in the first power only; there are no squares or cubes of things); and *constant coefficient* (because all the coefficients are constant).

We found, in each case, that the response of the network consists of two parts. One of these parts, which we call the natural response, has the form of a sum of exponential terms. The exponents are the complex natural frequencies of the network. They are either real and negative, corresponding to a simple decaying exponential, or complex with a negative real part, corresponding to a damped oscillation. As a limiting case, the natural frequencies may be imaginary, corresponding to an undamped oscillation. The natural response is also called the transient because it eventually dies down (except for the undamped oscillatory case).

The second part of the response depends on the forcing function, so it is called the forced response. In case the excitation is a constant, or a periodic function, such as a sinusoid, the forced response is called the steady state. In determining the forced response, we proceeded with the assumption that its waveform is the same as the waveform of the excitation. This assumption was verified in the cases which we considered. Our method of solution, the method of undetermined coefficients, involves assuming the waveform of the response and substituting into the differential equation in order to determine the coefficients. Clearly, this requires that the excitation waveform be differentiable as many times as the order of the equation. If this is not the case, the method cannot be used. We shall later introduce a method of solution, the Laplace transform, which will easily handle such excitations.

Problems

3-1 Let $A = 1 + j\sqrt{3}$ be a complex number. Compute the ratio of A to its conjugate. Comment on the general statement regarding the quotient of any complex number and its conjugate which is illustrated by this example.

3-2 With the complex numbers listed in this problem, compute the following quantities. Write the result in both rectangular and polar form.

$A = 3 + j4$	$A + BC$	$\dfrac{CA}{B}$	$\dfrac{1}{A} + \dfrac{B}{C}$
$B = 2 + j1$	$AB - C$	$B - \dfrac{C}{A}$	$A^* - BC^*$
$C = 4 + j5$	$A - \dfrac{B}{C}$	ABC	$\dfrac{A}{A^*} + \dfrac{BC}{B^*}$

3-3 Using the exponential forms of trigonometric functions, write the following in terms of real variables, either as exponential, trigonometric, or hyperbolic functions, or combinations of these: (*a*) $\cos jt$; (*b*) $\cos (t + j2t)$; (*c*) $\sin (3t + jt)$.

3-4 If A and B are two complex numbers, prove that

$$(AB)^* = A^*B^*$$

$$\left(\frac{A}{B}\right)^* = \frac{A^*}{B^*}$$

$$(A \pm B)^* = A^* \pm B^*$$

That is, the conjugate of the product, or quotient, or sum, or difference of two complex numbers is equal to the product, or quotient, or sum, or difference, respectively, of the conjugates.

3-5 Prove that addition and subtraction of complex numbers are commutative with the operation of taking the real part or the imaginary part; that is, if A and B are two complex numbers, then,

$$Re(A \pm B) = Re\,A \pm Re\,B$$

$$Im(A \pm B) = Im\,A \pm Im\,B$$

3-6 Prove that differentiation and integration of a function of a complex variable with respect to a real parameter are commutative with the operation of taking the real part or the imaginary part. That is, let $s = \sigma + j\omega$ be a complex variable, and let $f(s, t)$ be a function of s and a real variable t. This function is also complex and can be written $f = Re(f) + jIm(f)$. Then

$$\frac{d}{dt}(Re\,f) = Re\left(\frac{df}{dt}\right) \qquad \frac{d}{dt}(Im\,f) = Im\left(\frac{df}{dt}\right)$$

$$\int Re(f)dt = Re\left(\int f\,dt\right) \qquad \int Im(f)dt = Im\left(\int f\,dt\right)$$

Illustrate these properties for $f(s, t) = \epsilon^{st}$.

3-7 In the network of Fig. P3-7, let the initial voltage of the capacitor be $v(0) = V_0$ when the switch is closed at $t = 0$. Find the voltage v as a function of time by writing a loop equation instead of a node equation, and solving first for the current. Do this for the following source functions: (a) $v_g(t) = 0$; (b) $v_g(t) = V_1$ (a constant); (c) $v_g(t) = V_m \sin \omega t$; (d) $v_g(t) = V_m\epsilon^{at}$; (e) $v_g(t) = t$.

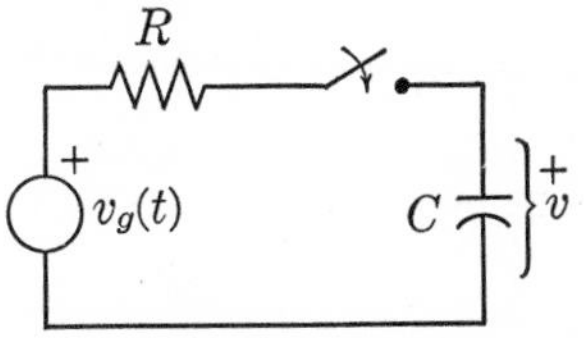

R
+
$v_g(t)$
C
+
v

FIG. P3-7

3-8 In the network of Fig. P3-8 the capacitor is charged to a voltage $v_C(0) = V_0$. At time $t = 0$, switch S_1 is closed. Find the voltage across the capacitor for $t > 0$. (Switch S_2 remains open.)

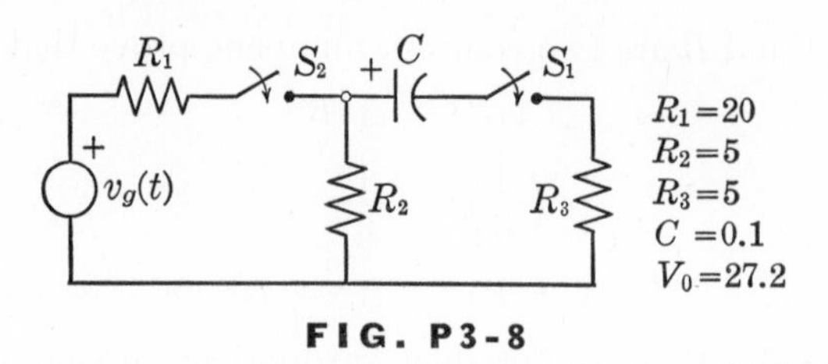

FIG. P3-8

3-9 In Fig. P3-8 switch S_2 is closed 1 sec after switch S_1. Find the voltage across the capacitor after this time. Do this first by writing node equations, choosing the datum node appropriately; then repeat, converting to a single loop by an equivalent source change. Use the following source functions: (*a*) $v_g(t) = 5$; (*b*) $v_g(t) = 10\epsilon^{2t}$; (*c*) $v_g(t) = 5t$; (*d*) $v_g(t) = 20 \cos 100t$. Identify the natural component and the forced component of the response.

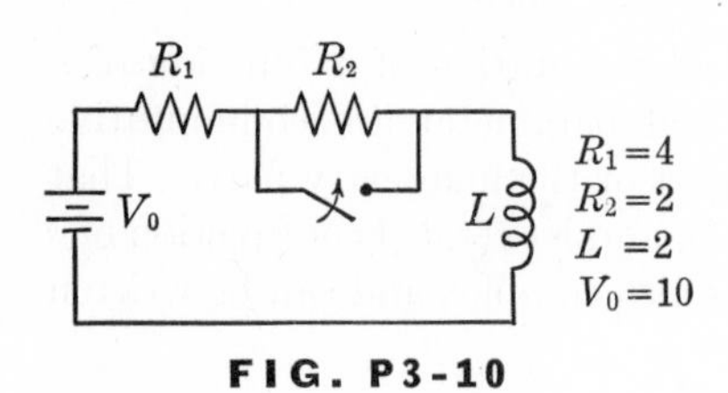

FIG. P3-10

3-10 In Fig. P3-10 the network has been in operation for a long time when the switch is closed at $t = 0$. Find the current after this time and sketch it for an interval of $-1 \leq t \leq 2$.

3-11 The network shown in Fig. P3-11 has been in operation for a long time when at $t = 0$ the switch is closed. (*a*) Find the value of the current in the capacitor immediately after the switch is closed. (*b*) Find the value of the capacitor voltage v following the closing of the switch. (*c*) Locate the "complex natural frequency" of the network in the complex plane.

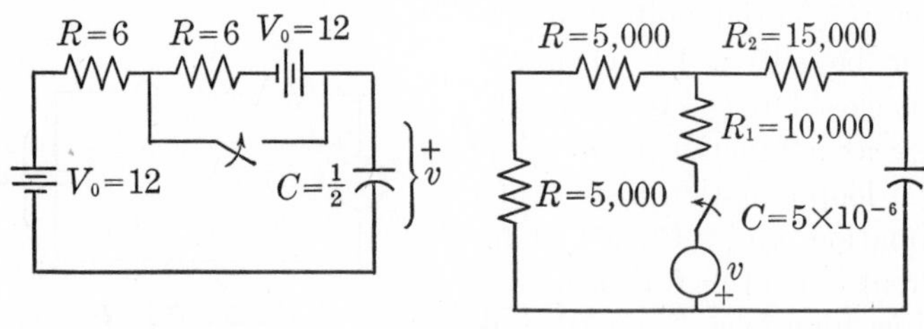

FIG. P3-11

3-12 In each of the networks of Fig. P3-12, the switch has been in position 1 for a long time. At $t = 0$ it is thrown to position 2, where it remains until $t = 1$, when it is thrown to position 3. Find the indicated response for $t > 0$ and sketch it.

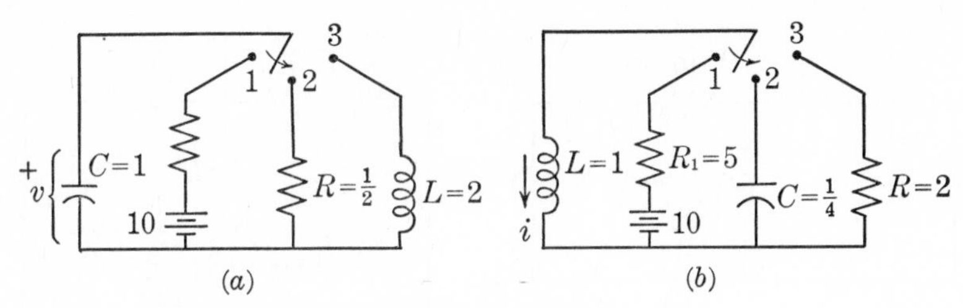

FIG. P3-12

3-13 In the network shown in Fig. P3-13, the capacitor is initially uncharged when the switch is closed at $t = 0$. Find the total current in the capacitor as a function of time following the closing of the switch. Avoid the appearance of integrals in the formulation of the problem. $v = 40 \sin 10t$.

3-14 In the network of Fig. P3-14, the switch is thrown from position 1 to position 2 at $t = 0$. Find the complete solution for the capacitor voltage v as a function of time. Locate the natural frequency on the complex-frequency plane. What is the initial value ($t = 0+$) of the current through R_3?

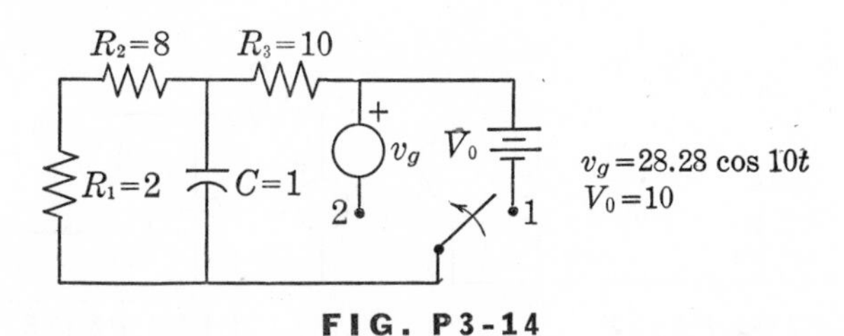

FIG. P3-14

3-15 In Figure P3-15 the network has been in operation for a long time when at $t = 0$ the switch is closed. (*a*) Find i, the current in the inductor, before $t = 0$ and immediately after $t = 0$. (*b*) Find the differential equation which must be satisfied by i after the switch is closed. (*c*) Find the complete solution of this equation. Indicate the natural component and the forced component of the solution. What is the value of the natural frequency?

3-16 In the network of Fig. P3-16, the switch has been open for a long time when at $t = 0$ it is closed. (*a*) Find the initial ($0+$) currents in L and C, and the initial voltage across C. (*b*) Write the loop equations after the switch is closed, choosing the loops shown in the figure. In addition to a derivative there will be an integral. (*c*) Using the v-i relationship of the capacitor, change one of the variables in the

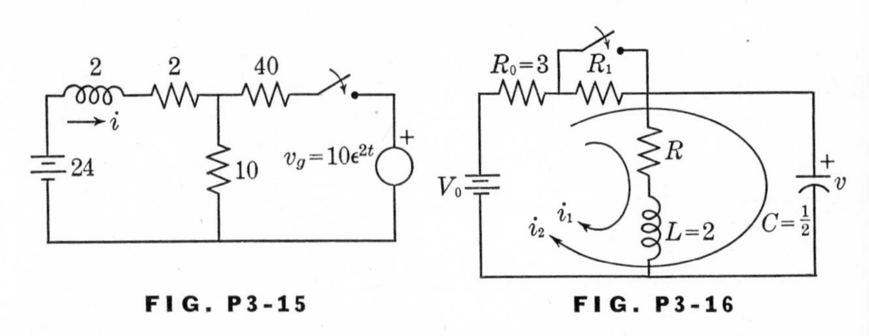

FIG. P3-15

FIG. P3-16

preceding equations to the capacitor voltage, thereby eliminating the integral. The two variables are now i_1 and v. (*d*) By elimination, reduce this pair of equations to a single second-order differential equation. (*e*) Find the natural frequencies for the following values of R: 0, 2, 3, 10.

3-17 In Fig. P3-17 switch S_2 has been closed and switch S_1 has been in position 1 for a long time. At $t = 0$, S_1 is changed from position 1 to position 2. At $t = 1$, S_2 is opened. Find the current in the inductor for $t > 0$. Sketch the pertinent sets of natural frequencies in the complex plane.

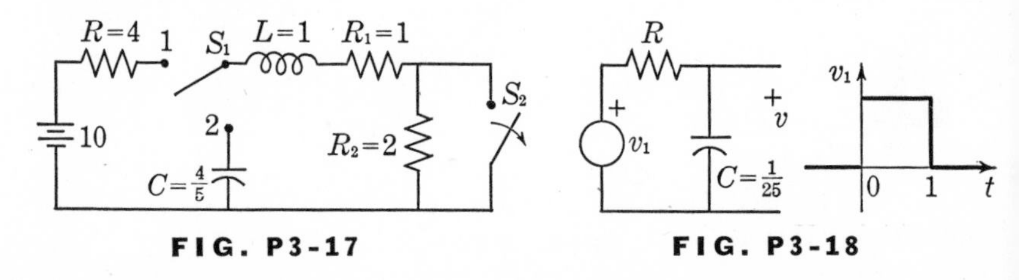

FIG. P3-17

FIG. P3-18

3-18 The waveform of the source voltage in Fig. P3-18 is a single pulse. If $R = 5$, find the voltage v across the capacitor for $t > 0$ assuming that it is initially zero. Repeat if $R = 25$. Sketch the output waveforms together with the input pulse. Comment on the waveforms in terms of the relative values of the pulse width and the RC time constant.

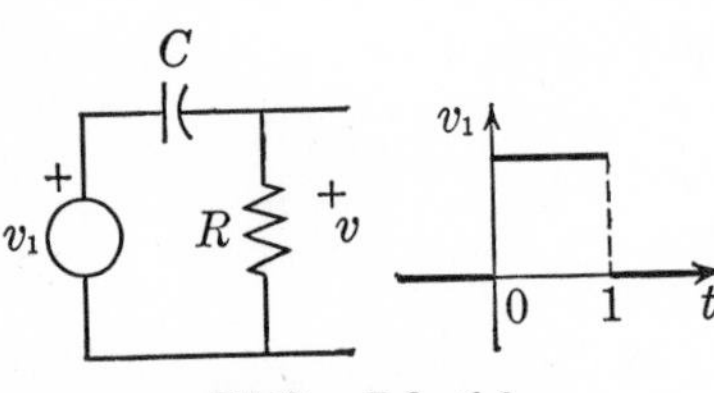

FIG. P3-19

3-19 In Fig. P3-19 find the output voltage in response to a single pulse. Assume that the capacitor is initially uncharged. Do this for $RC = 0.1$ and $RC = 10$.

capacator stores voltage ∴ $V_i = V_o$
inductor stores current $i_i = I_o$

4 SINUSOIDAL STEADY-STATE RESPONSE

WE are now ready to embark on the portion of network theory which carries the designation *a-c circuit analysis.* In the early applications of electricity, the rotating electrical machine was the major source of all electricity. (It still is the major electrical power source.) The waveshape generated by such a machine is periodic, and approximately sinusoidal. Hence, in the early studies of electrical circuits, it was enough to determine the behavior of circuits in response to sources which were sinusoidal in waveform. The sine wave reigned supreme for many years.

In present applications of electricity, many different waveshapes are to be found. Nevertheless, the sinusoid still plays a very important role, not only because it is important in its own right but because a very powerful tool of analysis, called the Fourier theorem, permits us to approximate almost any waveform as a summation of sinusoidal components. (You are not expected to appreciate the full significance of this statement at the present time.) Hence the ability to find the voltages and currents in a network when the sources have sinusoidal waveshapes is important to us.

In Chap. 3 we found that the response of a network to an excitation could be written in two parts—the natural response and the forced response. In case the excitation is a constant or is periodic, the forced response is called the steady state. We carried out a procedure for finding the steady-state response to a sinsusoidal excitation in terms of a simple RC network. By glancing back over the details of that procedure, you will note that it is extremely tedious and next to useless for networks containing more than two energy-storing elements. In this chapter we shall be concerned with developing a method of solution which will reduce the complexity of solution of the most general linear network to that of a resistive network.

4.1 USE OF THE EXPONENTIAL FUNCTION

Let us start this discussion by considering the network shown in Fig. 4-1. A voltage source having a sinusoidal waveform is connected to a series combination of a resistor and an inductor. There is but a single loop and the loop equation is

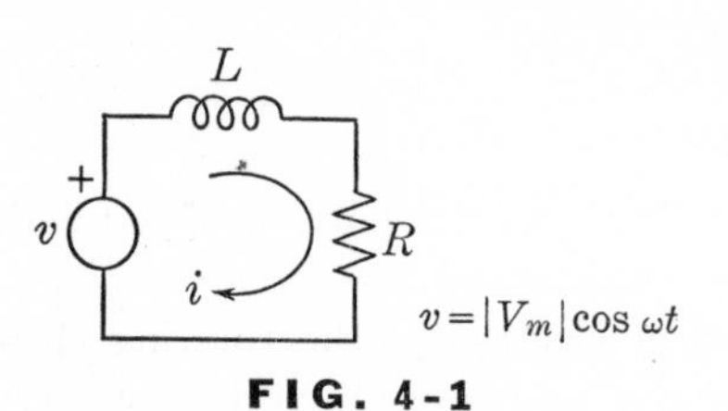

FIG. 4-1

Series R-L network fed by voltage source

$$L\frac{di}{dt} + Ri = |V_m| \cos \omega t \quad (4\text{-}1)$$

The origin of t is chosen so that the driving function is a cosine. Here we have a first-order differential equation. Our interest lies not in the complete solution of this equation but in the steady state. We know that the steady-state response will have the same waveform as the driving function, namely, a sinusoid of the same frequency. But its amplitude and angle are unknown. That is, we expect the steady-state current to have the form

$$i = |I_m| \cos (\omega t + \alpha) \quad (4\text{-}2)$$

in which $|I_m|$ and α are unknown.

Now, instead of substituting this expression into Eq. (4-1), let us first express the trigonometric function in exponential form. Referring back to Sec. 3-6, we find that $\cos \omega t$ can be written in either one of the following forms

$$\cos \omega t = \frac{\epsilon^{j\omega t} + \epsilon^{-j\omega t}}{2} \quad (a)$$

$$\cos \omega t = Re(\epsilon^{j\omega t}) \quad (b) \qquad (4\text{-}3)$$

Let us use the first of these and place both the driving function and the assumed steady-state response in exponential form. Thus

$$v = |V_m| \cos \omega t = \frac{|V_m|}{2} \epsilon^{j\omega t} + \frac{|V_m|}{2} \epsilon^{-j\omega t} \qquad (a)$$

$$i = |I_m| \cos (\omega t + \alpha) = \frac{|I_m|}{2} \epsilon^{j(\omega t+\alpha)} + \frac{|I_m|}{2} \epsilon^{-j(\omega t+\alpha)} \qquad (4\text{-}4)$$

$$= \frac{|I_m|\epsilon^{j\alpha}}{2} \epsilon^{j\omega t} + \frac{|I_m|\epsilon^{-j\alpha}}{2} \epsilon^{-j\omega t} \qquad (b)$$

Note that the coefficient of each of the exponential functions in the last line is a complex number; in fact the two are conjugates. We can therefore write the current as

$$i = \frac{I_m}{2} \epsilon^{j\omega t} + \frac{I_m^*}{2} \epsilon^{-j\omega t} \qquad (4\text{-}5)$$

where I_m is the complex quantity

$$I_m = |I_m| \epsilon^{j\alpha} \qquad (4\text{-}6)$$

Thus we see that a sinusoidal function of time with any amplitude and any angle can be written as the sum of two exponential functions of time with imaginary exponents and with complex coefficients. These complex coefficients give the pertinent information about the sinusoid, its amplitude and its angle.

If we compare Eq. (4-5) with Eq. [4-4(*a*)], we see that they have the same form, except that, in the latter case, the coefficients of the two exponentials are both real (and equal). Recall, however, that a real number is just a complex number with a zero imaginary part (or a zero angle). That is, $|V_m|$ can be looked upon as the magnitude of a complex number V_m whose angle is zero. Furthermore, the conjugate of a real number is simply itself. This is true because the conjugate has the same magnitude and the negative angle—and the negative of zero is still zero. Thus Eq. [4-4(*a*)] can be rewritten as

$$v = \frac{V_m}{2} \epsilon^{j\omega t} + \frac{V_m^*}{2} \epsilon^{-j\omega t} \qquad (a)$$

where V_m is the complex quantity (4-7)

$$V_m = |V_m| \epsilon^{j0} = |V_m| \qquad (b)$$

Let us now substitute into the differential equation both the excitation function and the assumed steady-state response in exponential form. At the same time, let us collect together all terms involving $\epsilon^{j\omega t}$

and $\epsilon^{-j\omega t}$. Remembering that the time derivative of $\epsilon^{j\omega t}$ is $j\omega\epsilon^{j\omega t}$, the result will be

$$[(R + j\omega L)I_m - V_m]\frac{\epsilon^{j\omega t}}{2} + [(R - j\omega L)I_m^* - V_m^*]\frac{\epsilon^{-j\omega t}}{2} = 0 \tag{4-8}$$

In this expression only the exponentials depend on t, and neither of them can ever be zero. The only way in which the entire expression on the left side can ever be zero is for each of the multipliers of the exponentials themselves to be zero. That is, in order to satisfy Eq. (4-8) for all time, we require that

$$I_m = \frac{V_m}{R + j\omega L} \qquad (a)$$

$$I_m^* = \frac{V_m^*}{R - j\omega L} \qquad (b) \tag{4-9}$$

Since $(R - j\omega L)$ is the conjugate of $(R + j\omega L)$, the second of these expressions will automatically be satisfied when the first one is. This fact leads to a very interesting conclusion.

Suppose we take only the positive exponential part of the driving function in Eq. (4-7) and assume only the positive exponential part of the response in Eq. (4-5); then we will get Eq. [4-9(a)] as the result. But, once I_m is known, we can easily take its conjugate. Thus, to find the steady-state response to an applied sinusoid, we need only find the response to the exponential $\epsilon^{j\omega t}$. From this we immediately write down the response to the exponential $\epsilon^{-j\omega t}$, it being simply the conjugate. The sum of these responses is the desired response to the sinusoid.

We still do not have I_m in the polar form given in Eq. (4-6), but this is a simple step from Eq. (4-9).

Let us write the complex quantity $(R + j\omega L)$ in polar form as

$$R + j\omega L = \sqrt{R^2 + (\omega L)^2}\epsilon^{j\theta} \qquad (a)$$

$$\theta = \tan^{-1}\frac{\omega L}{R} \qquad (b) \tag{4-10}$$

Then Eq. (4-9) becomes

$$I_m = \frac{V_m}{R + j\omega L} = \frac{|V_m|\epsilon^{j0}}{\sqrt{R^2 + (\omega L)^2}\epsilon^{j\theta}} = \frac{|V_m|\epsilon^{-j\theta}}{\sqrt{R^2 + (\omega L)^2}} \tag{4-11}$$

Comparing these expressions with Eq. (4-6), we see that the magnitude and angle of I_m are

FROM $|I| = \frac{|V|}{|Z|}$

$$|I_m| = \frac{|V_m|}{\sqrt{R^2 + (\omega L)^2}} \qquad (a)$$

$$\alpha = -\theta = -\tan^{-1}\frac{\omega L}{R} \qquad (b) \qquad (4\text{-}12)$$

Finally, substituting these values into Eq. (4-2), the desired steady-state response becomes

$$i_f = i_{ss} = i = \frac{|V_m|}{\sqrt{R^2 + (\omega L)^2}} \cos\left(\omega t - \tan^{-1}\frac{\omega L}{R}\right) \qquad (4\text{-}13)$$

It may appear that just as much work was required in arriving at this solution as in the method of undetermined coefficients, but this is not so. The power and generality of this method will become amply clear as we proceed.

In order to gain some additional insight, let us now repeat the solution of the problem under consideration, but this time express the sinusoidal function as the real part of an exponential in the form given by Eq. [4-3(*b*)]. Thus

$$v = |V_m| \cos \omega t = Re(V_m \epsilon^{j\omega t}) \qquad (a)$$

$$i = |I_m| \cos (\omega t + \alpha) = Re[|I_m| \epsilon^{j(\omega t + \alpha)}] = Re(I_m \epsilon^{j\omega t}) \qquad (b) \qquad (4\text{-}14)$$

where the complex quantities V_m and I_m are the same as before.

$$V_m = |V_m|\epsilon^{j0} \qquad (a)$$

$$I_m = |I_m|\epsilon^{j\alpha} \qquad (b) \qquad (4\text{-}15)$$

These expressions for v and i are to be substituted into the differential equation. When we do this, we get

$$L\frac{d}{dt} Re(I_m \epsilon^{j\omega t}) + R\, Re(I_m \epsilon^{j\omega t}) = Re(V_m \epsilon^{j\omega t}) \qquad (4\text{-}16)$$

At this point we meet a problem. There are two operations involved in the first term on the left: we are first to take the real part of a function, and then we are to differentiate it. In order to continue with our solution of the equation, it will be necessary to invert these two steps; that is, we want to differentiate the function first, and then take the real part. Is this inversion legitimate? The answer is yes, because the operation of differentiation is commutative with the operation of taking the real part. (See Prob. 3-6.) Hence, interchanging the two operations and carrying the right-hand side to the left, we get

$$\begin{aligned} &Re(j\omega LI_m\epsilon^{j\omega t}) + Re(RI_m\epsilon^{j\omega t}) - Re(V_m\epsilon^{j\omega t}) = 0 \\ \text{or} \quad &Re\{[(R + j\omega L)I_m - V_m]\epsilon^{j\omega t}\} = 0 \end{aligned} \tag{4-17}$$

In performing the last step, we used the fact that addition and subtraction are commutative with taking the real part. (See Prob. 3-5.) The last expression is in the form

$$Re(A\epsilon^{j\omega t}) = 0 \tag{4-18}$$

where A is constant with respect to time t. Remembering that this equation is to be satisfied identically for all values of t, we conclude that the only possibility is for the multiplier A to be zero. Hence

$$(R + j\omega L)I_m - V_m = 0$$

or

$$I_m = \frac{V_m}{R + j\omega L} \tag{4-19}$$

On comparison, we find that this is identical with Eq. [4-9(a)], which is certainly gratifying.

In order to gain more familiarity with this procedure and to determine if any difficulties are encountered by the introduction of a capacitor, let us now continue the discussion in terms of the network shown in Fig. 4-2. Again the source voltage is a sinusoid. This time let us suppose the time $t = 0$ is chosen so that the function is not a cosine but has an initial angle α; that is, $v_g = |V_{gm}| \cos(\omega t + \alpha)$. If we apply Kirchhoff's voltage law around the loop and substitute the v-i relation for R, we get

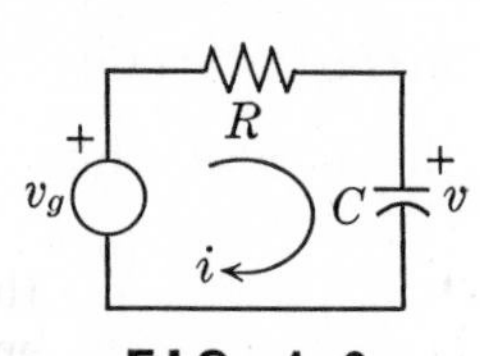

FIG. 4-2
Series RC network

$$Ri + v = |V_{gm}| \cos(\omega t + \alpha) \tag{4-20}$$

Here there are two unknowns v and i. Let us eliminate i by inserting $i = C\,dv/dt$, keeping v as the unknown. Then we have

$$RC\frac{dv}{dt} + v = |V_{gm}| \cos(\omega t + \alpha) \tag{4-21}$$

According to our conclusion following Eq. (4-9), it is necessary only to consider the response to a positive exponential; that is, instead of Eq. (4-21), we need to solve the equation

$$RC\frac{dv}{dt} + v = V_{gm}\epsilon^{j\omega t} \qquad (a) \qquad (4\text{-}22)$$

where

$$V_{gm} = |V_{gm}|\epsilon^{j\alpha} \qquad (b)$$

is a complex number formed from the amplitude and angle of the sinusoidal excitation. The forced solution of this equation will also be an exponential of the form

$$v = V_m\epsilon^{j\omega t} \qquad (a)$$
$$V_m = |V_m|\epsilon^{j\beta} \qquad (b) \qquad (4\text{-}23)$$

When this is substituted into the differential equation in Eq. (4-22), we get

$$(1 + j\omega RC)V_m\epsilon^{j\omega t} = V_{gm}\epsilon^{j\omega t} \qquad (4\text{-}24)$$

which can be satisfied if, and only if,

$$V_m = \frac{V_{gm}}{1 + j\omega RC} = |V_m|\epsilon^{j\beta} \qquad (4\text{-}25)$$

The right side of this equation expresses V_m in polar form. Having found the complex quantity V_m, we can certainly take its conjugate. Finally, as the solution of the original equation, Eq. (4-21), we can write

$$v = \frac{V_m}{2}\epsilon^{j\omega t} + \frac{V_m{}^*}{2}\epsilon^{-j\omega t} = |V_m|(\cos \omega t + \beta) \qquad (4\text{-}26)$$

As a matter of fact, there is no need to express the response in terms of V_m and its conjugate before writing the right side of this equation. V_m itself contains all the information we need. When V_m is put in polar form, as in the right side of Eq. (4-25), both the magnitude and the angle will be placed into evidence. Then the solution can immediately be written as the right side of Eq. (4-26).

So much for the capacitor voltage; let us now turn to the current. In terms of the voltage across the capacitor, given by Eq. (4-26), we can write

$$i = C\frac{dv}{dt} = C\frac{d}{dt}\left(\frac{V_m}{2}\epsilon^{j\omega t} + \frac{V_m{}^*}{2}\epsilon^{-j\omega t}\right)$$
$$= \frac{j\omega CV_m}{2}\epsilon^{j\omega t} - \frac{j\omega CV_m{}^*}{2}\epsilon^{-j\omega t}$$
$$= \frac{I_m\epsilon^{j\omega t}}{2} + \frac{I_m{}^*\epsilon^{-j\omega t}}{2} \qquad (4\text{-}27)$$

where

$$I_m = j\omega C V_m = j\omega C \frac{V_{gm}}{1 + j\omega RC} = \frac{V_{gm}}{R + \dfrac{1}{j\omega C}} \tag{4-28}$$

In the last equation we substituted Eq. (4-25) for V_m and then divided numerator and denominator by $j\omega C$. To put this in polar form, let us write

$$R + \frac{1}{j\omega C} = R - j\frac{1}{\omega C} = \sqrt{R^2 + (1/\omega C)^2}\epsilon^{j\theta} \quad (a)$$

$$\theta = \tan^{-1}\frac{-1}{\omega CR} \quad (b) \tag{4-29}$$

Then Eq. (4-28) can be written as

$$I_m = \frac{V_{gm}}{R + \dfrac{1}{j\omega C}} = \frac{|V_{gm}|\epsilon^{j\alpha}}{\sqrt{R^2 + (1/\omega C)^2}\epsilon^{j\theta}} = \frac{|V_{gm}|\epsilon^{j(\alpha-\theta)}}{\sqrt{R^2 + (1/\omega C)^2}} \tag{4-30}$$

The magnitude and angle of I_m are now placed in evidence. Hence the current as a function of time will be

$$i(t) = \frac{|V_{gm}|}{\sqrt{R^2 + (1/\omega C)^2}} \cos(\omega t + \alpha - \theta) \tag{4-31}$$

where α is the angle of the source voltage and θ is given by Eq. [4-29(*b*)]. We have now completed the steady-state solution for both the capacitor voltage and the current.

At this point let us return to Kirchhoff's voltage law expressed in Eq. (4-20). Instead of eliminating i from this expression, suppose we eliminate v. The result will be

$$Ri + \frac{1}{C}\int_0^t i(x)\,dx + v(0) = \frac{V_{gm}}{2}\epsilon^{j\omega t} + \frac{V_{gm}^*}{2}\epsilon^{-j\omega t} \tag{4-32}$$

Here, we find an equation which contains an integral whose integrand is the unknown function. Such an equation is called an *integral equation.* (When both integrals and derivatives are present, the equation is called an *integrodifferential* equation.) We are here using both the positive and the negative exponential, instead of taking short cuts.

One way to proceed at this point is to differentiate this equation, thus converting it to a differential equation. We shall leave this method for you to carry out.

As another alternative we can assume, as we have before, that the forced response alone (without adding the natural response) satisfies the equation. As a matter of fact, we have no firm basis for believing this to be true. We found it to be true in the case of a *differential* equation, but this does not mean that it must be true for an *integral* equation also. Nevertheless, let us proceed on this basis and see if we meet any difficulties.

We again assume a sinusoidal waveform for the current and express the sinusoid in exponential form, as in the last line of Eq. (4-27). (Of course, now I_m is not known.) When this is substituted into Eq. (4-32) and the terms are collected, the result will be

$$\left[\left(R + \frac{1}{j\omega C}\right) I_m - V_{gm}\right]\epsilon^{j\omega t} + \left[\left(R - \frac{1}{j\omega C}\right) I_m^* - V_{gm}^*\right]\epsilon^{-j\omega t} + \left[v(0) - \frac{I_m}{j\omega C} - \frac{I_m^*}{-j\omega C}\right] = 0 \qquad (4\text{-}33)$$

This time, in addition to the two exponential terms, there is a constant term which is made up of the initial voltage and the contributions from the lower limit of the integral. (Go through the details of calculating this expression.) In order for this equation to be satisfied for all values of t, we must require that the constant term and the multipliers of the two exponentials be individually zero. This requirement leads to

$$I_m = \frac{V_{gm}}{1 + \dfrac{1}{j\omega C}}; \quad I_m^* = \frac{V_{gm}^*}{1 - \dfrac{1}{j\omega C}} \qquad (4\text{-}34)$$

and

$$v(0) - \frac{I_m}{j\omega C} - \frac{I_m^*}{-j\omega C} = 0 \qquad (4\text{-}35)$$

The first of these agrees with the previous result given in Eq. (4-28), which is certainly a cause for joy. But the joy changes to sorrow when we look at Eq. (4-35). Remember that $v(0)$ is an arbitrary, independently specified, number. With I_m and I_m^* fixed from Eq. (4-34), how is it possible to satisfy Eq. (4-35)?

We must seek the answer to the question in the fact that the forced response alone *does not* satisfy the original integral equation. To the forced response we should add the natural response. In the previous chapter we found that, for the network under consideration here, the natural response is $K\epsilon^{-t/RC}$, with any value of K. If we now add this to the assumed forced response, before substituting into Eq. (4-32), there will be some additional terms on the left side of Eq. (4-33). These will be $RK\epsilon^{-t/RC}$ from the first term and $1/C$ times the integral of $K\epsilon^{-t/RC}$

from the second term. The contribution from the upper limit of the integral will cancel with $RK\epsilon^{-t/RC}$, leaving only the contribution from the lower limit, which will be simply KR. (Go through the details of this calculation.) Hence the constant term in Eq. (4-33) will be augmented by KR, which will lead to the following revision of Eq. (4-35).

$$v(0) - \frac{I_m}{j\omega C} - \frac{I_m^*}{-j\omega C} + KR = 0 \qquad (4\text{-}36)$$

Since the value of K is not restricted by any other requirement, this equation serves to determine it. In this way, all conditions are satisfied and the same steady-state solution is obtained from Eq. (4-34) as before.

The fact that the forced solution alone does not satisfy the integral equation is a disadvantage of the method of solution we are following. If we glance back over the discussion following Eq. (4-33), we notice that the initial capacitor voltage and the contribution of the lower limit of the integral do not influence the steady-state solution. Hence it would be justifiable henceforth to disregard these when seeking the steady-state solution to sinusoidal excitations. That is, when writing equations like Eq. (4-32), the initial voltage is simply omitted. Furthermore, only the upper limit of the integral is used. This amounts to writing the capacitor voltage as an antiderivative (indefinite integral) of the current.†

4.2 SINORS

Let us now pause and examine, somewhat critically, what we have done. The sinusoidal driving function is first expressed in exponential form. The assumed sinusoidal solution is also expressed in exponential form. For the purpose of discussion, let us assume that the response is the current. The two unknowns, the magnitude and phase of the response, appear together in the single complex number I_m. Once this complex number is determined, the solution is complete. Note that the complex number I_m is an algebraic quantity and does not depend on t. It contains all the information which is to be determined about the sinus-

† The question as to why this is not done from the start can be answered by remembering that the steady-state response is only part of the total response. Although it appears useful to treat the v-i relationship of a capacitor as an indefinite integral here, this is not true when the complete solution is sought, as by the Laplace transform, for example.

oidal current, its amplitude and phase. Once I_m is determined, the steady-state current is completely known. I_m is not the current; nevertheless, once we know it, then by a sequence of steps we can find the current. This can be done in one of two ways. In one of these methods, knowing I_m, we take its complex conjugate. Then we multiply I_m by $\epsilon^{j\omega t}/2$, and its conjugate by $\epsilon^{-j\omega t}/2$, and add. Alternatively, we multiply I_m by $\epsilon^{j\omega t}$, then take the real part.

To clarify these remarks, let us illustrate the underlying philosophy by means of a different mathematical procedure with which you are already familiar. Suppose we are to multiply together two real numbers x and y, their product being called w.

$$w = x \cdot y$$

One way of finding w is to take the logarithm of this expression. Let us designate the logarithm of a quantity by a capital letter; that is,

$$X = \log x$$

$$Y = \log y$$

$$W = \log w$$

Then, from the property of logarithms, we can write

$$\log w = \log x + \log y$$

or

$$W = X + Y$$

Capital W is not the quantity we want, but once we find capital W, we can easily find the quantity we want by taking the antilogarithm of it. We never deal with x and y and w; all operations are performed on X, Y, and W—the logarithms. At the very end, we convert to w by taking the antilogarithm of W. Note that the original operation of multiplication has been changed to a simpler one of addition in terms of the logarithms.

The steady-state response of linear networks is found by a completely analogous procedure.

In any linear electrical network when all the driving functions are sinusoidal and have the same frequency, the steady-state currents and voltages everywhere will also be sinusoidal, with the same frequency. The desired responses are functions of time, but the only quantities that remain to be determined are the amplitudes and phases of the sinusoids. If we express the sinusoids in exponential form, then the amplitude and phase information is conveniently given by the complex number which is the coefficient of the exponential. Hence we need not deal with the actual currents and voltages as functions of time, but only with these complex quantities. The point of view we shall adopt is that the actual time functions have been *transformed*, and that we deal with the transform quantities, just as we deal with the transform quantity which we call a logarithm when we wish to multiply, or divide, or raise a number to a power, or extract a root. All operations are performed on the transform quantities; at the very end, we convert back to the time functions.

These complex numbers, the coefficients of the exponentials, should be given a name, since we use them so often. Over the years a number of different names have been used. Among these are *vector*, *phasor*, *sinor*, and *complex amplitude*. The use of the word "vector" is unfortunate, since it is possible to confuse this with a three-dimensional space vector. We shall not use this word here. The word "phasor" seems to have found fairly wide acceptance, but, as this word is used, it appears to be a synonym for complex number. Although the quantities for which we seek a name are complex, they are also special, in that they are the coefficients of certain exponentials, a combination of which leads to a sinusoid. For this reason we shall use the word "sinor."

As we discussed in Chap. 1, one measure of a periodic function is its root-mean-square (rms), or effective value. There we found the effective value of a sinusoid to be $1/\sqrt{2}$ times its amplitude. It is common practice to introduce this numerical factor of $\sqrt{2}$ when defining a sinor. Thus, for the sinusoidal function $|A_m| \cos(\omega t + \alpha)$, we define the sinor A as

$$A = \frac{A_m}{\sqrt{2}} = \frac{|A_m|}{\sqrt{2}} \epsilon^{j\alpha} \tag{4-37}$$

The sinor is thus a complex quantity with a magnitude equal to the effective value of the sinusoid and an angle equal to the angle, or phase, of the sinusoid. Once the sinor is known, the sinusoid is determined. When the sinusoid represents a voltage or a current, we use the symbols V and I, respectively, for the sinors. Note that we are using capital letters for sinors, whereas lower-case letters are reserved for functions of time.

The following sinusoids and their sinors are given to illustrate this idea.

$$v(t) = 20 \cos(\omega t + \pi/3); \quad V = \frac{20}{\sqrt{2}} \epsilon^{j\pi/3}$$

$$v(t) = 5 \cos(\omega t - \pi/6); \quad V = \frac{5}{\sqrt{2}} \epsilon^{-j\pi/6}$$

$$i(t) = 10 \sin(\omega t + \pi/8); \quad I = ?$$

In the last case we encounter some difficulty, since the sinusoid is expressed as a sine rather than as a cosine. But, remembering that

$$\sin\left(x + \frac{\pi}{2}\right) = \cos x \qquad (a)$$
$$\cos\left(x - \frac{\pi}{2}\right) = \sin x \qquad (b) \qquad (4\text{-}38)$$

it is always possible to write a sine as a cosine with a different angle (and vice versa). Hence

$$i(t) = 10 \sin\left(\omega t + \frac{\pi}{8}\right) = 10 \cos\left(\omega t - \frac{3\pi}{8}\right)$$

$$I = \frac{10}{\sqrt{2}} \epsilon^{-j3\pi/8}$$

It is possible to give a simple graphical interpretation for the relationship between a sinusoidal function and its sinor. Since a sinor, say $V = |V|\epsilon^{j\alpha}$, is a complex number, it can be represented in the complex plane by a directed line making an angle α with the real axis, as shown in Fig. 4-3. (This representation of a sinor is the origin of the term vector which was used for such a long time in the past.)

When the sinor $|V|\epsilon^{j\alpha}$ is multiplied by the exponential $\epsilon^{j\omega t}$, we can look on the result as another sinor whose magnitude is $|V|$ and whose angle is now $(\alpha + \omega t)$. As time goes on, the angle of this new sinor increases at a uniform rate. The directed line representing it, then, appears to rotate in the counterclockwise direction at an angular velocity ω. As this line rotates, it will have a projection on the real axis which will be

$$\text{Projection on real axis} = Re[|V|\epsilon^{j(\omega t+\alpha)}]$$
$$= |V| \cos(\omega t + \alpha)$$

This projection differs from the sinusoid simply by the factor $\sqrt{2}$, which is a result of the way in which we defined the sinor, in terms of the effective value rather than the maximum value. But it is a simple job to supply the missing $\sqrt{2}$. The real-axis projection is shown in Fig. 4-3. The sinusoid is obtained by drawing a time axis vertically downward and choosing a convenient scale for t. For each value of t, the top of the rotating line is projected downward until it intersects a horizontal line drawn perpendicular to the t-axis at the same value of t. One such point is labeled t_1 in the figure.

Thus, to recover the sinusoid from its sinor, we should lay out the sinor in the complex plane at its initial angle, stretch it by a factor of

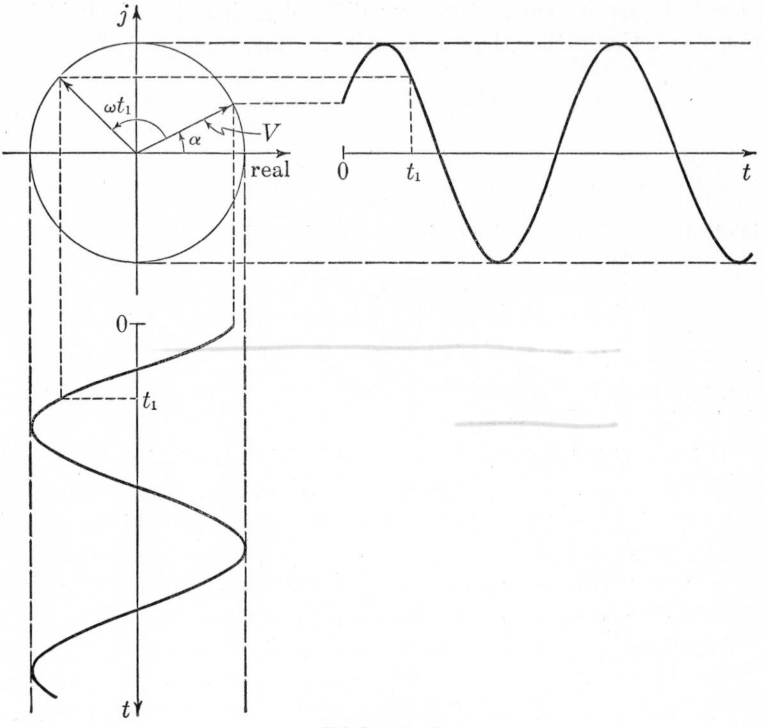

FIG. 4-3
Rotating sinor

$\sqrt{2}$, cause it to rotate counterclockwise at an angular velocity ω, and, finally, take its projection on the real axis. Of course, no one in his right mind would want to perform this sequence of steps graphically. It is much simpler for computational purposes to do it algebraically. The graphical representation gives just one more way of looking at things, however, and, as a wise man once said, "One purpose of education is to impart numerous viewpoints." (Anyway, some wise man should have said it!)

Note that the projection on the imaginary axis also generates a sinusoid. In this case

$$\begin{aligned}\text{Projection on imaginary axis} &= Im[|V|\epsilon^{j(\omega t+\alpha)}] \\ &= |V| \sin(\omega t + \alpha)\end{aligned}$$

This sinusoid is also shown in the figure. It is obtained in a similar graphical way.

You will recall that writing $Re(V\epsilon^{j\omega t})$ is but one way of representing sinusoids in terms of exponentials. An alternate way is to write

$$|V| \cos(\omega t + \alpha) = \frac{V}{2}\epsilon^{j\omega t} + \frac{V^*}{2}\epsilon^{-j\omega t}$$

We have seen that the first term on the right can be represented graphically in the complex plane by a line rotating counterclockwise with an angular frequency ω, starting at an angle α. In the present case the length of the line is half the previous length.

The second term on the left has the same magnitude as the first term but the opposite angle. This means that it can also be represented by a rotating line, but this time rotating clockwise, from an initial angle $-\alpha$. At each value of t, it will be in a position which is the conjugate of that of the first term, as shown in Fig. 4-4. Hence each rotating line will have the same projection on the real axis. The sinusoid is obtained by taking the sum of the two projections, as indicated in the figure.

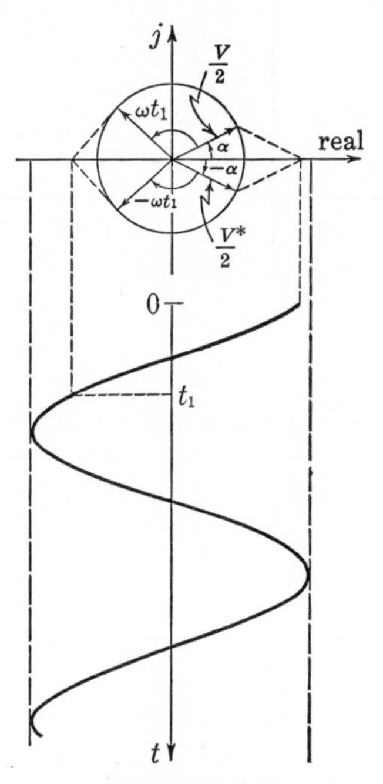

FIG. 4-4

Two rotating sinors

As mentioned before, these graphical constructions for recovering a sinusoid from its sinor are not useful for computational purposes. They serve, however, as an aid in the thought process of understanding the relationship between a sinusoid and its sinor.

In a given problem, if two sinusoids of the same frequency are involved, we can consider each of them to be generated by a rotating line. Since the two lines are rotating at the same angular velocity, their relative position does not change, but depends solely on the difference of initial angle, or *phase difference*, between the two sinusoids. Thus, if

$$v_1 = \sqrt{2}|V_1| \cos(\omega t + \alpha)$$

$$v_2 = \sqrt{2}|V_2| \cos(\omega t + \beta)$$

the sinors will be $|V_1|\epsilon^{j\alpha}$ and $|V_2|\epsilon^{j\beta}$. They can be represented in the complex plane by the lines shown in Fig. 4-5(*a*). These are the positions of the rotating lines at $t = 0$. The angle between them is labeled θ. At some later time the lines will have rotated as shown in part (*b*) of the figure, but their lengths, and the angle θ between them, will still be the

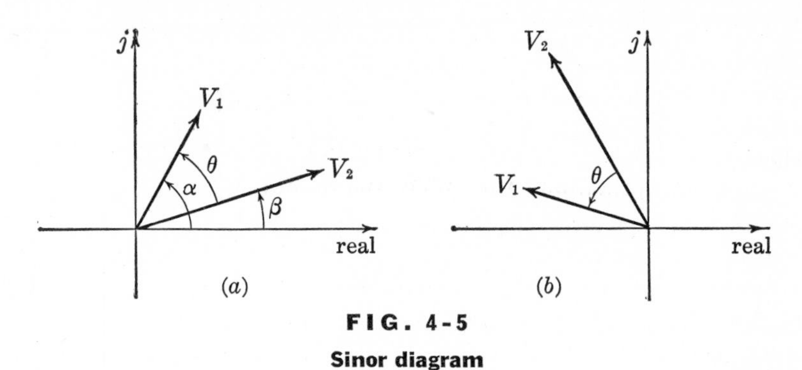

FIG. 4-5
Sinor diagram

same. A diagram such as the one in this figure is called a *sinor diagram*. Clearly, if there are other sinusoids involved in the same problem, the sinors can all be shown in the same diagram with the appropriate lengths and angles relative to each other. Since the absolute locations of the directed lines are not important, only their relative locations, it may be convenient to choose one of them in a horizontal position. This one is said to be taken as a reference for phase.

4.3 THE FUNDAMENTAL LAWS IN TERMS OF SINORS

In the preceding sections of this chapter, we laid the groundwork and the justification for finding the steady-state solution of linear networks to sinusoidal excitations in terms of the exponential function and sinors. Not much will be gained in the way of computational effort, however, if for each problem we must go through the entire process of writing the differential equations, substituting the exponentials for the sinusoids, then solving the equations. We shall now return to the fundamental laws themselves—Kirchhoff's laws and the v-i relationships—and examine these under sinusoidal steady-state conditions.

Consider first Kirchhoff's current law at any node of a network; it can be written

$$\sum_{k=1}^{n} i_k(t) = i_1 + i_2 + \cdots + i_n = 0 \tag{4-39}$$

(Since all the signs are positive, this assumes that all the current references have been taken away from the node.) We are assuming that any sources in the network have sinusoidal waveforms with the same frequency. In the steady state, all the currents will be sinusoids of the same frequency. Hence we can write for the kth current

$$i_k = Re(\sqrt{2}I_k\epsilon^{j\omega t}) = \sqrt{2}Re(I_k\epsilon^{j\omega t})$$

where I_k is a sinor. (Note that $\sqrt{2}|I_k|$ is the amplitude of the sinusoid.) If this is substituted into Eq. (4-39), the result will be

$$\begin{aligned}\sum Re(I_k\epsilon^{j\omega t}) &= Re[\sum I_k\epsilon^{j\omega t}] \\ &= Re[(I_1 + I_2 + \cdots + I_n)\epsilon^{j\omega t}] = 0\end{aligned} \tag{4-40}$$

The second step follows from the fact that the operations of taking the real part and addition are commutative. We now have an expression like that of Eq. (4-18) with $A = \sum I_k$ a complex constant. The only way in which the equation can be satisfied for all t is for the constant to be zero; that is,

$$\sum I_k = I_1 + I_2 + \cdots + I_n = 0 \tag{4-41}$$

Thus we find that, in the steady state, Kirchhoff's current law is satisfied not only by the instantaneous values but also by the sinors.

It is a simple matter to follow the same argument to see that Kirchhoff's voltage law is also satisfied in terms of sinors. Thus, under steady-state conditions, Kirchhoff's two laws hold when written in terms of the current and voltage sinors.

Finally, let us turn to the v-i relationships of the elements. When we use sinors instead of time functions, the v-i relationships take the form shown in Fig. 4-6. (Demonstrate these by expressing v and i in terms of exponentials and substituting into the v-i relationships.)

(a)	(b)	(c)
$+V \rightarrow I$	$+V \rightarrow I$	$+V \rightarrow I$
R	L	C
$V = RI$	$V = j\omega LI$	$V = \frac{1}{j\omega C} I$
$I = GV$	$I = \frac{1}{j\omega L} V$	$I = j\omega C V$

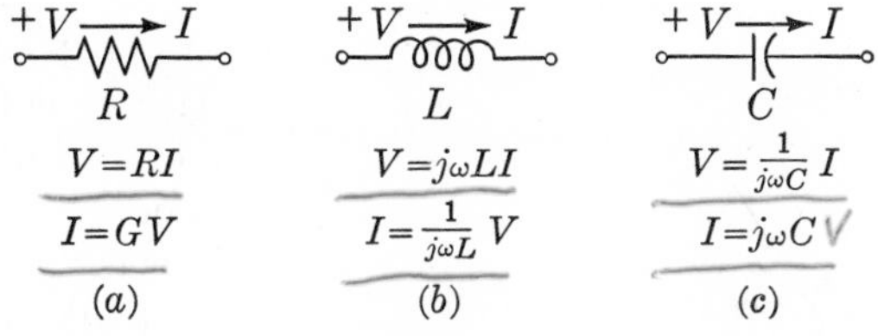

FIG. 4-6

Sinor V-I relationships

The implications which have been shaping up in this section have tremendous significance. All the methods of analysis which were carried out in Chap. 2 for resistive networks were based on Kirchhoff's two laws and the v-i relationships of the elements. If these have now been sucessfully expressed in terms of sinors in the steady state, surely it will be possible to extend the methods of analysis already discussed for resistive networks to more general networks in the steady state.

4.4 IMPEDANCE AND ADMITTANCE

In order to exploit the preceding ideas, let us define some new concepts and terminology. Consider Fig. 4-7(a). This shows an arbi-

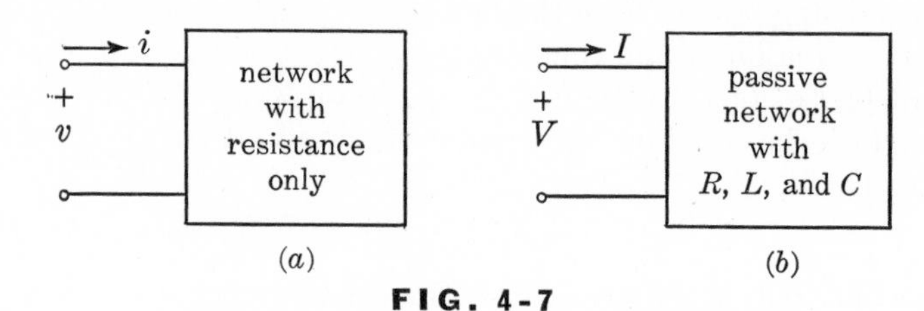

FIG. 4-7

Definition of impedance and admittance

trary network containing resistance only. No matter how complicated the network inside the box, the equivalent resistance at the terminals will be given very simply by the ratio of the voltage v and the current i. This is true, no matter what the wave form of v and i.

Now consider the network shown in Fig. 4-7(b) which may contain any number of R's, L's, and C's. We require only that it does not contain any sources in its interior. The only source is at the terminals and has a sinusoidal waveform. It may be either a current source or a voltage source which we represent by its sinor. We wish to define something similar to the equivalent resistance in the previous network. For the network of Fig. 4-7(b) we define the *impedance* Z and its reciprocal the *admittance* Y as

$$Z = \frac{V}{I} \qquad (a)$$

$$Y = \frac{I}{V} = \frac{1}{Z} \qquad (b) \qquad (4\text{-}42)$$

The impedance is the ratio of the terminal voltage sinor to the terminal current sinor; the admittance is the reciprocal of this ratio. Note very carefully that impedance and admittance are defined as ratios of sinors, not of time functions. Equation (4-42) is valid for the references shown in Fig. 4-7(*b*). If one of the references is reversed, then the sign of the impedance must be reversed.

If Eq. (4-42) is rewritten in the form $V = ZI$, then it becomes reminiscent of Ohm's law. In fact, this expression is often referred to as Ohm's law for sinors.

Suppose the network excitation is the voltage, the current being the response. Once the admittance is known, the steady-state response sinor is determined by the product of Y and the excitation sinor V. Similarly, if the current is the excitation, the response-voltage sinor is determined as the product of Z and I. That is, *the sinusoidal steady-state response of a network with one pair of terminals is completely determined from a knowledge of the impedance Z or the admittance Y.*

Since voltage and current sinors are generally complex, their ratio, the impedance or admittance, will also be complex. Hence, in rectangular form we can write

$$Z = R + jX \qquad (a)$$
$$Y = G + jB \qquad (b) \qquad (4\text{-}43)$$

These forms show the real and imaginary parts explicitly.

Of course, it is possible to refer to the component parts of Z and Y as the real part of Z, or the imaginary part of Y, etc. Certain other names, however, have come into rather general use. The real part of Z is referred to as the *resistance component, resistance part,* or simply *resistance,* and is given the symbol R; the imaginary part of Z is called the *reactance* and is given the symbol X; the real part of Y is called the *conductance* and is given the symbol G; and, finally, the imaginary part of Y is called *susceptance* and is given the symbol B.

The names "resistance" and "conductance," and the symbols R and G, for the real parts are unfortunate choices, because each of these names and symbols already represents a network parameter, but, since they are almost universally used, we shall also continue to use them, however reluctantly.

Let us now examine the impedances and admittances of some simple networks. The simplest ones are the single elements themselves. The resistor needs little comment. Its voltage and current sinors have the same angle; we say they are in phase. For the inductor we have

$$Z_L = \frac{V_L}{I_L} = j\omega L; \quad X_L = \omega L \qquad (a)$$

$$Y_L = \frac{1}{Z_L} = \frac{1}{j\omega L} = -j\frac{1}{\omega L}; \quad B_L = -\frac{1}{\omega L} \qquad (b) \qquad (4\text{-}44)$$

Both the impedance and the admittance are purely imaginary. The *inductive reactance* is $X_L = \omega L$, which is positive for positive ω. The inductive susceptance is $B_L = -1/\omega L$, which is negative for positive ω. The angle of the voltage sinor is 90 deg greater than that of the current sinor. We say that the voltage and current are 90 deg *out of phase,* or *in quadrature,* with the voltage leading the current. For a capacitor we have

$$Z_C = \frac{V_C}{I_C} = \frac{1}{j\omega C} = -j\frac{1}{\omega C}; \quad X_C = -\frac{1}{\omega C} \qquad (a)$$

$$Y_C = \frac{I_C}{V_C} = j\omega C; \quad B_C = \omega C \qquad (b) \qquad (4\text{-}45)$$

Again, both the impedance and the admittance are imaginary. The capacitive reactance is $X_C = -1/\omega C$, which is negative for positive ω. The capacitive susceptance is $B_C = \omega C$, which is positive for positive ω. The angle of the voltage sinor is 90 deg less than that of the current sinor. Again, the voltage and current are 90 deg out of phase, but this time the voltage lags the current.

Perhaps a better understanding of leading and lagging angles can be obtained by a consideration of the sinusoids shown in Fig. 4-8. Suppose the dashed one represents the current in an inductor. The voltage, being proportional to the derivative of the current, will be represented by the solid curve. A positive peak of the voltage curve occurs 90 deg before the closest positive peak of the current curve; hence it leads by

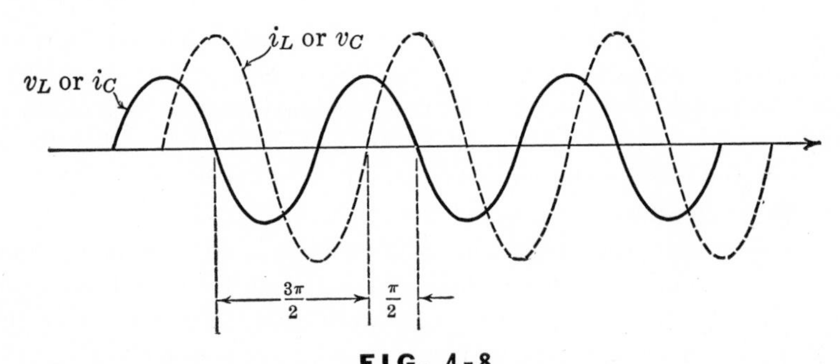

FIG. 4-8

Steady-state voltage and current in an inductor or capacitor

90 deg. It is also true that a positive peak of the voltage curve occurs 270 deg after the previous positive peak of the current curve. Hence we could say that the voltage lags the current by 270 deg.

In terms of a capacitor, the same discussion will apply but with the words voltage and current interchanged.

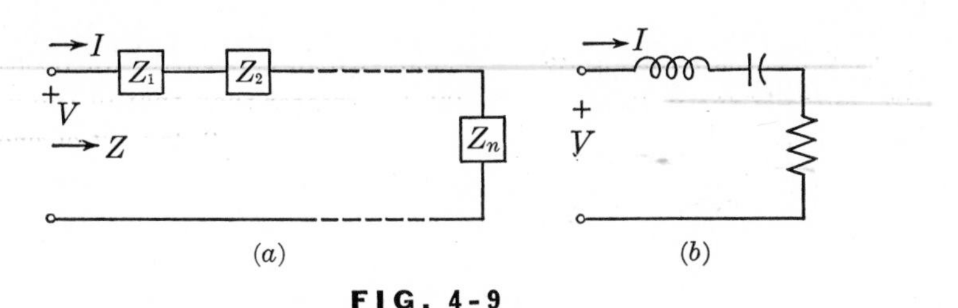

FIG. 4-9

Series-connected branches

Next let us consider a series connection of branches, as shown in Fig. 4-9(a). The branches have been labeled in terms of their impedances, and the terminal voltage and current have been specified by their sinors. If we write Kirchhoff's voltage law in terms of sinors and insert the v-i relationships of the branches (Ohm's law for sinors), the result will be

$$V = Z_1 I + Z_2 I + \cdots + Z_n I = (Z_1 + Z_2 + \cdots + Z_n)I = ZI \quad (4\text{-}46)$$

where

$$Z = Z_1 + Z_2 + \cdots + Z_n \quad (4\text{-}47)$$

The series connection of branches is characterized by the impedance Z which is simply the sum of each of the branch impedances. This is a general result for any number of impedances in series.

For the series-connected RLC network shown in Fig. 4-9(b), the impedance becomes

$$\begin{aligned} Z = \frac{V}{I} &= Z_R + Z_L + Z_C = R + j\omega L + \frac{1}{j\omega C} \\ &= R + j\left(\omega L - \frac{1}{\omega C}\right) = |Z|\epsilon^{j\theta} \end{aligned} \quad (4\text{-}48)$$

where

$$\begin{aligned} |Z| &= \sqrt{R^2 + \left(\omega L - \frac{1}{\omega C}\right)^2} \\ \tan\theta &= \frac{1}{R}\left(\omega L - \frac{1}{\omega C}\right) \end{aligned} \quad (4\text{-}49)$$

In this case we find that the impedance is complex. The real part is simply the resistance R. This fact is actually the origin of the use of "resistance" for the real part of an impedance. Note that in other networks the real part of the impedance will not be simply the value of one resistance; it will depend on the other elements and on the frequency as well.

To find the admittance of the series RLC network, we simply take the reciprocal of Z. Thus

$$Y = \frac{1}{Z} = \frac{1}{R + j\left(\omega L - \dfrac{1}{\omega C}\right)}$$

$$= \frac{1}{R^2 + \left(\omega L - \dfrac{1}{\omega C}\right)^2} - j\,\frac{\omega L - \dfrac{1}{\omega C}}{R^2 + \left(\omega L - \dfrac{1}{\omega C}\right)^2} \tag{4-50}$$

The rectangular form in the last step was obtained by rationalization. Note that the real part of the admittance depends on all three of the network elements as well as on the frequency. This is a rather involved expression; you should not feel it necessary to attempt to memorize it. It is only necessary to note the over-all form.

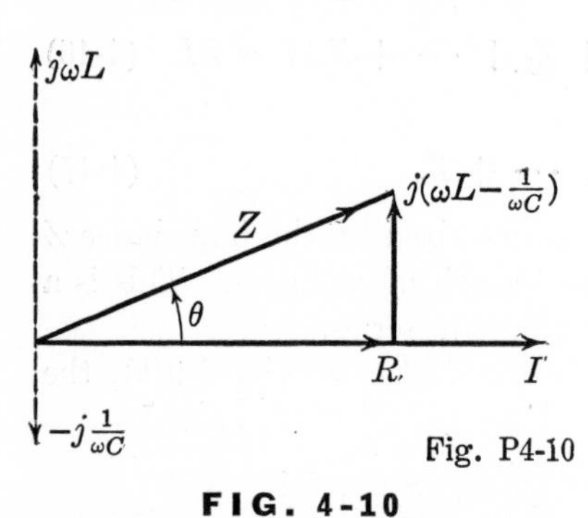

FIG. 4-10

Sinor diagram of series RLC network

Since the impedance and admittance are neither real nor imaginary but complex, the voltage and current of the series RLC network are neither in phase nor in quadrature, except in the special case when $X_L = X_C$. The exact phase relationship will depend on the relative sizes of the real part and the imaginary part of Z. The relationship is illustrated well by means of a sinor diagram, as shown in Fig. 4-10. Let us take the current sinor as a reference and lay it out horizontally. Not only is the angle of I taken as zero, its magnitude is also chosen as a unit for measuring all other magnitudes. Thus in units of $|I|$ the voltage across the resistor is simply R, and similarly for the voltages across L and C. The voltage across R is in phase with I; that across L leads I by 90 deg, while that across C lags I by 90 deg. For the case shown in Fig. 4-10, the capacitive reactance is smaller than the inductive reactance in magnitude. Hence the imaginary part of Z is positive, with the result that the total voltage leads the current but with an angle θ which is less than 90 deg.

As a final example, let us consider a parallel connection of branches as shown in Fig. 4-11(*a*). This time we have specified the branches in

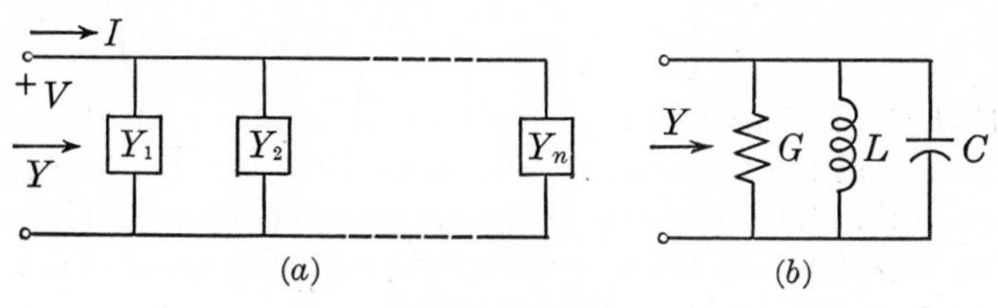

FIG. 4-11

Parallel-connected branches

terms of their admittances, strictly for convenience. This network is the dual of the series connection. If we write Kirchhoff's current law at the upper node in terms of sinors and insert the V-I relationships of the branches, the result will be

$$I = Y_1V + Y_2V + \cdots + Y_nV = (Y_1 + Y_2 + \cdots + Y_n)V = YV$$

where (4-51)

$$Y = Y_1 + Y_2 + \cdots + Y_n \tag{4-52}$$

Thus the parallel connection of branches is characterized by an admittance which is the sum of the admittances of the branches. This result could have been written directly by duality.

For the parallel-connected RLC network shown in Fig. 4-11(*b*), the admittance is

$$\begin{aligned} Y = Y_R + Y_C + Y_L &= G + j\omega C + \frac{1}{j\omega L} \\ &= G + j\left(\omega L - \frac{1}{\omega C}\right) = |Y|\epsilon^{j\phi} \end{aligned} \tag{4-53}$$

where

$$\begin{aligned} |Y| &= \sqrt{G^2 + \left(\omega C - \frac{1}{\omega L}\right)^2} \\ \tan\phi &= \frac{1}{G}\left(\omega C - \frac{1}{\omega L}\right) \end{aligned} \tag{4-54}$$

The admittance is seen to be complex with a real part which is simply the value of the conductance. The impedance of the network is found by taking the reciprocal of the admittance. Thus

$$Z = \frac{1}{Y} = \frac{1}{G + j\left(\omega C - \dfrac{1}{\omega L}\right)}$$

$$= \frac{G}{G^2 + \left(\omega C - \dfrac{1}{\omega L}\right)^2} - j\frac{\left(\omega C - \dfrac{1}{\omega L}\right)}{G^2 + \left(\omega C - \dfrac{1}{\omega L}\right)^2} \tag{4-55}$$

Note that the real part of the impedance depends on all the network parameters and the frequency. This network is the dual of the series *RLC* network. The admittance of the parallel network is identical with the impedance of the series network, and vice versa, but with an interchange of the dual quantities.

Again, the voltage and current of the parallel *RLC* network are neither in phase nor in quadrature. The voltage will lead or lag the current by an angle less than 90 deg, depending on the relative sizes of the real part and the imaginary part of Y. Again, a sinor diagram can be used to illustrate the phase relationship. Since the voltage is common to all branches, this time we choose the voltage sinor as a reference. In fact, because of the duality, Fig. 4-10 can represent the sinor diagram of the parallel *RLC* network simply by interchanging voltage with current and admittance with impedance.

4.5 MEDITATIONS ON GENERALITIES

It is now time to stand back from the details of the subject we have been discussing in order to get a perspective view before continuing. We have found that when excitations are sinusoidal (same frequency) and our interest lies in the steady-state response only, great simplicity can be achieved in the solution of network problems. We discussed two different points of view, both of which are but alternative paths to the same goal.

In one approach the network equations are formulated as differential equations, or integrodifferential equations, depending on the variables which are used. If loop equations are written with loop currents as the variables, each inductor in the network will lead to a derivative and each capacitor to an integral. Similarly, if node equations are written with node voltages as the variables, each inductor will lead to an integral and each capacitor to a derivative. To find the sinusoidal steady-state response of the network, it is only necessary to find the forced solution of these equations to the exponential function $\epsilon^{j\omega t}$. When this is attempted by the method of undetermined coefficients, we find that each derivative in the equations is replaced by $j\omega$ and each integral by $1/j\omega$, with the net result that the integrodifferential equations, whose variables are functions of time, are transformed into algebraic equations whose variables are sinors.

The solution of these algebraic equations now proceeds just as in the case of resistive networks, with the added complication that we are now dealing with complex numbers rather than real numbers. When a solution for the sinors has been obtained, the task is complete, since the sinors contain all the information needed to specify the sinusoids. The last step of actually writing the sinusoidal functions of time may or may not be carried out.

In the second approach the network equations are not first formulated as integrodifferential equations. Instead, Kirchhoff's laws and the V-I relationships of the branches are expressed in terms of sinors. The branches of the network are represented by their impedances and their admittances. Any network equations which are written are now algebraic equations from the start. The remaining work of solving the equations proceeds as before.

In this second point of view, the network itself is assumed to be transformed, and the transformed equations in terms of sinors are written directly. In all our subsequent work, this is the approach which we shall use. Always keep in mind, however, that what we are actually doing is finding the forced response to the exponential function $\epsilon^{j\omega t}$. This fact is actually the justification for the second approach.

4.6 VOLTAGE DIVIDERS, CURRENT DIVIDERS, AND SOURCE TRANSFORMATIONS

We are now ready to undertake the determination of the sinusoidal steady-state response of any electrical network. But, just as we did for resistive networks, we shall first consider certain simple, yet important, network configurations which are amenable to relatively simple solutions before we undertake the most general case.

Consider first the network shown in Fig. 4-12(a). The details of the network inside the rectangle on the left are unimportant; it is enough

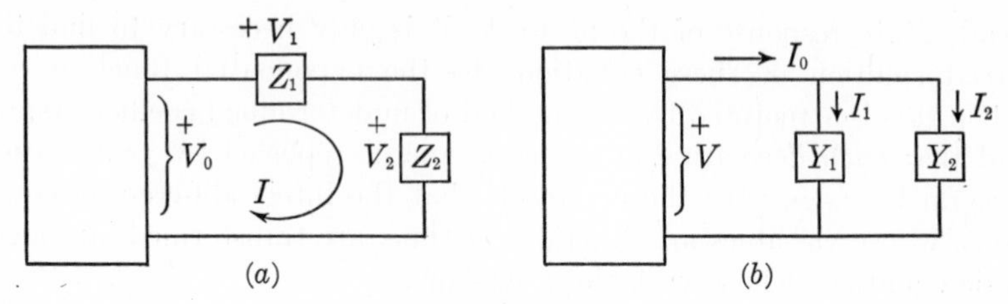

FIG. 4-12

Voltage divider and current divider

to know that all sources are sinusoidal (of the same frequency) and that the voltage sinor at the terminals is V_0. For simplicity you can assume that the network is simply a voltage source. We desire to find the voltage across each of the impedances Z_1 and Z_2.

This network structure is a voltage divider like the one first discussed in Sec. 2-2, except that the branches are now general ones rather than simply resistances. The current sinor is easily found as V_0 divided by $Z_1 + Z_2$, from which V_1 and V_2 follow. Thus

$$V_1 = Z_1 I = \frac{Z_1}{Z_1 + Z_2} V_0 \qquad (a)$$

$$V_2 = Z_2 I = \frac{Z_2}{Z_1 + Z_2} V_0 \qquad (b) \qquad (4\text{-}56)$$

$$\frac{V_1}{V_2} = \frac{Z_1}{Z_2} \qquad (c)$$

In words, this means that the voltage sinor across a series combination of two branches is distributed between them in the ratio of the impedance of each branch to the total impedance. Clearly, this result can be generalized to more than two branches in series.

The dual situation is shown in Fig. 4-12(*b*). Current sinor I_0 is known, and it is desired to find current sinors I_1 and I_2. The voltage sinor V is easily found as I_0 divided by the sum of the admittances. The individual branch-current sinors are now easily written in terms of V. Thus

$$I_1 = Y_1 V = \frac{Y_1}{Y_1 + Y_2} I_0 \qquad (a)$$

$$I_2 = Y_2 V = \frac{Y_2}{Y_1 + Y_2} I_0 \qquad (b) \qquad (4\text{-}57)$$

$$\frac{I_1}{I_2} = \frac{Y_1}{Y_2} \qquad (c)$$

In words, these equations say that the current sinor entering a parallel combination of two branches is distributed between them in direct proportion to the admittances of the branches (or inversely as the impedances). This result can also be generalized to more than two branches connected in parallel.

Note that we have been careful to say "voltage sinor" and "current sinor" rather than just "voltage" and "current." It becomes cumbersome, however, to say "sinor" every time a reference is made to voltage or current. Hence, in the sequel when we are discussing the sinusoidal steady state exclusively, we shall often omit saying sinor, except when it would otherwise be ambiguous, or if emphasis is desirable.

Next let us turn our attention to the transformation of sources. When discussing resistive networks in Chap. 2, we introduced the voltage-shift theorem and the current-shift theorem. These theorems simply refer to moving sources around in a network. They are obviously valid (as a glance back at the discussion will show) independently of the waveshapes of the source voltages or currents, and independently of the types of elements comprising the branches. Hence these theorems are valid in the sinusoidal steady state also. This means that no matter how voltage sources appear initially in a network, they can always be shifted so that each voltage source appears in series with a passive branch. Similarly, by means of the current-shift theorem, each current source will appear in parallel with a branch.

In the case of resistive networks, we found that the combination of a voltage source in series with a resistor can be equivalent to a current source in parallel with the same resistor if the values of the voltage and current are appropriately related. Let us now inquire whether such an equivalence still holds for sinors and *RLC* networks.

Consider the two networks shown in Fig. 4-13; a sinusoidal voltage source with sinor V_0 is in series with a branch having an impedance Z_0 in part (*a*), and in part (*b*) a sinusoidal current source of the same frequency, with sinor I_0, is in parallel with a branch having an impedance

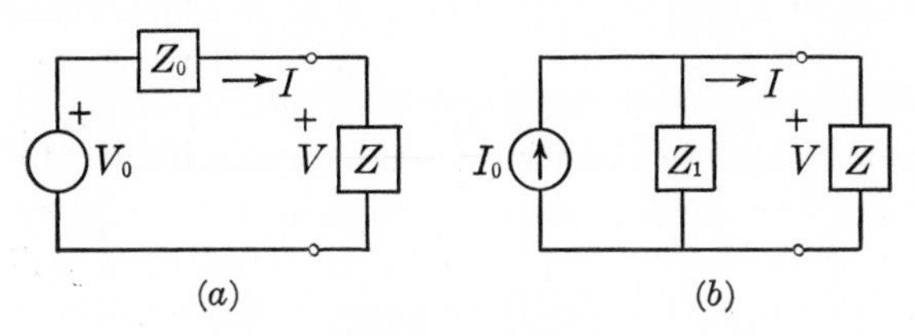

FIG. 4-13

Voltage-source and current-source equivalents

Z_1. The impedance Z can be considered as a load in both cases. The terminal V-I relationships for these networks will be

$$\text{For network } (a) \qquad V = V_0 - Z_0 I \qquad (a)$$

$$\text{For network } (b) \qquad I = I_0 - \frac{1}{Z_1} V \tag{4-58}$$

or

$$V = Z_1 I_0 - Z_1 I \qquad (b)$$

If these two networks are to be equivalent, the voltage V should be the same in both cases. Equating the two expressions and transposing terms, we get

$$(V_0 - Z_1 I_0) + (Z_1 - Z_0)I = 0 \tag{4-59}$$

This equation must be satisfied identically; that is, for any value of the current I, owing to any arbitrary value of the load Z. This is possible if, and only if,

$$Z_1 = Z_0 \qquad (a)$$

$$I_0 = \frac{V_0}{Z_0} \qquad (b) \tag{4-60}$$

Thus we find that the equivalence of the two networks is established, provided the series impedance in the one case is the same as the shunt impedance in the other, and provided that the two source sinors are related by Eq. [4-60(b)]. This is the same condition as for resistive networks, except that we are now dealing with impedances and sinors rather than with resistances and time functions.

In the case of resistive networks, we found an effective method of analysis to consist of simplifying the structure of a network. The steps involved in this process included the series combination of branches, the parallel combination of branches, and the voltage-source to current-source transformation (and vice versa). This procedure is valid in the sinusoidal steady state also.

Let us demonstrate this technique by finding the voltage V_2 in the network shown in Fig. 4-14. The source is sinusoidal and is represented by its sinor V; the angular frequency is ω. First of all, we replace the source V in series with the inductor by a current source $V/j\omega L$ in parallel with the same inductor. The inductor in parallel with the resistor is now considered as a single branch which is in parallel with the current source. The combination can then be replaced by a voltage-source equivalent. Figure 4-14 shows these steps. In the final network there is but a single loop. The current is obtained as the source voltage [the source in Fig. 4-14(c), not the original source] divided by the total im-

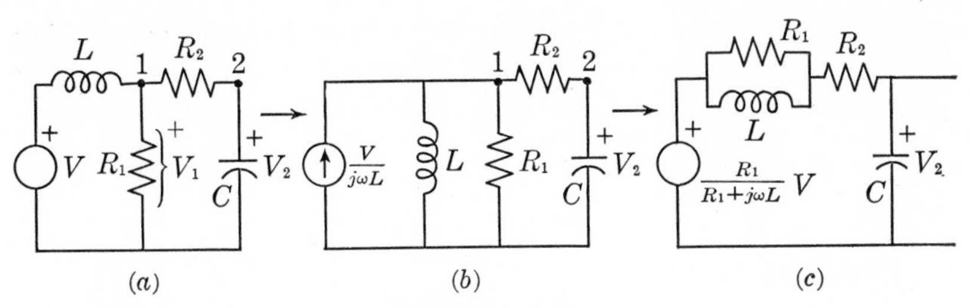

FIG. 4-14

Solution of problem by simplification of network structure

pedance at the source terminals. Finally, the desired voltage sinor V_2 is the product of this current sinor and the impedance of the capacitor. Alternatively, note that Fig. 4-14(c) is in the form of a voltage divider. Hence the final result is

$$V_2 = \frac{1}{j\omega C} I = \frac{1}{j\omega C} \frac{\dfrac{R_1 V}{R_1 + j\omega L}}{\dfrac{j\omega L R_1}{R_1 + j\omega L} + R_2 + \dfrac{1}{j\omega C}} \tag{4-61}$$

$$= \frac{R_1}{-\omega^2 L C R_1 + (1 + j\omega C R_2)(R_1 + j\omega L)} V$$

The last line is obtained by clearing the fractions in the numerator and denominator of the previous line. This is the desired result.

Note that the original structure of the network has been destroyed. If we would like to find the response in some other part of the network, we must work from the original network, not the modified one. Thus, suppose that we wish to find the current through R_1 in the original network. Now that we know V_2, this is an easy task. Suppose we apply Kirchhoff's current law at node 1. The currents away from the node in the three branches can be written

$$I_L + I_{R_1} + I_{R_2} = 0 \tag{4-62}$$

The current sinors can be expressed in terms of the branch-voltage sinors by Ohm's law for sinors. Doing this, the expression becomes

$$\frac{1}{j\omega L}(V_1 - V) + \frac{1}{R_1} V_1 + \frac{1}{R_2}(V_1 - V_2) = 0$$

or

$$\left(\frac{1}{j\omega L} + \frac{1}{R_1} + \frac{1}{R_2}\right) V_1 = \frac{1}{j\omega L} V + \frac{1}{R_2} V_2$$

or

$$I_{R_1} = \frac{V_1}{R_1} = \frac{j\omega L R_2}{R_1 R_2 + j\omega L(R_1 + R_2)} \left(\frac{V}{j\omega L} + \frac{V_2}{R_2}\right) \tag{4-63}$$

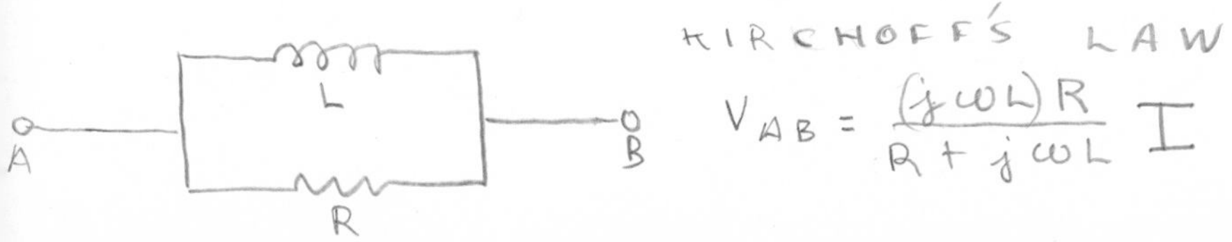

Substitution of V_2 from Eq. (4-61) into this expression causes everything on the right side to become known. Thus I_{R_1} is determined relatively easily, although the literal expression may appear formidable.

4.7 TRANSFER FUNCTIONS

At several places in the last section, we found expressions for one voltage sinor in terms of another, as in Eq. (4-56); or for one current sinor in terms of another, as in Eq. (4-57); or for a current sinor in terms of a voltage sinor, as in Eq. (4-63) with V_2 eliminated. In each of these cases, one of the sinors is a response while the other is an excitation. If we take the ratio of the response sinor to the excitation sinor, the result will depend only on the network parameters and the frequency; or, saying it another way, the result will depend only on the impedances in the network. Once this ratio of response to excitation sinors is known, the response to any given value of the excitation can be determined.

We shall use the general name *transfer function* to mean *the ratio of any response sinor to any excitation sinor*. Consider the general network shown in Fig. 4-15. One branch of the network is explicitly shown. There are no sources inside the rectangle, the only permissible source being at the left-hand terminals. Either a current source or a voltage source may be connected at these terminals; hence either V_1 or I_1 may be the excitation. We can define four possible ratios of response to excitation sinors:

$$G_{21} = \frac{V_2}{V_1}; \quad \text{voltage gain, or voltage-transfer ratio}$$

$$\alpha_{21} = \frac{I_2}{I_1}; \quad \text{current gain, or current-transfer ratio}$$

$$Z_{21} = \frac{V_2}{I_1}; \quad \text{transfer impedance}$$

$$Y_{21} = \frac{I_2}{V_1}; \quad \text{transfer admittance}$$

Note the order of the subscripts; the first one refers to the response, the second to the excitation. The voltage gain and current gain are dimensionless ratios. Z_{21} is dimensionally an impedance, and Y_{21} is dimensionally an admittance. Note carefully that these two are not reciprocals.

Recall that the impedance and admittance of a network at a pair of terminals are also a ratio of two sinors. One of these sinors may be considered an excitation, the other a response. Both the excitation and the response take place, however, at the same terminals. This is to be contrasted with the definition of the transfer functions, where the excitation and the response occur at different places in the network.

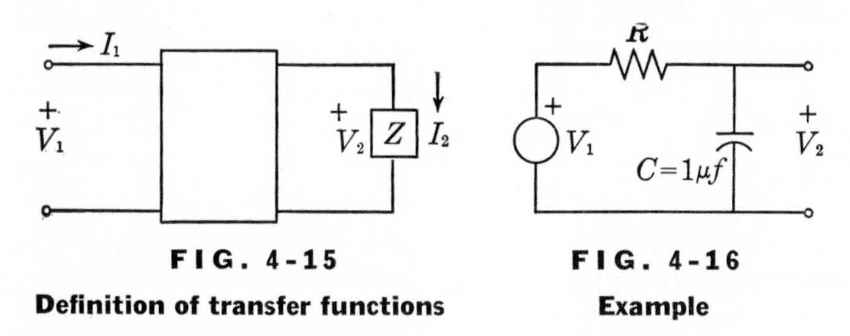

FIG. 4-15
Definition of transfer functions

FIG. 4-16
Example

As a simple illustration, consider the situation shown in Fig. 4-16. It is desired to find the steady-state voltage across the capacitor in response to a 10-volt sinusoidal source at a frequency of 10,000 radians per second, if the value of the resistance is such that the capacitor voltage lags the source voltage by $\pi/4$ radians when the source frequency is 5,000 radians per second.

The network is a simple voltage divider whose voltage-transfer function is obtained from Eq. (4-56) as

$$G_{21} = \frac{V_2}{V_1} = \frac{1/j\omega C}{R + 1/j\omega C} = \frac{1}{1 + j\omega CR} = \frac{1}{1 + j10^{-6}\omega R} \tag{4-64}$$

The last step is obtained by inserting the given value of C. Now the angle of the voltage gain G_{21} is simply the angle by which the output voltage leads the input voltage. The problem specifies that at $\omega =$ 5,000 this angle should be -45 deg. (Note that a negative angle of lead corresponds to a positive angle of lag.) Hence

$$\text{Angle of } G_{21} = -\tan^{-1} 10^{-6}\omega R = -\tan^{-1} 5 \times 10^{-3} R = -45°$$

$$R = \frac{1}{5 \times 10^{-3}} \tan 45° = 200 \text{ ohms}$$

Now we can write the output voltage sinor as

$$V_2 = G_{21}V_1 = \frac{1}{1 + j200 \times 10^{-6}\omega} V_1$$

It can be evaluated for any given value of the source voltage and frequency. Using the specified values $\omega = 10{,}000$ and $V_1 = 10$, we get

$$V_2 = \frac{1}{1 + j2} \times 10 \doteq 4.46\epsilon^{-j63.4°}$$

Notice that we chose the angle of V_1 as a reference. If the angle of V_1 is anything but zero, it will be added to the angle on the right of the last expression.

The value of the concept of transfer function and its utility will not be clearly evident from this brief discussion. As we proceed, we shall encounter many instances where this concept will take on added significance.

4.8 LADDER NETWORKS

The ladder network, which was first discussed in Sec. 2-5, is of such common occurrence that we shall devote some time in discussing techniques of its analysis. What was said in Chap. 2 for resistive networks is applicable here also. We shall illustrate the procedure in terms of the specific ladder shown in Fig. 4-17. The source voltage V is known,

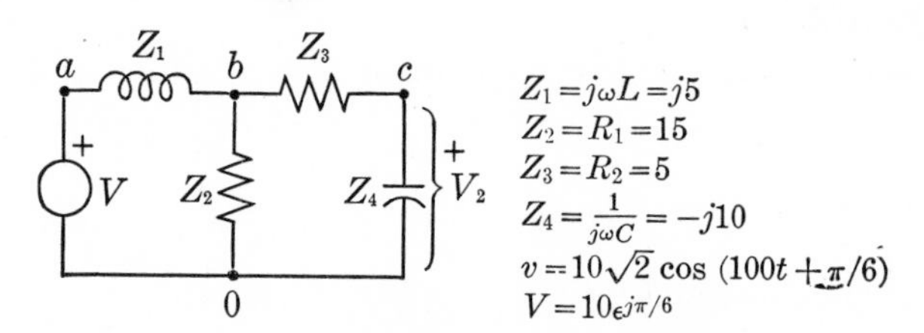

FIG. 4-17
Ladder network

and it is desired to find the voltages or currents of any of the branches.

The approach we take is to start with the right-hand end and label the voltage V_2. Now the current in the capacitor can be written in terms of V_2 by Ohm's law for sinors. This current is equal to the current in Z_3, and so the voltage across Z_3 follows from Ohm's law. The voltage across Z_2 now follows by applying Kirchhoff's voltage law around the right-hand loop. Knowing the voltage across Z_2, the current follows from Ohm's law. The current in Z_1 is now obtained by applying Kirchhoff's current law at node b, and so on. The process is continued, each step being the application of one of the three fundamental laws, until the source is reached. The following steps are self-explanatory.

$$
\begin{aligned}
I_{c0} &= Y_4V_2 && (a)\\
V_{bc} &= Z_3I_{bc} = Z_3Y_4V_2 && (b)\\
V_{b0} &= V_{bc} + V_2 = (Z_3Y_4 + 1)V_2 && (c)\\
I_{b0} &= Y_2V_{b0} = Y_2(Z_3Y_4 + 1)V_2 && (d) \qquad (4\text{-}65)\\
I &= I_{ab} = I_{b0} + I_{bc} = Y_2(Z_3Y_4 + 1)V_2 + Y_4V_2 && (e)\\
V &= Z_1I_{ab} + V_{b0} = [Z_1Y_2(Z_3Y_4 + 1) + Z_1Y_4]V_2 + (Z_3Y_4 + 1)V_2\\
&= [(Z_3Y_4 + 1)(Z_1Y_2 + 1) + Z_1Y_4]V_2 && (f)
\end{aligned}
$$

Hence

$$V_2 = \frac{1}{(Z_3Y_4 + 1)(Z_1Y_2 + 1) + Z_1Y_4}V \tag{4-66}$$

This is the desired answer for the response V_2 in terms of the excitation V; we have only to substitute for the Z's and Y's from the figure. The result will be

$$V_2 = \frac{1}{(1 + j\omega CR_2)\left(1 + j\dfrac{\omega L}{R_1}\right) - \omega^2 LC}V \tag{4-67}$$

Finally, we can substitute the numerical values and the job is completed. Thus

$$
\begin{aligned}
V_2 &= \frac{V}{\left[1 + 5\left(j\dfrac{1}{10}\right)\right]\left[1 + j5\left(\dfrac{1}{15}\right)\right] - 5\left(\dfrac{1}{10}\right)}\\
&= \frac{10\epsilon^{j30^\circ}}{\left(1 + j\dfrac{1}{2}\right)\left(1 + j\dfrac{1}{3}\right) - \dfrac{1}{2}}\\
&= \frac{60\epsilon^{j30^\circ}}{2 + j5} = \frac{60\epsilon^{j30^\circ}}{5.37\epsilon^{j68.4^\circ}} = 11.18\epsilon^{-j38.4^\circ}
\end{aligned}
$$

The corresponding sinusoid is

$$v_2 = 11.18\sqrt{2}\cos(1{,}000t - 38.4^\circ)$$

The method of solution for the output voltage in the ladder network is seen to be as simple in the sinusoidal steady state for RLC networks as it is for resistive networks, except for the added complication introduced by complex numbers.

With V_2 determined, it is now a simple matter to calculate any of the other voltages and currents from Eq. (4-65). This procedure then yields any response voltage or current in the ladder network. In particular, we can find the input current I by substituting Eq. (4-66) into Eq. [4-65(e)]. Thus

$$I = (Y_2 + Y_4 + Y_2Z_3Y_4)V_2 = \frac{Y_2 + Y_4 + Y_2Z_3Y_4}{(1 + Z_3Y_4)(1 + Z_1Y_2) + Z_1Y_4} V \qquad (4\text{-}68)$$

In the details of the steps leading to Eq. (4-65), certain operations—addition, multiplication—are performed on the sinor voltages and currents. Since sinors are complex numbers, these operations may sometimes be conveniently performed graphically, rather than algebraically, by means of a sinor diagram.

Let us illustrate this graphical procedure in terms of the steps given in Eq. (4-65). For convenience, we shall repeat these steps with the numerical values inserted.

$$\begin{aligned} I_{c0} &= j\frac{1}{10}V_2 && (a) \\ V_{b0} &= V_2 + 5I_{c0} && (b) \\ I_{b0} &= \frac{1}{15}V_{b0} && (c) \qquad (4\text{-}69) \\ I &= I_{b0} + I_{c0} && (d) \\ V &= j5I + V_{b0} && (e) \end{aligned}$$

The starting point is voltage sinor V_2. For convenience, V_2 can be chosen as a real number; that is, the angle of V_2 is chosen to be zero. We now choose a convenient scale and lay out a horizontal line whose length is the magnitude of V_2. In the present case the magnitude of V_2 is unknown. But this is no deterrent; we pick a convenient length and label it V_2, as shown in Fig. 4-18(a). All other magnitudes will be in terms of units of V_2.

In the first step I_{c0} is computed in terms of V_2; this is done by multiplying V_2 by $\frac{j}{10}$. Multiplication is most easily thought of in terms of polar coordinates. Thus the magnitude of V_2 is multiplied by $\frac{1}{10}$, and its angle is increased by 90 deg. The resulting sinor is also shown in Fig. 4-18(a). I_{c0} is in quadrature with V_2.

In the next step V_{b0} is found as the sum of V_2 and $5I_{c0}$. This addition is performed graphically, as illustrated in Fig. 4-18(b). From Eq. [4-69(c)] we see that I_{b0} has the same angle as V_{b0}; that is, they are in phase. Now I_{b0} is laid out along V_{b0} with a magnitude 1/15 that of V_{b0},

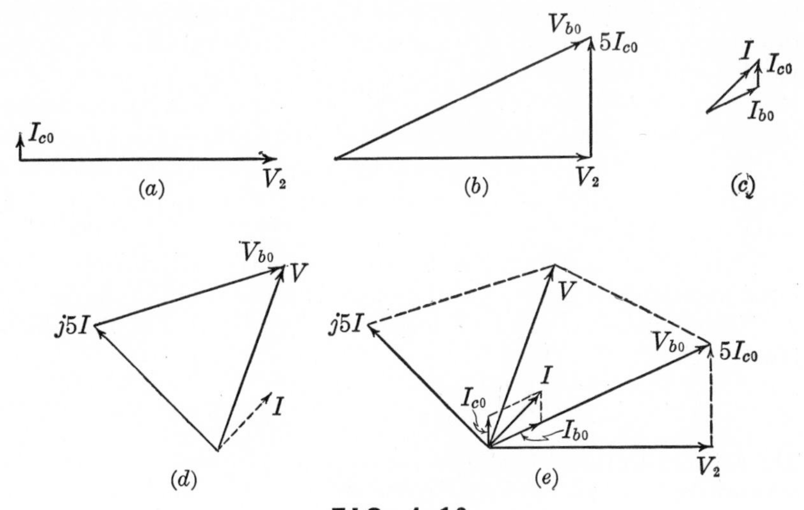

FIG. 4-18

Sinor diagram of example

as shown in Fig. 4-18(*c*). To this, I_{c0} is added graphically to get I. And, finally, to find V we first find $j5I$; this is done by laying out a line at 90 deg to sinor I with a length five times that of I, as shown in Fig. 4-18(*d*). When we now add V_{b0}, the result is V. Figure 4-18(*e*) shows the combined result. (The figure is not to scale; the smaller sinors have been exaggerated for the sake of clarity.)

The sinor diagram gives a graphical portrayal of the relative magnitudes and the relative angles of the sinors involved in a problem. In the present case the angles are all relative to the angle of V_2, which was chosen to be zero. The magnitudes are all relative to the magnitude of V_2.

A sinor diagram is not very useful for accurate numerical results. Nevertheless, it gives a quick method for obtaining an approximate answer. It also provides a method for checking on possible gross errors in arithmetic. In any numerical problem it is worthwhile to make at least a rough sketch of a sinor diagram as each step in the calculation is performed.

In case the desired response is the current I at the left-hand terminals in Fig. 4-17, an alternative procedure is available. In this case it is only necessary to compute the input impedance or admittance. This can be done with one of two approaches. We can start at the far end from the terminals and alternately combine impedances in series and in parallel until the input terminals are reached. Thus the impedance of R_2 in series with C is $R_2 + 1/j\omega C$. Its admittance is the reciprocal of this expression. This branch is in parallel with R_1; hence the admittance of the combination is the sum of the admittances. The combined branch is in series with Z_1; hence the impedance of this branch plus Z_1 is the desired impedance.

An alternative approach is to start at the left-hand terminals. The input impedance is Z_1 plus the impedance of the network to the right of Z_1. This impedance can be expressed as the reciprocal of the admittance. Thus

$$Z = Z_1 + \frac{1}{\text{admittance to right of } Z_1}$$

The network to the right consists of Y_2 in parallel with the rest of the network to the right of Y_2. Hence the last expression now becomes

$$Z = Z_1 + \cfrac{1}{Y_2 + \cfrac{1}{\text{impedance to right of } Y_2}}$$

The impedance to the right of Y_2 is simply $Z_3 + Z_4$ or $Z_3 + 1/Y_4$. Hence, finally,

$$Z = Z_1 + \cfrac{1}{Y_2 + \cfrac{1}{Z_3 + \cfrac{1}{Y_4}}} \tag{4-70}$$

The right-hand side of this expression is in a form which is called a *continued fraction.* The expression can be put in a more usual form by clearing the fractions. If we do this, the result will be

$$Z = Z_1 + \cfrac{1}{Y_2 + \cfrac{Y_4}{1 + Z_3Y_4}} = Z_1 + \frac{1 + Z_3Y_4}{Y_2(1 + Z_3Y_4) + Y_4}$$

$$= \frac{(1 + Z_1Y_2)(1 + Z_3Y_4) + Z_1Y_4}{Y_2 + Y_4 + Y_2Z_3Y_4} \tag{4-71}$$

This expression agrees with Eq. (4-68) in which the input current was calculated by another method.

It is not hard to appreciate that, had there been additional series and shunt branches in the ladder, these could have been easily accomodated in the continued fraction by continuing a little longer. Thus, for

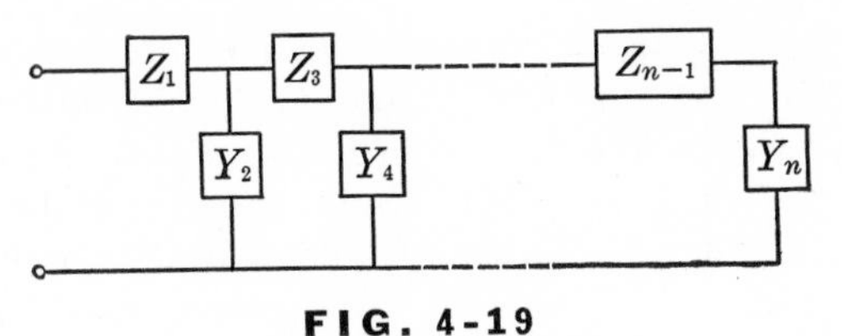

FIG. 4-19
General ladder network

the general ladder shown in Fig. 4-19, the input impedance is

$$Z = Z_1 + \cfrac{1}{Y_2 + \cfrac{1}{Z_3 + \cfrac{1}{Y_4 + \cdots + \cfrac{1}{Z_{n-1} + \cfrac{1}{Y_n}}}}} \tag{4-72}$$

In this way the impedance of a ladder network can be written down at a glance. After the continued fraction is obtained, however, there will be a considerable amount of labor in clearing it.

The methods of solution we have been describing in the last few sections are adequate for solving a large number of problems. In the case of resistive networks, we discussed more general methods of solution—loop equations and node equations—which could be used for arbitrary network structures. These methods are also applicable for *RLC* networks in the sinusoidal steady state. We shall, however, defer consideration of the details of these methods to a later chapter.

4.9 POWER AND ENERGY

In Chap. 1 we introduced the subject of electrical power and energy. There we found what influence the network elements have in dissipating or storing energy for any arbitrary waveshape of current or voltage.

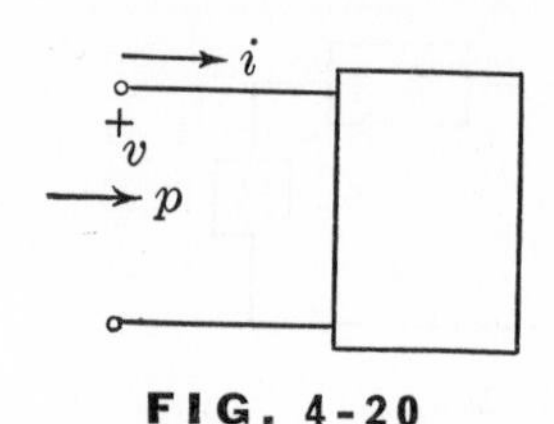

FIG. 4-20

Network for defining power

Let us now devote some time to a consideration of power and energy in the sinusoidal steady state. It would be worthwhile rereading Sec. 1-5 before proceeding.

Consider the network shown in Fig. 4-20. The voltage and current at the terminals are sinusoidal and are given by

$$v = \sqrt{2}|V| \cos (\omega t + \alpha); \quad V = |V|\epsilon^{j\alpha} \qquad (a)$$
$$i = \sqrt{2}|I| \cos (\omega t + \beta); \quad I = |I|\epsilon^{j\beta} \qquad (b) \qquad (4\text{-}73)$$

(The appropriate sinors have also been shown.) With the references for v and i shown in the figure, the reference for power is toward the right, as shown. The power at any instant of time is given by

$$\begin{aligned} p = vi &= 2|V|\,|I| \cos (\omega t + \alpha) \cos (\omega t + \beta) \\ &= |V|\,|I| \cos (\alpha - \beta) + |V|\,|I| \cos (2\omega t + \alpha + \beta) \end{aligned} \qquad (4\text{-}74)$$

The last form is obtained with the use of the identity $2 \cos x \cos y = \cos (x + y) + \cos (x - y)$. Whereas both the voltage and the current are sinusoidal, the power contains a constant term (independent of time) in addition to a sinusoidal term. Furthermore, the frequency of the sinusoidal term is twice that of the voltage or current. Plots of v, i, and p are shown in Fig. 4-21 for some specific values of α and β. The power

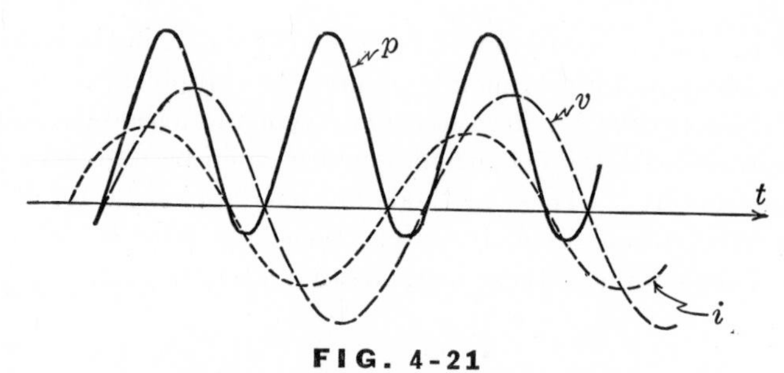

FIG. 4-21

Voltage, current, and power sinusoids

is sometimes positive and sometimes negative. This means that power does not always flow to the right in Fig. 4-20.

The energy which is transmitted into the network over an interval of time is found by integrating the power over this interval. If the area under the positive part of the power curve in Fig. 4-21 were the same as the area under the negative part, the net energy transmitted over one cycle would be zero. For the values of α and β chosen when plotting Fig. 4-21, however, the positive area is greater, so that there is a net transmission of energy toward the right. The energy flows back from the network to the source over part of the cycle. On the average, however, more energy flows toward the network than away from it. The amount of energy which is swapped back and forth between the network and the source is dependent on the relative values of the angles α and β.

Let us look at this from another point of view. Equation (4-74) shows the power to consist of a constant term and a sinusoid. The average value of a sinusoid is zero, so this term will contribute nothing to the net energy which is transmitted. Only the constant term will contribute. The constant term is the average value of the power. This can be seen either from Fig. 4-21 or by integrating Eq. (4-74) over a cycle. Denoting the average power by capital P and letting $\theta = \alpha - \beta$, we have

$$P = |V|\,|I| \cos \theta \tag{4-75}$$

Average power is measured in watts. This is a very important result and will be quite useful in subsequent work. Sometimes, however, other forms of this expression are more convenient. Hence, let us determine some alternative ways of expressing the average power.

With the voltage and current sinors given in Eq. (4-73), the impedance of the network in Fig. 4-20 will be

$$Z = \frac{V}{I} = \frac{|V|\epsilon^{j\alpha}}{|I|\epsilon^{j\beta}} = \frac{|V|}{|I|}\epsilon^{j(\alpha-\beta)} = |Z|\epsilon^{j\theta} \tag{4-76}$$

Thus the phase difference between the voltage and the current is the angle of the impedance. It is the cosine of this angle which appears in the expression for average power.

Remembering that the cosine can be expressed as the real part of an exponential, we can rewrite Eq. (4-75) as

$$\begin{aligned} P &= |V|\,|I| Re(\epsilon^{j\theta}) = Re[|V|\,|I|\epsilon^{j(\alpha-\beta)}] \\ &= Re[(|V|\epsilon^{j\alpha})(|I|\epsilon^{-j\beta})] \end{aligned} \tag{4-77}$$

In the last step the exponential in the previous step was broken up by the law of exponents and written as shown. We can now identify $|V|\epsilon^{j\alpha}$ as the voltage sinor and $|I|\epsilon^{-j\beta}$ as the conjugate of the current sinor. Thus Eq. (4-77) becomes

$$P = Re(VI^*) \tag{4-78}$$

In this form the average power is written in terms of the voltage and cur-

$V = |V|\epsilon^{\alpha j}$ $I = |I|\epsilon^{\beta j}$

$\theta = \alpha - \beta$

$W = P + Qj$

$P = |V||I| \cos\theta$ $Q = |V||I| \sin\theta$

rent sinors. This expression is sometimes more convenient, especially for theoretical developments.

We can express P in still other forms by substituting for either V or I^* its equivalent in terms of impedance or admittance. As a first step, let us set $V = ZI$. Then Eq. (4-78) becomes

$$P = Re(ZII^*) = |I|^2 Re(Z) = |I|^2 R \tag{4-79}$$

The final forms are obtained by noting that a complex number multiplied by its conjugate is equal to the square of the magnitude. On the right-hand side we have substituted R for the real part of Z. Remember that this is not necessarily the resistance of a single resistor. In this form the expression for the average power looks like the one for power in the case of a constant current.

An alternative expression for P can be obtained by substituting an equivalent form for I^* in Eq. (4-78). Since $I = YV$, I^* will by Y^*V^*. Thus Eq. (4-78) can be written as

$$P = Re(VV^*Y^*) = |V|^2 Re(Y^*) = |V|^2 Re(Y) = |V|^2 G \tag{4-80}$$

We have here used the fact that the real part of a complex number is the same as the real part of its conjugate. In the last form, G has been used for the real part of Y. This expression is quite similar to Eq. (4-79) except that voltage replaces current, and admittance replaces impedance.

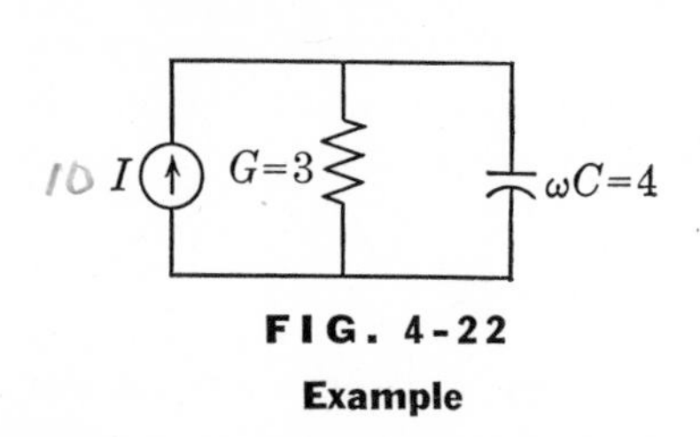

FIG. 4-22
Example

We have now obtained several different alternative expressions for the average power, all of which are equivalent. In any given situation one of them may be more convenient to apply than the others. As a simple illustrative example, consider the network shown in Fig. 4-22. A sinusoidal current having an effective value of 10 amp feeds the parallel combination of a resistor and a capacitor. The average power delivered by the source will be

$$P = Re(VI^*) = Re\left(\frac{I}{Y}I^*\right) = Re\frac{|I|^2}{G + j\omega C}$$

$$= |I|^2 Re\frac{G - j\omega C}{G^2 + (\omega C)^2} = \frac{G}{G^2 + (\omega C)^2}|I|^2$$

$$= \frac{3}{25} \times 100 = 12 \text{ watts}$$

Alternatively, we can use $|V|\,|I| \cos\theta$ to calculate the power, which means that we should first compute $|V|$ and $\cos\theta$. The impedance of the network is

$$Z = \frac{1}{Y} = \frac{1}{G + j\omega C} = \frac{1}{3 + j4} = \frac{3}{25} - j\frac{4}{25} = \frac{1}{5}\epsilon^{-j53.2°}$$

Hence

$$|V| = |ZI| = |2\epsilon^{-j53.2°}| = 2 \quad *$$

$$\cos\theta = \cos(-53.2°) = \frac{3}{5}$$

The average power is, then,

$$P = |V|\,|I| \cos\theta = 2 \times 10 \times \frac{3}{5} = 12 \text{ watts}$$

Let us return for a moment to Eq. (4-78). The average power is expressed as the real part of a complex quantity VI^*. This complex quantity also has an imaginary part. Let us designate VI^* as W and write it in rectangular form as follows:

$$\begin{aligned} W &= VI^* = |V|\,|I|\epsilon^{j\theta} = |V|\,|I| \cos\theta + j|V|\,|I| \sin\theta \\ &= P + jQ \end{aligned} \tag{4-81}$$

where

$$\begin{aligned} P &= |V|\,|I| \cos\theta & (a) \\ Q &= |V|\,|I| \sin\theta & (b) \end{aligned} \tag{4-82}$$

We already know P to be the average power. Since it is the real part of some complex quantity, it is also sometimes called the *real power*. The complex quantity W of which P is the real part is called the *complex power*. To be consistent, then, we should call Q the *imaginary power*. This, however, is not often done; Q is called the *reactive power*, and its unit is called a *var* (volt-amperes reactive). The magnitude of the complex power W is sometimes also given a name; it is called the *apparent power*, and its unit is called *volt-ampere* (va). (You are urged not to throw up your hands upon being introduced to all these names at once. Familiarity will eventually come and you will be able to associate the names with the "faces".)

It is possible to give a graphical interpretation of the quantities we have just named. Figure 4-23 showns a sinor diagram of V and I. The sinor voltage V can be resolved into two components, one parallel to sinor I (or in phase with I) and another perpendicular to I (or in quadrature). This is illustrated in Fig. 4-23(b). The average power P, then, is the magnitude of sinor I multiplied by the in-phase component of V; the reactive power Q is the magnitude of I multiplied by the quadrature component of V.

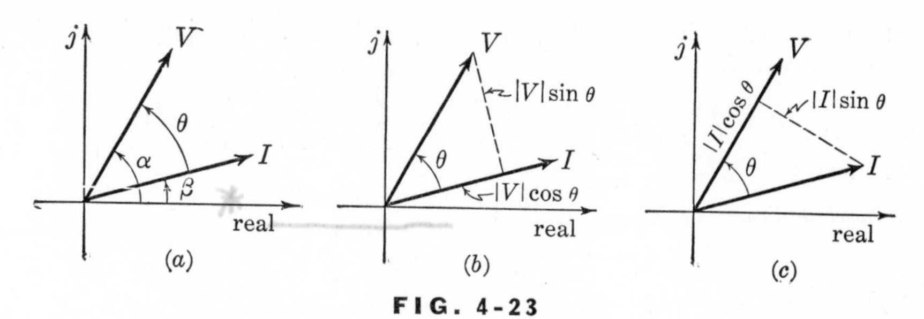

FIG. 4-23

In-phase and quadrature components of V and I

Alternatively, we can think of sinor I as being resolved into two components, one in phase with V and one in quadrature with it, as illustrated in Fig. 4-23(c). Then P is the product of the magnitude of V with the in-phase component of I, and Q is the product of the magnitude of V with the quadrature component of I. Real power is produced only by the in-phase components of V and I. The quadrature components contribute only to the reactive power.

Just as we found alternative expressions for P in Eqs. (4-79) and (4-80), it is also possible to find similar expressions for Q. Thus

$$Q = Im(VI^*) = Im(ZII^*) = |I|^2 Im(Z) = |I|^2 X \qquad (a)$$
$$Q = Im(VI^*) = Im(VV^*Y^*) = |V|^2 Im(Y^*) = -|V|^2 B \qquad (b) \qquad (4\text{-}83)$$

In the last expression we used the fact that the imaginary part of a complex number is the negative of the imaginary part of its conjugate.

Since the real and imaginary parts of the complex power are related to the real and imaginary parts of the impedance and admittance, we should be able to express W in terms of Z or Y. Thus, if we combine Eqs. (4-79) and [4-83(a)] on the one hand, and Eqs. (4-80) and [4-83(b)] on the other, we get

$$W = |I|^2 R + j|I|^2 X = |I|^2 Z \qquad (a)$$
$$W = |V|^2 G - j|V|^2 B = |V|^2 Y^* \qquad (b) \qquad (4\text{-}84)$$

Thus, if the excitation is a voltage (constant amplitude), the complex power is simply proportional to the conjugate of the admittance. Similarly, if the excitation is a current (again constant amplitude), the complex power is directly proportional to the impedance.

For any given network it is useful to know what part of the total complex power is real power and what part is reactive power. This is usually expressed in terms of the power factor F_p which is defined as the ratio of P to the magnitude of W. Thus

$$\text{Power factor} \equiv F_p = \frac{P}{|W|} \tag{4-85}$$

This is actually a general relationship, although we have developed it in terms of sinusoidal excitations. In the present case, with P given in Eq. (4-75) and W in Eq. (4-81), we see that the power factor is simply $\cos\theta$.

Note that since the cosine is an even function [that is, $\cos(-\theta) = \cos\theta$], specifying the power factor does not indicate the sign of the angle θ. Remember that θ is the angle of the impedance. If θ is positive, this means that the current lags the voltage, so we say that the power factor is a *lagging power factor*. On the other hand, if θ is negative, the current leads the voltage, and we say that the *power factor is leading.*

The power factor will reach its maximum value, unity, when the voltage and current are in phase. This will happen in a purely resistive network, of course. It can also occur in *RLC* networks for specific values of the elements and the frequency. For example, in the series *RLC* network of Fig. 4-9(*b*), we mentioned this to be the case when $X_L = X_c$. When the inductive reactance predominates, the power factor is lagging; and when the capacitive reactance predominates, the power factor is leading.

We can now obtain a physical interpretation for the reactive power. When the power factor is unity, the voltage and current are in phase, and $\sin\theta = 0$. Hence, the reactive power is zero. In this case the instantaneous power is always positive; the power curve in Fig. 4-21 never goes negative, and there is no exchange of energy between the source and the network. At the other extreme when the power factor is zero, the voltage and current are 90 deg out of phase, and $\sin\theta = 1$. Now the reactive power is a maximum. In this case the instantaneous power is positive over half a cycle and negative over the other half. All the energy which is delivered by the source over half the cycle is returned to the source by the network over the other half. The exchange of energy between the source and the network is complete.

It is clear, then, that the reactive power is a measure of the energy which is exchanged between the source and the network without being used by the network. Although none of this energy is retained by the network and it is returned unused to the source, nevertheless it is made available to the network by the source.†

† Power companies charge their large customers not only for the average power they use but also for the reactive power which the company must supply.

Let us now turn our attention to the individual network elements and determine their power and energy relationships. If the network in Fig. 4-20 is a single resistor with a resistance R, the complex power in Eq. (4-81) becomes

$$W = VI^* = RII^* = R|I|^2 \tag{4-86}$$

Thus there is no reactive power in this case. As a matter of fact, the instantaneous power, being $p = vi = Ri^2$, is always positive and never

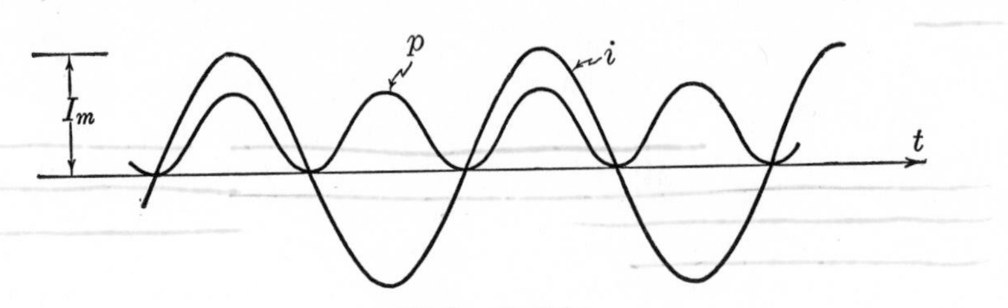

FIG. 4-24

Power in resistor

flows away from the resistor. Figure 4-24 shows the instantaneous power for a sinusoidal current in a resistor.

Let us next turn to an inductor. In this case the complex power will be

$$W = VI^* = j\omega L|I|^2 \tag{4-87}$$

That is, the average power in the inductor is zero, but there is reactive power; $Q = \omega L|I|^2$. In this case the reactive power is positive.

In terms of instantaneous values, the power flowing into an inductor is

$$p_L = Li\frac{di}{dt} = L\sqrt{2}|I|\cos\omega t(-\omega\sqrt{2}|I|\sin\omega t) = -\omega L|I|^2\sin 2\omega t \tag{4-88}$$

where the current has been taken as $i = \sqrt{2}|I|\cos\omega t$. The right side follows from the trigonometric identity $2\sin x\cos x = \sin 2x$. Thus, as time goes on, the power is alternately positive and negative. No net power flows into the inductor, in agreement with the previous result.

Let us now inquire into the energy stored in the inductor. Again taking $i = \sqrt{2}|I|\cos\omega t$, we have

$$w_L = \frac{1}{2}Li^2 = L|I|^2\cos^2\omega t = \frac{1}{2}L|I|^2(1 + \cos 2\omega t) \tag{4-89}$$

The right-hand side follows from the trigonometric identity $2\cos^2 x = 1 + \cos 2x$. Since the cosine varies between -1 and $+1$, the right-

watts vars

W = P + Qj

average power reactive power

hand side will vary between zero and $L|I|^2$. Since $\cos 2\omega t$ has an average value of zero, the average value of the stored energy, which we shall label W_L, will be

$$W_L = \frac{1}{2} L|I|^2 \tag{4-90}$$

Plots of the current, power, and stored energy are given in Fig. 4-25. The power and energy curves have twice the frequency of the current.

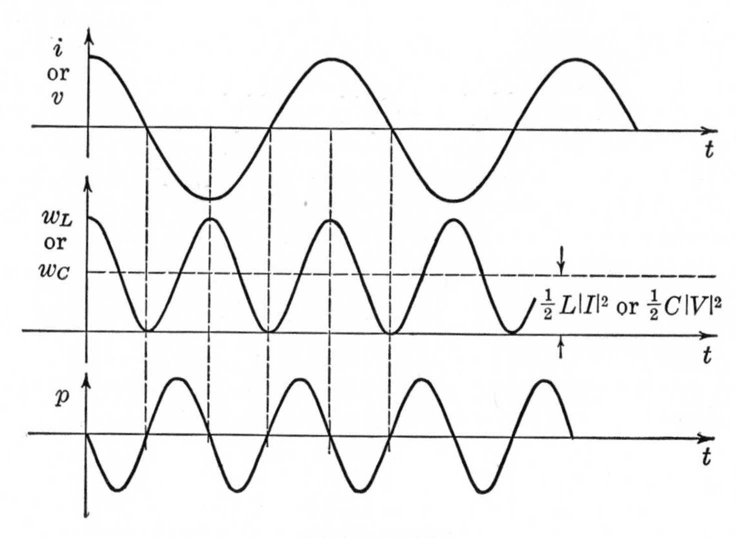

FIG. 4-25

Power and stored energy in inductor or capacitor

It is clear that, as the power flows in and out of the inductor, the stored energy rises to a peak value, then falls to zero. Although the power is negative over half the cycle, the stored energy is never negative. (A negative stored energy would mean that energy was actually being supplied by the inductor, which is not possible.)

Finally, let us consider a capacitor. In this case the complex power is

$$W = VI^* = \frac{1}{j\omega C} II^* = -j \frac{1}{\omega C} |I|^2 = -j\omega C|V|^2 \tag{4-91}$$

In this case also, there is no average power. Furthermore, the reactive power is negative.

In terms of instantaneous values, if we choose the voltage as $v = \sqrt{2}|V| \cos \omega t$, the power flowing into the capacitor will be

inductor

complex power = $W = \omega L|I|^2$

average energy stored = $W_L = \frac{1}{2}L|I|^2$

$$p_C = Cv\frac{dv}{dt} = C\sqrt{2}|V| \cos \omega t(-\omega\sqrt{2}|V| \sin \omega t) = -\omega C|V|^2 \sin 2\omega t \tag{4-92}$$

This expression is strikingly similar to that for the inductor power given in Eq. (4-88). It is only necessary to replace L by C and $|I|$ by $|V|$.

The energy stored in the capacitor will be

$$w_C = \frac{1}{2}Cv^2 = C|V|^2 \cos^2 \omega t = \frac{1}{2}C|V|^2(1 + \cos 2\omega t) \tag{4-93}$$

This equation, in turn, is identical with Eq. (4-89) except for a change in symbols. Thus Fig. 4-25 can be used to represent the power and stored energy of a capacitor as well, if the curve labeled current is now labeled voltage. The entire discussion regarding the flow of power in and out and the storage of energy need not be repeated; it applies in the present case as well (with appropriate changes of words). In this case, the average stored energy, which we will label capital W_C, will be

$$W_C = \frac{1}{2}C|V|^2 \tag{4-94}$$

In concluding this section let us give an example illustrating some of the relationships we have described. At the same time, the example will provide some additional ideas which we can then generalize.

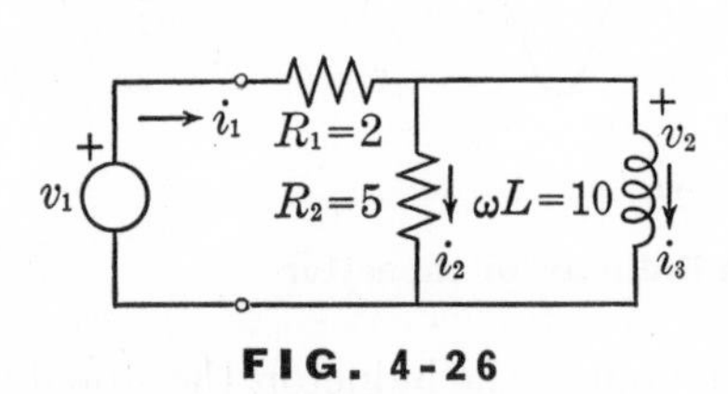

FIG. 4-26
Illustrative example

Consider the network shown in Fig. 4-26. A sinusoidal voltage source v_1 having an effective value of 10 volts supplies the network. It is required to find the average power and the reactive power supplied by the source. One procedure would be to compute the current supplied by the source, thus determining the power factor. Equation (4-82) now determines P and Q.

A more convenient procedure is to use Eq. [4-84(*b*)]. Since $|V|$ is already given, it remains only to find the admittance of the network. This can be written as

$$Y = \frac{1}{Z} = \frac{1}{R_1 + \dfrac{1}{\dfrac{1}{R_2} + \dfrac{1}{j\omega L}}} = \frac{1}{2 + \dfrac{1}{\dfrac{1}{5} - j\dfrac{1}{10}}} = \frac{1}{2 + \dfrac{10}{2 - j1}} = \frac{2 - j1}{14 - j2}$$

$$\doteq \frac{2.24\epsilon^{-j26.5°}}{14.2\epsilon^{-j8.1°}} = 0.158\epsilon^{-j18.4°}$$

Hence

$$W = |V|^2 Y^* = 100(0.158\epsilon^{j18.4°}) = 15 + j5$$
$$P = 15 \text{ watts}$$
$$Q = 5 \text{ vars}$$

These are the required values.

The average power supplied by the source is dissipated in the resistors. An alternative way of computing the average power is to find the power dissipated in each resistor and add. This requires finding the branch current in each resistor. In the present case

$$I_1 = \frac{V_1}{Z}$$

$$I_2 = \frac{j\omega L}{R_2 + j\omega L} I_1$$

where Z is the impedance at the terminals. The second expression results from the current-divider relationship. Hence

$$P = R_1|I_1|^2 + R_2|I_2|^2 = \frac{R_1|V_1|^2}{|Z|^2} + R_2 \frac{|V_1|^2}{|Z|^2} \frac{(\omega L)^2}{R_2^2 + (\omega L)^2}$$

$$= |V_1|^2|Y|^2 \left(R_1 + \frac{R_2(\omega L)^2}{R_2^2 + (\omega L)^2} \right) = 100(0.158)^2 \left(2 + \frac{500}{125} \right) = 15 \text{ watts}$$

The result, of course, agrees with the previous answer.

This is an illustration of a general result. The average power which is dissipated in a network can be obtained by adding the power dissipated in each resistor of the network. Usually, this would be a more complicated calculation than the alternative procedures.

A similar result is true of the reactive power. One method of finding it is to add the reactive power in each reactive element in the network. In the present example there is only a single reactive element. The reactive power in the inductor is

$$Q = \omega L|I_3|^2 = \omega L \left| \frac{R_2}{R_2 + j\omega L} \right|^2 |I_1|^2 = \frac{\omega L R_2^2 |V_1|^2 |Y|^2}{R_2^2 + (\omega L)^2} = 5 \text{ vars}$$

which agrees with the previous result.

In the preceding example the impedance of the network is not purely real; it is complex. That is, the power factor is less than unity. Suppose we place a capacitor across the terminals of the network, as shown in Fig. 4-27. The total reactive power will be the reactive power of the original network plus the reactive power of the capacitor. Since the latter is negative, there will be a cancellation of reactive power, the extent of which will depend on the reactance of the capacitor. Let us inquire what value of ωC will cause the power factor to become unity. The total reactive power will be

$$Q_{\text{total}} = 5 - \omega C|V|^2$$

For unity power factor, this should vanish. Hence

$$\omega C = \frac{5}{|V|^2} = \frac{5}{100}$$

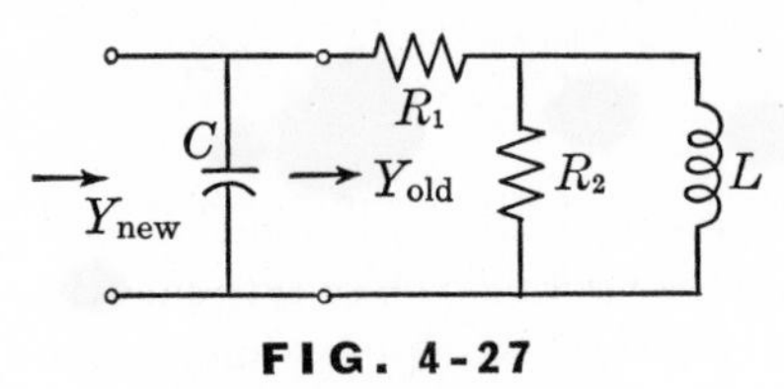

FIG. 4-27

Power factor correction

This procedure of utilizing a capacitor (which often takes the form of a so-called synchronous condenser in a practical situation) to balance out inductive reactive power is called *power-factor correction.*

An alternative point of view is to think in terms of impedance or admittance rather than reactive power. The admittance of the modified network is equal to the admittance of the original network plus that of the capacitor. Thus

$$Y_{\text{new}} = Y_{\text{old}} + j\omega C = 0.158\epsilon^{-j18.4°} + j\omega C = 0.05 - j0.05 + j\omega C$$

In order to make the new admittance purely real, we require that $\omega C = 0.05$, as before.

4.10 SUMMARY

In this chapter we started, but did not complete, the study of the sinusoidal steady-state response of electrical networks. We showed that it is sufficient to find the forced response to the exponential function $\epsilon^{j\omega t}$, since a sinusoid can be expressed in terms of this exponential. The voltage and current variables are represented by sinors, which are transforms of the time functions, and the branches of the network by impedances or admittances. Time is completely eliminated from the formulation and solution of problems. The sinors contain all the unknown information about the responses, so that often it may not be necessary to convert the solution back to the time domain.

We introduced the concept of a network function, defined as the ratio of a response sinor to an excitation sinor. When the excitation and response are at the same terminals, the network function is an impedance or admittance; otherwise, a transfer function. Knowledge of

for unity power factor

$Q = 0$

power factor $= \frac{P}{|W|}$

the network function will completely determine the response once the excitation is given.

We discussed some simple analytic techniques which apply for certain specific network configurations. A large number of network problems can be solved by these methods. We introduced the concept of a sinor diagram as a graphical aid in the solution of problems.

Finally, we concluded with a consideration of energy and power in the sinusoidal steady state. We discussed average power, reactive power, complex power, and power factor.

Problems

4-1 Write sinors for the following sinusoids: (*a*) $v = 32 \cos (\omega t - 30°)$; (*b*) $v = 18 \sin \omega t$; (*c*) $i = 20 \cos (\omega t + 52°)$; (*d*) $i = 100 \sin (\omega t + 70°)$.

4-2 Write the sinusoidal function of time corresponding to each of the following sinors, both as cosine functions at an angle and as sine functions at an angle: (*a*) $I = 10\epsilon^{-j20°}$; (*b*) $I = 3 + j4$; (*c*) $V = 20\epsilon^{j140°}$; (*d*) $V = -2 + j5$.

4-3 Laboratory measurements on a coil of wire in series with a resistor, as shown in Fig. P4-3, give the following values: $|V_1| = 50$, $|V_2| = 60$, and $|V| = 90$ (all volts). The frequency of the sinusoid used in the measurements is 60 cycles per second. Assume that the coil of wire can be represented by a series connection of R and L. Find the numerical values of R and L.

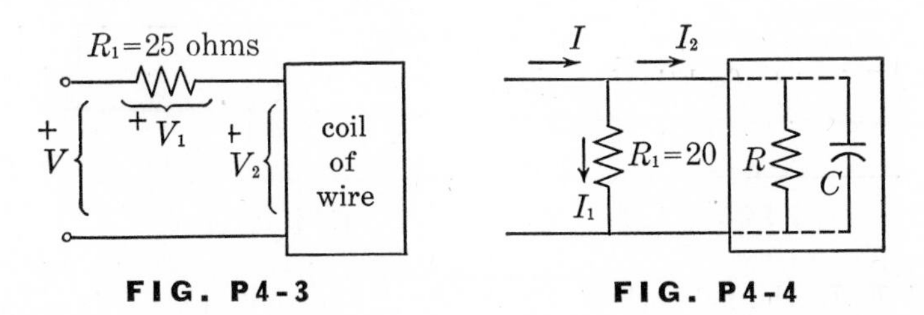

FIG. P4-3 **FIG. P4-4**

4-4 The physical device inside the box shown in Fig. P4-4 is to be represented by a model consisting of the parallel combination of a resistor and a capacitor. Laboratory measurements using a 400-cycle-per-second sinusoidal source give the following data: $|I_1| = 100$, $|I_2| = 200$, $|I| = 250$ (all milliamperes). Find the numerical values of R and C.

4-5 In the network of Fig. P4-5, find the current sinor as R takes on the values $R = 10$, 20, and 50 ohms. Do this both algebraically and graphically by means of a sinor diagram. Tabulate the variation of the magnitude and angle with R. The source voltage is $v_g = 2{,}200\sqrt{2}\cos 400t$.

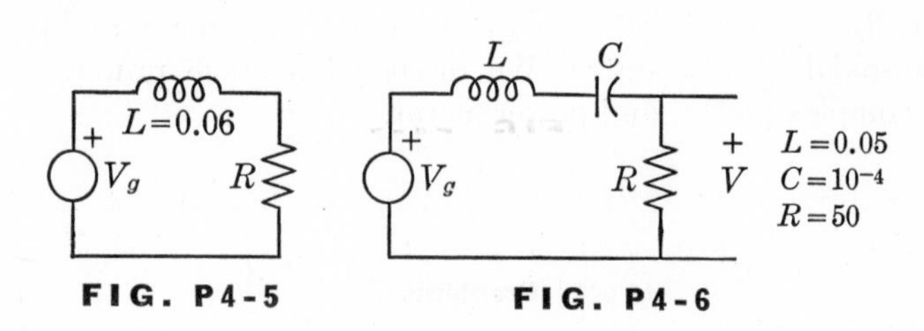

FIG. P4-5 FIG. P4-6

4-6 In the network of Fig. P4-6, find the magnitude and angle of the voltage across the resistor as frequency varies. Take $\omega = 300$, 400, 500. Use the magnitude of the source voltage as a unit of measure, and measure angles relative to that of V_g.

4-7 In the network of Fig. P4-7, the sinusoidal voltage v_1 has an effective value of 10 volts. (*a*) Find voltage v_2. (*b*) Find current i. (*c*) Compute the average power and reactive power at the terminals in terms of the input impedance or admittance. (*d*) Compute the average power by finding the power dissipated by each resistor and adding. (*e*) Compute the reactive power by finding the reactive power of the inductor.

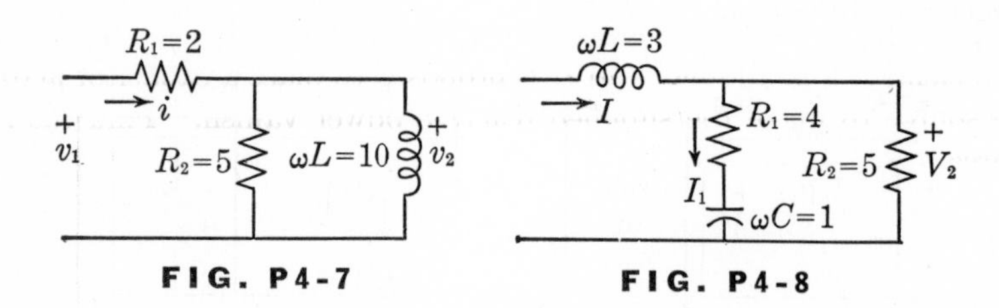

FIG. P4-7 FIG. P4-8

4-8 In the network of Fig. P4-8, the current sinor I_1 is given as $I_1 = j10$. (*a*) Find the input current sinor I. (*b*) Find the voltage sinor V_2. (*c*) Find the input impedance. (*d*) Find the power input to the network by calculations at the input, and by calculating the power dissipated in each resistor.

4-9 In the networks of Fig. P4-9, find the output voltage sinor V_2.

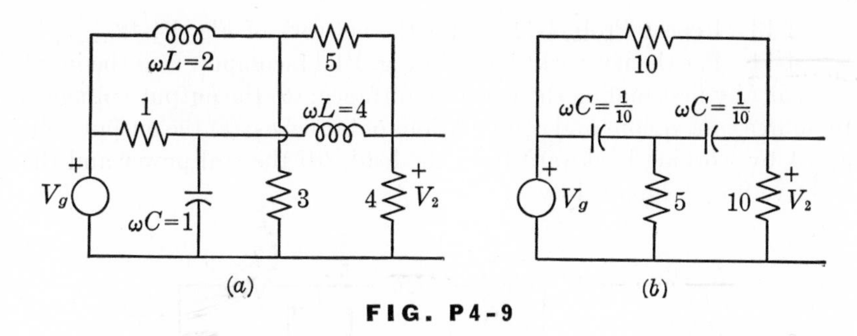

FIG. P4-9

4-10 In Fig. P4-10 find the steady-state current in the output branch (R_2): (*a*) by source conversions and combinations of branches until a simple network structure is obtained; (*b*) by applying the ladder-network technique. (*c*) With the procedure of part (*b*), find the current in the source. (*d*) Find the input admittance of the network looking toward the right from the source terminals and again determine the source current.

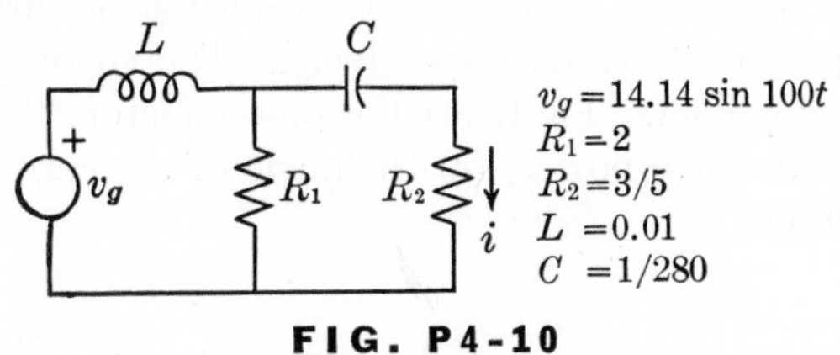

FIG. P4-10

4-11 (*a*) In the previous problem find the real power and the reactive power supplied by the source. Do this once by calculations at the terminals of the source; repeat by calculating the pertinent power for each element and adding. (*b*) It is proposed to place a capacitor across the source to make the supplied reactive power vanish. Find the required value of C.

4-12 Repeat Prob. 4-10 for the network of Fig. P4-12. (The output branch is now L.)

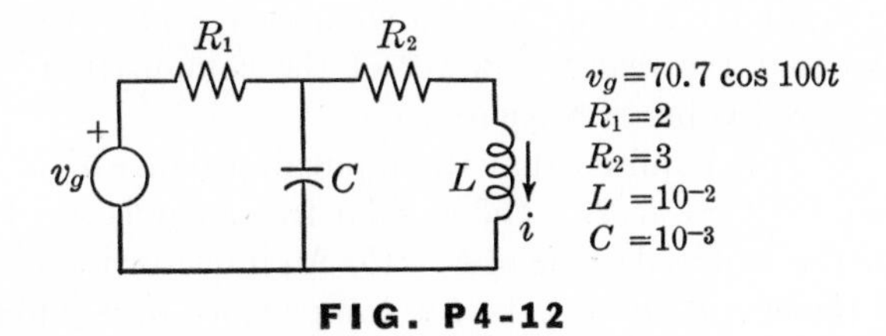

FIG. P4-12

4-13 Repeat Prob. 4-11 (*a*) for the network of Fig. P4-12.

4-14 For the network shown in Fig. P4-14, compute: (*a*) the input current i by first finding the input admittance; (*b*) the output voltage v_2 by source conversions and combinations of branches; (*c*) the output voltage v_2 by working backward from the load; (*d*) the real power and the reactive power supplied by v_1.

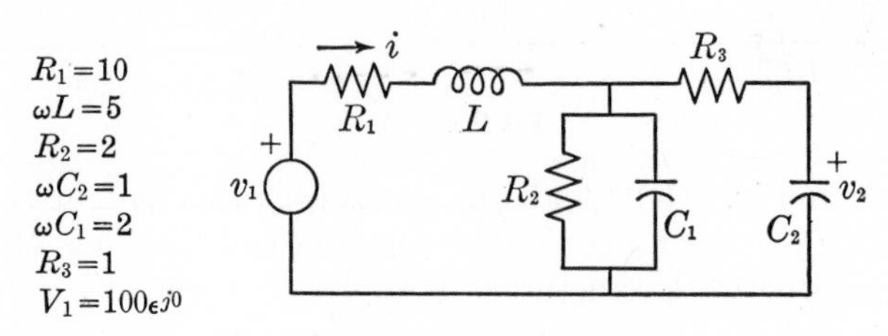

FIG. P4-14

4-15 Figure P4-15 shows a network branch which is receiving average power $P = 500$ watts. The effective values of the current and voltage are 6 amp and 120 volts, respectively. The angular frequency is 400 radians per second. Find: (*a*) the power factor; (*b*) the complex power and the reactive power; (*c*) the angle of the current relative to that of the voltage.

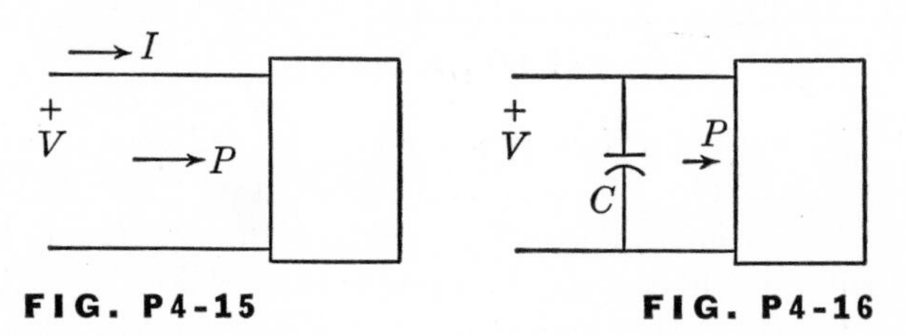

FIG. P4-15 **FIG. P4-16**

4-16 The network in the rectangle in Fig. P4-16 is receiving an average power $P = 2{,}000$ watts. The effective value of the voltage is 250 volts, the power factor is 0.8, and the angular frequency is $\omega = 400$ radians per second. It is desired to connect a capacitor C across the terminals so that the reactive power of the combination is 500 vars. Find the required value of capacitance.

4-17 (*a*) It is required that the input impedance Z in Fig. P4-17 be such that $Re(Z) = R/2$. The angular frequency is ω. Find an expression for the required value of C. (*b*) An inductor is to be placed in series with the circuit so that the total impedance Z_1 is purely real and equal to $R/2$. ($Z_1 = R/2$). Find the required value of L.

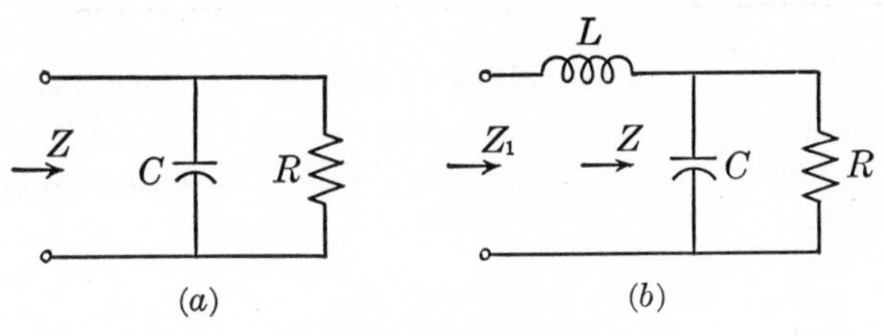

FIG. P4-17

4-18 In the network of Fig. P4-18, the average power dissipated in R_3 is 25 kw. The sources are adjusted so that $I_1 = j2I_2$. Assume that I_3 is real. (*a*) Find I_1 and I_2. (*b*) Find the voltage and the average power supplied by each source.

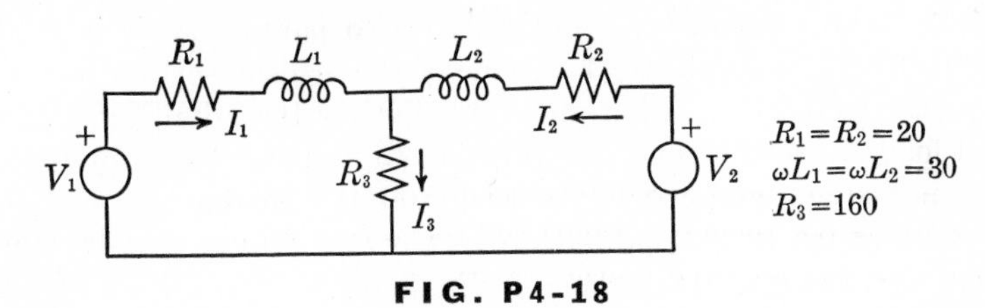

FIG. P4-18

4-19 When $i(t) = 30\sqrt{2}\cos(100t + 20°)$ in Fig. P4-19, the complex power is found to be $W = 2{,}700 + j4{,}500$. (*a*) Find the input impedance Z. (*b*) Find the voltage $v(t)$. (*c*) Find the value of the capacitance or inductance (as the case may be) which must be placed in series with Z in order to make the reactive power vanish. How is the real power affected?

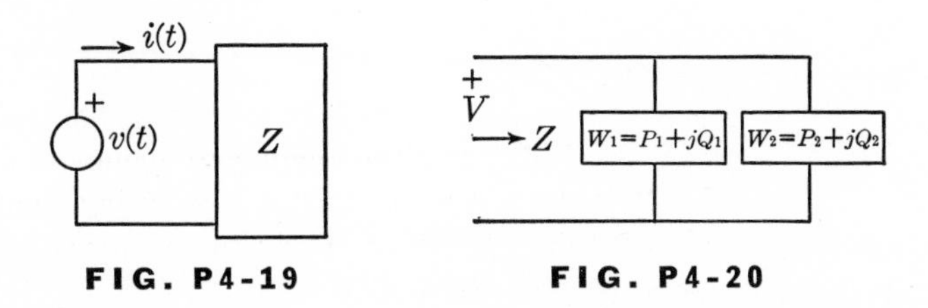

FIG. P4-19 FIG. P4-20

4-20 In Fig. P4-20 the complex power taken by each of the two parallel branches is known at the given voltage. The complex power is specified either in terms of the real and reactive powers or in terms of the real power and the power factor. Find the impedance of the parallel combination in terms of the known quantities.

4-21 The power in the two parallel branches in Fig. P4-21 is specified as follows: branch 1, 1,000 watts at a leading power factor of 0.6; branch 2, 1,500 watts at a lagging power factor of 0.8. The effective value of the voltage across the parallel branches is 100 volts. Find the source voltage V_g and the impedance of the parallel combination.

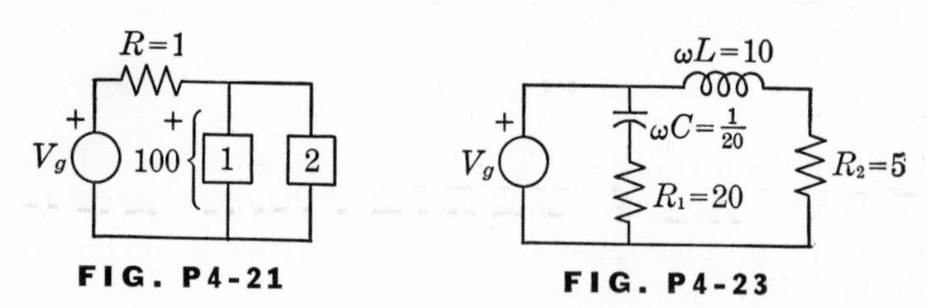

FIG. P4-21

FIG. P4-23

4-22 The total power absorbed by two parallel loads is P at a power factor F_p and voltage $|V|$. One of the loads absorbs a power P_1 at a power factor F_{p1}. Find an expression for the admittance of the second load. Use the general expression to find the result for the following numerical cases: (*a*) $P = 10$ kw at 0.8 lagging power factor, $P_1 = 8$ kw at 0.6 lagging power factor; (*b*) $P = 20$ kw at unity power factor, $P_1 = 20$ kw at 0.8 lagging power factor.

4-23 In Fig. *P*4-23 the power dissipated in R_2 is 500 watts. Find the voltage V_g and the complex power supplied by the source.

FREQUENCY VARIATION AND RESONANCE

5

IN our study of the sinusoidal steady state, we found that the response sinor can always be expressed in terms of the excitation sinor multiplied by a network function—an impedance or a transfer function. These functions depend on the values of the network elements and on the frequency.

In Chap. 4 we assumed that the network elements and the frequency are fixed. It is possible, however, for either a network element or the frequency to vary over some range. For example, the signal involved in a telephone conversation varies in the audio range from about 100 to 3,000 cycles per second.

In this chapter we shall study how the network functions of some simple networks vary with frequency. We shall introduce some concepts which can later be applied to more elaborate networks. Since the functions are complex, both the magnitudes and the angles will vary with frequency.

5.1 FREQUENCY RESPONSE

To start the discussion, let us consider the dependence on frequency of the impedances or admittances of the individual network elements.

In the case of the resistor, the impedance is independent of frequency.

For the inductor we have

$$Z_L = j\omega L = \omega L \epsilon^{j90°} \qquad (a)$$

$$Y_L = \frac{1}{j\omega L} = -j\left(\frac{1}{\omega L}\right) = \frac{1}{\omega L}\epsilon^{-j90°} \qquad (b) \qquad (5\text{-}1)$$

$$X_L = \omega L; \quad B_L = -\frac{1}{\omega L} \qquad (c)$$

The impedance and admittance are purely imaginary. The angle of the impedance is constant at 90 deg for all frequencies. (The angle of the admittance is the negative of this.) The inductive reactance $X_L = \omega L$ and the susceptance $B_L = -1/\omega L$ are the only things that vary with

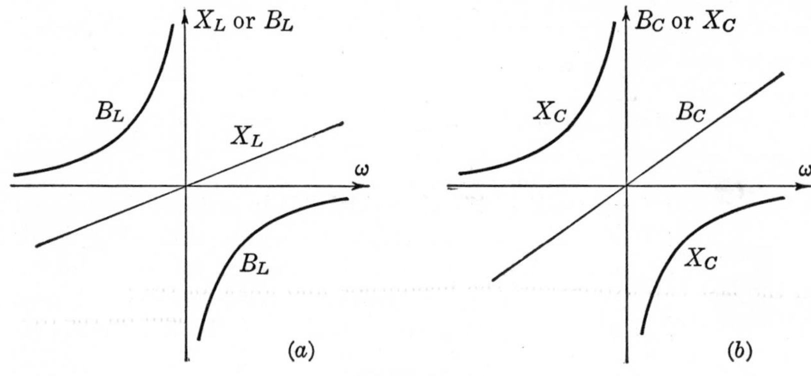

FIG. 5-1

Variation of inductor and capacitor reactance and susceptance with frequency

frequency. These are shown plotted in Fig. 5-1(*a*) as a function of ω. We see that X_L is a straight line with a slope L, while B_L is a hyperbola.

Similarly, for a capacitor the corresponding relationships are

$$Z_C = \frac{1}{j\omega C} = -j\left(\frac{1}{\omega C}\right) = \frac{1}{\omega C}\epsilon^{-j90°} \qquad (a)$$

$$Y_C = j\omega C = \omega C \epsilon^{j90°} \qquad (b) \qquad (5\text{-}2)$$

$$X_C = -\frac{1}{\omega C}; \quad B_C = \omega C \qquad (c)$$

Here the behavior of Z_C is similar to that of Y_L and the behavior of Y_C is similar to that of Z_L. The reactance and susceptance are shown plotted against ω in Fig. 5-1(*b*). As a matter of fact, the same curves can be used for X_L and B_C, and for B_L and X_C. This is a manifestation of the dual nature of inductance and capacitance.

Consider next the network shown in Fig. 5-2. A sinusoidal voltage whose frequency can be varied is applied at the input terminals. The amplitude and phase of this voltage are assumed to stay fixed as the frequency varies. That is, sinor V_1 is a constant complex number.

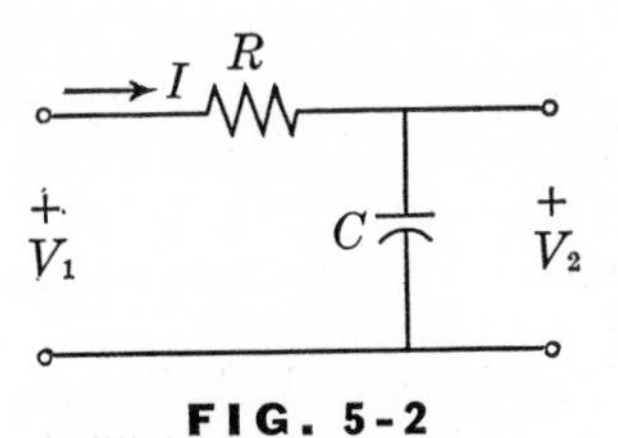

FIG. 5-2

Simple frequency-selective network

Suppose that the desired response is the voltage across the capacitor. It is sufficient to deal with the voltage transfer ratio $G_{21} = V_2/V_1$. By the voltage-divider relationship, we find the voltage ratio to be

$$G_{21} = \frac{V_2}{V_1} = |G_{21}|\epsilon^{j\phi} = \frac{1/j\omega C}{R + 1/j\omega C} = \frac{1}{RC}\frac{1}{1/RC + j\omega} \tag{5-3}$$

$$|G_{21}| = \frac{1}{RC}\frac{1}{\sqrt{\omega^2 + (1/RC)^2}} \quad (a)$$

$$\phi = -\tan^{-1}\omega CR \quad (b) \tag{5-4}$$

In the last two expressions the magnitude and angle of G_{21} have been explicitly shown. It is seen that both of these are dependent on the frequency. As the frequency varies, the magnitude and angle both vary.

Sketches of the magnitude of G_{21} and its angle against ω are shown in Fig. 5-3. Either from the figures or from Eq. (5-4), it is seen that the

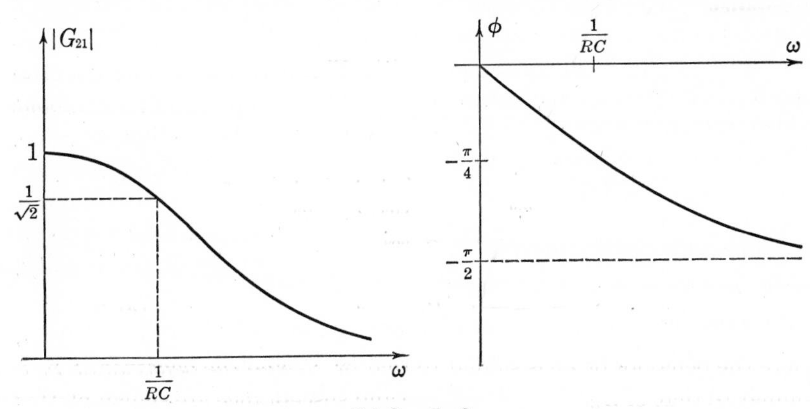

FIG. 5-3

Variation of magnitude and angle functions with frequency

frequency - response curves

magnitude starts with a value of unity at zero frequency and gradually falls as the frequency increases. For very high frequencies it approaches -90 deg as the frequency approaches infinity.

These variations with frequency are also evident from a glance at the network. At low frequencies the reactance of the capacitor has a large magnitude. Hence most of the input voltage appears across the capacitor. As the frequency increases, $1/\omega C$ decreases relative to R, so that more and more of the input voltage appears across the resistor, and less and less across the capacitor.

Prob

Plots of the response magnitude and angle (per unit excitation) against ω are called the *frequency-response curves.* Individually, they are called the *amplitude response* and the *phase response,* respectively.

If we look upon Fig. 5-2 as a signal-transmission network, we can see that sinusoidal signals having a relatively low frequency are transmitted with not much change in amplitude and angle. High-frequency sinusoids, on the other hand, have their amplitudes appreciably reduced and their angles reduced by almost 90 deg. The network is, thus, *frequency selective.* It passes certain sinusoids relatively unchanged, and stops others from passing to a large extent. Since the signals which are passed are low-frequency sinusoids, we call the network a *low-pass* network. (This particular network constitutes a very simple low-pass filter.) Sinusoidal signals which are "passed" are said to lie in the *pass band,* and those which are "stopped" are said to be in the *stop band.*

It is clear from Fig. 5-3 that there is no particular value of ω to which we can point and say that sinusoids having a lower frequency will be passed while those having a higher frequency will be stopped. The variation of the magnitude of G_{21} with ω is smooth and continuous. Hence, in order to give some quantitative measure of the pass band, we arbitrarily pick a value of ω and say that it forms the edge of the pass band. It is often convenient to choose for this purpose the value of ω at which the square of the magnitude is one-half its maximum value. Thus, we define the bandwidth β of the network as the *angular frequency interval* over which the square of the magnitude of the response sinor is one-half its maximum value, per unit excitation.

From Eq. [5-4(a)] we note that the maximum value of $|G_{21}|$ is unity, and it occurs at $\omega = 0$. To find the value of ω at which the square of the magnitude of G_{21} is 1/2 (or at which the magnitude itself is $1/\sqrt{2}$), we square Eq. [5-4(a)] and set it equal to 1/2. The result gives $\omega = 1/RC$. The pass band runs from $\omega = 0$ to $\omega = 1/RC$. Hence, the bandwidth is $\beta = 1/RC$ radians per second. Note from Eq. [5-4(b)]

that when $\omega = 1/RC$, the angle of G_{21} is -45 deg. The real and imaginary parts of G_{21} are equal at this frequency.

Keep in mind, however, that the definition of the bandwidth which we have given is not unique. At times we may wish to define the limits of the pass band differently. Hence, when referring to "the bandwidth," it is necessary to specify exactly what is meant.

Another point should be mentioned here. The bandwidth we have defined is in terms of the angular frequency ω. Very often we may wish to use the actual frequency instead. This will simply introduce a factor of $1/2\pi$. Thus a bandwidth of 6,283 radians per second corresponds to $6{,}283/2\pi = 1{,}000$ cycles per second.

5.2 SERIES TUNED RLC NETWORK

Let us now turn to a consideration of the series *RLC* network shown in Fig. 5-4, and discuss the variation of the response with frequency. A sinusoidal voltage whose frequency can be varied is applied at the terminals. The amplitude and phase of the voltage are assumed to stay unchanged as the frequency is varied. The sinor V is thus a constant complex number.

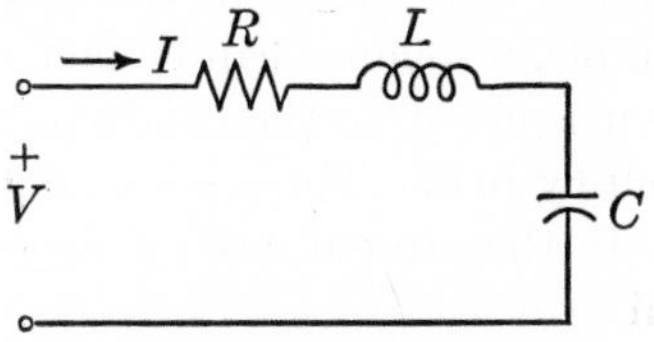

FIG. 5-4

Series tuned circuit

The desired response is the current. But since V is constant, it is sufficient to deal with the admittance $Y = I/V$, which is

$$Y = \frac{I}{V} = \frac{1}{R + j\left(\omega L - \dfrac{1}{\omega C}\right)} \tag{5-5}$$

Let us temporarily consider the impedance Z rather than the admittance, as a matter of convenience. It is

$$Z = R + j\left(\omega L - \frac{1}{\omega C}\right) \tag{5-6}$$

The real part is constant and does not vary with frequency. The imaginary part is dependent on frequency, and Fig. 5-5 gives a plot of the reactance X against ω. Shown on the same figure are plots of $X_L = \omega L$ and $X_C = -1/\omega C$. At low frequencies, the capacitive reactance predominates, and X is negative. At high frequencies the inductive reactance predominates, and X is positive. Since the two reactances vary in opposite ways with ω, there will be one frequency at which the two are equal (and of opposite sign) so that they cancel, making $X = 0$. At this frequency the impedance, and hence also the admittance, is purely real. The voltage and current at the terminals will be in phase. We say that a condition of *resonance* exists; the network is *in resonance.* at ω_0

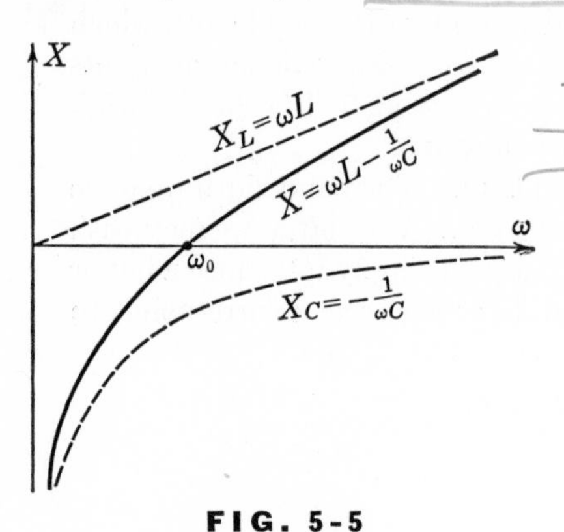

FIG. 5-5

Reactance curve

More precisely, we shall say that *a network is in resonance at a particular frequency if the terminal voltage and current are in phase at that frequency. The frequency in question is called the resonant frequency.* Alternative statements of the condition of resonance can be made. For example, resonance can be defined as the condition for which the power factor is unity, or the condition for which the impedance or admittance is purely real. These statements are all equivalent.

In the present case, if we label the resonant angular frequency ω_0, then

$$\omega_0^2 = \frac{1}{LC} \tag{5-7}$$

Let us now return to the admittance in Eq. (5-5) and consider its magnitude. It will be given by

$$|Y| = \frac{1}{\sqrt{R^2 + (\omega L - 1/\omega C)^2}} \tag{5-8}$$

This is clearly a function of frequency. It is also clear that the magnitude of the admittance will be a maximum when $\omega L = 1/\omega C$, since the denominator will then have its smallest possible value. But this is precisely the point at which resonance occurs. At the resonant frequency $|Y| = 1/R$. If we use the notation Y_m to designate the maximum value of the magnitude of Y, then Eqs. (5-5) and (5-8) will become

$$\frac{Y}{Y_m} = \frac{1}{1 + j\frac{1}{R}(\omega L - 1/\omega C)} \qquad (a)$$

$$\frac{|Y|}{Y_m} = \frac{1}{\sqrt{1 + \frac{1}{R^2}(\omega L - 1/\omega C)^2}} \qquad (b) \tag{5-9}$$

These expressions can be put in alternative forms by performing some algebraic manipulations. In the first place, using $\omega_0^2 = 1/LC$, let us substitute for C in Eq. [5-9(a)]. The result will be

$$\frac{Y}{Y_m} = \frac{1}{1 + j\frac{1}{R}\left(\omega L - \frac{\omega_0^2 L}{\omega}\right)} = \frac{1}{1 + j\frac{\omega_0 L}{R}\left(\frac{\omega}{\omega_0} - \frac{\omega_0}{\omega}\right)}$$

$$\frac{Y}{Y_m} = \frac{1}{1 + jQ_0\left(\frac{\omega}{\omega_0} - \frac{\omega_0}{\omega}\right)} \tag{5-10}$$

In the second step $\omega_0 L$ is factored from each term in the parentheses. In the last step a quantity Q_0 is introduced which is defined as

$$Q_0 \equiv \frac{\omega_0 L}{R} = \frac{1}{\omega_0 CR} = \frac{1}{R}\sqrt{\frac{L}{C}} \tag{5-11}$$

The alternative forms for Q_0 are obtained by using $\omega_0^2 = 1/LC$. Shortly, we shall have much more to say about this quantity. For the present, let us think of it as just a shorthand way of writing the combination of things $\omega_0 L/R$. Note that Q_0 is the ratio of the reactance of the inductor or the capacitor at resonance to the resistance.

Let us comment on what these algebraic manipulations have accomplished. In Eq. (5-5) the admittance is written in terms of the network elements R, L, and C. We wish to look upon the entire series circuit as a single unit, however, with emphasis on the terminal behavior. For this purpose the network parameters individually are not the best things to use for describing the operation of the network as frequency is varied. What we have done in Eq. (5-10) is to use a different set of three parameters to describe the network instead of R, L, and C; namely Y_m, ω_0, and Q_0.

The magnitude of the relative admittance can now be written in terms of the newly defined parameters as

$$\frac{|Y|}{Y_m} = \frac{1}{\sqrt{1 + Q_0^2\left(\frac{\omega}{\omega_0} - \frac{\omega_0}{\omega}\right)^2}} \tag{5-12}$$

Let us consider plotting this function against ω. First of all, note that, since we are dividing $|Y|$ by its maximum value, the curve will always have a value of unity at its highest point. Second, if, instead of ω, we take ω/ω_0 as the variable, then the shape of the curve will be dependent on only one parameter, Q_0. A sketch of Eq. (5-12) is shown in Fig. 5-6. When Q_0 is low, the curve is relatively broad and has a relatively flat top. When Q_0 is high, the curve is sharper and narrower. In Fig. 5-6(a) both curves have the same maximum value, since we are sketching the relative admittance magnitude. If the actual magnitude is sketched, rather than the relative value, as shown in Fig. 5-6(b), perhaps a better picture is obtained of the effect of Q_0. For the same inductive reactance, a low Q_0 means a large resistance, and vice versa. In the limiting situation of zero resistance, Q_0 will approach infinity, and the curve will become very sharp and narrow indeed.

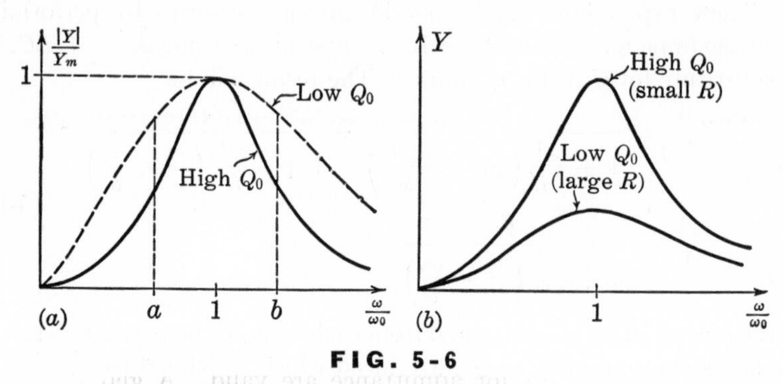

FIG. 5-6
Variation of admittance magnitude with ω; (a) relative, (b) actual

The most interesting part of the curve is in the immediate vicinity of the resonant frequency. In this vicinity the curve is controlled by the second term in the denominator of Eq. (5-12). This term involves the difference between two terms which are both approximately the same. Not much numerical accuracy can be obtained when subtracting two quantities which are nearly equal. Hence, near resonance, it would be more useful to use as a variable the change in frequency from the resonant value, rather than frequency itself. Let us, therefore, define the *fractional frequency deviation*

$$\delta = \frac{\omega - \omega_0}{\omega_0} = \frac{\omega}{\omega_0} - 1 \tag{5-13}$$

This quantity is very close to zero when ω is near the resonant value.

When ω is greater than ω_0, δ is positive; when ω is less than ω_0, δ is negative. In terms of δ, we can now write

$$\frac{\omega}{\omega_0} - \frac{\omega_0}{\omega} = \delta + 1 - \frac{1}{\delta + 1} = \delta\left(\frac{2 + \delta}{1 + \delta}\right) \qquad (5\text{-}14)$$

When ω is near ω_0, δ is small. In this case the right side of the last equation can be written approximately as

$$\frac{\omega}{\omega_0} - \frac{\omega_0}{\omega} \doteq 2\delta \qquad (5\text{-}15)$$

With this approximation, the expressions for the admittance and its magnitude become

$$\frac{Y}{Y_m} = \frac{1}{1 + j2Q_0\delta} \qquad (a)$$

$$\frac{|Y|}{Y_m} = \frac{1}{\sqrt{1 + (2Q_0\delta)^2}} \qquad (b) \qquad (5\text{-}16)$$

At this point let us inquire as to the conditions under which the approximate expressions for admittance are valid. A glance at Fig. 5-6(a) shows that the variation in the curve for a given fractional change in frequency from the resonant value is much greater for the case of high Q_0 than it is for low Q_0. If we let δ vary from zero to some value, say to the point labeled a or b in Fig. 5-6(a), most of the variation in the high Q_0 curve will be covered, but very little variation of the low Q_0 curve will be covered. Saying this in another way, to cover most of the variation of the curve from its maximum value will require a relatively small excursion from the resonant frequency for the case of high Q_0, but it will require a relatively large excursion from the resonant frequency for the case of low Q_0.

Thus, in order to cover the important part of the curve, a larger value of δ will be required for the case of low Q_0 than for high Q_0. Since the error made in the approximation in Eq. (5-15) gets progressively worse as δ increases, we see that the approximation is best for the case of high Q_0. For this reason the approximation is called the *high-Q approximation.*

For Q_0 of the order of 20 or greater, the approximate expression for the admittance is quite adequate for almost all values of δ. For much lower values of Q_0, the error is tolerable only for smaller values of δ.

Let us now plot the relative magnitude of $|Y|$ as given by Eq. [5-16(b)]. If we use $Q_0\delta$ as the abscissa, the resulting curve will be applicable to all tuned circuits having a high value of Q_0. This curve is shown in Fig. 5-7. It is called the *universal resonance curve.* A scale for ω/ω_0 is also shown, but this is not a linear scale. It should be repeated that the universal resonance curve applies to highly resonant circuits only.

The series tuned circuit is another example of a frequency-selective circuit. The magnitude of the current-response sinor (which is propor-

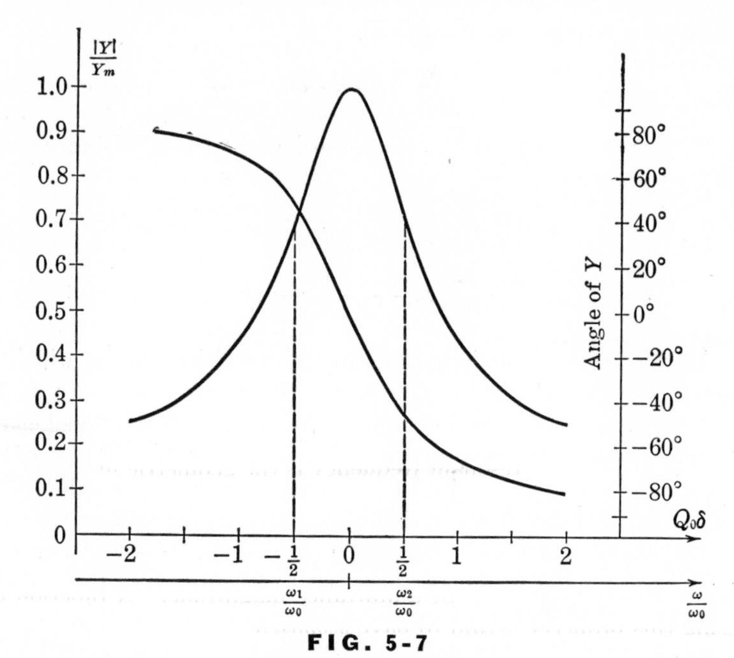

FIG. 5-7

Universal resonance curve

tional to Y) is relatively large over a frequency interval in the vicinity of the resonant frequency. In the present case both low-frequency sinusoids and high-frequency sinusoids are "stopped." Frequencies which are "passed" are in the vicinity of ω_0. We call such a network a *band-pass* network. Again, a measure of the selectivity is obtained by defining the bandwidth. As before, we shall define the bandwidth as the interval of angular frequency over which the magnitude squared of the response sinor is no less than one half its maximum value.

In order to find the frequency values at the edges of the band, we should square $|Y|/Y_m$ and set it equal to 1/2. For both the exact expression in Eq. (5-12) and the high-Q approximation in Eq. [5-16(b)], this means that the quantity under the radical should be 2. Let us first deal with the exact expression. Then we shall have

$$Q_0^2\left(\frac{\omega}{\omega_0}-\frac{\omega_0}{\omega}\right)^2=1$$
$$\frac{\omega}{\omega_0}-\frac{\omega_0}{\omega}=\pm\frac{1}{Q_0} \tag{5-17}$$

The last expression can be rewritten as a quadratic in ω. From it we get two values of frequency—the lower band edge ω_1 and the upper band edge ω_2. They will be given by

$$\omega_1=\omega_0\sqrt{1+(1/2Q_0)^2}-\frac{\omega_0}{2Q_0} \qquad (a)$$
$$\omega_2=\omega_0\sqrt{1+(1/2Q_0)^2}+\frac{\omega_0}{2Q_0} \qquad (b) \tag{5-18}$$

The bandwidth β will then be

$$\beta=\omega_2-\omega_1=\frac{\omega_0}{Q_0} \tag{5-19}$$

This is a very useful expression relating the resonant frequency, the bandwidth, and Q_0.

Suppose we now form the product of ω_1 and ω_2 as given in Eq. (5-18). We get

$$\omega_1\omega_2=\omega_0^2 \tag{5-20}$$

Thus we see that the resonant frequency is the geometric mean between the two band edges. As a matter of fact, by glancing back at Eq. (5-12), we see that if ω is replaced by the value ω_0^2/ω, the value of $|Y|$ is not changed. Thus any two frequencies whose geometric mean is ω_0 lead to the same value of the admittance magnitude. A function possessing this property is said to have *geometric symmetry* around ω_0.

Let us now repeat the preceding discussion in terms of the high-Q approximation. The band edges are found by setting the quantity under the radical in Eq. [5-16(*b*)]equal to 2. Then

$$(2Q_0\delta)^2=1$$
$$\delta_{1,2}=\pm\frac{1}{2Q_0} \tag{5-21}$$

On the scale of $Q_0\delta$ in Fig. 5-7, these two points are the ones labeled $+\frac{1}{2}$ and $-\frac{1}{2}$. The corresponding values of ω are formed by inserting these values of δin Eq. (5-13). The results are

$$\omega_2=\omega_0(1+\delta_2)=\omega_0+\frac{\omega_0}{2Q_0} \qquad (a)$$
$$\omega_1=\omega_0(1+\delta_1)=\omega_0-\frac{\omega_0}{2Q_0} \qquad (b) \tag{5-22}$$

These are to be compared with Eq. (5-18). The two expressions are identical if $(1/2Q_0)^2$ under the radical in Eq. (5-18) is neglected relative to 1.

If we now compute the bandwidth, we again get Eq. (5-19). Thus this equation relating β, ω_0, and Q_0 holds for the high-Q approximation also.

The two band edges ω_1 and ω_2 are centered arithmetically about ω_0, as shown in Eq. (5-22). Thus the universal resonance curve shown in Fig. 5-7 has *arithmetic symmetry* about ω_0. If Q_0 is large, the exact curve having geometric frequency very closely approximates the approximate curve having arithmetical symmetry.

Very often, it is more useful to express the bandwidth as a fraction of the resonant frequency. It is not very informative, for example, to state simply that the bandwidth of a network is 1,000 radians per second. This could represent a wide band network if the resonant frequency happened to be 5,000 radians per second; on the other hand, it would be a narrow band network if the resonant frequency were 10^8 radians per second. From Eq. (5-19) we find the *fractional bandwidth* to be

$$\frac{\beta}{\omega_0} = \frac{1}{Q_0} \tag{5-23}$$

Thus the value of Q_0 is a clear indication of the fractional bandwidth.

Figure 5-7 also shows a plot of the angle of Y, which is $-\tan^{-1} 2Q_0\delta$. This curve is also symmetrical about the point $\delta = 0$. The angle goes from $+90$ deg at zero frequency to -90 deg at infinite frequency. At the lower band edge it is $+45$ deg, whereas at the upper band edge it is -45 deg.

Up to this point we have considered the response to be the current. Let us now give a brief consideration to the voltages across each of the elements. The resistor voltage needs no comment, since it is proportional to the current. Choosing appropriate references, the capacitor voltage and inductor voltage sinors will be

$$V_C = -j\frac{1}{\omega C} I \qquad (a)$$

$$V_L = j\omega L I = j\frac{\omega}{\omega_0^2 C} I = j\left(\frac{\omega}{\omega_0}\right)^2 \frac{I}{\omega C} = -\left(\frac{\omega}{\omega_0}\right)^2 V_C \qquad (b) \tag{5-24}$$

In the case of the inductor voltage, we substituted for L using relationship $LC = 1/\omega_0^2$. We see that the two voltage sinors differ by an angle of 180 deg and a factor (ω/ω_0^2). At the resonant frequency this factor is 1, so that the two voltage sinors have the same magnitude at resonance.

To carry the analysis further, let us substitute for I its value from Eq. (5-10), remembering that $I = YV$ and that $Y_m = 1/R$. Thus the capacitor voltage becomes

$$V_C = -j \frac{V/\omega CR}{1 + jQ_0\left(\frac{\omega}{\omega_0} - \frac{\omega_0}{\omega}\right)} \tag{5-25}$$

The factor in the numerator can be further rearranged as follows:

$$\frac{1}{\omega CR} = \frac{1}{\omega_0 CR}\frac{\omega_0}{\omega} = Q_0\frac{\omega_0}{\omega} \tag{5-26}$$

Hence the capacitor and inductor voltages now become

$$V_C = \frac{(-jQ_0\omega_0/\omega)V}{1 + jQ_0\left(\frac{\omega}{\omega_0} - \frac{\omega_0}{\omega}\right)} \qquad (a)$$

$$V_L = \frac{(jQ_0\omega/\omega_0)V}{1 + jQ_0\left(\frac{\omega}{\omega_0} - \frac{\omega_0}{\omega}\right)} \qquad (b) \tag{5-27}$$

Finally, at the resonant frequency $\omega = \omega_0$, and these become

$$V_C = -jQ_0V \qquad (a)$$

$$\text{at resonance} \tag{5-28}$$

$$V_L = jQ_0V \qquad (b)$$

We observe that, at the resonant frequency, the capacitor and inductor voltage magnitudes are larger than the input voltage magnitude by a factor Q_0. For circuits with a high value of Q_0, these voltages will be extremely high indeed. The capacitor voltage lags the input by 90 deg, and the inductor voltage leads the input by 90 deg.

5.3 POWER AND ENERGY IN THE TUNED CIRCUIT

Let us now turn to a study of the energy storage and dissipation in a series RLC network. This will give us some additional insight into the operation of the network and into the phenomenon of resonance. The network is redrawn in Fig. 5-8.

Let us initially look at the variables as functions of time and write the loop equation as a differential equation. The result will be

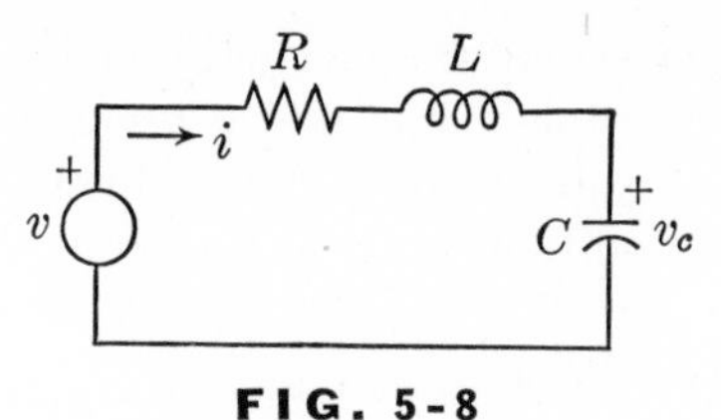

FIG. 5-8

Series RLC network

$$v = Ri + L\frac{di}{dt} + v_C \quad (5\text{-}29)$$

The power input to the network at any instant of time is the product of v and i. Let us then multiply both sides of the equation by i. The result will be

$$iv = Ri^2 + Li\frac{di}{dt} + v_C i \quad (5\text{-}30)$$

In the last term on the right, we can replace i by Cdv_C/dt. Then

$$iv = Ri^2 + Li\frac{di}{dt} + Cv_C\frac{dv_C}{dt}$$

$$= Ri^2 + \frac{d}{dt}\left(\frac{1}{2}Li^2 + \frac{1}{2}Cv_C^2\right) = Ri^2 + \frac{d}{dt}(w_L + w_C) \quad (5\text{-}31)$$

The last line is obtained by noting that $2ydy/dx = d(y^2)/dx$. w_L and w_C are the instantaneous stored energies in L and C, respectively.

This equation makes a very important statement; $w_L + w_C$ is the total stored energy in the network. The left-hand side is the power entering the network. Thus the equation expresses the fact that the power entering the network is equal to the power dissipated plus the rate of change of the stored energy. If we were to integrate this equation over an interval of time, the result would be a statement of the conservation of energy.

The preceding discussion is valid no matter what the waveforms of the variables. Let us now specify that they are sinusoidal. It is convenient to choose the capacitor voltage as the point of departure. Hence let us write

$$v_C = \sqrt{2}|V_C| \cos \omega t \quad (a)$$

$$i = C\frac{dv_C}{dt} = -\sqrt{2}\omega C|V_C| \sin \omega t \quad (b) \qquad (5\text{-}32)$$

With these expressions we can find the energy stored in the capacitor and the inductor at any instant of time. They are

$$w_C = \frac{1}{2}Cv_C^2 = C|V_C|^2 \cos^2 \omega t \quad (a)$$

$$w_L = \frac{1}{2}Li^2 = L\omega^2C^2|V_C|^2 \sin^2 \omega t = \left(\frac{\omega}{\omega_0}\right)^2 C|V_C|^2 \sin^2 \omega t \quad (b) \qquad (5\text{-}33)$$

The total energy stored in the circuit at any instant of time is the sum of these two expressions. Thus

$$w_C + w_L = C|V_C|^2 \left(\cos^2 \omega t + \left(\frac{\omega}{\omega_0}\right)^2 \sin^2 \omega t\right) \tag{5-34}$$

Observe that $\cos^2 \omega t$ and $\sin^2 \omega t$ have the same maximum value (one). Hence Eq. (5-33) shows that the maximum value of w_L is $(\omega/\omega_0)^2$ times the maximum value of w_C. That is, when the frequency is below the resonant frequency, the maximum value of the stored energy in the inductor is less than that in the capacitor, whereas for frequencies higher than the resonant frequency, these maxima are reversed. Exactly at resonance, the energy stored in the inductor has the same maximum value as the energy stored in the capacitor. As a matter of fact, from Eq. (5-34) we find the total energy stored at resonance to be

$$w_C + w_L = C|V_C|^2 = CQ_0{}^2|V|^2 \qquad \text{at resonance} \tag{5-35}$$

To get the right-hand side, we used Eq. (5-28).

This expression states that the sum of the stored energies is a constant for all time. The energy surges between the inductor and the capacitor, and none is supplied from the source (except to supply the power lost in the resistor). This agrees with the fact that, at resonance, the admittance is real, so that the voltage and current at the terminals are in phase; there is no reactive power at the terminals.

Further information can be obtained on this point by repeating the analysis, this time in terms of sinors. The complex power entering the network can be found by first writing the loop equation in terms of sinors and then multiplying both sides by I^*. Thus

$$V = RI + j\omega LI - j\frac{1}{\omega C}I$$

$$W = VI^* = R|I|^2 + j\left(\omega L|I|^2 - \frac{1}{\omega C}|I|^2\right) \tag{5-36}$$

$$= R|I|^2 + j\omega(L|I|^2 - C|V_C|^2)$$

The last step is obtained by noting that $|I| = \omega C|V_C|$. Finally, recalling the expressions in Eqs. (4-90) and (4-94) for the average energy stored in an inductor (W_L) and a capacitor (W_C), the complex power becomes

$$W = P + j2\omega(W_L - W_C) \tag{5-37}$$

From this expression we see that reactive power is proportional to the difference between the average energy stored in the inductor and the average energy stored in the capacitor. Note that the average stored energies are related by

$$W_L = \left(\frac{\omega}{\omega_0}\right)^2 W_C \tag{5-38}$$

This can be verified by finding the average values from Eq. (5-33) or by recalling the expressions for average stored energy from Sec. 4-9. Inserting this expression into Eq. (5-37), the complex power becomes

$$W = R|I|^2 + j2\omega W_C\left[\left(\frac{\omega}{\omega_0}\right)^2 - 1\right] \tag{5-39}$$

When the frequency is above resonance, the reactive power is positive; below resonance, it is negative. Exactly at resonance the reactive power vanishes, as we observed before.

Glance back at Eq. (5-37) for the complex power. This expression was derived for the series RLC network under study here. Let us state parenthetically, however, that this is a general relationship valid for all networks, if R is interpreted as the real part of the impedance, W_L as the average energy stored in all the inductances, and W_C as the average energy stored in all the capacitances.

Now let us look at the average power dissipated in the resistor, which is $P = R|I|^2$. Using the value of $|I|$ from Eq. [5-32(*b*)], the power becomes

$$P = R\omega^2C^2|V_C|^2 = \frac{\omega_0}{Q_0}\left(\frac{\omega}{\omega_0}\right)^2 C|V_C|^2 \tag{5-40}$$

The right-hand side is obtained by using $RC = 1/Q_0\omega_0$. Note from Eq. [5-33(*b*)] that $C|V_C|^2(\omega/\omega_0)^2$ is the maximum value of the energy stored in the inductor. Hence, using this fact, Eq. (5-40) becomes

$$P = \frac{\omega_0}{Q_0} w_L(\max) \tag{5-41}$$

We have now arrived at an expression which involves Q_0 and some fundamental quantities. Remember that, up to this point, Q_0 has simply been a symbol used to replace a combination of other symbols, such as $\omega_0 L/R$. If we now solve Eq. (5-41) for Q_0, we get

$$Q_0 = \omega_0 \frac{\text{maximum energy stored in inductor}}{\text{average power dissipated}} \tag{5-42}$$

This expression can be written alternatively as

$$Q_0 = 2\pi \frac{\text{maximum energy stored}}{\text{energy dissipated in one cycle}} \tag{5-43}$$

since ω_0 is 2π divided by the period, and the energy dissipated in one cycle is the average power times the period.

These expressions for Q_0 which we have derived are quite general and apply to any network, not only to the series-tuned one. They give a measure of the quality of a network in storing energy versus its ability to dissipate it. We could have adopted Eq. (5-43) as the definition of a quality factor Q (not to be confused with reactive power), of which Q_0 is the value at resonance. We could then have derived the expressions for Q_0 with which we started in Eq. (5-11). We shall henceforth assume that this has been done, and we shall make the following fundamental definition of the quality factor Q of a network

$$\begin{aligned} Q &= \omega \frac{\text{maximum energy stored in network}}{\text{average power dissipated}} \\ &= 2\pi \frac{\text{maximum energy stored in network}}{\text{energy dissipated per cycle}} \end{aligned} \tag{5-44}$$

From these expressions it is clear that Q is a function of frequency. At the resonant frequency Q takes on the value Q_0.†

Let us again look at the dissipated power from another point of view. In $P = R|I|^2$, let us substitute $I = YV$. Then

$$P = R|V|^2|Y|^2 \tag{5-45}$$

At resonance the dissipated power is a maximum (since the current is a maximum) which we can designate P_m. Also, at resonance $|Y| = Y_m$. Hence

$$P_m = R|V|^2Y_m^2 \tag{5-46}$$

The ratio of these two values of power will be

$$\frac{P}{P_m} = \frac{|Y|^2}{Y_m^2} \tag{5-47}$$

The edges of the pass band were defined to make the right side of this expression equal to 1/2. Thus we see that, at the band edges, the power supplied by the source is one half its maximum value, which occurs at resonance. For this reason, the edges of the band are called the *half-power points.*

5.4 NORMALIZATION

When discussing the series resonant circuit in Sec. 5-2, we found it convenient and useful to talk about the value of the admittance magni-

† We have used the symbol Q to stand for quality factor and also for reactive power. It is also used for electric charge. These three uses are not usually made in the same discussion, so there should be no reason for excessive confusion on this score.

tude relative to its maximum value. Thus the ordinate used in plotting the curve in Fig. 5-7 is $|Y|/Y_m$. Similarly, we found it convenient to express the frequency relative to its resonant value. Thus one of the scales of the abscissa in Fig. 5-7 is ω/ω_0. This process of expressing a quantity relative to some standard is called normalization.

Normalizing permits us to deal with convenient small numbers. For example, if the frequencies of interest are in the megacycle range, we can define a new frequency variable which is the old variable divided by 10^6. Then the new frequency variable will vary about unity.

Consider the impedance of the series RLC network.

$$Z = R + j\omega L + \frac{1}{j\omega C} = R + j\left(\frac{\omega}{\omega_1}\right)(\omega_1 L) + \frac{1}{j\left(\frac{\omega}{\omega_1}\right)(\omega_1 C)} \tag{5-48}$$

On the right side we have multiplied and divided the frequency-dependent terms by ω_1, which maneuver leaves the value of the expression unchanged. Suppose we now divide both sides of the equation by a constant, say R_1. Then we will have

$$\frac{Z}{R_1} = \frac{R}{R_1} + j\left(\frac{\omega}{\omega_1}\right)\left(\frac{\omega_1 L}{R_1}\right) + \frac{1}{j\left(\frac{\omega}{\omega_1}\right)(\omega_1 RC)}$$

or

$$Z_n = R_n + j\omega_n L_n + \frac{1}{j\omega_n C_n} \tag{5-49}$$

where

$$Z_n = \frac{Z}{R_1} \qquad L_n = \frac{\omega_1 L}{R_1}$$
$$R_n = \frac{R}{R_1} \qquad C_n = \omega_1 RC \tag{5-50}$$

The subscript n stands for "normalized."

Let us inquire into the net effect of this normalization. The form of Eq. (5-49) is identical with that of the original impedance in Eq. (5-48). That is, the variation of Z_n with the frequency ω_n is the same as that of Z with the frequency ω. But the scale of values of Z_n is $1/R_1$ times the values of Z. We say that the *impedance level* is changed, the normalizing factor being R_1.

Normalization often avoids tedious numerical computation. We have anticipated this fact in previous numerical work by usually specifying convenient small values for the elements and for the frequency. At the time it may have seemed strange to deal with capacitance values of 1 and radian frequencies of 2, for example. On the other hand, in the illustrative example given in Fig. 4-16, more usual numerical values were specified: $C = 10^{-6}f$, $R = 200$ ohms, $\omega = 10^4$ radians per second. Suppose, instead, that in this example we choose normalizing factors $\omega_1 = 10^4$ and $R_1 = 100$. Then, according to Eq. (5-50), the normalized values will be $C_n = 1$, $R_n = 2$, and $\omega_n = 1$. If you compute the voltage-transfer function of the example with the normalized numerical values, you will find that it is exactly the same as before. That is, frequency normalization and impedance level have no influence on the voltage-transfer function. In this simple example there is not much computational advantage in using normalized values. In more extensive networks, however, dealing with numbers such as 1 and 2 can be decidedly more attractive than manipulating less convenient numbers.

What we have illustrated by means of the series RLC network culminating in the expressions for normalized element values in Eq. (5-50) is valid for all networks. We shall not prove this statement here, but you will undoubtedly agree that it is reasonable.

5.5 LOGARITHMIC MEASURE

Turn back to some of the amplitude-response curves which were shown in the last few sections, such as Fig. 5-3 and 5-6. The scales of both coordinates are linear. It is often more convenient to use a logarithmic scale for either the response variable or the frequency variable or both. (We shall note some reasons for this shortly.) We shall therefore take some time to discuss logarithmic variables and logarithmic plots.

Let us initially turn back to Fig. 5-2 and the voltage gain given in Eq. (5-3). The logarithm of G_{21} can be written

$$\ln G_{21} = \ln\left(|G_{21}|\epsilon^{j\phi}\right) = \ln \epsilon^{\alpha' + j\phi} = \alpha' + j\phi$$

where

$$\alpha' = \ln |G_{21}| \qquad (a)$$
$$|G_{21}| = \epsilon^{\alpha'} \qquad (b) \qquad (5\text{-}51)$$

and ln designates the natural logarithm (logarithm to the base ϵ). The quantity α' is the natural logarithm of $|G_{21}|$; its unit is the *neper*, named in honor of John Napier who invented the natural logarithms. (The spelling of his name is still the subject of controversy; hence the difference between neper and Napier.) We shall call α' the *logarithmic voltage gain*, to distinguish it from the plain, ordinary voltage gain G_{21}.

For numerical work the common logarithm (logarithm to the base 10) is much more convenient to use than the natural logarithm. If we use the common logarithm in Eq. (5-51), however, the resulting unit will be quite large. Hence it is common practice to make the following definition:

$$\begin{aligned} \alpha &= 20 \log |G_{21}| && (a) \\ |G_{21}| &= 10^{\alpha/20} && (b) \end{aligned} \qquad (5\text{-}52)$$

where "log" designates the common logarithm. (We shall consistently use "ln" to designate the natural logarithm, and "log" to designate the common logarithm, in order to avoid writing the appropriate base each time.) The unit of α is called the *decibel* (db). By inserting Eq. [5-51(*b*)] into Eq. [5-52(*a*)], we find the relationship between the decibel and the neper to be

$$\alpha = 20 \log \epsilon^{\alpha'} = 20\alpha' \log \epsilon \doteq 8.686\alpha' \qquad (5\text{-}53)$$

That is, the number of decibels is obtained by multiplying the number of nepers by approximately 8.686.

Historically, the decibel was introduced as a unit for the logarithmic measure of a ratio of two values of power. Thus, if P_1 and P_2 refer to two values of power, then $\log P_2/P_1$ was defined as a logarithmic power ratio with the *bel* as a unit (named after Alexander Graham Bell). For convenience of size, a unit one tenth of a bel was chosen and called a decibel Thus the original definition of the logarithmic measure in decibels was

$$\alpha(\text{in decibels}) = 10 \log \frac{P_2}{P_1} \qquad (5\text{-}54)$$

In electrical networks, power is proportional to the square of a voltage-sinor magnitude or a current-sinor magnitude; that is,

$$\begin{aligned} P_1 &= |V_1|^2 Re(Y_1) = |I_1|^2 Re(Z_1) && (a) \\ P_2 &= |V_2|^2 Re(Y_2) = |I_2|^2 Re(Z_2) && (b) \end{aligned} \qquad (5\text{-}55)$$

where Z_1 is the impedance and Y_1 the admittance of the network in which the power P_1 is dissipated, and Z_2 is the impedance and Y_2 the admittance of the network in which P_2 is dissipated. When these expressions are substituted into Eq. (5-54), we get

$$\alpha(\text{in decibels}) = 10 \log \left|\frac{V_2}{V_1}\right|^2 \frac{Re(Y_2)}{Re(Y_1)}$$
$$= 10 \log \left|\frac{I_2}{I_1}\right|^2 \frac{Re(Z_2)}{Re(Z_1)} \tag{5-56}$$

In the particular case in which P_2 and P_1 are the powers dissipated in the same network and $Re(Z_2) = Re(Z_1)$, these expressions reduce to

$$\alpha(\text{in decibels}) = 10 \log \left|\frac{V_2}{V_1}\right|^2 = 10 \log \left|\frac{I_2}{I_1}\right|^2 \tag{5-57}$$

Only in such a case is the historical definition of the decibel consistent with our definition in Eq. (5-52). If we wish to be entirely consistent, we should used a different name for the logarithmic unit defined in Eq. (5-52). In some cases this unit would be the same as a decibel. It has become common practice among electrical engineers, however, to define the decibel as in Eq. (5-52). We shall continue this practice in this book.

In the network now under consideration, V_2 is the voltage sinor across a capacitor; hence there is no average power involved, and the historical definition of the logarithmic unit would not be applicable. Nevertheless, the definition we have adopted is still valid.

Let us now turn to a consideration of the logarithmic measure of the frequency. The first task is to normalize the variable by choosing an appropriate reference frequency and referring all others to this one. What frequency is chosen as a reference is arbitrary. In the present case the frequency that suggests itself is the value $\omega = 1/RC$ which defines the bandwidth. Let us call this value ω_1 and divide all frequencies by ω_1. We then define the following logarithmic frequency variable:

$$u = \log \frac{\omega}{\omega_1} \tag{5-58}$$

To get an idea of the unit of u, note that $u = 0$ when $\omega = \omega_1$. To make $u = 1$ requires that $\omega = 10\omega_1$. Similarly, when $u = 2$, $\omega = 100\omega_1$. Thus each unit change of u requires the frequency to change by a factor of 10. For this reason, the unit of u is named a *decade*.

Another unit for the logarithmic frequency variable is in common use. It is based on the use of the logarithm to the base 2. Thus, if we write

$$u' = \log_2 \frac{\omega}{\omega_1} \tag{5-59}$$

then, in order for u' to change from zero to 1 requires ω to change from ω_1 to $2\omega_1$. Each time the frequency doubles, u' changes by 1 unit. Hence the unit of u' is called an *octave*.

To find the relationship between the number of octaves and the number of decades, it is only necessary to note that the logarithm of a quantity to the base 2 is equal to the logarithm of the quantity to the base 10 times the logarithm of 10 to the base 2. Thus

$$u' = \log_2 \frac{\omega}{\omega_1} = (\log_2 10) \log \frac{\omega}{\omega_1} \doteq \frac{10}{3} u \tag{5-60}$$

That is, the number of octaves is approximately $3\frac{1}{3}$ times the number of decades.

Let us now turn to the question of plotting an amplitude-response curve. Instead of plotting $|G_{21}|$ against ω as we have done in Fig. 5-3, we plot the logarithmic value of $|G_{21}|$ (α in decibels) against the logarithmic value of ω (u in decades), suitably normalized. This is done on ordinary linear graph paper. Alternatively, we can use logarithmic graph paper. In this case the decades on the frequency axis are already all laid out. We can conveniently show the actual frequency scale here also. Whichever method is used, the same curve will be obtained.

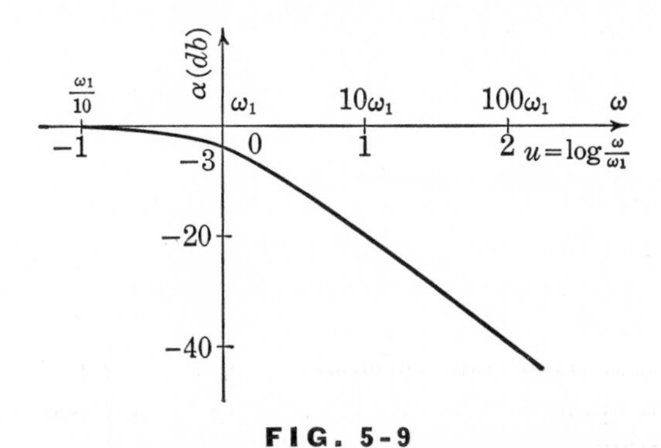

FIG. 5-9

Logarithmic frequency-response curve

Figure 5-9 shows the amplitude response of Fig. 5-3 plotted against logarithmic variables. The abscissa is labeled in terms of frequency as well as u in decades. Negative values of u correspond to frequencies below ω_1. This method of plotting tends to expand the curve in the neighborhood of $\omega = \omega_1$ in which the variation is large. At the frequency $\omega = \omega_1 = 1/RC (u = 0)$ the value of $|G_{21}|$ from Eq. (5-4) is $1/\sqrt{2}$. Its logarithmic measure, therefore, is

$$\alpha(\text{at band edge}) = 20 \log \frac{1}{\sqrt{2}} = -20 \log (2)^{1/2} = -10 \log 2 \doteq -3 \text{ db} \tag{5-61}$$

This information provides us with another way of referring to the bandwidth. The edge of the band is called the *3-db point* and the bandwidth is the *3-db bandwidth.*

It is clear that what we have discussed in terms of the voltage gain applies to any response function. For example, the normalized admittance of the tuned circuit given in Eq. (5-12) can also be plotted logarithmically. Again, the band edges are called the 3-db points.

Before terminating this section, let us make a comment. No material has been presented here which is of fundamental importance. That is to say, nothing of basic value would be lost if we were never to mention logarithmic units of measurement and logarithmic plots of curves. These units, however, are convenient to use for numerical work in many areas of engineering. The terminology we have introduced here is the common language in engineering circles. For these reasons, it is worthwhile getting acquainted with logarithmic units and logarithmic plots.

5.6 PARALLEL TUNED CIRCUIT

Let us now turn our attention to the parallel tuned circuit shown in Fig. 5-10. It is desired to find the voltage response at the terminals when a sinusoidal current of constant amplitude and phase but variable frequency is applied. The sinor I is thus constant. Hence, to find V, it is sufficient to deal with the impedance Z. For the impedance we have

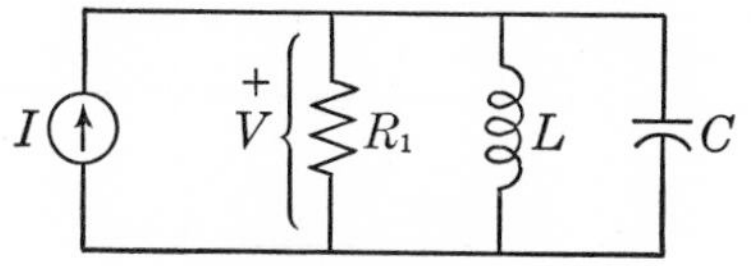

FIG. 5-10
Parallel tuned circuit

$$Z = \frac{1}{G_1 + j\left(\omega C - \dfrac{1}{\omega L}\right)} = \frac{R_1}{1 + jR_1\left(\omega C - \dfrac{1}{\omega L}\right)} \tag{5-62}$$

Note that the parallel circuit under consideration here is the dual of the series circuit. Hence the entire discussion which was carried out in Sec. 5-2 applies here also, with appropriate changes in the words current, voltage, admittance, impedance, etc.

The resonant frequency is again defined as the frequency at which the response and excitation are in phase, leading to $\omega_0^2 = 1/LC$, just as before. The response, which is now the voltage, again has a maximum magnitude at this frequency. The bandwidth is again defined as the angular frequency interval over which the input power exceeds one half its maximum value, or the interval over which the logarithmic value of the response is down no more than 3 db. At resonance the stored energy surges back and forth between the capacitor and the inductor, as in the series circuit.

To find the Q of the circuit at resonance, let us use the basic definition in Eq. (5-44). The maximum energy stored in the inductor will be the same as that in the capacitor at resonance. In the capacitor this will be $C|V|^2$. Hence, letting Q_p be the Q of the parallel resonant circuit at resonance, we get

$$Q_p = \omega_0 \frac{C|V|^2}{G_1|V|^2} = \omega_0 C R_1 = \frac{R_1}{\omega_0 L} = R_1 \sqrt{\frac{C}{L}} \tag{5-63}$$

In the case of the parallel resonant circuit, we find that the expression for Q_P in terms of the elements is just the reciprocal of what it is for the series resonant circuit. Actually, we could have anticipated this result by recalling that we are dealing with the dual of the series circuit and, hence, the dual quantities should be substituted into the expression for Q_0. Thus, in $\omega_0 L/R$, we should replace L by C and R by G_1. This will lead to Eq. (5-63).

Having defined ω_0 and Q_p, let us rewrite the expression for impedance in Eq. (5-62). The result will be

$$\begin{aligned} \frac{Z}{R_1} &= \frac{1}{1 + jR_1\left(\omega C - \dfrac{1}{\omega L}\right)} = \frac{1}{1 + jR_1\sqrt{\dfrac{C}{L}}\left(\dfrac{\omega}{\omega_0} - \dfrac{\omega_0}{\omega}\right)} \\ &= \frac{1}{1 + jQ_P\left(\dfrac{\omega}{\omega_0} - \dfrac{\omega_0}{\omega}\right)} \end{aligned} \tag{5-64}$$

The form of this expression is identical with that of the admittance in Eq. (5-10). Hence the universal resonance curve shown in Fig. 5-7 will apply to the parallel resonant circuit as well as to the series resonant circuit. It is necessary only to label the ordinate impedance rather than admittance.

Let us now consider the variation of the parallel tuned circuit which is shown in Fig. 5-11. The branch consisting of the series L and R is a

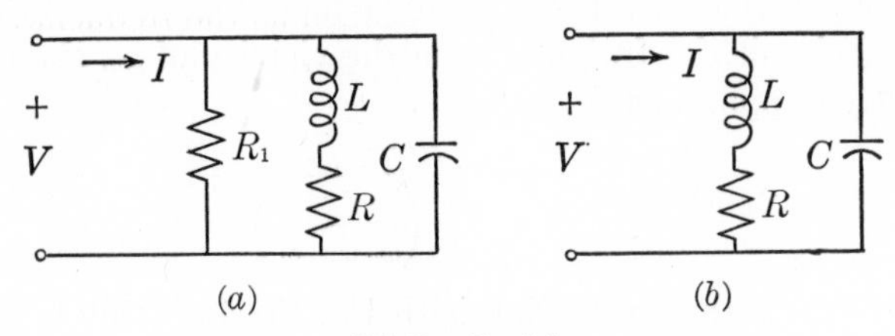

FIG. 5-11

Variations of the parallel tuned circuit

model of a coil of wire in which R represents the resistance of the wire. The branch R_1 does not add anything essential to the behavior of this network. Hence we shall remove it and consider the further modification shown in Fig. 5-11(*b*). We shall refer to this as the *two-branch parallel resonant network*.

The impedance of this network is readily found to be

$$Z = \frac{1}{j\omega C + \dfrac{1}{R + j\omega L}} = \frac{R + j\omega L}{(j\omega)^2 LC + j\omega CR + 1}$$
$$= \frac{1 + R/j\omega L}{\dfrac{RC}{L} + j\omega C + \dfrac{1}{j\omega L}} = \frac{L}{RC} \frac{1 - j\dfrac{R}{\omega L}}{1 + j\dfrac{1}{R}\left(\omega L - \dfrac{1}{\omega C}\right)} \tag{5-65}$$

The next-to-last step was obtained by dividing both numerator and denominator by $j\omega L$. To get the final form, we factored RC/L from the denominator.

Let us again define ω_0 as $1/LC$ and Q_0 as $\omega_0 L/R$. In terms of these quantities, Eq. (5-65) can be written

$$Z = \frac{RQ_0^2\left(1 - j\dfrac{\omega_0}{Q_0\omega}\right)}{1 + jQ_0\left(\dfrac{\omega}{\omega_0} - \dfrac{\omega_0}{\omega}\right)} \tag{5-66}$$

We can now make some observations concerning the two-branch parallel tuned circuit based on this expression. First of all, we note that the impedance is not purely real when $\omega = \omega_0$, the frequency at which the imaginary part of the denominator vanishes. The numerator still has an imaginary part which is $-1/Q_0$ at that frequency. If Q_0 is large, say $Q_0 > 20$, this imaginary part will be quite small relative to the real part, so that the impedance is almost real at $\omega = \omega_0$. Hence, for a high-Q circuit, for frequencies in the vicinity of ω_0, the impedance can be written approximately as

$$Z \doteq \frac{RQ_0^2}{1 + jQ_0\left(\frac{\omega}{\omega_0} - \frac{\omega_0}{\omega}\right)} \tag{5-67}$$

Now compare this expression with Eq. (5-64), which is the impedance of the three-branch parallel resonant circuit. They are similar, except that Q_0 now replaces Q_P and RQ_0^2 appears in the numerator where R_1 was. This fact suggests that the two-branch circuit can be replaced by a three-branch circuit, provided the resistance R in the two-branch case is replaced by a resistance $R_1 = Q_0^2R$ in the three-branch circuit. If we follow through with this suggestion, and compute the value of Q_P with $R_1 = Q_0^2R$, we get

$$Q_P = R_1\sqrt{\frac{C}{L}} = Q_0^2R\sqrt{\frac{C}{L}} = \left(\frac{1}{R}\sqrt{\frac{L}{C}}\right)^2 R\sqrt{\frac{C}{L}} = \frac{1}{R}\sqrt{\frac{L}{C}} = Q_0 \tag{5-68}$$

That is, the Q of a two-branch circuit is the same as the Q of a three-branch circuit having the same L and C, provided the resistances are related by $R_1 = Q_0^2R$.

Let us emphasize that this equivalence is valid only for high-Q circuits in the vicinity of the resonant frequency. For example, near zero frequency, for any value of Q, the impedance of Fig. 5-10 is approximately zero, whereas that of Fig. 5-11(*b*) is approximately R.

We should briefly consider the question of exactly what the resonant frequency is in the two-branch parallel resonant circuit. It is clear from Eq. (5-66) that the angle of the impedance is not zero at $\omega = \omega_0$. Hence, at this frequency, the terminal voltage is not in phase with the current; the power factor is not unity. It is a relatively simple job to find the frequency at which the power factor is unity, which requires that Z be real. In terms of Eq. (5-66), we require that the angle of the numerator be equal to the angle of the denominator; the total angle of the impedance will then be zero. If the angles of the numerator and denominator are to be equal, their tangents will be also. But the tangents are the ratios of imaginary to real parts. Hence, for unity power factor, we require that

$$-\frac{\omega_0}{Q_0\omega} = Q_0\left(\frac{\omega}{\omega_0} - \frac{\omega_0}{\omega}\right) \tag{5-69}$$

Solving this expression for ω leads to

$$\omega\ (\textit{for unity power factor}) = \omega_0\sqrt{1 - 1/Q_0^2} \qquad (5\text{-}70)$$

For values of Q_0 greater than about 10, we see that the frequency of unity power factor differs from ω_0 by a very small amount.

In the case of the series resonant circuit, we found that, at the frequency of unity power factor, the magnitude of the response is a maximum. Let us investigate this question for the present case. Conceptually, it is quite simple to find the frequency at which the magnitude of Z is a maximum. We should write an expression for $|Z|$, differentiate it with respect to ω, and then set the derivative equal to zero. The solution of the resulting equation will give the desired value of ω. Although easy conceptually, the actual mathematical manipulations are tedious. We shall here give the result of this process and leave for you the task of carrying out the details.

For maximum value of $|Z|$

$$\omega = \omega_0 \sqrt{\sqrt{1 + 2/Q_0^2} - 1/Q_0^2} \doteq \omega_0\sqrt{1 - 1/Q_0^4} \qquad (5\text{-}71)$$

The last step is obtained by expanding the inner radical by means of the binomial theorem and retaining only the first three terms.

This expression shows that, for a given value of Q_0, the frequency of maximum response is even closer to ω_0 than the frequency of unity power factor. For engineering purposes, these three frequencies can be considered identical in the case of high-Q circuits. For Q_0 as low as 4, for example, the last two equations give

For unity power factor $\qquad \dfrac{\omega}{\omega_0} = 0.969$

For maximum $|Z|$ $\qquad \dfrac{\omega}{\omega_0} = 0.998$ approx.

$\qquad = 0.999$ exact

5.7 SUMMARY

In this chapter we introduced the study of the variation of network functions with frequency. This is a very important and extensive subject, and by no means have we yet exhausted it. We introduced the major concepts of frequency response and resonance. Resonance was defined as the condition in which the voltage and current are in phase. It was interpreted as the condition in which stored energy is traded back and forth by an inductor and a capacitor.

We discussed frequency selectivity, pass bands, stop bands, bandwidth, quality factor, resonant frequency, and their interrelationships. Since frequency-response curves are smooth, there is no unique way of specifying the limits of the pass band. Customarily, however, the half-power points, or 3-db points, are taken as the edges of the band.

We discussed frequency normalization and impedance normalization. We introduced logarithmic units; decibels and nepers, decades and octaves.

Problems

5-1 Sketch the indicated frequency-response curves of the networks shown in Fig. P5-1 against ω. Repeat, using log log paper.

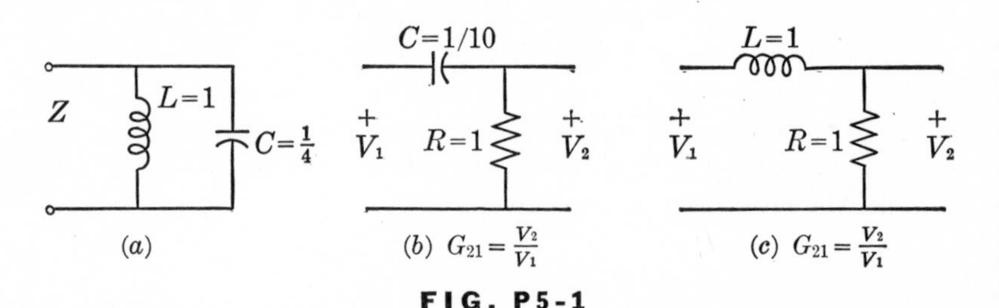

FIG. P5-1

5-2 In a series *RLC* circuit the element values are $R = 1$, $L = 10^{-2}$, $C = 10^{-6}$. Find the bandwidth between half-power points. Would you describe this as a wide-band or a narrow-band circuit?

5-3 The resonant frequency of a series *RLC* circuit is 2 megacycles per second. The half-power bandwidth is $\frac{1}{2}$ per cent of the resonant frequency. At resonance the input impedance has a magnitude of 10. Find the element values.

5-4 The resonant frequency of a three-branch parallel resonant circuit is to be 1 megacycle per second and its half-power bandwidth is to be 2 kc per second. The value of the inductance is $L = 1\ mh$. Find the values of C and R. Is this a high-Q or a low-Q circuit?

5-5 A coil of wire is to be represented by a model consisting of a series L and R. A known capacitance $C = 100\mu\mu f$ is connected across the coil, as shown in Fig. P5-5. The circuit is excited by a variable-frequency, constant-amplitude current source. Two measurements of frequency are made. When the voltage across the capacitor has maximum magnitude, the angular frequency is found to be $\omega_1 = 10^7$ radians per second. When the voltage magnitude reaches half the maximum value, the angular frequency is $\omega_2 = 99 \times 10^5$ radians per second. Find L and R.

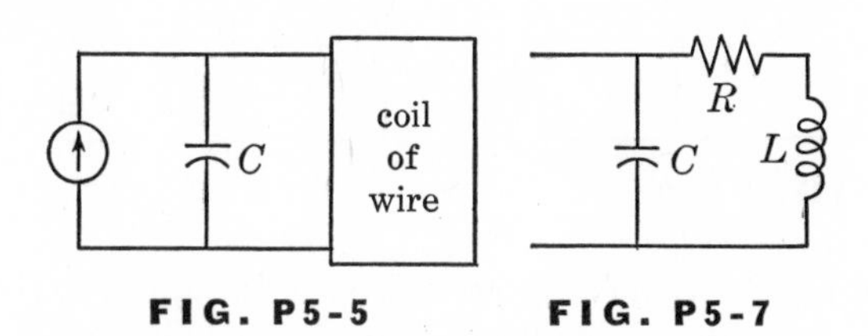

FIG. P5-5 FIG. P5-7

5-6 A coil of wire is to be represented by a series L and R. A 1-μf capacitor is connected in series with it, and the resonant frequency is found to be 1.5 kc. This is done by exciting the series combination with a variable-frequency voltage source and determining the frequency at which the voltage and current are in phase. The capacitor is now placed in parallel with the coil, and the combination is excited with a variable-frequency current source. The frequency at which the voltage and current are in phase is now found to be 1,455 cycles per second. Determine L and R.

5-7 A coil of wire is to be used at such a high frequency that an adequate model to represent it takes the form shown in Fig. P5-7. It is desired to make some measurements similar to the ones in the last two problems in order to determine the elements of the model. The highest frequency of the variable-frequency current source which is available, however, is not high enough for the peak of the response to be reached. The only other equipment available consists of a few capacitors. Suggest a way to determine the element values and obtain formulas for these elements in terms of measured values. Will your scheme always work?

5-8 A series resonant circuit is excited with a 10-volt variable-frequency source. At an angular frequency of 100 radians per second, it is found that the magnitude of the inductor voltage and that of the capacitor voltage are both 100 volts. At this frequency the current magnitude is found to be 2 amp. Find the L, R, and C element values.

5-9 A variable-frequency voltage source is applied to the network of Fig. P5-9. At an angular frequency of 2 radians per second, it is desired that the current in the resistor be zero. At 1 radian per second it is desired that the current in the resistor be a maximum. (*a*) Determine the value of C. (*b*) Specify whether the branch marked Z is a capacitor or an inductor, and determine its value.

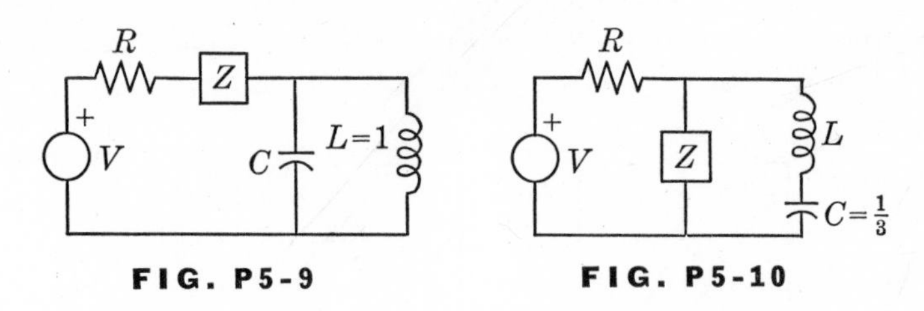

FIG. P5-9 FIG. P5-10

5-10 A variable-frequency voltage source is applied to the network of Fig. P5-10. At an angular frequency of 3 radians per second, it is required that the current in the resistor be a maximum. At 2 radians per second the current should be zero. (*a*) Determine the value of L. (*b*) Specify whether the branch marked Z should be a capacitor or an inductor, and determine its value.

5-11 A variable-frequency voltage source is applied to the network shown in Fig. P5-11. It is desired that the output voltage V_2 have a maximum magnitude at an angular frequency of 2 radians per second. It is also desired that the current supplied by the source be a maximum at an angular frequency of 5 radians per second. (*a*) Determine the value of C. (*b*) Specify whether the branch marked Z should be a capacitor or an inductor, and determine its value.

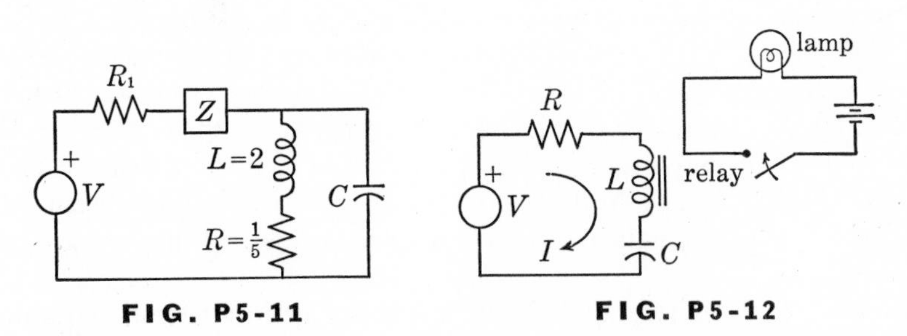

FIG. P5-11 FIG. P5-12

5-12 The circuit shown in Fig. P5-12 is used as a frequency-sensing device. As the frequency of the source is varied, whenever the magnitude of the current I exceeds $1/\sqrt{5}$ times its maximum value the relay closes, thereby turning on the indicator lamp. Find the band of frequencies over which the lamp is on. Assume that the relay coil can be represented by an inductor. Take $Q_0 = 100$ and $\omega_0 = 1{,}000$ radians per second.

5-13 The circuit in Fig. P5-13 is the dual of the two-branch parallel resonant circuit. Find the frequency at which the voltage and current at the terminals are in phase. Compare with the frequency at which L and C alone would be resonant. Find the frequency band over which the magnitude of the admittance is no less than 80 per cent of its value at unity power factor.

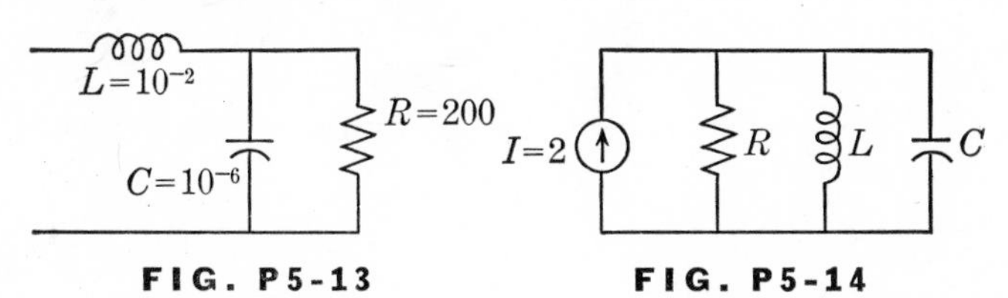

FIG. P5-13 FIG. P5-14

5-14 A 2-amp current source excites the three-branch parallel resonant circuit shown in Fig. P5-14. The resonant angular frequency is $\omega_0 = 1{,}000$. At this frequency the voltage across the circuit is 10 volts. The 3-db bandwidth is $\beta = 10$ radians per second. (*a*) Determine the element values. (*b*) Determine the band of frequencies over which the voltage exceeds 6 volts.

5-15 In a three-branch parallel resonant circuit the resonant angular frequency is $\omega_0 = 5 \times 10^6$ and $Q_p = 200$. Find the frequencies at which the power is one quarter the maximum. At these frequencies find the ratio of the capacitor current to the inductor current.

5-16 The network inside the box in Fig. P5-16 is to be represented by a three-branch parallel resonant circuit as shown. C_1 is a known external capacitor. With C_1 disconnected, it is determined that the resonant angular frequency is ω_0 and the half-power bandwidth is β. With C_1 connected, the resonant angular frequency shifts to ω_1. Determine the values of R, L, and C in terms of C_1, ω_0, β, and ω_1.

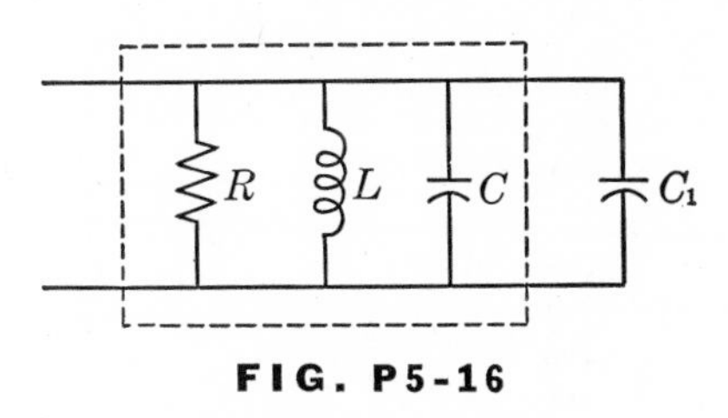

FIG. P5-16

5-17 The magnitude of the voltage gain of a low-pass RC filter can be written $|G_{21}| = 1/\sqrt{1 + \omega^2/\omega_1^2}$. [See Eq. (5-4) in the text.] Defining the logarithmic gain α as in Eq. (5-52), and the logarithmic frequency variable u as in Eq. (5-58), prove that the slope of the α versus u curve is given by

$$\frac{d\alpha}{du} = \frac{-20}{1 + 10^{-2u}} \text{ db per decade}$$

which approaches -20 db per decade as $u \to \infty$. What is the value of the slope if the logarithmic frequency variable is measured in octaves?

5-18 A series *RLC* circuit is excited with a variable-frequency, constant-amplitude voltage source. It is found that at 3,200 cycles per second the rms value of the current is 13 amp and the power input of 845 watts is a maximum. At 2,800 cycles per second the power input is 125 watts. Find the *RLC* element values and the source voltage.

5-19 For each of the circuits shown in Fig. P5-19, compute an expression for the complex power supplied by the source and verify that it takes the form of Eq. (5-37) in the text.

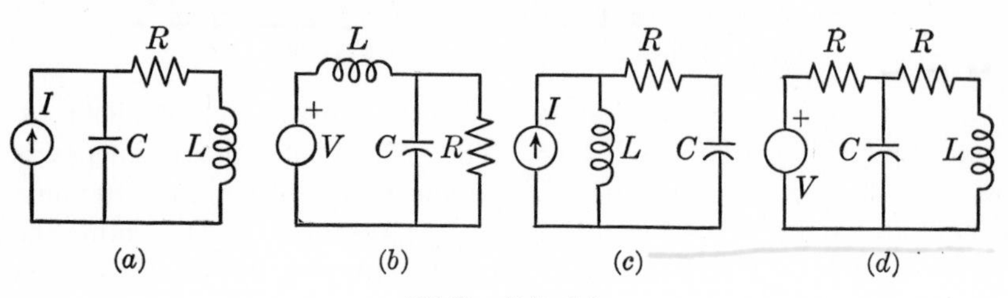

FIG. P5-19

THE COMPLEX-FREQUENCY VARIABLE

6

IN the last two chapters we have been studying certain aspects of the problem of determining the steady-state response of electrical networks to excitations which are sinusoidal functions of time. The method of solution which has been developed consists of four parts: First of all, the problem is formulated in the time domain as one or more integrodifferential equations. Then a transformation is carried out so that the voltage and current variables are no longer sinusoidal functions of time but sinors, and the equations are no longer integrodifferential but algebraic. The transformed problem is then solved by algebraic methods identical with those used for resistance networks. Finally, the solution is converted back to the time domain. As a matter of fact, the first and last steps are very often omitted, and the problem is phrased from the beginning in terms of sinors and impedances.

The fundamental idea on which this procedure is based is Euler's theorem, which permits us to express a sinusoid in terms of the exponential $\epsilon^{j\omega t}$. In this chapter we shall exploit this relationship even further by generalizing the exponent.

6.1 COMPLEX FREQUENCY

In the exponential function $\epsilon^{j\omega t}$ the exponent is an imaginary quantity. Now an imaginary quantity is but a special case of something more general, namely, a complex quantity. Let us, therefore, define the complex quantity

$$s = \sigma + j\omega \tag{6-1}$$

whose imaginary part is the usual angular frequency ω. We shall call s the *complex-frequency variable*, or simply the *complex frequency*. When the real part of s is zero, then s is simply $j\omega$.

Let us now consider the exponential function ϵ^{st}. By the law of exponents and Euler's theorem, this can be written

$$\epsilon^{st} = \epsilon^{\sigma t}\epsilon^{j\omega t} = \epsilon^{\sigma t}(\cos \omega t + j \sin \omega t) \tag{6-2}$$

This is a complex function of time.

A graphical interpretation of this function is readily obtained. We have already interpreted the exponential $\epsilon^{j\omega t}$ as a rotating sinor in the complex plane, its projections on the real and imaginary axes generating the cosine and sine functions. Now, however, $\epsilon^{j\omega t}$ is multiplied by $\epsilon^{\sigma t}$. Thus $\epsilon^{\sigma t}$ is the magnitude of the sinor. When σ is zero, the magnitude of the sinor is constant. When σ is not zero, however, the magnitude will be changing. Figure 6-1 shows the path traced by

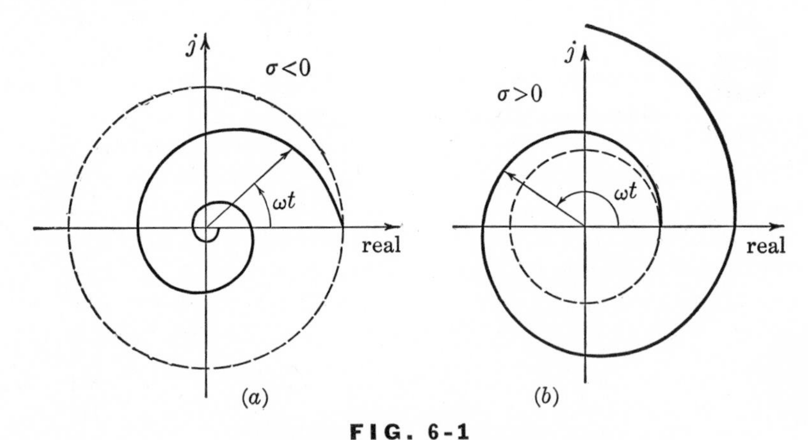

FIG. 6-1

Spiral paths generated by $\epsilon^{\sigma t}\epsilon^{j\omega t}$

the tip of the sinor as it rotates, for both negative and positive values of σ. Both parts of the figure are spirals. When σ is negative, the magnitude of the sinor is decreasing so that the path spirals inward; when σ is positive, the magnitude of the sinor is increasing so that the path spirals outward.

The projections of these spirals on the real and imaginary axes will be cosine and sine functions multiplied by the real exponential $\epsilon^{\sigma t}$. When σ is negative, the resulting waveform is a damped sinusoid; when σ is positive, the waveform is an exponentially increasing sinusoid, as shown in Fig. 6-2.

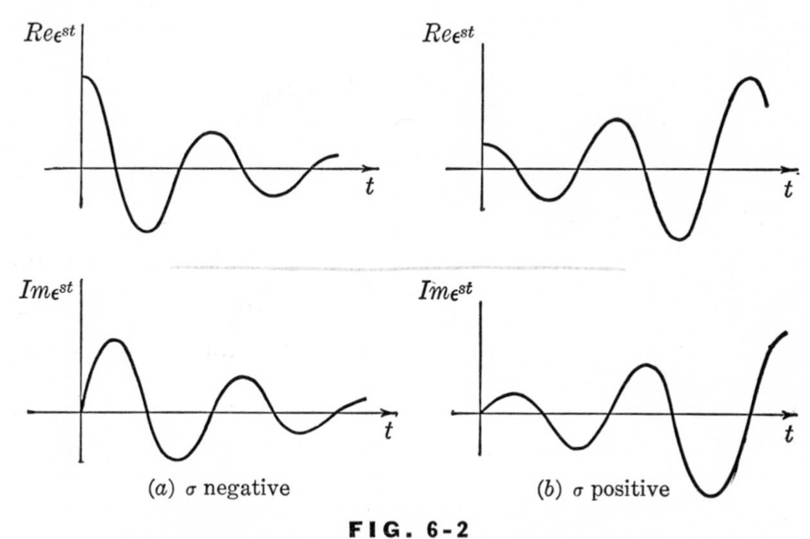

FIG. 6-2

Waveforms of real and imaginary parts of $\epsilon^{\sigma t}\epsilon^{j\omega t}$

The same conclusions follow from Eq. (6-2) by taking the real and imaginary parts; thus

$$\begin{aligned} Re(\epsilon^{st}) &= \epsilon^{\sigma t}\cos\omega t & (a) \\ Im(\epsilon^{st}) &= \epsilon^{\sigma t}\sin\omega t & (b) \end{aligned} \qquad (6\text{-}3)$$

It appears, then, that particular values of s in the exponential function ϵ^{st} lead to sinusoidal waveforms with amplitudes which are constant, exponentially decreasing, or exponentially increasing. But this is not all. When the imaginary part of s is zero, so that $s = \sigma$, ϵ^{st} becomes simply a real exponential. In this case the possible waveforms of ϵ^{st} are shown in Eq. (6-3) for positive, negative, and zero values of σ.

A word is also in order about the units of s. The units of ω are expressed in radians per second; but a radian is really a dimensionless quantity, since it is defined as the ratio of an arc length of a circle to its radius. Hence, dimensionally, ω is the reciprocal of time. The same should be true of s, of which the imaginary part is ω, and of σ. The unit of the exponent of a real exponential has acquired the name neper, however, as discussed in the last chapter. Thus, if σt has the

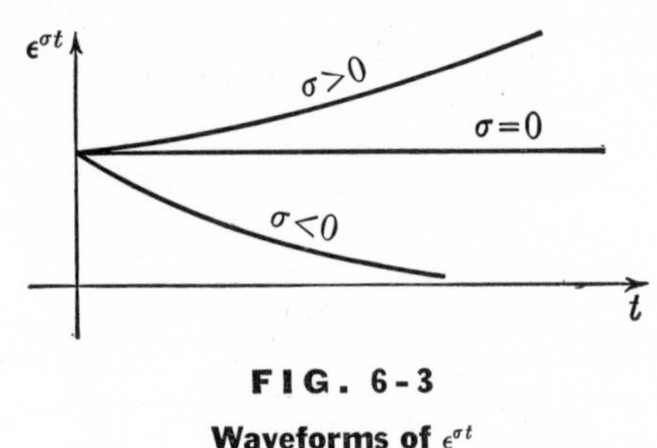

FIG. 6-3

Waveforms of $\epsilon^{\sigma t}$

unit neper, then σ should have the unit *neper per second*. In fact, on the basis of this unit for σ, the name *neper frequency* has been given to σ, in contrast to the names *angular frequency* or *radian frequency* for ω.

The concept of the complex exponential which we have introduced is a very powerful one. We see that the function ϵ^{st} serves to bring under the same roof the real exponential function of time, the constant-amplitude sinusoid, the exponentially damped and increasing sinusoids, and even the steady constant. If we find the forced response of a network to an excitation of the form ϵ^{st}, we shall at the same time be finding it for all the functions just mentioned. Thus the function ϵ^{st} unifies d-c and a-c network analysis! There is no need to treat these as two distinct studies.

The sinusoidal steady-state response of a network can be found by first obtaining the forced response to the exponential excitation ϵ^{st}, then setting $\sigma = 0$, which amounts to setting $s = j\omega$. Let us inquire into the changes that might be introduced by the use of the complex frequency into the methods of steady-state analysis which were developed in Chap. 4.

First of all, note that Kirchhoff's two laws in terms of sinors will remain unchanged. You can demonstrate this fact by repeating the development in Sec. 4-3, using ϵ^{st} instead of $\epsilon^{j\omega t}$. Next, let us look at the *v-i* relationships of the elements. Since that of the resistor does not depend on frequency, it will remain unchanged. The differentiation and integration in the *v-i* relationships of inductors and capacitors lead to factors of s and $1/s$ instead of $j\omega$ and $1/j\omega$, respectively. Thus the impedances of an inductor and a capacitor will be $Z_L = Ls$ and $Z_C = 1/Cs$, respectively. All that is necessary is to replace $j\omega$ by s in everything that we have done, and we shall have the forced response to the exponential ϵ^{st}.

6.2 NETWORK FUNCTIONS AND COMPLEX FREQUENCY

We are now ready to exploit the concept of complex frequency introduced in the last section. Let us start by determining the impedances and admittances of some simple networks in terms of the complex frequency s.

Figure 6-4 shows a series RL network excited by a voltage source whose time variation is of the form ϵ^{st}. In terms of sinors, the equation relating the voltage excitation and the current response is

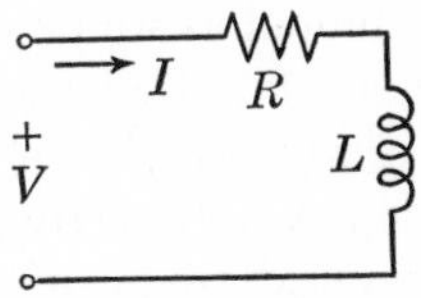

FIG. 6-4
Series RL network

$$V = RI + LsI$$

The admittance of the network is the ratio of the response to excitation sinors. Thus

$$Y = \frac{I}{V} = \frac{1}{Ls + R} = \frac{1}{L(s + R/L)} \tag{6-4}$$

Of course, the impedance is the reciprocal of this, namely

$$Z = L(s + R/L) \tag{6-5}$$

This expression is in the form of a polynomial in which s is the variable. It is a known fact that an nth-degree polynomial has n roots and that the polynomial can be written in terms of its factors. (This fact is referred to as the fundamental theorem of algebra.) In Eq. (6-5) the polynomial is of the first order and is already in factored form. Thus it can be written as

$$Z(s) = L(s - s_1) \qquad (a)$$

where (6-6)

$$s_1 = -R/L \qquad (b)$$

is the root. Of course the admittance can also be written in this form.

$$Y = \frac{1/L}{s - s_1} \tag{6-7}$$

The quantity s_1 is the particular value of s at which the impedance is zero, and we see that it depends only on the network parameters R and L. Aside from the constant L, which can be considered simply as a normalizing factor, the impedance or admittance depends on two things: the complex-frequency variable s, which is contributed by the source, and the quantity s_1, which is contributed by the network.

Let us here digress temporarily and return to a consideration of the natural response of the RL network which was discussed in Chap. 3. There we found that the natural response is characterized by the natural frequency, which, for the series RL network, is nothing but $-R/L$, the same quantity which determines the steady-state response! This is a truly remarkable fact of the utmost significance; its implications will unfold as we proceed.

Thus we find that the impedance is proportional to the difference between s, which is the complex frequency of the source, and s_1, which is the complex natural frequency of the network. In the present case these "complex" frequencies are not complex at all. For the source, s is simply $j\omega$, and hence is imaginary; for the network, s_1 is real (and negative). Nevertheless, if we look upon s as a general complex variable, then $j\omega$ and s_1 are simply two particular values of this complex variable. This viewpoint is extremely productive. It permits the application of a branch of mathematics (known as the theory of functions of a complex variable) to circuit problems, in terms of which it is possible to unify various aspects of network theory. We shall see more and more of the utility of this approach as we go along, even though we shall not attain a level of accomplishment in this book which will permit extensive exploitation of the mathematics.

Since s is a complex quantity, we can represent it geometrically in the complex plane. Figure 6-5 shows a complex plane which can be designated the *complex-frequency plane,* or simply the *s-plane.* The

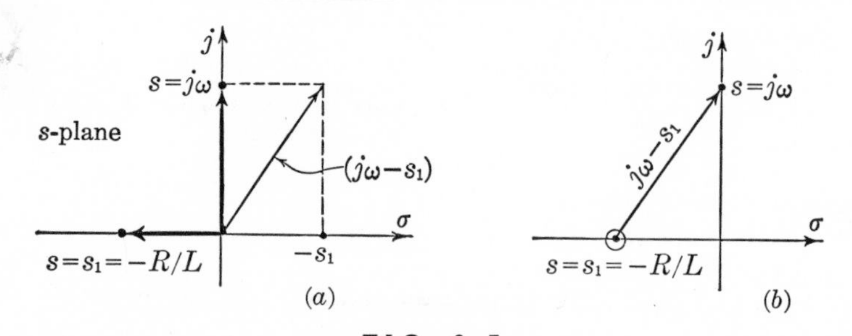

FIG. 6-5

Graphical representation of impedance of RL network

quantities $s = j\omega$ and $s = s_1 = -R/L$ are represented by directed lines from the origin. The difference $j\omega - s_1$ is found graphically by changing the direction of s_1 and adding. Now if the resulting directed line, which is labeled $j\omega - s_1$ in Fig. 6-5(*a*), is translated parallel to itself, it still represents the same quantity. Thus Fig. 6-5(*b*) shows that the difference $j\omega - s_1$ is a directed line from s_1 to $j\omega$. Aside from the multiplying factor L in Eq. (6-6), this directed line completely describes the impedance Z. Since the admittance is the reciprocal of the impedance, the line completely describes the admittance also.

Let us continue the development of these ideas in terms of some additional simple networks. For the series RC network of Fig. 6-6, the V-I relationship and the admittance are

$$V = IR + \frac{1}{sC} I$$

$$Y = \frac{I}{V} = \frac{1}{R + \dfrac{1}{sC}} = \frac{1}{R} \frac{s}{s + 1/RC} = \frac{1}{R} \frac{s}{s - s_1} \tag{6-8}$$

where

$$s_1 = -\frac{1}{RC} \tag{6-9}$$

In this case neither the impedance nor the admittance is a simple polynomial. But we notice that Y is the ratio of s to a first-degree

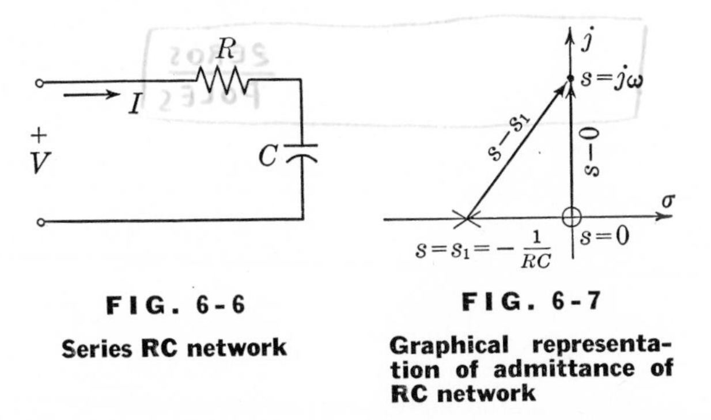

FIG. 6-6

Series RC network

FIG. 6-7

Graphical representation of admittance of RC network

polynomial. Now s alone can be looked upon as a factor $(s - s_2)$ in which s_2 is zero; that is, we can write $s = s - 0$. Thus both the numerator and the denominator of the right side of Eq. (6-8) have the same form. The denominator can be represented geometrically by a line directed from $s_1 = -1/RC$ to $s = j\omega$ in the complex plane, as shown in Fig. 6-7. Similarly, the numerator can be represented by the line directed from zero (the origin in the s-plane) to $s = j\omega$. Aside from the multiplying factor $1/R$ in Eq. (6-8), these two directed lines completely describe the admittance.

Consider again the natural response by glancing back at Chap. 3. The complex natural frequency of the series RC network is none other than $-1/RC$, the same quantity which characterizes the steady-state response. Again we find this intimate relationship between the natural response (the transient) and the steady state.

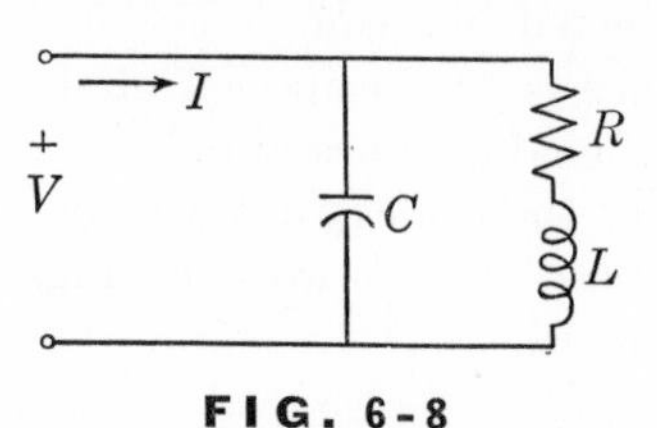

FIG. 6-8

Parallel resonant network

As another example let us consider the two-branch parallel resonant network shown in Fig. 6-8. The impedance of this network is found by parallel and series combination of branches to be

$$Z = \frac{1}{Cs + \dfrac{1}{sL + R}}$$

$$= \frac{sL + R}{LCs^2 + RCs + 1} = \frac{1}{C}\,\frac{(s + R/L)}{s^2 + \dfrac{R}{L}s + \dfrac{1}{LC}}$$

$$= \frac{1}{C}\,\frac{(s - s_1)}{(s - s_2)(s - s_3)} \tag{6-10}$$

In the final form we have set $s_1 = -R/L$. Also, since the denominator is a quadratic, it will have two roots which we have labeled s_2 and s_3. For our present considerations it is unimportant what the values may be. Since there are two roots, either they are both real or they form a complex conjugate pair. For purposes of discussion, let us assume that the latter is the case.

Note again that the numerator and denominator of Z are both polynomials. Each of the factors can be represented in the complex plane by a line directed from one of the points s_1, s_2, or s_3 to the point $s = j\omega$, as shown in Fig. 6-9(a). Aside from the multiplying factor $1/C$ in Eq. (6-10), the impedance is completely determined by these directed lines.

In a numerical example it may be required to find the magnitude and angle of Z. This can be done graphically by means of a diagram

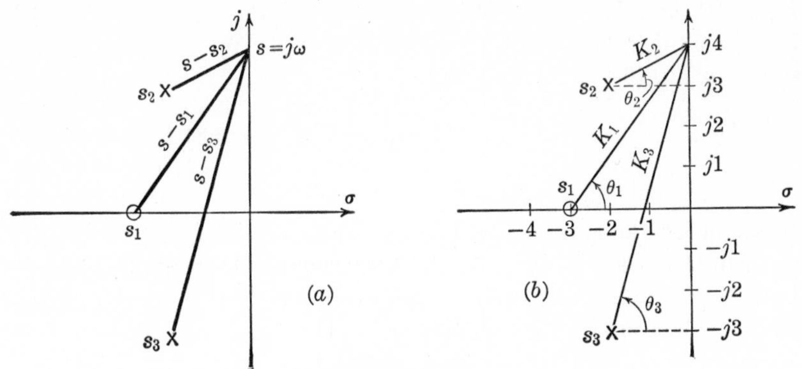

FIG. 6-9

Complex-plane representation

such as that in Fig. 6-9. As a first step let us express each of the factors of Eq. (6-10) in polar form.

$$
\begin{aligned}
s - s_1 &= K_1\epsilon^{j\theta_1} \\
s - s_2 &= K_2\epsilon^{j\theta_2} \\
s - s_3 &= K_3\epsilon^{j\theta_3}
\end{aligned}
\tag{6-11}
$$

The quantities K_1, K_2, and K_3 are the lengths of the directed lines in Fig. 6-9, and θ_1, θ_2, and θ_3 are the angles which they make with the horizontal. The magnitude and angle of Z will be obtained by inserting these expressions into Eq. (6-10). Thus

$$
|Z| = \frac{1}{C}\frac{K_1}{K_2K_3} \qquad (a)
$$

$$
\text{angle of } Z = \theta_1 - \theta_2 - \theta_3 \qquad (b) \qquad (6\text{-}12)
$$

That is, the magnitude of Z is obtained by taking the lengths of the lines representing the denominator factors and dividing into the length of the line representing the numerator factor (aside from the multiplying factor $1/C$). Also, the angle of Z is the angle of the numerator factor minus the angles of the denominator factors. These lengths and angles can be measured directly from the diagram.

As a numerical example let us take the values

$$
\omega = 4 \qquad s_2 = -2 + j3
$$

$$
s_1 = -3 \qquad s_3 = -2 - j3
$$

These values are laid out approximately to scale in Fig. 6-9(b). By means of a divider, the lengths of the lines are found to be $K_1 = 4$,

$K_2 = 2.2$, and $K_3 = 7.3$. With a protractor, the angles are found to be $\theta_1 = 53$ deg, $\theta_2 = 26$ deg, and $\theta_3 = 74$ deg. With these values inserted in Eq. (6-12), we find that

$$|Z| = \frac{1}{C}\frac{K_1}{K_2K_3} = \frac{1}{C}\frac{5}{2.2 \times 7.3} = 0.31\left(\frac{1}{C}\right)$$

$$\text{angle of } Z = \theta_1 - \theta_2 - \theta_3 = -47°$$

If you check the actual values of the magnitude and angle of Z arithmetically from Eq. (6-10), you will find that the values obtained graphically agree to a fair degree of approximation. If greater graphical accuracy is desired, it will be necessary to make a larger diagram.

In each of the examples discussed in this section we have dealt with the impedance or admittance of a network. That is, we have considered the response to be located at the same terminals as the excitation. There is no reason to be so restrictive.

As a final illustration let us take the network shown in Fig. 6-10(a).

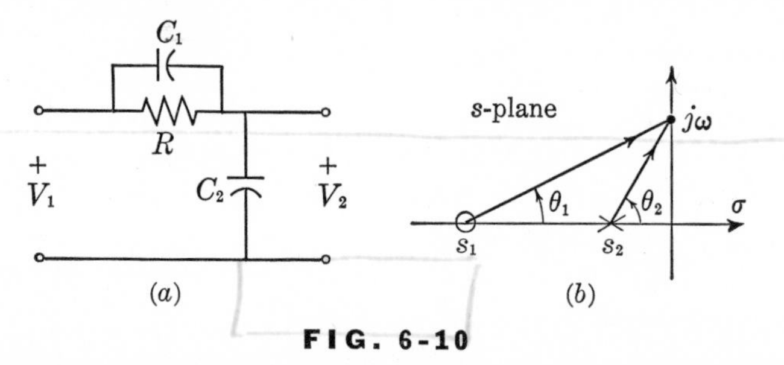

FIG. 6-10

Voltage-divider network and its complex-plane representation

(When used in control systems, this network is called a lag network.) Again we assume a voltage-source excitation varying as ϵ^{st}, and we wish to find the voltage-gain function V_2/V_1. The structure is in the form of a simple voltage divider. Hence we can write

$$V_2 = \frac{\dfrac{1}{C_2s}}{\dfrac{1}{C_2s} + \dfrac{1}{Cs_1 + \dfrac{1}{R}}}V_1 = \frac{C_1}{C_1 + C_2}\,\frac{s + \dfrac{1}{RC_1}}{s + \dfrac{1}{R(C_1 + C_2)}}V_1$$

Finally,

$$G_{21} = \frac{V_2}{V_1} = K\frac{s - s_1}{s - s_2} \tag{6-13}$$

where

$$K = \frac{C_1}{C_1 + C_2}$$
$$s_1 = -\frac{1}{RC_1} \tag{6-14}$$
$$s_2 = -\frac{1}{R(C_1 + C_2)}$$

Again we find that the function of interest, in this case the voltage gain, is the ratio of two polynomials. Aside from the multiplying factor K, the voltage-gain function is completely determined by the factors $s - s_1$ and $s - s_2$. In the present case both s_1 and s_2 are negative real numbers, s_2 having a smaller magnitude than s_1, as is evident from Eq. (6-14). In the complex plane these two points can be marked as shown in Fig. 6-10(b). Graphically, the voltage gain is represented by the two lines from s_1 to $j\omega$ and from s_2 to $j\omega$.

The origin of the name "lag network" is clear from a consideration of Fig. 6-10(b). The angle of the voltage-gain function is $\theta_1 - \theta_2$, as you can see from Eq. (6-13). It is clear from the figure that, at any angular frequency ω, the angle θ_1 is smaller than the angle θ_2. Hence $\theta_1 - \theta_2$ is negative, which means that the angle of the response V_2 is less than that of the excitation V_1.

Let us now glance back over the expressions for the network functions (impedance, admittance, or voltage gain) of the simple networks we have considered in this section. We observe that the form of each expression consists of the ratio of two polynomials in s. [Equation (6-7) is also of this form, with a numerator polynomial which is of zero degree.]

It should be clear that this result is true for any of the functions which express the ratio of a response to an excitation sinor: transfer admittance, current gain, etc. A function which is the ratio of two polynomials is called a *rational function.* As a conclusion we can now make the statement that *each of the network functions is a rational function of the complex-frequency variable s.*

When s is purely imaginary (that is, $\sigma = 0$ and $s = j\omega$), there may be no values of s at which any of the factors in the numerator or denominator of a network function can become zero. If s is allowed to take on any complex value whatsoever, however, then both the numerator and the denominator will go to zero at some specific values of s.

When the denominator goes to zero, the whole function goes to infinity.

We define some words to name the values of s for which these things happen. Thus a value of s at which a function vanishes is called a *zero* of the function. Similarly, a value of s at which the function becomes infinite is called a *pole* of the function. Note that, since impedance is the reciprocal of admittance, the poles of the impedance of a network will be the zeros of the admittance, and the zeros of the impedance will be the poles of the admittance.

In terms of this terminology, it is clear that the value of any one of the network functions of a network depends only on the locations in the complex-frequency plane of its poles and zeros, and on the frequency of the source (aside from a scale factor). This is an extremely significant fact whose full implications you cannot yet appreciate. We shall have much more to say about this topic in subsequent chapters.

When displaying the poles and zeros of a function in the complex plane, we shall conventionally use a small circle to show the location of a zero, and a small x to show the location of a pole. Such a display of the poles and zeros is called a *pole-zero diagram*. If you glance back at the figures, you will notice that we have already adhered to this convention.

In the case of the series RL and RC networks, at least, we found that the impedance has a zero (and the admittance has a pole) at the natural frequency of the network. Thus there is a very close tie between the natural response of a network and the steady-state response to a sinusoid. We found previously that the form of the natural response is determined completely by the natural frequencies. We now find that the natural frequencies have an intimate connection with the steady-state response also. In a later section we shall explore this relationship somewhat further.

6.3 POLE-ZERO DIAGRAMS AND RESONANCE

In discussing the pole-zero diagrams and the graphical method of computing the steady-state response from such diagrams, it was assumed that the frequency is constant. No thought was given to varying the frequency. However, by glancing back at Fig. 6-9, for example, you can appreciate that the complete frequency response can be determined from the pole-zero diagram simply by allowing the point representing the source frequency to move up the $j\omega$ axis from zero to infinity. If line lengths and angles are measured at a number of frequencies, sketches of the magnitude and angle as a function of frequency can be drawn rapidly.

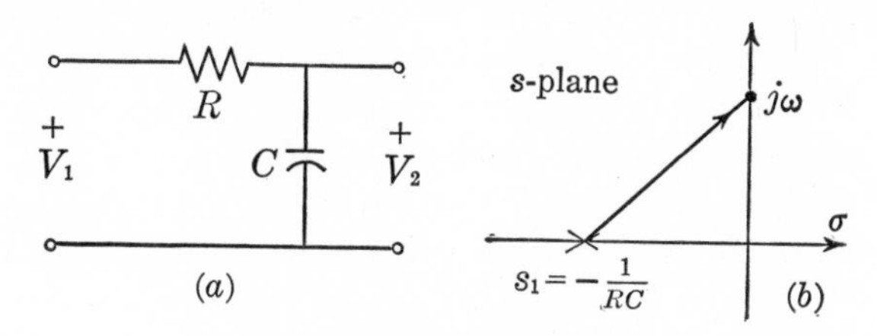

FIG. 6-11

RC network and its pole-zero diagram

To illustrate these remarks consider the simple low-pass filter network shown in Fig. 6-11(a). The transfer voltage ratio can be written from the voltage-divider relationship as

$$G_{21} = \frac{V_2}{V_1} = \frac{1/Cs}{R + 1/Cs} = \frac{1}{RC} \frac{1}{s + 1/RC} = \frac{1}{RC} \frac{1}{s - s_1} \quad (6\text{-}15)$$

where

$$s_1 = -1/RC$$

There is only one pole, as shown in Fig. 6-11(b), and there are no finite zeros. The magnitude of G_{21}, aside from the factor $1/RC$, is the reciprocal of the line length from s_1 to $j\omega$. As ω increases, the length of this line will increase; its minimum length will occur when $\omega = 0$. Thus $|G_{21}|$ has a maximum at $\omega = 0$ and decreases monotonically for increasing ω. Our present reasoning confirms the shape of the curve found in Fig. 5-3. We can determine the 3-db bandwidth by noting that the line from s_1 to $j\omega$ is the hypotenuse of a triangle. It will have a length $\sqrt{2}$ times its length at zero frequency when the triangle becomes isosceles; that is, when $\omega = 1/RC$. Hence the 3-db bandwidth is $1/RC$, a fact which we already know.

At the close of the last section, we observed the intimate relationship between the steady-state response and the natural response in terms of the poles and zeros of the pertinent network function. For the simple network under consideration here, we can demonstrate how the natural response can be determined from a knowledge of the steady-state response.

Assume that measurements of the output voltage are made on a physical circuit, a model of which is the RC network under consideration here, using a variable frequency source. With the source voltage held constant, the frequency at which the output voltage amplitude is $1/\sqrt{2}$ of its value at zero frequency is measured. We know that this frequency should be $\omega = 1/RC$. We also know that the natural frequency of the network is $-1/RC$ and that the natural response has the form $\epsilon^{-t/RC}$. Thus, by measurement of the frequency response, we are able to determine the natural response.

In the particular network under consideration here, only one pole is involved, and the location of this pole is easily determined by measurement. Of course, in other more complicated networks such a simple approach will not be available to us, but the principle is nevertheless valid.

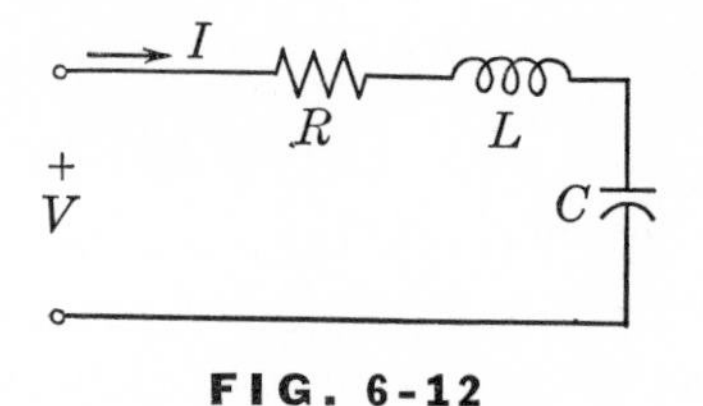

FIG. 6-12
Series resonant circuit

The example which we have just discussed is relatively simple, but it illustrates the ideas which apply in more complicated cases as well. Let us now turn to a consideration of the series resonant circuit shown in Fig. 6-12 and see what we can learn about resonance from the pole-zero diagram. In terms of the complex-frequency variable, the admittance is

$$Y = \frac{1}{sL + R + \dfrac{1}{Cs}} = \frac{1}{L}\,\frac{s}{s^2 + \dfrac{R}{L}s + \dfrac{1}{LC}} \tag{6-16}$$

Note that the denominator is a quadratic and hence has two roots; that is, the admittance has two poles, which may be either real or complex, depending on the relative values of the coefficients. Notice also that the admittance has a zero at the origin of the s-plane.

In discussing resonance in the last chapter, we introduced the concepts of Q, the resonant frequency ω_0, and the bandwidth β. But certain other parameters were also introduced in Chap. 3 in connection with the natural response of this network, namely, the damped and undamped frequencies ω_d and ω_0, and the damping constant α. Let us collect all these expressions.

$$\omega_0{}^2 = \frac{1}{LC} \qquad \alpha = \frac{R}{2L} = \frac{\omega_0}{2Q_0}$$

$$Q_0 = \frac{\omega_0 L}{R} \qquad \omega_d{}^2 = \omega_0{}^2 - \alpha^2 \tag{6-17}$$

$$\beta = \frac{\omega_0}{Q_0}$$

Notice that the damping constant and the bandwidth differ simply by a constant multiplier, $\beta = 2\alpha$.

In terms of these parameters, the expression for the admittance in Eq. (6-16) can be written in several different forms, as follows:

$$Y = \frac{1}{L} \frac{s}{s^2 + 2\alpha s + \omega_0^2} = \frac{1}{L} \frac{s}{s^2 + \dfrac{\omega_0}{Q_0} s + \omega_0^2}$$

$$= \frac{1}{L} \frac{s}{s^2 + \beta s + \omega_0^2} = \frac{1}{L} \frac{s}{(s - s_1)(s - s_2)} \qquad (6\text{-}18)$$

In the last line the factors of the denominator have been shown explicitly. The poles can also be expressed in several alternative forms. The most appropriate form will depend on whether or not the poles are complex. For the moment, let us assume that the poles are complex. Then we find that

$$s_{1,2} = -\alpha \pm j\sqrt{\omega_0^2 - \alpha^2} = -\alpha \pm j\omega_d \qquad (a)$$

$$s_{1,2} = -\frac{\beta}{2} \pm j\omega_0\sqrt{1 - (1/2Q_0)^2} \qquad (b) \qquad (6\text{-}19)$$

From the last equation it is clear that the poles will be complex if $Q_0 > 1/2$. The pole-zero diagram is shown in Fig. 6-13(a).

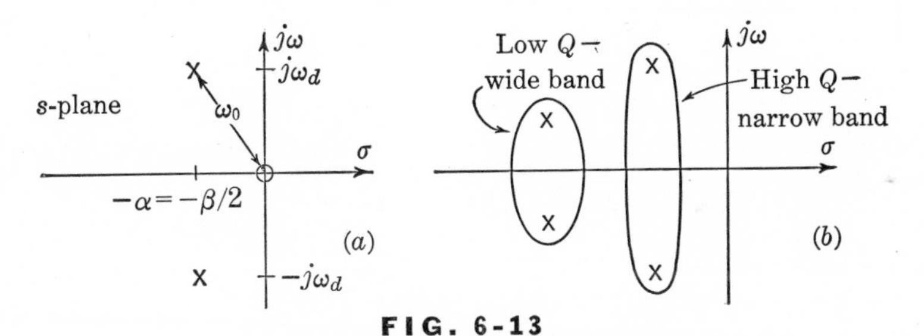

FIG. 6-13

Pole-zero diagram of resonant circuit

Notice that the size of the real part of the poles (namely, α) is half the 3-db bandwidth, and the distance from each pole to the origin (which is the magnitude of the pole) is the resonant angular frequency ω_0. Remembering that the bandwidth is inversely proportional to Q_0 (for a given ω_0), this means that, the closer the poles are to the $j\omega$ axis, the narrower will be the bandwidth and the larger will be Q_0. Conversely, when the poles are far from the $j\omega$ axis, Q_0 will be small and the bandwidth will be large. These comments are illustrated by the two sets of poles in Fig. 6-13(b).

Let us now determine graphically, from the pole-zero diagram, what will happen to the magnitude of the admittance as the frequency is increased from zero to some high value. Consider Fig. 6-14. The magnitude of the admittance will be

$$|Y| = \frac{|s|}{|s - s_1|\,|s - s_2|} = \frac{K_1}{K_2 K_3} \tag{6-20}$$

where the K's are the lengths of the lines from the poles and zero to the point on the $j\omega$ axis corresponding to the particular source frequency, as illustrated in the figure.

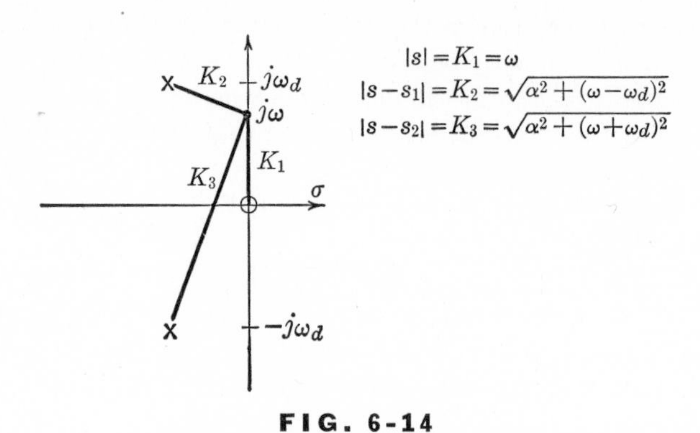

FIG. 6-14

Calculation of resonance curve

As the frequency increases from zero, K_1 and K_3 increase monotonically. However, K_2 at first decreases until ω equals ω_d; that is, until the varying point takes a position opposite the pole at s_1. From this point on, K_2 begins to increase again. Thus K_2 has a minimum at $\omega = \omega_d$. Since K_2 is in the denominator of Eq. (6-20), we might expect $|Y|$ to have a maximum at this frequency. This is not quite an accurate conclusion, however, since it is not necessarily true that a function which depends on more than one factor will reach a maximum when one of the factors reaches a maximum.†

Nevertheless, if the poles are very close to the $j\omega$ axis (high Q, low α) and we restrict ourselves to the range of frequencies in the vicinity of ω_d, the lengths of the lines K_1 and K_3 will be varying at approximately the same rate. Hence the variation of $|Y|$ will depend almost entirely on the variation of K_2. But in the case of high Q (low damping), ω_d is approximately the same as ω_0, the frequency at which

† Suppose $y = f_1(x) \cdot f_2(x)$. Then $y' = f_1 f_2' + f_2 f_1'$. That is, y' will not generally be zero where either f_1' or f_2' is zero.

we know that the magnitude of Y is a maximum. [See Eq. (6-19).]

So we see that the visual picture presented by the pole-zero diagram gives a clear interpretation of the phenomenon of resonance, and, in the high-Q case at least, it yields quantitative results which are approximately correct to quite a high degree of approximation. Very often, such an approximate estimate of network behavior may be all that is required.

The lower the Q, the worse will be the approximate result obtained by neglecting the variations of the lines from $s = 0$ and $s = s_2$. When Q_0 becomes less than 1/2, the poles become real, and the preceding detailed analysis does not apply. The s-plane description in terms of line lengths from the poles and zero, however, is still a valid representation.

We can again demonstrate the relationship between the steady-state response and the natural response in terms of the high-Q resonant circuit. With a variable-frequency voltage source, we again make measurements of the input current. We determine the maximum current and the frequency at which it occurs, which we know is ω_0. Then we find the two frequencies ω_1 and ω_2 on either side of ω_0 at which the current has a value $1/\sqrt{2}$ of its maximum. The bandwidth is then $\beta = \omega_2 - \omega_1$, and the damping constant is $\alpha = (\omega_2 - \omega_1)/2$. The damped angular frequency ω_d is then calculated from Eq. (6-17).

If you recall the discussion of the oscillatory natural response of the series resonant circuit given in Sec. 3-7, the form of this response, $K\epsilon^{-\alpha t} \cos (\omega_d t + \theta)$, is determined completely by α and ω_d. Thus we find that the natural response is determined (except for amplitude and phase) by steady-state measurements.

6.4 POLES, ZEROS, AND NATURAL FREQUENCIES

When examining the simple series RC and RL networks in Sec. 6-2, we found that in each case the admittance function has a pole at the precise point in the s-plane corresponding to the transient (natural) frequency. In more extensive networks the admittance has more than one pole, and it also has zeros. It is our purpose now to inquire into the question of the relationship among the poles and zeros of the admittance or impedance, and the natural frequencies. We shall not be able to carry out this inquiry in a completely general manner, but shall restrict ourselves to some illustrations, which nevertheless will point to some general conclusions.

Consider again the two-branch parallel resonant circuit shown in Fig. 6-15. In the first case assume that a current source is the excitation. The response is the voltage. The ratio of the response sinor to the excitation sinor is the impedance $Z = V/I_g$, an expression for which was found in Eq. (6-10). It is repeated here for convenience.

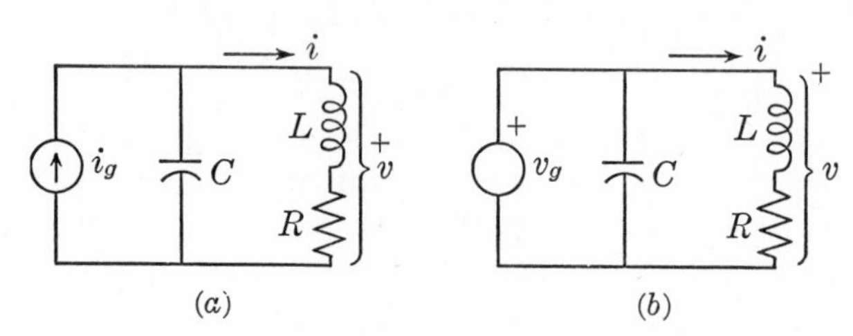

FIG. 6-15

Parallel resonant circuit

$$Z = \frac{1}{C} \frac{(s + R/L)}{s^2 + \frac{R}{L}s + \frac{1}{LC}} \tag{6-21}$$

The poles of the impedance are the roots of the polynomial in the denominator.

Let us next turn to the task of determining the natural frequencies of the same network. We still assume that a current source is the excitation. We must now deal with the time functions and write the appropriate differential equations. One equation is obtained by applying Kirchhoff's current law at the upper node. Another one is obtained by writing the V-I relationship of the RL branch. These will be

$$C\frac{dv}{dt} + i = i_g \qquad (a)$$

$$L\frac{di}{dt} + Ri = v \qquad (b) \tag{6-22}$$

Substituting the second one into the first yields

$$\frac{d^2i}{dt^2} + \frac{R}{L}\frac{di}{dt} + \frac{1}{LC}i = \frac{i_g}{LC} \tag{6-23}$$

Recall that the transient is obtained by setting the excitation equal to zero in this equation and assuming $i = K\epsilon^{st}$, where s is the unknown natural frequency. The result of this process is the following characteristic equation:

$$s^2 + \frac{R}{L} s + \frac{1}{LC} = 0 \tag{6-24}$$

(Go through the details.) The roots of this equation are the natural frequencies.

We immediately notice that the polynomial in Eq. (6-24) is the same one in the denominator of the impedance in Eq. (6-21). Thus *the poles of the impedance are the same as the natural frequencies when the network is excited by a current source.*

Having established this point, let us temporarily abandon this aspect of the discussion and now assume that the same network is excited by a voltage source, as shown in Fig. 6-15(b). The current through the capacitor will be $i_C = C\, dv/dt$. With a known v, this current is found by differentiation. There will be no transient, no natural frequency associated with the capacitor. For the RL branch we shall have

$$v = L\frac{di}{dt} + Ri \tag{6-25}$$

To find the transient, we again set the excitation to zero and assume $i = K\epsilon^{st}$. The result will be

$$s + R/L = 0 \tag{6-26}$$

The solitary root of this equation is the natural frequency under the present conditions. Returning to Eq. (6-21), we see that this root is precisely the zero of the impedance. Thus *the zero of the impedance is the same as the natural frequency when the network is excited by a voltage source.*

It is possible to give another interpretation of the two results we have found. Observe that the natural frequencies in any situation are determined by setting the excitation equal to zero. When the excitation is a current source, this implies opening the terminals, since a current in a branch is made to go to zero by open-circuiting the branch. The natural frequencies under this condition are called the *open-circuit natural frequencies.* Remember also that, when the excitation is a current source, it is the impedance which is the ratio of the response sinor (voltage) to the excitation sinor (current). Hence we say that the *poles of the impedance are the open-circuit natural frequencies.*

On the other hand, when the excitation is a voltage source, this implies short-circuiting the terminals, since the voltage of a branch is made to vanish by shorting the branch. The natural frequencies under this condition are called the *short-circuit natural frequencies.* Furthermore, when the excitation is a voltage source it is the admittance which is the ratio of the response sinor (current) to the excitation sinor (voltage). Hence *the poles of the admittance are the short-circuit natural frequencies.*

It appears that natural frequencies, whether they are open circuit or short circuit, are the same as the poles of some function. The appropriate function is the ratio of response to excitation sinors. When the excitation is a current and the response is the voltage at the same terminals, the appropriate function is the impedance $Z = V/I$. When the excitation is a voltage and the response is the current in the same terminals, the appropriate function is the admittance $Y = I/V$.

But is it necessary that the response be at the same terminals as the excitation? In order to answer this question, let us return to Fig. 6-15(*a*) and assume that the response of interest is the voltage across the inductor. The function expressing the ratio of response to excitation sinors is the transfer impedance $Z_{21} = V_L/I_g$. The voltage sinor V_L can be found in terms of V by the voltage-divider law. Thus

$$V_L = \frac{sL}{sL + R} V \tag{6-27}$$

The transfer impedance is V_L/I_g; it is obtained by dividing both sides of this expression by I_g. Finally, the resulting ratio V/I_g on the right is replaced by the impedance from Eq. (6-21). The result will be

$$Z_{21} = \frac{V_L}{I_g} = \frac{1}{C} \frac{s}{s^2 + \frac{R}{L} s + \frac{1}{LC}} \tag{6-28}$$

Observe that the poles of the transfer impedance are the same as the poles of the input impedance. To complete the story, we should now write the differential equation satisfied by the inductor voltage v_L. This is easily done by replacing di/dt in Eq. (6-23) by v_L/L. From the result, we shall find that the characteristic equation obtained is again the same as the one in Eq. (6-24). (Go through the details.) We conclude that the open-circuit natural frequencies are the same as the poles of the transfer impedance as well as of the input impedance.

We still have not exhausted all possibilities. How about the possibility of asking for a current response with a current excitation? Suppose, for example, that we ask for the current I as the response. An expression for I in terms of I_g can be found from Fig. 6-15(*a*), using the current-divider relationship. Thus

$$\frac{I}{I_g} = \frac{1/Cs}{sL + R + 1/Cs} = \frac{1/LC}{s^2 + \dfrac{R}{L}s + \dfrac{1}{LC}} \tag{6-29}$$

The function expressing the ratio of response to excitation sinors is the current-transfer ratio. Again we find that the poles of this function are the same as the poles of the other functions which express the ratio of a response sinor to the same excitation sinor. We begin to suspect that this is a result whose validity is general and not restricted to the example we have been discussing. This is indeed the case, but we shall not be able to prove it here. Similar conclusions are true for a voltage-source excitation.

Let us here summarize the results of this section. Consider a network excited by a voltage source V_1 at a pair of terminals. The possible responses are the current I_1 in the same terminals, the current I_k somewhere else in the network, and the voltage V_k somewhere else in the network. The network functions which express the ratio of response to excitation sinors are the input admittance $Y = I_1/V_1$, the transfer admittance $Y_{k1} = I_k/V_1$, and the transfer voltage ratio $G_{k1} = V_k/V_1$. These functions all have the same poles, which are the same as the short-circuit natural frequencies—the natural frequencies of the network resulting when the terminals at which the source is connected are shorted.

Next consider a network excited by a current source I_1 at a pair of terminals. The possible responses are the voltage V_1 at the same terminals, the voltage V_k somewhere else in the network, and the current I_k somewhere else. The network functions which express the ratio of response to excitation sinors are the input impedance $Z = V_1/I_1$, the transfer impedance $Z_{k1} = V_k/I_1$, and the transfer current ratio $\alpha_{21} = I_k/I_1$. These functions all have the same poles, which are the same as the open-circuit natural frequencies—the natural frequencies of the network resulting when the terminals at which the source is connected are open.

6.5 SUMMARY

In this chapter we introduced the concept of complex frequency s. This concept permits the extension of the concept of impedance, as well as of the other network functions, which now become functions of a complex variable. For the lumped networks under consideration in this book, these functions are all rational functions of s. As such, aside from a constant multiplier, they are completely characterized by their poles and zeros.

We discussed a graphical procedure for determining the magnitude and angle of the network functions from a knowledge of the positions of the poles and zeros in the complex plane. This graphical procedure gives a clear picture of the variation of the magnitude and angle with source frequency. This point of view was found to be useful in describing the phenomenon of resonance, at least for highly oscillatory (high-Q) networks.

Finally, we discussed the relationships between the poles of a network function and the natural frequencies. We found that the poles of the impedance are the open-circuit natural frequencies and the zeros of the impedance are the short-circuit natural frequencies.

Problems

6-1 Using the fundamental laws of network theory, find a differential equation with current i as independent variable in the network of Fig. P6-1. Assuming that the source voltage has a waveform given by $V\epsilon^{st}$, find the forced solution of this equation. Compare the multiplier of ϵ^{st} in this solution with the admittance of the network.

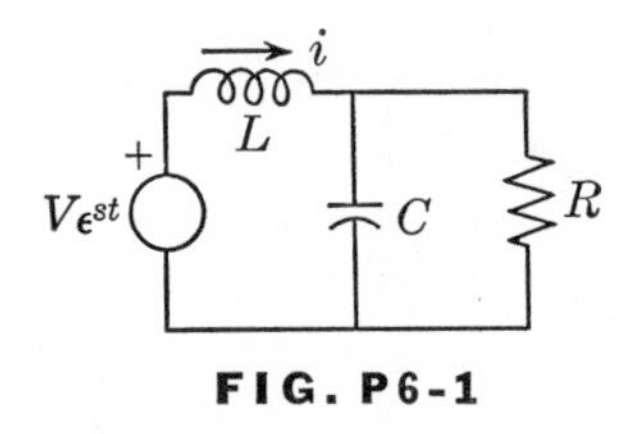

FIG. P6-1

6-2 In each of the networks in Fig. P6-2: (*a*) Find the indicated ratio of response to excitation sinors in terms of the complex frequency variable s. (*b*) Locate the positions of the poles and zeros in the complex-frequency plane. (*c*) Sketch the frequency-response curves utilizing the geometry of the pole-zero diagram.

6-3 In each of the networks in the previous problem, find the ratio of one other response sinor to the excitation sinor and show that the poles are the same.

6-4 The first three networks in Fig. P6-2 are excited by a voltage source, the last three by a current source. Show that the short-circuit natural frequencies of the first three networks are the same as the poles of the corresponding functions determined in the last two

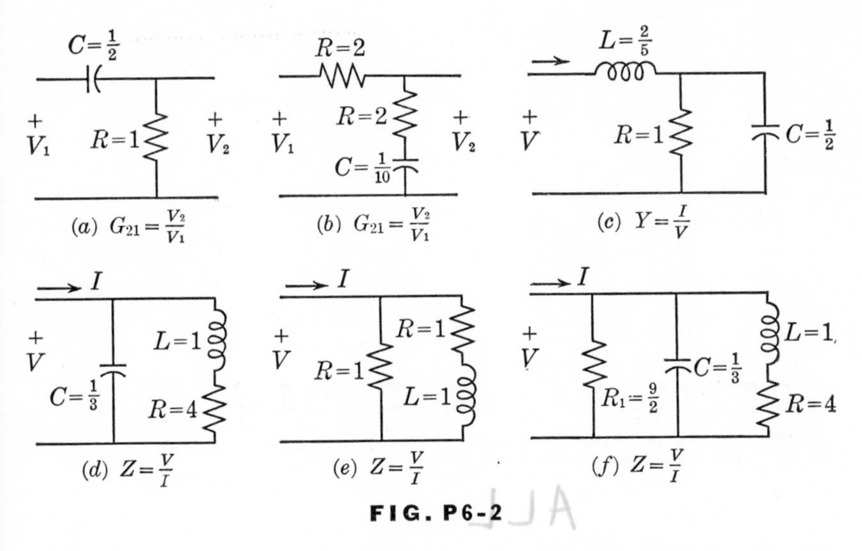

FIG. P6-2

problems; also show that the open-circuit natural frequencies of the last three networks are the same as the poles of the corresponding functions.

6-5 Find the locations of the poles and zeros of the impedance of the parallel resonant circuit shown in Fig. P6-5. A resistor R_1 is to be placed in parallel with the circuit. Find the value of R_1 which will cause the poles to move horizontally one unit to the left in the complex-frequency plane.

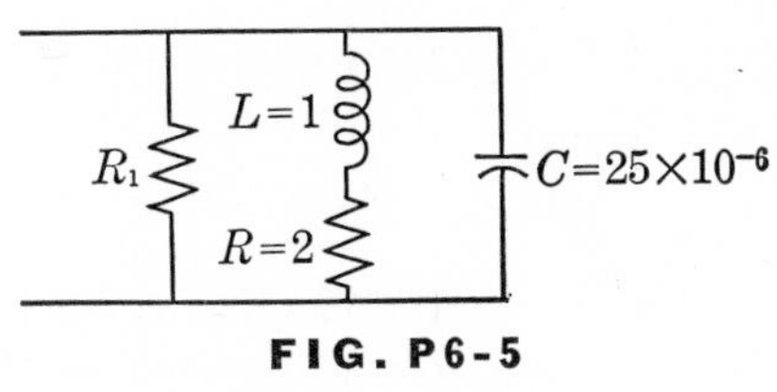

FIG. P6-5

ADDITIONAL METHODS OF ANALYSIS

ALL

7

IN Chap. 4 the sinor method was developed for finding the sinusoidal steady-state response of electrical networks. This procedure starts by transforming the time variables representing voltages and currents into complex variables, the sinors. In terms of the sinors, the *V-I* relationships of all the branches reduce to the form of Ohm's law, the ratio of the voltage sinor to current sinor of any branch being the impedance of that branch.

Following the preliminary discussion of this general technique, we studied some simple methods of solution. These methods fall into two broad categories. When it is desired to find only one particular voltage or current as a response, one technique consists of a step-by-step transformation of the network into equivalent configurations, until, finally, the network is reduced to a simple form, such as a single loop. This is done by alternate uses of the voltage-shift and the current-shift theorems, the equivalences of sources, and series combinations and parallel combinations of branches. Another technique consists of starting from the desired response as an unknown and, by a step-by-step application of the three fundamental laws (Kirchhoff's voltage and current laws and the *V-I* relationships of the branches, which reduce to Ohm's law for sinors), relating the response sinor to the excitation sinor.

In this chapter we shall discuss additional methods of solution of steady-state problems. In particular, we shall extend the methods of loop equations and node equations, which were first introduced for resistive networks, to the sinusoidal steady state. It is worthwhile to review these topics in Chap. 2 before proceeding.

At this point we have reached a degree of sophistication which will permit us to deal habitually with the complex-frequency variable s; that is, we shall write branch impedances and admittances in terms of the complex frequency. When we wish to write these quantities in terms of their real and imaginary parts, or when we wish to substitute numerical values, or when it is necessary to be explicit for any other reason, we can then replace s by $j\omega$, and proceed.

7.1 LOOP EQUATIONS

Let us start by recalling that the fundamental laws on which the analysis of electrical networks is based are Kirchhoff's two laws and the V-I relationships of the elements. For the sinusoidal steady state the V-I relationships in terms of sinors are all of the form $V = ZI$, which we have called Ohm's law for sinors. The systematic solution of steady-state problems, then, will involve the application of these three fundamental laws.

Let us also review the question of the number of nodes, the number of branches, the number of independent equations, etc. It is possible to consider each individual resistor, capacitor, and inductor as a branch. A node is then formed whenever two or more elements are connected. When counting branches, we do not include voltage sources and current sources. Voltage sources are assumed to be short-circuited and current sources open-circuited. In Fig. 7-1, for example, if we count each element (excluding the source) as a branch, there will be seven branches, and five nodes which are labeled A, B, C, D, and E. (The upper end of the voltage source is assumed to be shorted to node E.)

It was previously established that there are $N_v - 1$ independent Kirchhoff current-law equations and $N_m = N_b - (N_v - 1)$ independent Kirchhoff voltage-law equations; where N_b, N_v, and N_m are the numbers of branches, nodes, and independent voltage-law equations, respectively. In the present case, then, there are $5 - 1 = 4$ independent current-law equations and $7 - (5 - 1) = 3$ independent voltage-law equations.

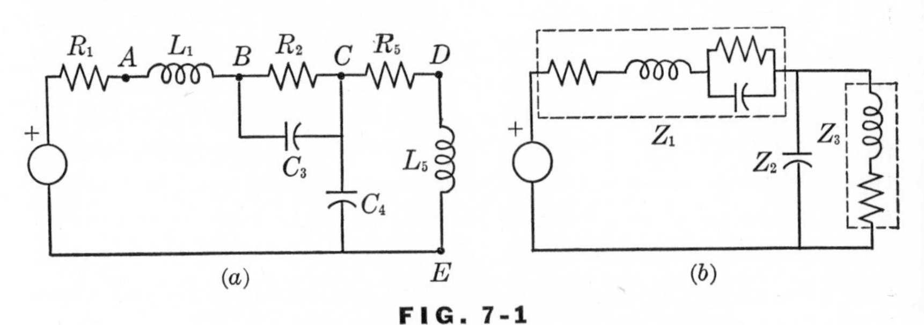

FIG. 7-1

Example for counting branches, nodes, and loops

Although permissible, it is not necessary to consider each element as a branch. Any series combination of elements can be considered as a single branch, such as R_1 in series with L_1, or R_5 in series with L_5. The points between these elements, labeled A and D, will no longer be nodes. Note that, by considering R_1 in series with L_1 as a single branch, the number of branches is reduced by one, but so is the number of nodes. Hence N_m remains the same.

Similarly, any parallel connection of elements can be considered as a single branch, such as C_3 in parallel with R_2. This will reduce the number of branches by one, but it will also reduce the number of loops by one. Hence $N_v - 1$ will stay the same. As a matter of fact, any combination of branches having only two terminals can be counted as a single branch. Any nodes or loops internal to this combination will be destroyed as far as the remainder of the analysis is concerned. Thus it is possible to consider that the network in question consists of only three branches, as illustrated in Fig. 7-1(*b*). There are now only two nodes, C and E. The number of independent voltage-law equations is $N_m = 3 - (2 - 1) = 2$. It is obvious, from an inspection of the network, that this is correct. Thus it does not matter how the elements are grouped into branches; the fundamental relationship among N_b, N_v, and N_m remains valid.

In writing loop equations, the following sequence of steps is taken.

1. Choose an appropriate set of N_m loop currents and express the branch currents in terms of these. For simple networks the

most convenient choice is obvious from an inspection of the network.

2. Choose an appropriate set of N_m loops for writing Kirchhoff's voltage-law equations and write these equations. These loops need not be the same as those chosen in the first step. It is simplest to do so, however, and the resulting equations are more desirable. Hence we shall always do so.
3. Into the voltage equations substitute the V-I relationships of the branches. The variables are now all branch-current sinors.
4. Finally, substitute the loop-current equivalents for the branch currents. The result is a set of N_m equations in the N_m loop-current sinors.

Although we have outlined four distinct steps in arriving at the loop equations, it is not necessary to go through each of these individually, as previously discussed in Chap. 2. After some experience you will find that steps 3 and 4 can be carried out simultaneously while writing the Kirchhoff voltage-law equations in step 2. We shall later discuss a formal procedure for immediately writing the equations in a standard form which avoids all of these steps (except for choosing the loop currents). This formal procedure, however, should not be emphasized until you have a firm grasp of the fundamental process, which is simply an application of the fundamental laws.

Let us illustrate the method of loop equations by means of the network shown in Fig. 7-2. This is the same as that of Fig. 7-1, except

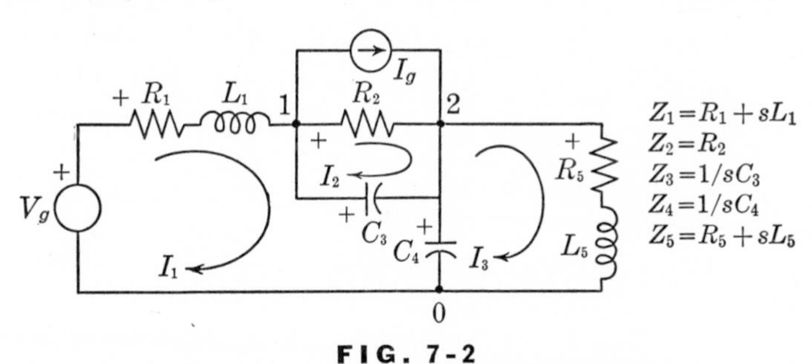

FIG. 7-2

Illustrative example for writing loop equations

that a current source has been added. With the choice of branches indicated in the figure, we see that $N_b = 5$ and $N_v = 3$. Hence $N_m = 5 - (3 - 1) = 3$. The choice of loop currents is shown in the figure. In terms of these loop currents, the branch currents are given by

$$
\begin{aligned}
I_{b1} &= I_1 \\
I_{b2} &= I_2 - I_g \\
I_{b3} &= I_1 - I_2 \\
I_{b4} &= I_1 - I_3 \\
I_{b5} &= I_3
\end{aligned}
\tag{7-1}
$$

We have used the subscript b to distinguish the branch currents from the loop currents. These equations are written assuming the branch-current reference arrows have their tails at the branch-voltage reference plus signs. The reference arrows are omitted for the sake of clarity of the diagram.

The next step involves writing Kirchhoff's voltage law around the three loops. With the references on the diagram, these equations are

$$
\begin{aligned}
V_{b1} + V_{b3} + V_{b4} &= V_g & (a) \\
V_{b2} - V_{b3} &= 0 & (b) \\
-V_{b4} + V_{b5} &= 0 & (c)
\end{aligned}
\tag{7-2}
$$

There are five unknowns in these three equations.

The next step is to substitute the branch V-I relationships. These are all of the form $V = ZI$. The result will be

$$
\begin{aligned}
Z_1 I_{b1} + Z_3 I_{b3} + Z_4 I_{b4} &= V_g & (a) \\
Z_2 I_{b2} - Z_3 I_{b3} &= 0 & (b) \\
-Z_4 I_{b4} + Z_5 I_{b5} &= 0 & (c)
\end{aligned}
\tag{7-3}
$$

Here we still have five unknowns, the branch currents. As the final step, we substitute Eq. (7-1) into the last set. The result is

$$
\begin{aligned}
Z_1 I_1 + Z_3(I_1 - I_2) + Z_4(I_1 - I_3) &= V_g & (a) \\
Z_2(I_2 - I_g) - Z_3(I_1 - I_2) &= 0 & (b) \\
-Z_4(I_1 - I_3) + Z_5 I_3 &= 0 & (c)
\end{aligned}
\tag{7-4}
$$

These are the desired loop equations. Notice that there are now only three unknowns, the loop currents.

After some practice this final set of equations can be written immediately. The intervening steps are not written explicitly but carried out mentally. Thus, adding voltages around the first loop, we would say $V_{b1} = Z_1$ times the current through it, which is I_1. To this is added $V_{b3} = Z_3$ times the current through it, which is $(I_1 - I_2)$. To this is added $V_{b4} = Z_4$ times the current through it, which is $(I_1 - I_3)$.

To this is added $-V_g$ giving a total of zero. In this way Eq. [7-4(a)] is written in one step, and similarly for the other loops.

In preparation for solving the loop equations, let us collect all the terms involving the same current. Since I_g is a known source current, it is transferred to the right side. The result of this step will be

$$
\begin{aligned}
(Z_1 + Z_3 + Z_4)I_1 - Z_3I_2 - Z_4I_3 &= V_g \qquad (a)\\
-Z_3I_1 + (Z_2 + Z_3)I_2 &= Z_2I_g \qquad (b)\\
-Z_4I_1 + (Z_4 + Z_5)I_3 &= 0 \qquad (c)
\end{aligned}
\tag{7-5}
$$

Here we have a set of three simultaneous linear algebraic equations. Note that in the second equation Z_2I_g appears on the right side just as a voltage source would. This is not surprising, since in the original network we could have replaced I_g in parallel with Z_2 by a voltage source Z_2I_g in series with Z_2.

One method of solving algebraic equations is in terms of determinants and Cramer's rule, as first discussed in Chap. 2. The determinant Δ of this set of equations is

$$
\Delta = \begin{vmatrix} (Z_1 + Z_3 + Z_4) & -Z_3 & -Z_4 \\ -Z_3 & (Z_2 + Z_3) & 0 \\ -Z_4 & 0 & (Z_4 + Z_5) \end{vmatrix} \tag{7-6}
$$

The solution of the loop currents will now be

$$
I_1 = \frac{\Delta_1}{\Delta} \qquad I_2 = \frac{\Delta_2}{\Delta} \qquad I_3 = \frac{\Delta_3}{\Delta} \tag{7-7}
$$

where the three determinants in the numerators are

$$
\begin{aligned}
\Delta_1 &= \begin{vmatrix} V_g & -Z_3 & -Z_4 \\ Z_2I_g & (Z_2 + Z_3) & 0 \\ 0 & 0 & (Z_4 + Z_5) \end{vmatrix} \qquad (a)\\
\Delta_2 &= \begin{vmatrix} (Z_1 + Z_3 + Z_4) & V_g & -Z_4 \\ -Z_3 & Z_2I_g & 0 \\ -Z_4 & 0 & (Z_4 + Z_5) \end{vmatrix} \qquad (b)\\
\Delta_3 &= \begin{vmatrix} (Z_1 + Z_3 + Z_4) & -Z_3 & V_g \\ -Z_3 & (Z_2 + Z_3) & Z_2I_g \\ -Z_4 & 0 & 0 \end{vmatrix} \qquad (c)
\end{aligned}
\tag{7-8}
$$

Each of these is the same as Δ, except that one column has been replaced by the column of driving functions on the right of Eq. (7-5).

In order to place each of the source quantities into evidence, let us expand these determinants; Δ_1 is expanded by the first column, Δ_2 by the second column and Δ_3 by the third column, yielding

$$\begin{aligned} \Delta_1 &= \Delta_{11}V_g + \Delta_{21}Z_2I_g & (a) \\ \Delta_2 &= \Delta_{12}V_g + \Delta_{22}Z_2I_g & (b) \\ \Delta_3 &= \Delta_{13}V_g + \Delta_{23}Z_2I_g & (c) \end{aligned} \tag{7-9}$$

In these expressions, Δ_{11}, Δ_{12}, etc., are cofactors of Δ. Their values are

$$\begin{aligned} \Delta_{11} &= (Z_2 + Z_3)(Z_4 + Z_5) & (a) \\ \Delta_{12} &= \Delta_{21} = Z_3(Z_4 + Z_5) & (b) \\ \Delta_{22} &= (Z_1 + Z_3 + Z_4)(Z_4 + Z_5) - Z_4^2 & (c) \\ \Delta_{13} &= Z_4(Z_2 + Z_3) & (d) \\ \Delta_{23} &= Z_3Z_4 & (e) \end{aligned} \tag{7-10}$$

Finally, substituting Eq. (7-9) into Eq. (7-7), the solution for the loop currents takes the form

$$\begin{aligned} I_1 &= \frac{\Delta_{11}}{\Delta} V_g + \frac{\Delta_{21}Z_2}{\Delta} I_g & (a) \\ I_2 &= \frac{\Delta_{12}}{\Delta} V_g + \frac{\Delta_{22}Z_2}{\Delta} I_g & (b) \\ I_3 &= \frac{\Delta_{13}}{\Delta} V_g + \frac{\Delta_{23}Z_2}{\Delta} I_g & (c) \end{aligned} \tag{7-11}$$

Having found the loop currents, all the branch currents can now be found from Eq. (7-1). From these the branch voltages follow directly. Thus all unknowns in the network are found in a systematic manner.

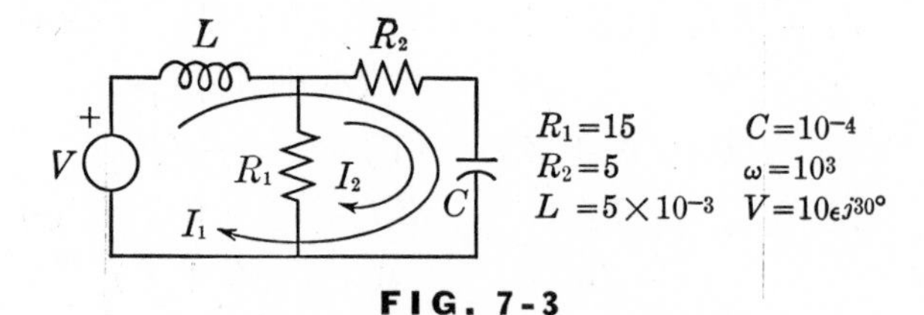

FIG. 7-3

Numerical example for loop analysis

Let us now further illustrate the method of loop analysis by means of a numerical example. In the network of Fig. 7-3 it is desired to find the voltage across the resistor R_1. This network is the same one given in Fig. 4-14; it was there solved by the step-by-step method.

The first step is to choose the loop currents. Counting the number of branches and nodes, we find that two loop currents are required. (For such a simple network the required number of loop currents is obvious at a glance. It is best, however, to get into the habit of checking the formula to make sure.) If we choose the most obvious set of loops—the windows—then the current in R_1 will be the difference (or the sum, depending on the references chosen) of two loop currents. This means that we must find the numerical values of two loop currents. On the other hand, if we choose the loops so that the branch current in R_1 coincides with a loop current, then only one loop current need be found. Hence let us choose the loop currents indicated in the figure.

The next step is to write Kirchhoff's voltage law around these two loops. As we reach each branch voltage, let us mentally replace the voltage by the branch impedance times the branch current, and then immediately express the branch current in terms of the loop currents. Since the branch references have no influence on the loop equations, we mentally take each branch-voltage reference to be at the tail of the corresponding branch-current arrow, and each branch-current reference to coincide with the loop-current reference of the loop around which we are writing the loop equation. This means that the branch references may be reversed when writing equations for different loops. We should emphasize that it is not necessary to follow such a rule; any set of branch references will lead to the same loop equations. Such a procedure, however, will be very convenient.

Following this plan, then, the following equations are obtained.

$$\begin{aligned} sLI_1 + R_2(I_1 + I_2) + \frac{1}{sC}(I_1 + I_2) &= V \qquad (a) \\ R_2(I_1 + I_2) + \frac{1}{sC}(I_1 + I_2) + R_1 I_2 &= 0 \qquad (b) \end{aligned} \tag{7-12}$$

In this case the two loop-current references coincide in branches R_2 and C, so that these branch currents are the sums of the two loop currents. If we again collect terms, Eq. (7-12) becomes

$$\begin{aligned} \left(sL + R_2 + \frac{1}{sC}\right) I_1 + \left(R_2 + \frac{1}{sC}\right) I_2 &= V \qquad (a) \\ \left(R_2 + \frac{1}{sC}\right) I_1 + \left(R_1 + R_2 + \frac{1}{sC}\right) I_2 &= 0 \qquad (b) \end{aligned} \tag{7-13}$$

Let us now substitute the numerical values given in the figure. To do this, we must first replace s by $j\omega$. The equations now become

$$\begin{aligned} (5 - j5)I_1 + (5 - j10)I_2 &= 10\epsilon^{j30°} \qquad (a) \\ (5 - j10)I_1 + (20 - j10)I_2 &= 0 \qquad (b) \end{aligned} \tag{7-14}$$

We are, at present, interested only in current I_2. By Cramer's rule the solution will be

$$I_2 = \frac{\begin{vmatrix} (5 - j5) & 10\epsilon^{j30°} \\ (5 - j10) & 0 \end{vmatrix}}{\begin{vmatrix} (5 - j5) & (5 - j10) \\ (5 - j10) & (20 - j10) \end{vmatrix}} = \frac{-50(1 - j2)\epsilon^{j30°}}{(5 - j5)(20 - j10) - (5 - j10)^2}$$

$$= \frac{-50(1 - j2)\epsilon^{j30°}}{25(5 - j2)} = \frac{-2(2.24\epsilon^{-j63.5°})\epsilon^{j30°}}{(5.4\epsilon^{-j21.8°})} = -0.83\epsilon^{-j11.7°}$$

Finally, taking the reference plus of the voltage across R_1 to be at the upper terminal, we find

$$V_{R1} = -15I_2 = 12.4\epsilon^{-j11.7°}$$

The preceding discussion gives the essence of the method of loop equations for finding the sinusoidal steady-state solution of network problems. The method is applicable to a network having any structure whatsoever, and any number and type of elements, including current sources. Once an appropriate set of the correct number of loops is chosen, the procedure is straightforward. For planar networks, an appropriate set of loops is usually obvious. The question of having a sufficient number of loops is important. It might intuitively appear that, once you have chosen enough loops so that every branch of the network falls on at least one loop, this number should be sufficient. This, however, is not the case.

Consider either of the networks shown in Fig. 7-4. Two loops are shown, which, together, account for every element in each network. Yet, in either case, an incorrect answer will be obtained with this choice

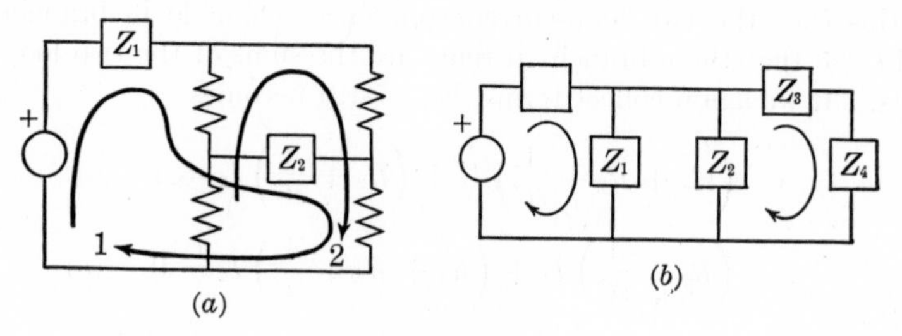

FIG. 7-4

Incomplete choice of loops

of loops. A somewhat closer scrutiny will disclose an inconsistency in each case. Thus, in Fig. 7-4(a), the branch currents in Z_1 and Z_2 are forced to be the same, since loop current 1 is the only current in each of these branches. Furthermore, this result must be true no matter what types of elements there are in any of the branches, or what values they might have. This is an impossible requirement.

Similarly, in Fig. 7-4(b), the choice of loops requires that the same current must flow in Z_2 and in Z_3, no matter what values the impedances have. The three branches consisting of Z_1, Z_2, and the series connection of Z_3 and Z_4, however, are in parallel. The total current entering this combination divides among the branches according to a definite relationship involving the values of the impedances (current-divider law). If any one of the impedances is changed, the current in Z_2 will change relative to the current in Z_3. Hence, the choice of loops forces an impossible situation.

By counting branches and nodes for either of these networks, we see that $N_m = 3$. Thus one more loop is required in each case. Whereas our intuition is not correct in expecting an appropriate set of loops if all elements are included in at least one loop, it is true that (1) *if all elements are included in at least one loop* and (2) *if there are* $N_b - (N_v - 1)$ *loops*, this set of loops will be appropriate. Thus, in Fig. 7-4, any one other loop, together with the ones shown, will be an adequate set of loops.

We promised earlier to discuss a formal procedure for writing loop equations; this we shall now do. To start the discussion, look back at the loop equations given in Eq. (7-5). These equations have the following form.

$$\begin{aligned} a_{11}I_1 + a_{12}I_2 + a_{13}I_3 &= V_1 && (a) \\ a_{21}I_1 + a_{22}I_2 + a_{23}I_3 &= V_2 && (b) \qquad (7\text{-}15) \\ a_{31}I_1 + a_{32}I_2 + a_{33}I_3 &= V_3 && (c) \end{aligned}$$

The unknowns in these equations are the loop currents. The known quantities appear on the right. They include all the source quantities, whether current source or voltage. For ease of making a general statement, it would be convenient to imagine that all current sources are converted to voltage-source equivalents. Then V_1 has the significance of being the algebraic sum of all the source voltages on loop 1; V_2 is the algebraic sum of all the source voltages on loop 2; V_3 is the algebraic sum of all the source voltages on loop 3. In the present case $V_1 = V_g$

and V_2 is the equivalent voltage owing to the current source, namely Z_2I_g. Since there are no sources on loop 3, $V_3 = 0$.

Now let us examine the coefficients a_{ik}. These are of two varieties: those whose two subscripts are the same, like a_{11}, a_{22}, and those whose two subscripts are different. We see that a_{11}, the coefficient of I_1 in the equation for loop 1, is equal to $Z_1 + Z_3 + Z_4$. This is the sum of all the impedances on loop 1; it is sometimes called the *self-impedance* of loop 1. Similarly, $a_{22} = Z_2 + Z_3$. This is the sum of all the impedances on loop 2 (the self-impedance of loop 2), and $a_{33} = Z_4 + Z_5$ is the sum of all the impedances on loop 3 (the self-impedance of loop 3).

Then there are the coefficients with different subscripts. Thus $a_{12} = -Z_3$. We note that Z_3 is the impedance of the branch common between loops 1 and 2, sometimes called the *mutual impedance* between loops 1 and 2. Similarly, $a_{13} = -Z_4$, and Z_4 is the impedance common between loops 1 and 3. Of course, the impedance common between loops 1 and 2 will also be common between loops 2 and 1. Thus $a_{12} = a_{21}$, $a_{13} = a_{31}$, and, in general, $a_{ik} = a_{ki}$ for all values of the subscripts i and k. Since there is no branch common between loops 2 and 3, then $a_{23} = a_{32}$ should be zero; and we see from Eq. (7-5) that this is true.

Let us also check the loop equations of the numerical example of Fig. 7-3 to see if they fit the pattern we have set up. They are given in Eq. (7-13). We notice that, again, a_{11} and a_{22} correctly give the self-impedances of the two loops. Now, however, the mutual-impedance term is preceded by a plus sign rather than a minus sign, as in the previous example. It is not hard to find the condition for the appearance of a plus sign or a minus sign before the mutual-impedance terms. In Fig. 7-3 the two loop-current references coincide in the branch common between the two loops; hence the plus sign. On the other hand, in Fig. 7-2, the loop-current references are in opposite directions in the common branches, both between loops 1 and 2 and between loops 1 and 3; hence the negative signs.

With these examples in front of us, it is easy to conclude that the loop equations for any network will have the following form, which we call the *standard form* of the loop equations.

$$\begin{aligned} a_{11}I_1 + a_{12}I_2 + \cdots + a_{1n}I_n &= V_1 \\ a_{21}I_1 + a_{22}I_2 + \cdots + a_{2n}I_n &= V_2 \\ \cdots\cdots\cdots\cdots\cdots\cdots\cdots\cdots \\ a_{n1}I_1 + a_{n2}I_2 + \cdots + a_{nn}I_n &= V_n \end{aligned} \qquad (7\text{-}16)$$

Each of the voltages appearing on the right-hand side is the algebraic sum of all source quantities appearing on the contour of the corresponding loop. Thus, suppose the loop shown in Fig. 7-5 is loop 2 of an extensive network. The voltage V_2 will involve all three of the source quantities. Because of the source references relative to that of loop current I_2, we have

$$V_2 = V_{g1} - V_{g2} - Z_1 I_g \quad (7\text{-}17)$$

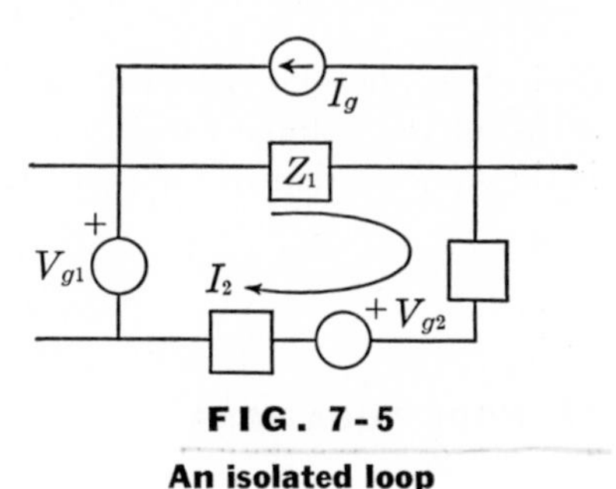

FIG. 7-5

An isolated loop

Note that, if the same sources appear on the contours of other loops in addition to loop 2, they will form part of the equations corresponding to these other loops also.

The a_{kk} coefficients (the ones having two subscripts alike) are the self-impedances of each loop. They always have a positive sign. The a_{ik} coefficients (the ones having unlike subscripts) are plus or minus the impedance common between loops i and k. The sign is plus if the references of loop currents i and k coincide in the common branch; it is minus if the loop-current references are opposite in the common branch.

Note also the ordering on the subscripts of the coefficients. The first subscript in each case refers to the loop for which the equation is written; it coincides with the subscript of the voltage on the right. The second subscript is the same as that of the corresponding loop current.

After this discussion it is a relatively easy job to write the loop equations of any network in standard form, once the loop currents are properly chosen. This mechanistic, formal procedure is not recommended, however, until a thorough grasp is obtained of the fundamental ideas on which the loop equations are based.

The solutions of the loop equations can be written by Cramer's rule as follows.

$$I_1 = \frac{\Delta_{11}}{\Delta} V_1 + \frac{\Delta_{21}}{\Delta} V_2 + \cdots + \frac{\Delta_{n1}}{\Delta} V_n \quad (a)$$

$$I_2 = \frac{\Delta_{12}}{\Delta} V_1 + \frac{\Delta_{22}}{\Delta} V_2 + \cdots + \frac{\Delta_{n2}}{\Delta} V_n \quad (b) \qquad (7\text{-}18)$$

$$\cdots\cdots\cdots\cdots\cdots\cdots\cdots\cdots\cdots$$

For the kth loop current the solution is

$$I_k = \frac{\Delta_{1k}}{\Delta} V_1 + \frac{\Delta_{2k}}{\Delta} V_2 + \cdots + \frac{\Delta_{nk}}{\Delta} V_n \tag{7-19}$$

Note that the first subscript on each cofactor corresponds to the subscript of the voltage which it multiplies, whereas the second subscript is the same as that of the corresponding loop current.

7.2 NODE EQUATIONS

In the last section we discussed one systematic method for finding the steady-state solution of network problems. This method is based on applying Kirchhoff's two laws and Ohm's law for sinors in a specific order. By modifying the order in which these laws are applied, we can arrive at a second systematic method for solving problems. This method leads to the node equations.

To write the node equations, the following sequence of steps is taken.

1. Choose one node of a network as a datum, or reference, node. The voltages of every other node relative to the datum are the node voltages. Express all branch voltages in terms of these. Since each branch is connected to two nodes, each branch voltage will be the difference between two node voltages. For those branches with one end connected to the datum node, the branch voltage is identical with the corresponding node voltage.
2. Choose $N_v - 1$ nodes for writing Kirchhoff's current-law equations, and write these equations. The node which is omitted in this step need not be the datum node chosen in the previous step. The simplest set of equations follows with this choice, however, and we shall always do this.
3. Into the current equations, substitute the V-I relationships of the branches. The variables are now all branch voltages.
4. Finally, substitute for the branch voltages their equivalents in terms of node voltages. The result is a set of $N_v - 1$ equations in the $N_v - 1$ node voltages.

Since each branch voltage is always the difference between two node voltages (the node voltage of the datum node being zero), the last two steps can be performed simultaneously; that is, when substituting the branch V-I relationships into the current-law equations, immediately express the branch voltages in terms of the node voltages.

Let us illustrate the method of node equations by the network of Fig. 7-2 which is redrawn in Fig. 7-6. With the same choice of branches

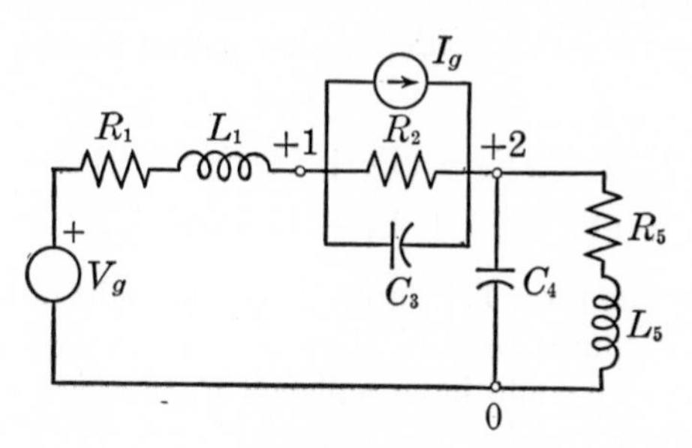

FIG. 7-6

Illustrative example for writing node equations

and nodes as before, we see that $N_v - 1$ is 2. Thus, there are only two independent node equations. The node labeled 0 has been chosen as datum; V_1 and V_2 are the node voltages with references at nodes 1 and 2. The expressions for the branch voltages in terms of the node voltages are

$$\begin{aligned} V_{b1} &= V_g - V_1 && (a) \\ V_{b2} &= V_{b3} = V_1 - V_2 && (b) \\ V_{b4} &= V_{b5} = V_2 && (c) \end{aligned} \qquad (7\text{-}20)$$

For the sake of clarity of the diagram, we have not shown the branch-voltage references. The references are the ones implied by these equations; they are the same as shown in Fig. 7-2. The branch-current references are taken so that the tail of each is at the corresponding voltage-reference plus sign.

Applying Kirchhoff's current law at nodes 1 and 2, we get

$$\begin{aligned} -I_{b1} + I_{b2} + I_{b3} &= -I_g && (a) \\ -I_{b2} - I_{b3} + I_{b4} + I_{b5} &= I_g && (b) \end{aligned} \qquad (7\text{-}21)$$

The next step is to substitute the V-I relationships of the branches, which are all in the form $I = YV$.

$$\begin{aligned} -Y_1V_{b1} + Y_2V_{b2} + Y_3V_{b3} &= -I_g && (a) \\ -Y_2V_{b2} - Y_3V_{b3} + Y_4V_{b4} + Y_5V_{b5} &= I_g && (b) \end{aligned} \qquad (7\text{-}22)$$

The variables are now the branch voltages. When we express these in terms of the node voltages by means of Eq. (7-20), the result will be

$$\begin{aligned} Y_1(V_1 - V_g) + Y_2(V_1 - V_2) + Y_3(V_1 - V_2) &= -I_g \quad (a) \\ -Y_2(V_1 - V_2) - Y_3(V_1 - V_2) + Y_4V_2 + Y_5V_4 &= I_g \quad (b) \end{aligned} \tag{7-23}$$

These are the node equations; there are two equations in the two node-voltage variables. The source voltage V_g is known, and this term can be transferred to the right side. If we collect terms, these equations will become

$$\begin{aligned} (Y_1 + Y_2 + Y_3)V_1 - (Y_2 + Y_3)V_2 &= Y_1V_g - I_g \quad (a) \\ -(Y_2 + Y_3)V_1 + (Y_2 + Y_3 + Y_4 + Y_5)V_2 &= I_g \quad (b) \end{aligned} \tag{7-24}$$

Notice that the term Y_1V_g appears on the right of the first equation just as a current source would. This is to be expected because, in the original network, we could have replaced V_g in series with Y_1 by a current source Y_1V_g in parallel Y_1.

The solution of this set of equations is effected very simply in terms of determinants by Cramer's rule. If the determinant of the set of equations is labeled Δ, then the solution becomes

$$\begin{aligned} V_1 &= \frac{\Delta_1}{\Delta} = \frac{\Delta_{11}}{\Delta}(Y_1V_g - I_g) + \frac{\Delta_{21}}{\Delta}I_g \\ &= \frac{\Delta_{11}Y_1}{\Delta}V_g + \left(\frac{\Delta_{21} - \Delta_{11}}{\Delta}\right)I_g \\ V_2 &= \frac{\Delta_2}{\Delta} = \frac{\Delta_{12}}{\Delta}(Y_1V_g - I_g) + \frac{\Delta_{22}}{\Delta}I_g \\ &= \frac{\Delta_{12}Y_1}{\Delta}V_g + \left(\frac{\Delta_{22} - \Delta_{12}}{\Delta}\right)I_g \end{aligned} \tag{7-25}$$

where

$$\begin{aligned} \Delta &= \begin{vmatrix} (Y_1 + Y_2 + Y_3) & -(Y_2 + Y_3) \\ -(Y_2 + Y_3) & (Y_2 + Y_3 + Y_4 + Y_5) \end{vmatrix} \quad (a) \\ \Delta_1 &= \begin{vmatrix} (Y_1V_g - I_g) & -(Y_2 + Y_3) \\ I_g & (Y_2 + Y_3 + Y_4 + Y_5) \end{vmatrix} \quad (b) \\ \Delta_2 &= \begin{vmatrix} Y_1 + Y_2 + Y_3 & (Y_1V_g - I_g) \\ -(Y_2 + Y_3) & I_g \end{vmatrix} \quad (c) \end{aligned} \tag{7-26}$$

and the Δ_{ik}'s are the cofactors of Δ, in terms of which Δ_1 and Δ_2 are expanded.

Equation (7-25) constitutes the solution for the node voltages. The branch voltages now follow from Eq. (7-20). Finally, the branch currents are obtained from the V-I relationships. The solution is then complete. Note that, in these expressions, Δ is the determinant of node admittances. This is not the same as the Δ used in the solution of the loop equations.

As pointed out before, it is not necessary to follow explicitly all the steps leading to the node equations in Eq. (7-23). Once the datum node is chosen, the node equations can be written immediately in the form of Eq. (7-23). Consider writing the current equation at node 1. Except for sources, whose references already appear on the diagram, we take all branch-current references to be away from the node, and the branch-voltage references to be at the tails of the current arrows. Thus, starting at branch 1, we say $I_{b1} = Y_1$ times the voltage across it, which is $(V_1 - V_g)$. To this is added $I_{b2} = Y_2$ times the voltage across it, which is $(V_1 - V_2)$. To this is added $I_{b3} = Y_3$ times the voltage across it, which is $(V_1 - V_2)$. To this is added I_g, giving a total of zero. Equation [7-23(*a*)] is the result. The same plan is followed at the other nodes. Note that when writing the equation at each node the branch-current references are all taken away from the node. Hence, the references change as we go from node to node. But this causes no difficulty since the branch references have no influence on the node equations.

Let us now solve the problem given in Fig. 7-3, this time by node equations. The network is redrawn in Fig. 7-7. There are three nodes

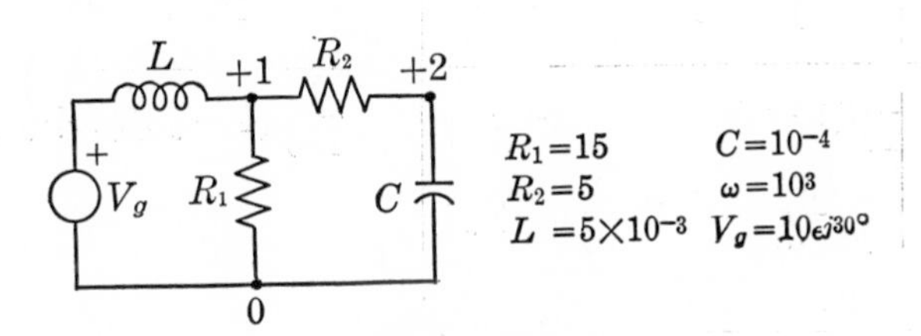

FIG. 7-7

Numerical example for node analysis

and hence two independent node equations. The node labeled 0 is chosen as a datum.

The next step is to apply Kirchhoff's current law at nodes 1 and 2. For each branch we mentally replace the branch current by the admittance times the branch voltage, which we then immediately express in terms of the node voltages. Following this plan, and choosing the branch references as discussed above, we obtain the following equations.

$$\frac{1}{sL}(V_1 - V_g) + \frac{1}{R_1}V_1 + \frac{1}{R_2}(V_1 - V_2) = 0 \quad (a)$$
$$\frac{1}{R_2}(V_2 - V_1) + sCV_2 = 0 \quad (b) \qquad (7\text{-}27)$$

Collecting terms and transposing the term involving V_g to the right, these become

$$\left(\frac{1}{R_1} + \frac{1}{R_2} + \frac{1}{sL}\right)V_1 - \frac{1}{R_2}V_2 = \frac{1}{sL}V_g \quad (a)$$
$$-\frac{1}{R_2}V_1 + \left(\frac{1}{R_2} + sC\right)V_2 = 0 \quad (b) \qquad (7\text{-}28)$$

Let us now substitute the numerical values, first replacing s by $j\omega$. We get

$$\frac{1}{15}(4 - j3)V_1 - \frac{1}{5}V_2 = -j2\epsilon^{j30°} \quad (a)$$
$$-\frac{1}{5}V_1 + \frac{1}{10}(2 + j1)V_2 = 0 \quad (b) \qquad (7\text{-}29)$$

Our interest is in finding V_1. By Cramer's rule we get

$$V_1 = \frac{\begin{vmatrix} -j2\epsilon^{j30°} & -\frac{1}{5} \\ 0 & \frac{1}{10}(2 + j1) \end{vmatrix}}{\begin{vmatrix} \frac{1}{15}(4 - j3) & -\frac{1}{5} \\ -\frac{1}{5} & \frac{1}{10}(2 + j1) \end{vmatrix}} = \frac{\frac{-j(2 + j1)}{5}\epsilon^{j30°}}{\frac{1}{150}(4 - j3)(2 + j1) - \frac{1}{25}}$$

$$= \frac{30(1 - j2)\epsilon^{j30°}}{5 - j2} = 12.4\epsilon^{-j11.7°} \qquad (7\text{-}30)$$

This answer agrees with the one we found previously, using loop equations.

We have now discussed the essentials of the method of node equations for finding the sinusoidal steady-state solution of network problems. In the present case we shall encounter no difficulty corresponding to the problem of choosing an adequate set of loops in the loop method of analysis. Any one of the nodes of a network is chosen as the datum, and node equations are written at all the other nodes.

Just as we were able to write the loop equations in a standard form, so also it is possible to write the node equations in a standard form. Look back at Eq. (7-24). These are in the form

$$\begin{aligned} b_{11}V_1 + b_{12}V_2 &= I_1 & (a) \\ b_{21}V_1 + b_{22}V_2 &= I_2 & (b) \end{aligned} \tag{7-31}$$

The variables are the node voltages. The quantities appearing on the right side include the sources in the network, both voltage and current. It is convenient to imagine that all voltage sources have been converted to their current-source equivalents. Then I_1 has the significance of being the algebraic sum of all source currents connected to node 1 and similarly for I_2.

Now look at the coefficients. b_{11} is equal to $Y_1 + Y_2 + Y_3$, which is the sum of all admittances connected to node 1, sometimes called the *self-admittance* of node 1. Similarly, $b_{22} = Y_2 + Y_3 + Y_4 + Y_5$ is the sum of all the admittances connected to node 2. On the other hand, $b_{12} = b_{21} = -(Y_2 + Y_3)$; and $(Y_2 + Y_3)$ is the admittance of the branches common between nodes 1 and 2, the *mutual admittance.*

With a little thought, it will be clear that the node equations for any network will have the following form, which we call the *standard* form for node equations.

$$\begin{aligned} b_{11}V_1 + b_{12}V_2 + b_{13}V_3 + \cdots + b_{1n}V_n &= I_1 \\ b_{21}V_1 + b_{22}V_2 + b_{23}V_3 + \cdots + b_{2n}V_n &= I_2 \\ \cdots\cdots\cdots\cdots\cdots\cdots\cdots\cdots & \\ b_{n1}V_1 + b_{n2}V_2 + b_{n3}V_3 + \cdots + b_{nn}V_n &= I_n \end{aligned} \tag{7-32}$$

Each of the currents appearing on the right is the algebraic sum of all source currents connected to the corresponding node. (Any voltage source is assumed to be replaced by its current-source equivalent for this purpose.) If a particular current source is connected between two nodes, this source current will appear in the equations for both nodes, once with a positive sign and once with a negative sign.

The coefficient b_{kk} is the sum of all admittances connected to node k. The sign is always positive. The coefficient b_{ik} is minus the admittance common between nodes i and k, and $b_{ik} = b_{ki}$. The sign here is always negative, in contrast with the case of the a_{ik} coefficients in the loop equations. In that case we saw that a plus sign is appropriate when the two loop-current references coincide in the common branch. The corresponding situation for node equations would occur if we chose one of the node-voltage references to be positive at the datum node. Since we have agreed never to do this, the sign of the b_{ik} coefficients will always be negative.

The solution of the node equations for any one of the node voltages, say V_k, will be

$$V_k = \frac{\Delta_{1k}}{\Delta} I_1 + \frac{\Delta_{2k}}{\Delta} I_2 + \cdots + \frac{\Delta_{nk}}{\Delta} I_n \tag{7-33}$$

We should again emphasize that, although this formal way of writing the node equations may save a step or two, you should use it only after you have a solid grasp of the foundations.

Let us now glance briefly at the over-all features of the loop method and the node method of analysis. The result of either method is a set of simultaneous algebraic equations; of these there are $N_v - 1$ in the node method and $N_b - (N_v - 1)$ in the loop method. In a given problem the choice between the use of the two methods might very well depend on the number of simultaneous equations required to be solved. Expanding a low-order determinant is much easier than expanding a high-order one. For some networks $N_v - 1$ will be less than $N_b - (N_v - 1)$; for others, the opposite will be true.

If the objective is a complete solution for all branch variables, there probably is no other criterion for choosing the method, other than number of equations. If only one or more specific voltages are required, however, it might be most advantageous to use the node method. Similarly, if one or more currents are desired, the loop method might be better.

7.3 THE PRINCIPLE OF SUPERPOSITION

Let us recall that the sinusoidal steady-state responses in which we are interested are the forced solutions of linear differential equations with constant coefficients. When discussing the forced solution of such equations in Chap. 3, in response to excitations of any arbitrary waveshape, we made the following observation. The forced response to a combination of different excitations is the same as the sum of the forced responses to these same excitations when they are applied individually, all others being removed. This result is a consequence of the linearity of the equations which, in turn, is a result of our choice of a network model. Such a statement can be made in any kind of relationship whenever the effect (the response) is linearly related to the cause (the excitation).

A very general statement of the principle of superposition can be made in the following way. *Whenever an effect is linearly related to its cause, then the effect owing to a combination of causes is the same as the sum of the effects owing to each cause acting alone, all other causes being inoperative.* We have already seen this principle demonstrated when the "effect" is the forced response of a linear electric network and the "causes" are excitations of any arbitrary waveshapes. We now wish to restrict it further and consider all the excitations to be of sinusoidal waveforms.

In the first place, suppose all sources in a network are sinusoidal and of the same frequency. Refer to Eq. (7-19) which gives the solution of the loop equations for any of the loop currents. If all the sources except V_i are set equal to zero, the steady-state response for I_k will be $\Delta_{ik}V_i/\Delta$. This is the same as the partial response of source V_i when all the sources are operating as shown by Eq. (7-19), which demonstrates the truth of the superposition principle for the case under consideration here. The same result can be established by considering the node equations instead.

For actually computing numerical results in the sinusoidal steady state, the superposition principle often has very little utility. It is often just as easy to find a numerical solution with all sources operating as it is to find the partial solutions with one source at a time, and then adding. As a conceptual tool, however, the superposition principle is of first rank in importance.

The principle of superposition is also of fundamental value in finding the response to a waveform which is not sinusoidal. Consider, for example, the situation illustrated in Fig. 7-8. A voltage source v_g, which consists of the sum of two sinusoidal voltages of different frequencies ω_1 and ω_2, is connected to a network. It is desired to find the current response in a branch which we have shown explicitly in the

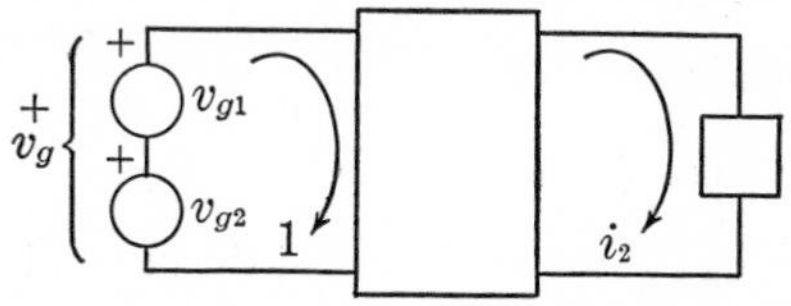

v_g v_{g1} v_{g2} 1 i_2

$$v_{g1}=\sqrt{2}\,|V_{g1}|\cos(\omega_1 t+\alpha)$$
$$v_{g2}=\sqrt{2}\,|V_{g2}|\cos(\omega_2 t+\beta)$$

FIG. 7-8

Use of superposition for nonsinusoidal waveforms

figure. Let us choose the loop currents in such a way that there is only one loop current i_2 in this branch, so that i_2 is also the branch current. The source appears in loop 1. The source voltage v_g, being the sum of v_{g1} and v_{g2}, is not sinusoidal. Hence we cannot use the sinor method for finding the steady-state response directly.

The superposition principle, however, tells us that the steady-state response to v_g consists of the sum of the steady-state responses to v_{g1} and to v_{g2} when these are acting alone. The steady-state response to each of these alone can certainly be found by sinor methods. We must convert back to the time responses, however, before we add the parts to get the total response.

As an illustration of this process, consider the situation illustrated in Fig. 7-9. A voltage source v_g, which is the sum of two sinusoids of

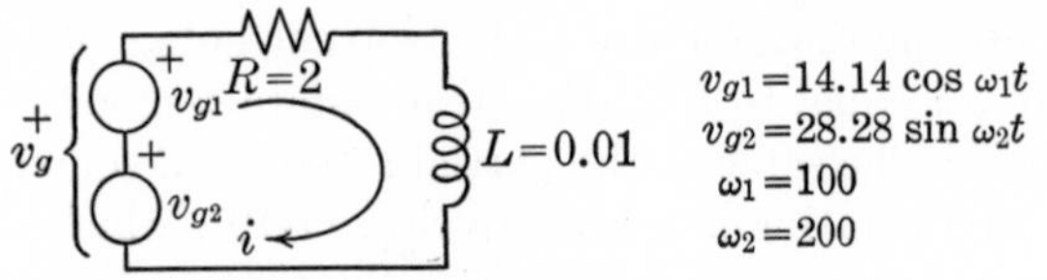

FIG. 7-9

Illustration of principle of superposition

different frequency, is applied to a series RL circuit. It is desired to find the steady-state current i. We shall solve the problem by finding the partial currents i_1 and i_2 owing to each source term acting alone.

To find i_1, we first find the sinor I_1 which is given by

$$I_1 = \frac{V_{g1}}{R + j\omega_1 L} = \frac{10\epsilon^{j0°}}{2 + j100(0.01)} = 4.46\epsilon^{-j26.5°} \tag{7-34}$$

Then the corresponding time function is

$$i_1 = 4.46\sqrt{2} \cos (100t - 26.5°)$$

Next we turn to i_2. This is found by first finding the sinor I_2. It is

$$I_2 = \frac{V_{g2}}{R + j\omega_2 L} = \frac{20\epsilon^{-j90°}}{2 + j200(0.01)} = \frac{10\epsilon^{-j90°}}{1 + j1} = 7.07\epsilon^{-j135°} \tag{7-35}$$

The corresponding time function is

$$i_2 = 7.07\sqrt{2} \cos (200t - 135°)$$

By the principle of superposition, the total steady-state current is the sum of i_1 and i_2

$$i = i_1 + i_2 = 6.31 \cos (100t - 26.5°) + 10 \cos (200t - 135°) \tag{7-36}$$

If you attempt to find the total current i in this simple example without using the superposition principle, you will appreciate its great value in the case under consideration here.

Although we have shown the two sinusoidal sources of different frequency to be connected at the same place in the network, the superposition principle applies no matter where the sources are connected. It should also be clear that the principle is applicable for any number of sources and for either type—voltage or current. The superposition principle is the basis for some of the advanced techniques for finding the response of linear networks (or any kind of linear systems) to excitations of arbitrary waveshape. We shall discuss these techniques in later chapters.

It should be remembered that, if an effect is not linearly related to its cause, then the superposition principle does not apply. A common error is to apply the superposition principle to the calculation of power. For example, the instantaneous power dissipated in the resistor in Fig. 7-9 is

$$p = Ri^2 = R(i_1 + i_2)^2 = R(i_1^2 + i_2^2 + 2i_1i_2)$$

If we compute the power due to each component current and add the two resulting expressions, erroneously believing the validity of the superposition principle, we shall get only the first two terms on the right side of the last equation, which is incorrect.

7.4 THÉVENIN'S AND NORTON'S THEOREMS

SEE LAB

The loop and node methods of analysis are two systematic ways of finding all the branch current and voltage responses of networks. When the response in only one branch is required, however, perhaps the power and generality of these methods is wasted. It would be useful to have simpler methods to deal with these cases. We have already discussed some simple methods in Chap. 4.

One such method of solution consists of simplifying the network in stages until the entire network, excluding the branch in question, consists of a voltage source in series with a branch. This is illustrated in Fig. 7-10. The question arises as to whether or not this process can always be carried out, no matter what the structure of the network might be. Furthermore, if it *is* always possible to find an equivalent in the form under discussion, must we always go through the step-by-

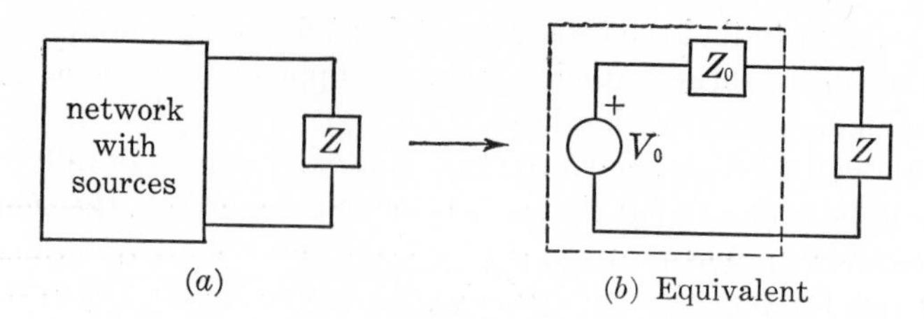

FIG. 7-10

Equivalent network obtained by source conversions

step process, or is there some other formulation that will also lead to the result?

To answer these questions, consider Fig. 7-11(*a*). One of the branches of a network which contains any number of sinusoidal voltage and current sources, all of the same frequency, is singled out for observation. The current sinor in the branch is labeled I. Suppose, now, that we add a sinusoidal voltage source, having the same frequency, in series with the branch, as shown in Fig. 7-11(*b*), and adjust the sinor V

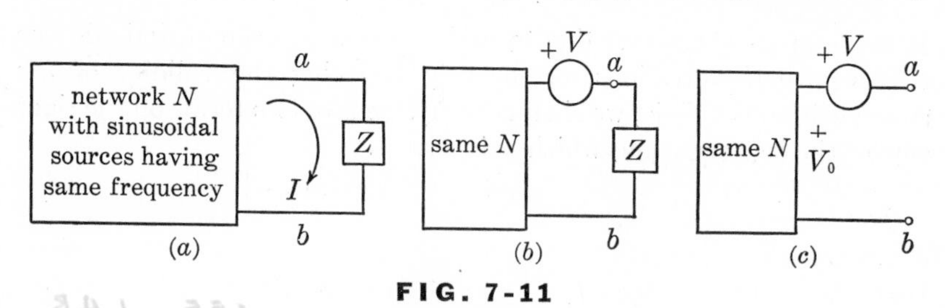

FIG. 7-11

Deriving Thévenin's theorem

until the current in the branch becomes zero. Now, according to the principle of superposition, the total current in the branch is the sum of the currents owing to all the sources, including the one we added. The current owing to all the sources in the original network is I. Since the total current is zero, the current owing to the added source, with all the other sources removed, is $-I$. If we reverse the reference of V, then the current it produces in the branch Z will be the negative of what it was before, namely $-(-I) = I$. It follows that, if we reverse the reference for V in Fig. 7-11(*b*), and remove all the sources in the original network, the same current will flow in branch Z as before. Hence this configuration will be equivalent to the original network, as far as branch Z is concerned. It remains only to find what the value of V must be.

To find the required value of V, note that no current flows in branch Z, and no voltage appears across it, under the conditions of Fig. 7-11(*b*). Hence we might as well open-circuit it, as shown in Fig. 7-11(*c*). The voltage at these open terminals is still zero. Let us call the voltage sinor at the terminals of network N alone (exclusive of the source) V_0. Then, from Fig. 7-11(*c*), $V - V_0 = 0$. Hence the desired value of V is $V = V_0$. But V_0 is simply the open-circuit voltage, since the terminals of the network are open in Fig. 7-11(*c*).

In conclusion, then, we have obtained the network of Fig. 7-12(*a*)

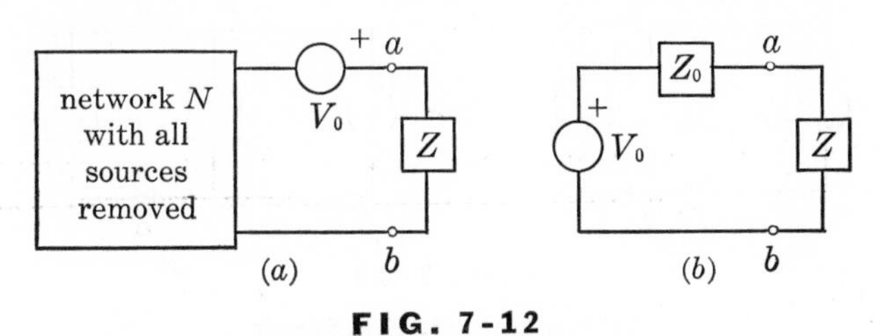

FIG. 7-12

Thévenin equivalent network

as the equivalent of the original network at the terminals *a-b*. The equivalent consists of a voltage source whose voltage V_0 is the voltage which will appear at the terminals *a-b* of the original network, if these terminals are open-circuited, in series with the original network but with all the sources removed. A completely passive network can be characterized by a single quantity—its impedance (or admittance). Hence it is usual to draw the equivalent circuit as shown in Fig. 7-12(*b*), where Z_0 is the impedance of the original network at the terminals *a-b*, with all the sources removed.

A word is in order about removing sources. To remove a voltage source means to make the voltage zero. This is done by short-circuiting it. Similarly, to remove a current source means to make its current zero. This is done by open-circuiting it.

We now have the result we were after. Any network consisting of R, L, and C elements, and voltage and current sources, can be replaced at any pair of terminals by a voltage source in series with a branch consisting of R's, L's, and C's. This theorem was first advanced by M. L. Thévenin, in France, about three quarters of a century ago, and it is known as *Thévenin's theorem*. The network is called the *Thévenin equivalent network*. (The theorem can be extended to include other elements of our model as we introduce them later.) It is not necessary to carry out the step-by-step conversions we did previously. Any method by which the open-circuit voltage and the impedance at the terminals, with all the sources removed, can be calculated is permissible.

Before we give an example illustrating the use of this theorem, let us make the following observation. We have already found that a voltage source in series with a passive branch can be replaced by a current source of the correct value, in parallel with the branch. Thus the Thévenin equivalent of Fig. 7-12(*b*) can be replaced by still another equivalent, as shown in Fig. 7-13. This equivalence was first pointed

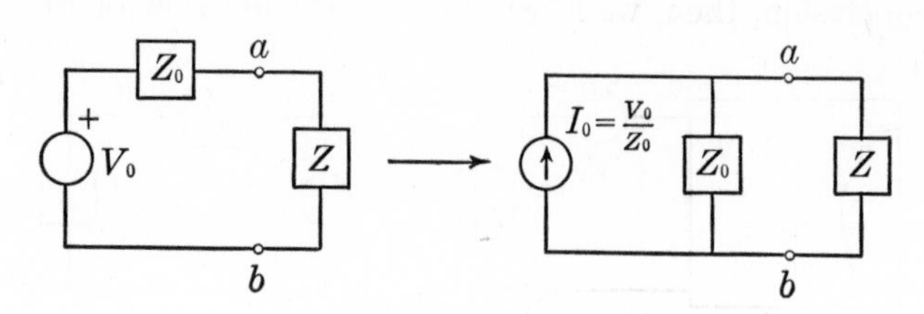

FIG. 7-13

Norton equivalent network

out by E. L. Norton; the resulting network is called the *Norton equivalent network*.

The current I_0 in the Norton equivalent is the ratio of V_0 and Z_0 in the Thévenin equivalent. An interpretation for I_0 can be obtained by noting from the Thévenin equivalent that, if the load impedance Z is shorted, the current in the short circuit will be $V_0/Z_0 = I_0$. Thus I_0 is interpreted as the short-circuit current at the terminals *a-b* of the original network. Observe that, to get either the Thévenin or the Norton equivalent, it is necessary to know only two of the three quantities V_0, I_0, Z_0. In some cases it might be more convenient to calculate V_0 and I_0; then Z_0 is found by taking their ratio.

Let us now illustrate the use of these theorems by means of an example. Consider the network of Fig. 7-14. It is required to find the

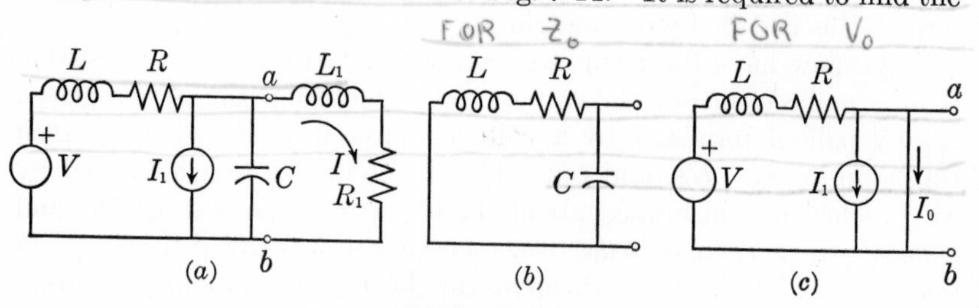

FIG. 7-14

Illustrative example

current in R_1. We shall solve the problem by replacing the network to the left of terminals *a-b* by a Thévenin equivalent. To do this, it is necessary to find the two quantities V_0 and Z_0. To find Z_0 we short-circuit the voltage source and open-circuit the current source. The resulting network is shown in Fig. 7-14(*b*). It consists of two branches in parallel. Hence the Thévenin impedance is

$$Z_0 = \frac{1}{sC + \dfrac{1}{sL + R}} = \frac{sL + R}{s^2LC + sCR + 1} \qquad (7\text{-}37)$$

We now turn to the open-circut voltage V_0 at terminals *a-b*. Finding this is much more difficult. Suppose we consider, however, finding the short-circuit current instead. When terminals *a-b* are shorted, the result is shown in Fig. 7-14(*c*). The short-circuit current I_0 is easily found by applying Kirchhoff's current law at node *a*. We find

$$I_0 = \frac{V}{R + sL} - I_1 \qquad (7\text{-}38)$$

Knowing I_0 and Z_0, the open-circuit voltage is now calculated to be

$$V_0 = Z_0 I_0 = \frac{V - (sL + R)I_1}{s^2LC + sCR + 1} \qquad (7\text{-}39)$$

The resulting network is shown in Fig. 7-15. The desired current is now readily found to be

$$I = \frac{V_0}{R_1 + sL_1 + Z_0} \qquad (7\text{-}40)$$

with Z_0 and V_0 given by Eqs. (7-37) and (7-39).

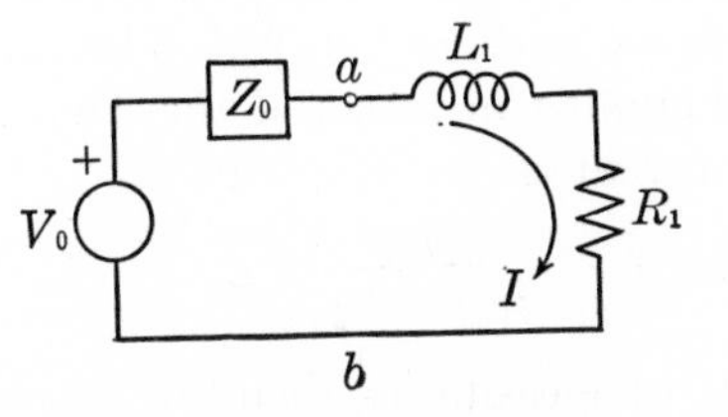

FIG. 7-15
Illustrative example continued

In discussing the Thévenin and Norton theorems, we have assumed that there is a network with a known structure and elements and sources, so that we can compute the open-circuit voltage or the short-circuit current, and we can remove the sources to calculate the impedance. Suppose, rather, that we have a physical electrical circuit which has a pair of terminals to which external connections can be made. We wish to represent this circuit by a model having the form of the Thévenin equivalent or the Norton equivalent. The internal sources may or may not be available for us to open-circuit or short-circuit. It is possible to find the desired equivalent circuit by measurements at the terminals.

Such a physical circuit with terminals a and b is shown in Fig. 7-16. To find the Thévenin or the Norton equivalent, we need to determine two of the three quantities V_0, I_0, and Z_0. The first thought that probably comes to mind is to measure both the open-circuit voltage and

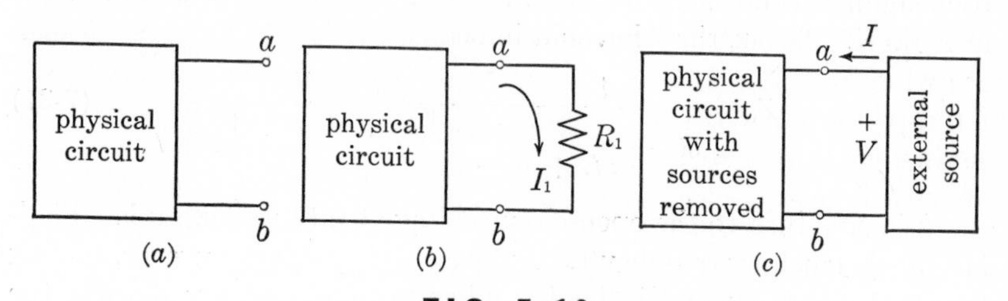

FIG. 7-16

Thévenin equivalent of a physical circuit

the short-circuit current. Conceptually, this may be the simplest thing to do. There will sometimes, however, be practical difficulties associated with such a step. Placing a short circuit across the terminals of a physical circuit may produce such large currents as to cause damage to the circuit.

It is not actually necessary to measure V_0 and I_0. Suppose, instead, we place a known load at terminals a-b, say a resistance R_1, as shown in Fig. 7-16(b). The current under this condition is labeled I_1. Thinking in terms of the Thévenin equivalent, an expression for this current is

$$I_1 = \frac{V_0}{Z_0 + R_1} \tag{7-41}$$

If we measure I_1, then this equation will give us a relationship between the quantities V_0 and Z_0. If we can also measure the open-circuit voltage V_0, then Z_0 can be calculated. If V_0 cannot be measured, another resistance R_2 can be placed at terminals a-b and the resulting current I_2 measured. Again, in terms of the Thévenin equivalent, an expression for I_2 is

$$I_2 = \frac{V_0}{Z_0 + R_2} \tag{7-42}$$

The two equations can now be solved for V_0 and Z_0.

In case the internal sources are available and can be removed, then it is possible to measure Z_0 in the following way. With the sources removed, we apply an external physical source at the terminals a-b, as shown in Fig. 7-16(*c*), and measure the voltage V and current I. The impedance Z_0 is then the ratio of V and I.

We should make one comment concerning the measurements we have mentioned. When we say "measure I_1," this means measure both the magnitude and the angle, since I_1 is a complex number. This holds true for all the other voltage or current sinors which we have suggested measuring. It is not our purpose here to discuss the techniques for carrying out these measurements; this is outside our scope. At some stage in your technical education you must learn about the theory of measurements and proper measurement technique under different conditions. Here, the desire is simply to become aware that the concept of the Thévenin and the Norton equivalents is useful in obtaining a model for an actual physical circuit with the help of measurements made at the terminals of the circuit.

Thévenin's theorem is useful in explaining another concept which is quite commonly used. Consider the network containing sources shown in Fig. 7-17. A load Z is to be connected at the terminals, and

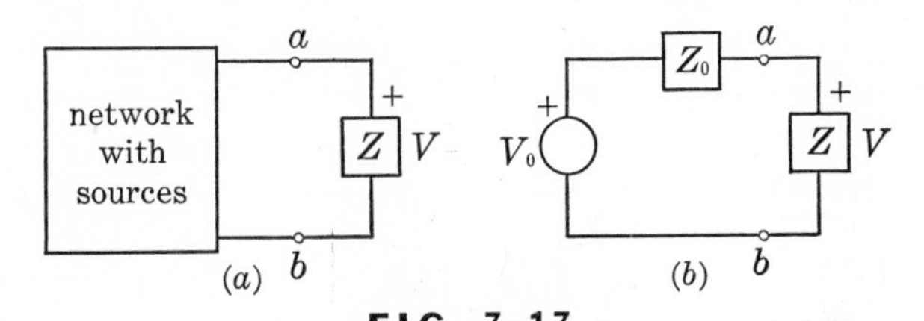

FIG. 7-17

Voltage regulation

it is desired to know how the voltage magnitude at the load changes as the load is varied. A measure of this variation is called the *voltage regulation* defined as

$$\text{Voltage regulation} = \frac{(\text{no-load voltage}) - (\text{loaded voltage})}{\text{no-load voltage}} \tag{7-43}$$

(Magnitudes are implied here.) Thus, if there is no variation of the voltage from the no-load condition to the loaded condition, then the network behaves like a voltage source, and the regulation is zero.

Let us replace the network by its Thévenin equivalent as shown in Fig. 7-17(*b*). The no-load voltage magnitude is simply $|V_0|$. The regulation can now be calculated as

$$\text{Voltage regulation} = \frac{|V_0| - \left|\dfrac{Z}{Z_0 + Z} V_0\right|}{|V_0|}$$

$$= 1 - \left|\frac{Z}{Z_0 + Z}\right| = 1 - \frac{|Z/Z_0|}{|1 + Z/Z_0|} \qquad (7\text{-}44)$$

that is, the voltage regulation depends simply on the relative magnitudes of the load and the Thévenin equivalent impedance. When $|Z_0|$ is small relative to $|Z|$, then the regulation is small. It is common to refer to regulation as being *good* (small) or *poor* (large). The closer a network with sources approximates a voltage source, the better will be the regulation.

7.5 TEE-PI TRANSFORMATION

In order to motivate the subject discussed in this section, let us contemplate determining the Thévenin equivalent of the network shown in Fig. 7-18 at the terminals *a-b*. When discussing source trans-

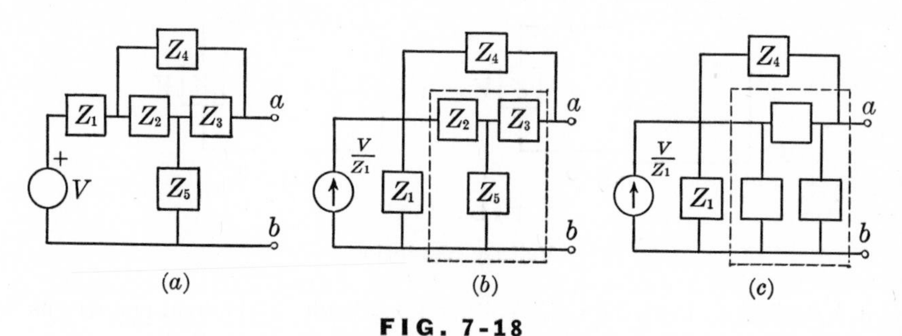

FIG. 7-18

Thévenin equivalent of a bridged tee

formations in Chap. 2, we described a step-by-step procedure for finding the Thévenin equivalent of a similar network, but without impedance Z_1. (We did not call it a Thévenin equivalent at the time.) To find the Thévenin equivalent, we must find two of the three quantities V_0, I_0, and Z_0.

Suppose we try to find V_0 by the method of loop equations or of node equations. By counting branches and nodes, we see that three node equations are required, but only two loop equations are needed. On the other hand, if loop equations are used, the direct solution will be for the loop currents; to find the voltage at terminals *a-b*, we shall need to solve for both loop currents and multiply by appropriate impedances. Either of these procedures involves a considerable amount of work.

All is not lost, however. We can still try finding I_0 and Z_0 directly, which may be easier. Suppose we short-circuit the terminals and try solving for the current. If we do this, we shall create still another loop, so that now three loop equations must be solved simultaneously. Finding I_0 first does not seem to be very promising.

Perhaps we can still avoid extra labor, since finding Z_0 might involve combining branches alternately in series and parallel, and this is relatively easy. So we remove the voltage source V (by shorting it); but we find that no simplifications are possible by series or parallel combinations of branches. Perhaps the task of finding Z_0 will be facilitated if we fall back on the definition of impedance. We can apply a voltage source at terminals *a-b*, compute the current, and take the ratio of the two sinors. But this again involves solving three simultaneous loop equations.† Alternatively, we can apply a current source at *a-b* and compute the resulting voltage. This time we require the solution of three simultaneous node equations.‡ We conclude that the direct calculation of V_0, I_0, and Z_0 all involve a considerable amount of work.

Instead of finding the quantities in Thévenin's equivalent directly, perhaps we can convert voltage to current-source equivalents, and vice versa, step by step, as we did before. Thus V in series with Z_1 can be converted to a current source V/Z_1 in parallel with Z_1. If we do this, we shall get the network shown in Fig. 7-18(*b*). But now we find that there are still no branches that can be combined in series or parallel in order to continue the step-by-step process. When confronted with a similar situation before, we resorted to the voltage-shift or current-shift theorems, but if we employ either of these theorems here, no simplifications will take place. (Try it.) We are therefore at an impasse, confronted with the unpleasant task of computing V_0 and Z_0 directly, with all the labor involved.

Let us pause and reflect here for a moment. Several times, now, we have found that replacing one network by an equivalent has resulted in a simplification of some sort. Thus, when two branches are in

† Node equations may be used, and only two would be required. But, after finding the node voltages, an application of Kirchhoff's current law would be required to find the current.

‡ This time, loop equations can be used, and only two will be required. But, after finding the loop currents, an application of Kirchhoff's voltage law would be required around the loop including the current source.

parallel, we add their admittances and think of the combination as a single branch. Similarly, the conversion of a voltage source to a current-source equivalent, or vice versa, leads to simplifications in many instances. If you look over all the "equivalent" networks we have discussed so far, you will note that the equivalence applies only at a single pair of terminals. In fact, whenever we say that two networks are equivalent, we must specify the terminals at which they are equivalent. When deciding whether or not two networks are equivalent at a pair of terminals, the criterion is this: If the relationship between the voltage and current at the terminals is the same in both networks, then they are equivalent. One of them can be substituted for the other, if this is connected in a more extensive network, and the rest of the network will not know the difference.

Let us now extend this idea of equivalence to include networks with more than two terminals to which external connections can be made. For example, the structure formed by Z_2, Z_3, and Z_5 in Fig. 7-18 has three terminals to which external connections can be made. The structure has the form of a letter T; hence it is called a *tee network*. Similarly, the structure formed by Z_1, Z_2, and Z_5 in Fig. 7-18(*b*) has three external terminals. (It looks as if there are four, but remember that the bottom ends of Z_1 and Z_5 are connected together so that only one terminal is involved here.) The structure has the form of the Greek letter π, hence it is called a *pi network*.

Now, suppose that it were possible to replace the tee formed by Z_2, Z_3, and Z_5 in Fig. 7-18(*b*) by an equivalent in the form of a pi, as shown inside the dashed lines in Fig. 7-18(*c*). Then Z_1 could be combined in parallel with one of the branches of the pi, and Z_4 in parallel with another, and our log jam would be broken. Further step-by-step conversions would now be possible to obtain the Thévenin equivalent.

Let us, therefore, consider the conditions under which the tee and the pi shown in Fig. 7-19 will be equivalent at the terminals labeled

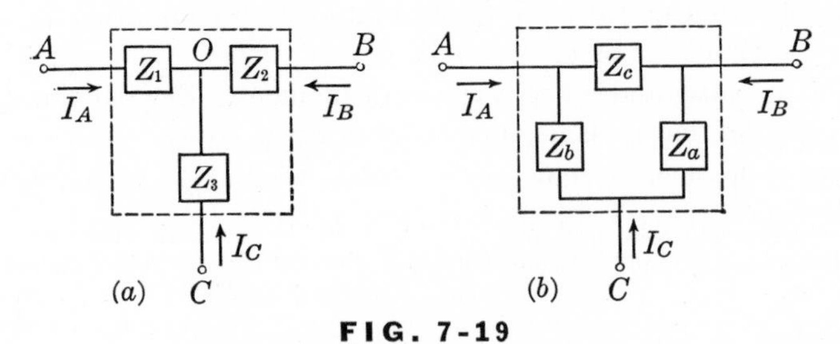

FIG. 7-19

Tee and pi equivalent networks

A, B, and C. As far as the outside world is concerned, the networks inside the two dashed rectangles are to be completely interchangeable. For this to be true, we should write the expressions for I_A, I_B, and I_C in terms of the terminal voltages for the tee and for the pi, and then demand that they be the same. To satisfy this demand, the branches of the tee and the pi must be related in certain ways. We want to determine these relationships.

For the tee network, let us apply Kirchhoff's voltage law around the three loops formed by the terminals taken in pairs. We shall get

$$\left.\begin{aligned} V_{AB} &= Z_1 I_A - Z_2 I_B \qquad (a) \\ V_{BC} &= Z_2 I_B - Z_3 I_C \qquad (b) \\ V_{CA} &= Z_3 I_C - Z_1 I_A \qquad (c) \end{aligned}\right\} \text{ for the tee} \tag{7-45}$$

If we add these three equations together, the terms on the right cancel in pairs, yielding zero for the sum. Hence these three equations are not independent; any one of them is the negative sum of the other two. To see further how they are related, note that

$$I_A + I_B + I_C = 0 \tag{7-46}$$

as a consequence of Kirchhoff's current law applied to node 0 in Fig. 7-19(a).

Now let us turn to the pi network and apply Kirchhoff's current law at each of the junctions. The result will be

$$\left.\begin{aligned} I_A &= Y_c V_{AB} - Y_b V_{CA} \qquad (a) \\ I_B &= Y_a V_{BC} - Y_c V_{AB} \qquad (b) \\ I_C &= Y_b V_{CA} - Y_a V_{BC} \qquad (c) \end{aligned}\right\} \text{ for the pi} \tag{7-47}$$

Again, adding these equations together yields zero, since the terms cancel in pairs. Hence the equations are not independent. The relationship is evident by noting that

$$V_{AB} + V_{BC} + V_{CA} = 0 \tag{7-48}$$

as a consequence of Kirchhoff's voltage law applied around the loop formed by the three branches in Fig. 7-19(b).

In order to compare the V-I relationships of the tee and the pi, we should either (1) solve Eq. (7-45) for the currents and compare with Eq. (7-47), or (2) solve Eq. (7-47) for the voltages and compare with Eq. (7-45). In each case, remember that the equations are dependent, and so the auxiliary relationships in either Eq. (7-46) or Eq. (7-48) must be used. We shall not go through the details of the algebra here but shall leave it to you.† The result of the algebra gives the following relationships among the branch impedances and admittances of the tee and pi which must be satisfied if the two are to be equivalent.

For converting from tee to pi:

$$Y_a = \frac{Y_2 Y_3}{Y_1 + Y_2 + Y_3} = \frac{Z_1}{Z_1 Z_2 + Z_2 Z_3 + Z_3 Z_1} \quad (a)$$

$$Y_b = \frac{Y_3 Y_1}{Y_1 + Y_2 + Y_3} = \frac{Z_2}{Z_1 Z_2 + Z_2 Z_3 + Z_3 Z_1} \quad (b) \qquad (7\text{-}49)$$

$$Y_c = \frac{Y_1 Y_2}{Y_1 + Y_2 + Y_3} = \frac{Z_3}{Z_1 Z_2 + Z_2 Z_3 + Z_3 Z_1} \quad (c)$$

For converting from pi to tee:

$$Z_1 = \frac{Z_b Z_c}{Z_a + Z_b + Z_c} = \frac{Y_a}{Y_a Y_b + Y_b Y_c + Y_c Y_a} \quad (a)$$

$$Z_2 = \frac{Z_c Z_a}{Z_a + Z_b + Z_c} = \frac{Y_b}{Y_a Y_b + Y_b Y_c + Y_c Y_a} \quad (b) \qquad (7\text{-}50)$$

$$Z_3 = \frac{Z_a Z_b}{Z_a + Z_b + Z_c} = \frac{Y_c}{Y_a Y_b + Y_b Y_c + Y_c Y_a} \quad (c)$$

Let us make some comments about these formulas. First of all, they are not simple enough to remember after a casual glance. As a matter of fact, you should make no attempt to memorize them. They are not fundamental to an understanding of network theory. When they are required in the solution of a problem, you can consult a reference.

Second, we have given the formulas in terms of both the impedance and the admittance. Note that the branches of the tee are labeled with numerical subscripts and those of the pi with literal subscripts, in order to avoid confusion. The formulas show a marked degree of symmetry. In fact, two of the ones in Eq. (7-49) can be obtained from the other merely by a cyclical permutation of the subscripts; and similarly for Eq. (7-50).

In the third place, the tee and the pi are dual networks. The equations for converting from a pi to a tee can be obtained from those for converting a tee to a pi by changing impedances to admittances. (The subscripts should also be changed, to avoid confusion.)

The tee and the pi networks are sometimes drawn in the form

† An alternative method for finding the conditions for equivalence of a tee and a pi will be developed in the next chapter.

shown in Fig. 7-20. Because of the appearances of these structures, they are called a wye (tee) and a delta (pi), respectively.

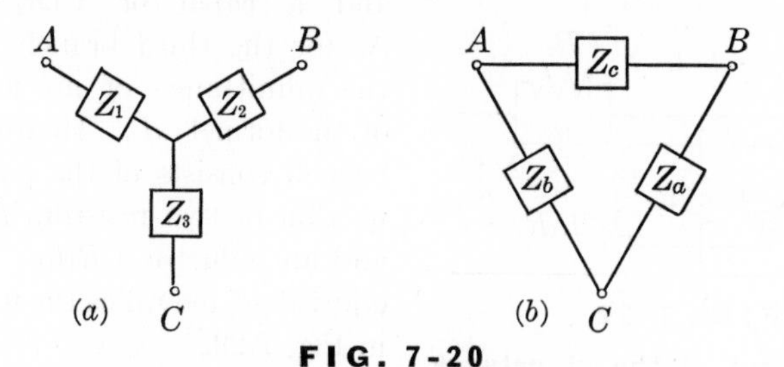

FIG. 7-20

Redrawing of tee and pi as wye and delta

Let us now illustrate the use of the tee-pi or pi-tee transformations by means of some examples. Consider the pi network shown in Fig. 7-21, for which it is desired to find a tee equivalent. Applying Eq. (7-50), the branch impedances of the equivalent tee will be

$$Z_1 = \frac{R_1/sC}{R_1 + R_2 + 1/sC} = \frac{1}{s\left(\dfrac{R_1 + R_2}{R_1}C\right) + \dfrac{1}{R_1}} = \frac{1}{Y_1} \qquad (a)$$

$$Z_2 = \frac{R_2/sC}{R_1 + R_2 + 1/sC} = \frac{1}{s\left(\dfrac{R_1 + R_2}{R_2}C\right) + \dfrac{1}{R_2}} = \frac{1}{Y_2} \qquad (b) \qquad (7\text{-}51)$$

$$Z_3 = \frac{R_1R_2}{R_1 + R_2 + 1/sC} = \frac{1}{\dfrac{1}{R_1} + \dfrac{1}{R_2} + \dfrac{1}{sCR_1R_2}} = \frac{1}{Y_3} \qquad (c)$$

It remains to see what types of branch are represented by these three expressions. Each impedance is expressed as "one over something."

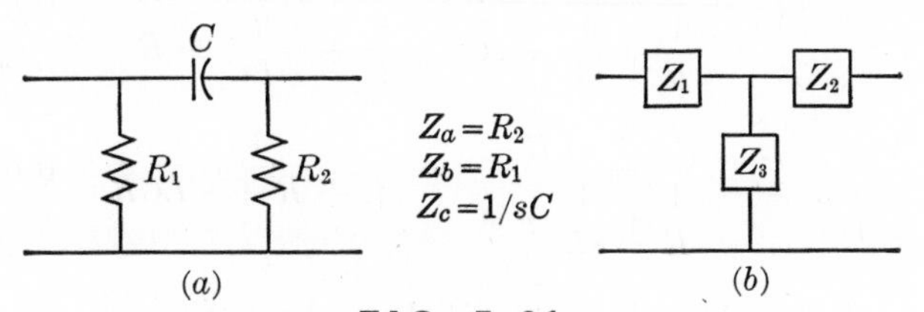

FIG. 7-21

Example of pi-tee transformation

The something, hence, is the branch admittance. Since the admittance of a capacitor is of the form sC, we recognize the first branch as the parallel connection of a resistor R_1 and a capacitor $C(R_1 + R_2)/R_1$. Similarly, the second branch is the parallel connection of a resistor R_2 and a capacitor $C(R_1 + R_2)/R_2$. As for the third branch, note that the admittance of an inductor is of the form $1/sL$. Hence the third branch consists of the parallel connection of two resistors R_1 and R_2, and an inductor CR_1R_2. Thus the equivalent tee takes the form shown in Fig. 7-22.

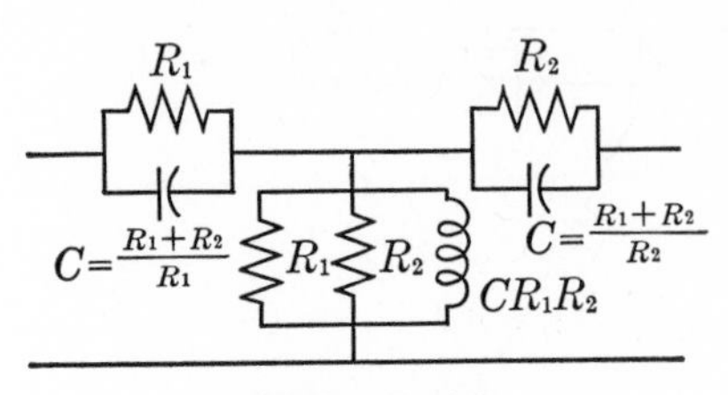

FIG. 7-22

Tee equivalent of the pi network shown in Fig. 7-21

In this example we were able to find a tee equivalent of the given pi which is valid at all frequencies. The number of elements in the tee, however, is more than twice as many as in the original pi. The "simplification" which the transformation is supposed to provide is not apparent in this example. Actually, simplification will result if this pi forms part of a more extensive network, such as the one in Fig. 7-18 with which we started this discussion, and this transformation permits further combinations. Simplification will also result when numerical values are involved. No matter how complicated the branches, each impedance will simply be a complex number.

Let us next turn to the tee network shown in Fig. 7-23, for which it is desired to find an equivalent pi. Using Eq. (7-49), the branch admittances of the pi will be

$$Y_a = \frac{C/L}{\dfrac{1}{R} + \dfrac{1}{sL} + sC} = \frac{1}{\dfrac{L}{RC} + \dfrac{1}{sC} + sL} \qquad (a)$$

$$Y_b = \frac{sC/R}{\dfrac{1}{R} + \dfrac{1}{sL} + sC} = \frac{1}{\dfrac{1}{sC} + \dfrac{R}{s^2LC} + R} \qquad (b) \qquad (7\text{-}52)$$

$$Y_c = \frac{1/sLR}{\dfrac{1}{R} + \dfrac{1}{sL} + sC} = \frac{1}{sL + R + s^2LCR} \qquad (c)$$

The first of these can be recognized as the series connection of an inductor L, a capacitor C, and a resistor L/RC. We cannot, however, recognize what type and combination of elements constitute the other two branches, at least with our present knowledge. As a matter of fact, there is no network consisting of the elements we have so far

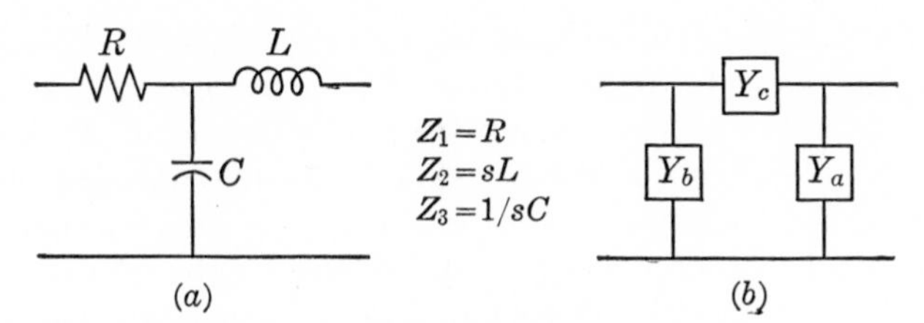

FIG. 7-23

Example of tee-pi transformation

introduced in our model whose admittance can be Y_b and Y_c as given in these equations.

To carry on further, let us choose some numerical values. Let $R = 2$, $L = 0.05$, and $C = 10^{-3}$. Also let the frequency be $\omega = 100$. (That is, $s = j\omega = j100$.) Then the branch admittances become

$$Y_a = \frac{1}{25 - j10 + j5} = \frac{5}{130} + j\frac{1}{130} \qquad (a)$$

$$Y_b = \frac{1}{-j10 - 4 + 2} = \frac{1}{-2 - j10} = -\frac{1}{52} + j\frac{5}{52} \qquad (b) \qquad (7\text{-}53)$$

$$Y_c = \frac{1}{j5 + 2 - 1} = \frac{1}{26} - j\frac{5}{26} \qquad (c)$$

We now have the numerical values of the branch admittances of the pi which is equivalent to the given tee. We had already recognized Y_a as the admittance of a parallel *R-L-C* branch. At the given frequency we see that this branch is capacitive. Now look at Y_c. Whereas there is no network having Y_c in Eq. (7-52) as its admittance (for all frequencies), Eq. [7-53(*c*)] shows that the branch can be represented by a parallel *RL*, at the given frequency. But, looking at the expression for Y_b, we see that even this is not possible. Even at the single given frequency, Y_b does not represent the admittance of any branch consisting of elements in our model. It has a negative real part, implying a negative conductance.

Let us say that a network is *realizable* if it consists of elements in our network model. (As we introduce additional elements into the model later on, these will also be included.) In this terminology we can state that the pi equivalent of the tee network given in Fig. 7-23 is not realizable for all frequencies. It is not even realizable at a single frequency. Notwithstanding these comments, the tee network in the figure can be replaced by the pi, and correct results will be obtained as far as voltages and currents at the terminals are concerned. You should not, however, expect to find a pi with positive R, L, and C elements, which is equivalent to the tee even at this one frequency.

As a general conclusion about tee-pi or pi-tee transformations, let us state the following. In some cases it is possible to find a physically realizable pi (tee) equivalent to a given tee (pi) which is valid for all frequencies. In other cases it may be possible to find a realizable equivalent only at a given frequency. In still other cases no realizable equivalent may be possible, even at a single frequency. Nevertheless, it is always possible to compute a mathematical equivalent which will give valid results as far as external connections are concerned.

7.6 POWER TRANSFER

One of the most common problems of electrical engineering is that of transmitting power from a source to a load through some intervening circuits. The conditions which influence the power flow are of great interest. We shall now turn to a consideration of this problem. A typical situation is shown in Fig. 7-24 in which Z represents the load

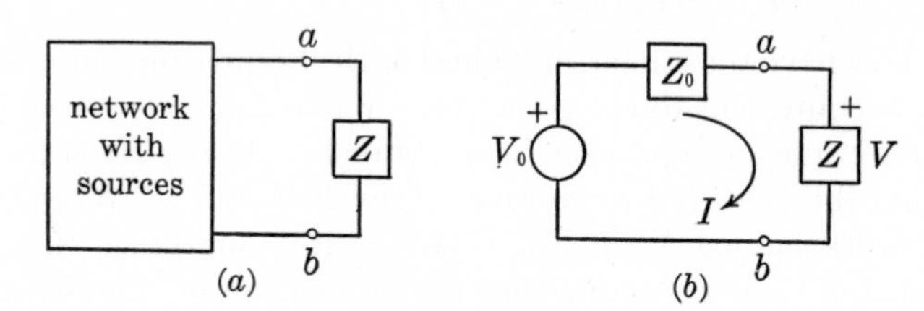

FIG. 7-24

Transfer of power to a load

impedance. The network can be replaced by its Thévenin equivalent as shown in Fig. 7-24(*b*). The real power delivered to the load is

$$P = |I|^2 Re(Z) = \frac{|V_0|^2}{|Z + Z_0|^2} Re(Z) \qquad (7\text{-}54)$$

Let us assume that the network is fixed but that the load impedance can be varied. We wish to find the value of Z for which the maximum power will be delivered to the load.

Now Z and Z_0 are complex quantities. They can be expressed either in rectangular form or in polar form, as follows:

$$Z = R + jX = |Z|\epsilon^{j\theta} \qquad (a)$$
$$Z_0 = R_0 + jX_0 = |Z_0|\epsilon^{j\theta_0} \qquad (b) \tag{7-55}$$

Let us first use the rectangular form in Eq. (7-54); the result will be

$$P = \frac{|V_0|^2 R}{(R + R_0)^2 + (X + X_0)^2} \tag{7-56}$$

The variables in this expression are R and X. As they vary, the power will vary. If a quantity which depends on two variables is to have a maximum value, the partial derivatives of the quantity with respect to each of the variables must be zero. Hence, to find the conditions under which the power is a maximum, we should differentiate P partially with respect to R and to X and set both derivatives equal to zero. The details will be left for you to work out. The resulting conditions on the load impedance will be

$$R = R_0$$
$$X = -X_0$$

or

$$Z = Z_0^* \tag{7-57}$$

That is, in order to have the maximum power delivered from the network to the load, the load impedance must be adjusted until it is the conjugate of the Thévenin equivalent impedance of the network. In such a case we say there is a *conjugate match.* The load is called a *matched load.* If we call the maximum power P_m, then, using Eq. (7-57) in Eq. (7-56), we find

$$P_m = \frac{|V_0|^2}{4R_0} \tag{7-58}$$

When the Thévenin impedance is fixed, this value of power is the maximum which is available from the network.

Sometimes it is possible that the angle of the load impedance is fixed and only the magnitude is variable. Such is the case, for example, if the load is to be a resistor. In such a case it is more useful to express Z and Z_0 in terms of their magnitudes and angles. If we do this, Eq. (7-54) for the power becomes

$$P = \frac{|V_0|^2|Z| \cos\theta}{\left||Z| \cos\theta + j|Z| \sin\theta + |Z_0| \cos\theta_0 + j|Z_0| \sin\theta_0\right|^2}$$
$$= \frac{|V_0|^2|Z| \cos\theta}{|Z|^2 + |Z_0|^2 + 2|ZZ_0| \cos(\theta - \theta_0)} \tag{7-59}$$

In this expression only $|Z|$ is variable. To find the condition for a maximum value of P in this case, we differentiate with respect to $|Z|$ and equate the derivative to zero. The resulting condition is

$$|Z| = |Z_0| \tag{7-60}$$

(Go through the details.)

Thus, when the magnitude only of the load is variable, it must be made equal to the magnitude of the Thévenin impedance. With this value of $|Z|$, the power in Eq. (7-59) will be a maximum but under the constraint that θ is fixed. Let us call this maximum P_{m1}. Then

$$P_{m1} = \frac{|V_0|^2|Z_0| \cos \theta}{2|Z_0|^2 + 2|Z_0|^2 \cos (\theta - \theta_0)}$$
$$= \frac{|V_0|^2 \cos \theta \cos \theta_0}{2|Z_0| \cos \theta_0[1 + \cos (\theta - \theta_0)]} \tag{7-61}$$

In this expression we multiplied numerator and denominator by $\cos \theta_0$. Note that $|Z_0| \cos \theta_0 = R_0$ and that $2 \cos \theta \cos \theta_0 = \cos (\theta + \theta_0) + \cos (\theta - \theta_0)$. Using these expressions, the power becomes

$$\begin{aligned} P_{m1} &= \frac{|V_0|^2 [\cos (\theta + \theta_0) + \cos (\theta - \theta_0)]}{4R_0[1 + \cos (\theta - \theta_0)]} \\ &= P_m \frac{\cos (\theta + \theta_0) + \cos (\theta - \theta_0)}{1 + \cos (\theta - \theta_0)} \end{aligned} \tag{7-62}$$

Examine the fraction involving the cosines on the right. Since $\cos (\theta + \theta_0)$ is always less than unity, this fraction has a numerator which is always less than the denominator. Hence P_{m1} is less than P_m. We knew this already, since P_m is absolutely the maximum power. But, in terms of this expression, we can find out how radically P_{m1} differs from P_m.

Suppose, for example, that the load is a resistor ($\theta = 0$). Let us find how much θ_0 can differ from θ before P_{m1} differs from P_m by more than 1 per cent. That is, set the factor multiplying P_m in Eq. (7-62) equal to 0.99 and calculate the resulting value of θ_0.

$$\left.\frac{\cos (\theta + \theta_0) + \cos (\theta - \theta_0)}{1 + \cos (\theta - \theta_0)}\right|_{\theta=0} = \frac{2 \cos \theta_0}{1 + \cos \theta_0} = 0.99$$

$$\cos \theta_0 = \frac{0.99}{1.01}; \quad \theta_0 \doteq 11.5°$$

This result shows that, although the maximum power is transmitted to the load under conditions of conjugate match, not much less than the maximum is transmitted, even when the angles of the load and the Thévenin impedance differ by about 10 deg, as long as the two magnitudes are equal.

The constraint which was placed on the load impedance, namely that of fixed magnitude, is not the only possible constraint. It is possible that the angle, or the real part, or the imaginary part is fixed. For each of these constraints, a different condition can be found for maximizing the power transfer.

It is not enough for you to memorize any of the results developed here without associating therewith the specific conditions under which the results are valid. In fact, you should make it a habit throughout your career to note carefully the conditions under which a theorem is proved to be true, and to apply the theorem only when the required conditions are satisfied. In common language, "Don't apply formulas blindly."

In the case of the power-transfer formulas under discussion here, it is known that misleading results have been obtained for certain electronic circuits by the application of these formulas without taking into account the specific constraints under which the particular circuit operates. Be warned!

7.7 THE COMPENSATION THEOREM

Several times we have observed that replacing a network or a subnetwork by something equivalent may lead to a simplification of some sort or other. In this section we shall consider a particular instance of this substitution of an equivalent.

Consider the network of Fig. 7-25. Suppose that a solution for the voltage $V = 20$ across Z_3 and the current $I = 10$ through Z_2 has been obtained by one method or another. It is then possible to place a

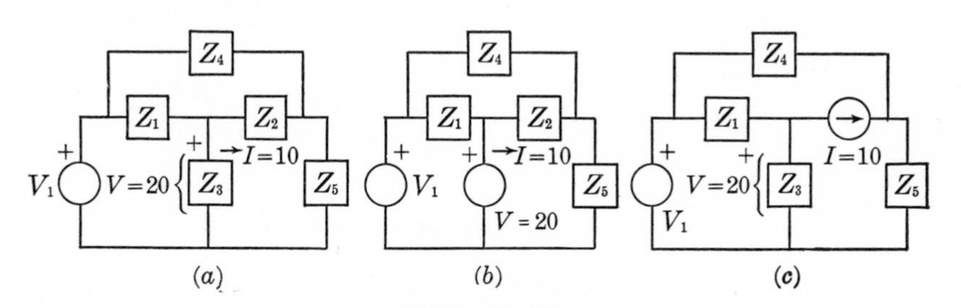

FIG. 7-25

Compensation theorem

voltage source $V = 20$ in parallel with Z_3 without disturbing the rest of the network, since the voltage across Z_3 is already 20. The impedance Z_3 can now be removed, since it is in parallel with a voltage source, and the rest of the network will not be affected. The net result is to replace a network branch by a voltage source whose value is equal to the original voltage across the branch, as shown in Fig. 7-25(*b*).

A similar discussion applies with respect to the known current. Suppose a current source $I = 10$ is placed in series with Z_2. This source will have no influence on the network behavior, since the current in Z_2 is already $I = 10$. But now Z_2 is redundant as far as the remaining network is concerned; its removal will not influence the remaining currents and voltages. The net result is to replace a network branch by a current source whose value is equal to the original current through the branch, as shown in Fig. 7-25(*c*).

The results we just described are dignified by the name *compensation theorem*, or sometimes *substitution theorem*. Notice that, in order to apply the theorem, it is first necessary to know the voltage across a branch or the current through it. Although we introduced the theorem in terms of a specific example, the argument used in establishing it can be used for any branch in any network. Thus the result is a general one. We shall discuss two particular situations in which the compensation theorem can be applied.

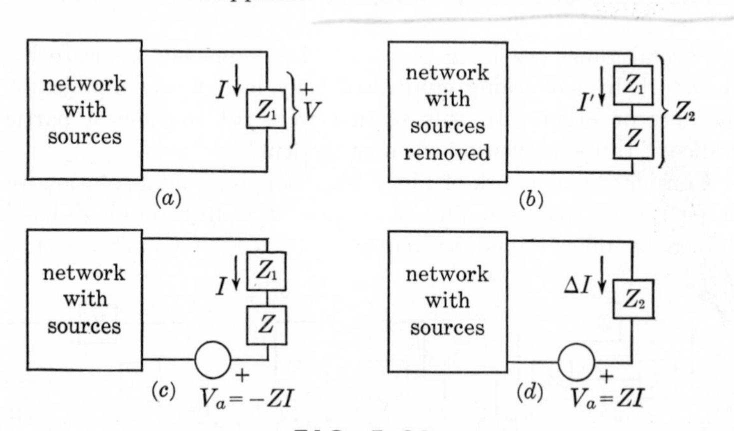

FIG. 7-26

Influence of an impedance variation

The first problem we shall discuss is the following. Consider the network of Fig. 7-26(*a*). One branch of the network having an impedance Z_1 is shown explicitly. By analyzing the network, the voltage

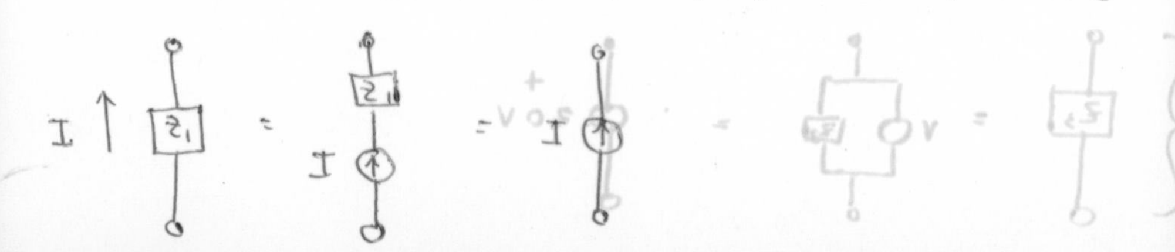

across the branch has been found to be V, and the current I. Now suppose that the impedance Z_1 changes to a new value Z_2, the change being $Z = Z_2 - Z_1$. This is illustrated in Fig. 7-26(b). The current in this branch will now be I', different from the original current. It is desired to assess the change in any voltage or current in the network owing to this change in the impedance.

One approach, of course, is to repeat the solution of the network in all its details. This is a valid approach, but we should like to find a means of avoiding this labor.

Suppose that a voltage source V_a is inserted in series with the branch with the reference shown in Fig. 7-26(c). Its value is then adjusted until it is just equal in magnitude to the voltage across Z, but opposite in sign; that is, the voltage across the combination of Z and the source is zero. Hence this situation is equivalent to the original one in Fig. 7-26(a), before the change in the impedance. This means that the current in the branch is the original current I, so that the voltage of the source is $V_a = -ZI$.

The net effect of the voltage source is to cancel the influence of the change in the impedance; that is, the effect of the source $-IZ$ will be just the negative of the effect due to the change in impedance. Finally, if we change the sign of the voltage, making it $V_a = +ZI$, the effect will be identical with that of changing the impedance. Note that the reference of the voltage source is at the tail of the original current reference.

In conclusion, in order to compute the changes in any current or voltage in the network after a change of Z in the original impedance, it is necessary to find the corresponding current or voltage owing to a voltage source $V_a = ZI$, placed in series with the changed impedance and with the appropriate reference, where I is the current in the branch before the impedance change. According to the superposition principle, this is done by removing all the original sources, as illustrated in Fig. 7-26(d).

It is possible to arrive at these results alternatively by an analytical argument. Return to Fig. 7-26(b) and contemplate writing loop equations after the change of the impedance. All variables after the change are labeled with a prime, those before the change being unprimed. For convenience, let us assume that the loops are chosen so that the branch in question appears in only one loop and we call this loop 1.

The loop equations before the change of impedance will look like Eq. (7-16). After the change, the equation for loop 1 will become

$$(a_{11} + Z)I' + a_{12}I_2' + a_{13}I_3' + \cdots + a_{1n}I_n' = V_1 \qquad (7\text{-}63)$$

where a_{11} is the self impedance of loop 1 (not including Z) and the other a's are the mutual impedances between loop 1 and the other loops. All the other loop equations will remain unchanged, except that the variables will be primed. Suppose we now subtract the original loop equations from the new ones. The result will be

$$\begin{aligned} ZI' + a_{11}(I' - I) + a_{12}(I_2' - I_2) + \cdots + a_{1n}(I_n' - I_n) &= 0 \\ a_{21}(I' - I) + a_{22}(I_2' - I_2) + \cdots + a_{2n}(I_n' - I_n) &= 0 \\ \cdots\cdots\cdots\cdots\cdots\cdots\cdots\cdots\cdots\cdots\cdots\cdots \\ a_{n1}(I' - I) + a_{n2}(I_2' - I_2) + \cdots + a_{nn}(I_n' - I_n) &= 0 \end{aligned} \tag{7-64}$$

The differences between the primed and unprimed variables are the changes in the loop currents owing to the change in impedance. Notice the term ZI' in the first equation. If we subtract ZI from both sides of the first equation, this term will become $Z(I' - I)$ and can then be combined with $a_{11}(I' - I)$. The resulting set of equations now becomes

$$\begin{aligned} a_{11}'\Delta I + a_{12}\Delta I_2 + a_{13}\Delta I_3 + \cdots + a_{1n}\Delta I_n &= -ZI \\ a_{21}\Delta I + a_{22}\Delta I_2 + a_{23}\Delta I_3 + \cdots + a_{2n}\Delta I_n &= 0 \\ \cdots\cdots\cdots\cdots\cdots\cdots\cdots\cdots\cdots\cdots\cdots\cdots \\ a_{n1}\Delta I + a_{n2}\Delta I_2 + a_{n3}\Delta I_3 + \cdots + a_{nn}\Delta I_n &= 0 \end{aligned} \tag{7-65}$$

where $a_{11}' = a_{11} + Z$ is the sum of the impedances on the first loop, including the changed impedance, and the ΔI's are the changes in the loop currents owing to the impedance change. Notice that ZI appears as a voltage source in loop 1.

This set of equations can now be solved; according to Eq. (7-19) the solution for the change in the kth loop current will be

$$\Delta I_k = \frac{\Delta_{1k}}{\Delta}(-ZI) = Y_{1k}(-ZI) \tag{7-66}$$

where Δ is the determinant of the system of equations, and Δ_{1k} is a cofactor. Notice that the determinant is evaluated after the change in impedance. This expression is precisely what would be obtained from Fig. 7-26(d). Y_{1k} is the transfer admittance, including the effect of the change.

What has been done in terms of a change of impedance can be repeated in terms of a change of admittance in a completely dual manner. The compensating source will now be a current source in parallel with the changed admittance. The preceding discussion will apply in its entirety simply by replacing impedance by admittance, voltage by current and vice versa, loop current by node voltage, loop equations by node equations, etc. You are urged to go through the steps of the development.

Let us illustrate the preceding discussion by means of the network shown in Fig. 7-27. The resistance R_1 can vary by 10 per cent. It is

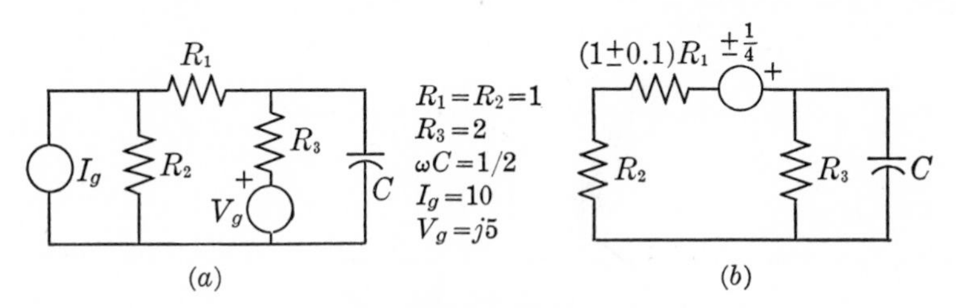

FIG. 7-27

Numerical illustrative example

desired to know how this variation influences the voltages across all the other elements. The actual voltages may or may not be required, but the changes are. With the numerical values given in the figure, the current in R_1 is computed to be $I = 5/2$. (Carry out the computation.) The change in impedance is $\pm 1/10$. Hence the compensating source is $V_a = (\pm 5/2)(1/10) = \pm 1/4$. To find the changes in the voltages, we insert this source and remove the original ones (open-circuit I_g and short-circuit V_g), as shown in Fig. 7-27(*b*). We see that the change in voltage across R_3 is the same as that across C. By the voltage-divider relationship, we easily find

$$\Delta V_{R2} = \frac{R_2(\pm 1/4)}{R_2 + (1 \pm 0.1)R_1 + \dfrac{R_3}{j\omega C R_3 + 1}} \doteq 0.077\epsilon^{j17.9°} \quad \text{or} \quad -0.0813\epsilon^{j19.1°}$$

$$\Delta V_C = \Delta V_{R_3} = \frac{\dfrac{R_3}{1 + j\omega R_3 C}}{R_2} \Delta V_{R_2} \doteq 0.109\epsilon^{-j27.1°} \quad \text{or} \quad -0.115\epsilon^{-j25.9°}$$

It may not appear that the procedure we have used saves much work. You can convince yourself of the substantial saving, however, even in this relatively simple example, by recomputing all the voltages with the modified value of R_1 and then subtracting to find the change.

Note carefully that there is no restriction on the magnitude of the impedance variation. The result we have established is not an approximate relationship which holds for small variations. It is exact and applies to an impedance variation of any size.

As a second application of the compensation theorem, we shall show how the computation of a response in an extensive network can be simplified. In the network of Fig. 7-28(a), it is required to find the voltage V_2. This can be done, of course, by straightforward loop or node analysis. It is also possible to apply Thévenin's theorem. But

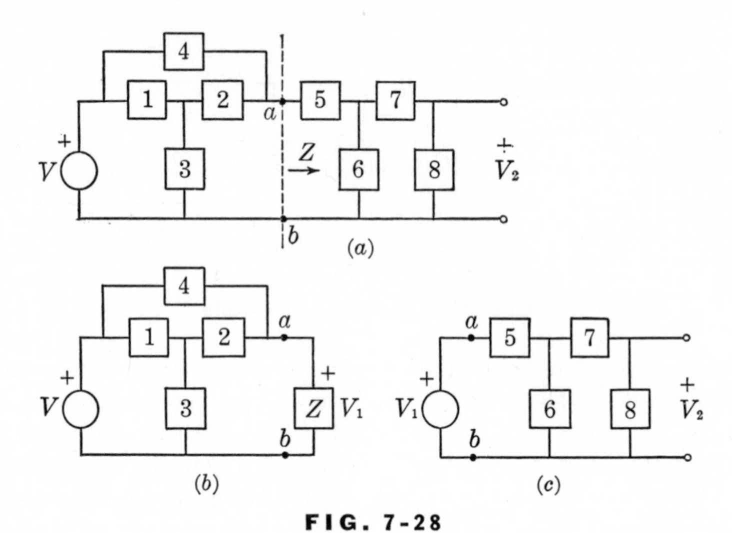

FIG. 7-28

Application of the compensation theorem

suppose we proceed by first determining the impedance of the network to the right of the dashed line at a-b. The resulting network will have the form shown in Fig. 7-28(b). This is a simpler network than the original one. In the next step we solve for the voltage V_1 across the terminals a-b. The entire network to the left of a-b can be considered as a single branch, the voltage across which is V_1. Hence, by the compensation theorem, everything to the left of a-b can be replaced by a voltage source V_1. If we now restore the network to the right of a-b, Fig. 7-28(c) will result. It is now a relatively easy task to determine V_2.

Notice that in Fig. 7-28(c) it might appear as if the network to the right of a-b has been replaced by a Thévenin equivalent. This is not the case; the voltage V_1 is not the open-circuit voltage at terminals a-b.

Looking over the solution of the preceding problem, you will observe that the original problem was broken down into three simpler ones. First we found the impedance of the right-hand part of the network; then we found a response in the resulting simpler left-hand network; and, finally, the compensation theorem permitted a simplification of the right-hand network. Notice that the absence of sources to the right of *a-b* is not an essential part of this approach. If sources are present there, we simply find the Thévenin equivalent to the right of *a-b* as the first step. In Fig. 7-28(*b*), then, the Z which appears will be the Thévenin impedance, and it will be in series with the Thévenin voltage. The procedure from here on is not modified at all.

7.8 COMPLEX LOCUS PLOTS

The steady-state response of a network to a sinusoidal excitation is described completely in terms of the appropriate network function. If the network has one pair of terminals, this function is the impedance or admittance. If there is more than one pair of terminals, the function may be a transfer impedance, or a transfer voltage ratio, etc. The more ways we can find for representing network functions, the more ways we shall have for describing a network response.

One method of representing a network function which we have discussed has to do with the locations of the poles and zeros in the complex plane. Once the poles and zeros are located, the response is completely determined—not only the steady-state response, but the total response, including the transient. We have seen how the variation of the steady-state response with frequency can be determined from the pole-zero diagram.

Another method of representing a network function which we have mentioned involves plotting parts of the function against frequency. Being complex, a network function can be expressed either in terms of its real and imaginary parts or in terms of its magnitude and angle. Thus the function is represented by plotting its real and imaginary parts as functions of frequency, or by plotting its magnitude and angle as functions of frequency. Examples of such curves are the reactance curves in Fig. 5-1 and the resonance curve in Fig. 5-7.

In this section we shall discuss yet a third way of representing a network function. The value of any network function depends on the frequency and on the network parameters—the R's, L's, and C's. If all of these are fixed, the function will have a fixed complex value. Hence it can be represented by a point in the complex plane. Suppose, however, that either the frequency or one of the network parameters is variable. For each value taken on by the varying quantity, the network function will take on a different value. Hence the point representing the function in the complex plane will move, thereby tracing out a curve. This curve is called the complex locus of the function.

Let us illustrate this concept in terms of the input impedance of the network of Fig. 7-29. The reactance of the capacitor is assumed

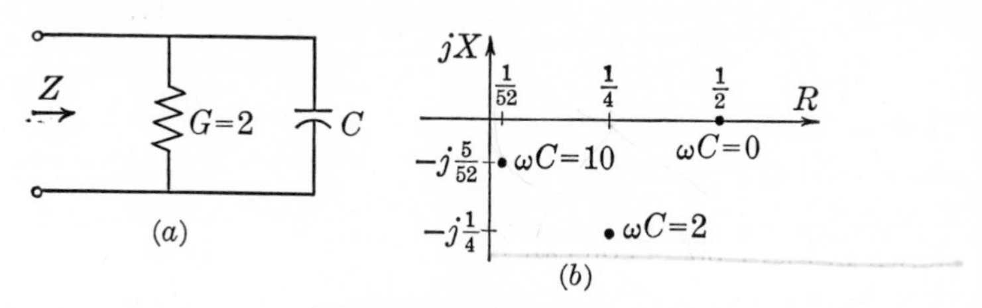

FIG. 7-29

Impedance locus of parallel RC network

to be variable. This can be achieved by having either a variable capacitance or a variable frequency. The impedance is

$$Z = \frac{1}{G + j\omega C} = \frac{2}{4 + (\omega C)^2} - j\frac{\omega C}{4 + (\omega C)^2} = R + jX \qquad (7\text{-}67)$$

where R and X are the real and imaginary parts of Z, respectively. If we now let ωC take on several different values, we can calculate the corresponding values of Z and locate these points in the complex plane. The values of Z corresponding to $\omega C = 0$, 2, and 10 are shown in Fig. 7-29(*b*).

From these three points alone, it is not possible to tell what the complete locus will be in the Z plane. We could, of course, get a better idea by calculating more points for different values of ωC. On the other hand, for the present example it is possible to find an analytical expression for the locus curve. In order to do this, let us first write the real and imaginary parts of Z from Eq. (7-67).

$$R = \frac{2}{4 + (\omega C)^2} \qquad (a)$$

$$X = \frac{-\omega C}{4 + (\omega C)^2} \qquad (b) \qquad (7\text{-}68)$$

Here we have two quantities, R and X, expressed in terms of a third quantity ωC. Let us now eliminate ωC from these equations, thereby getting a single equation involving R and X. One sequence of steps to accomplish this is to take the ratio of the second to the first equation, obtaining $\omega C = -2X/R$. Substituting this expression back into the first equation leads to

Z = R + Xj

impedance locus is graph of R Vs X

jX R

$$R^2 - \frac{R}{2} + X^2 = 0 \tag{7-69}$$

Completing the square on the left-hand side finally leads to

$$\left(R - \frac{1}{4}\right)^2 + X^2 = \left(\frac{1}{4}\right)^2 \tag{7-70}$$

This we now recognize as the equation of a circle of radius 1/4 in the plane having coordinates R and X. The center of the circle is at $R = 1/4$, $X = 0$. The locus is shown in Fig. 7-30. You can verify that the previously computed points fall on the locus.

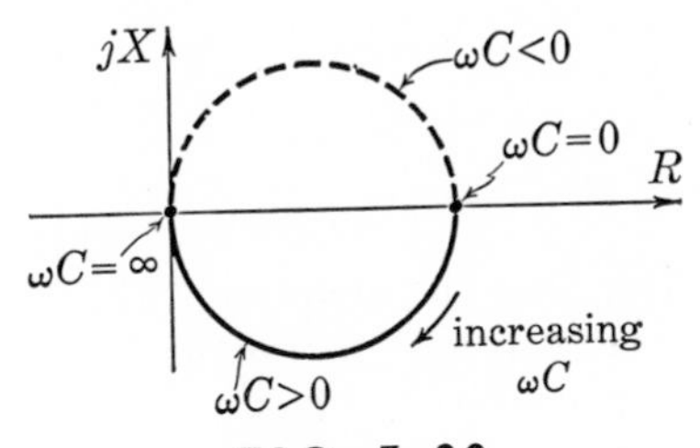

FIG. 7-30
Circular locus

The locus is seen to fall in the first and fourth quadrants, which guarantees that the real part of Z is positive for all values of ωC, as it should be. Equation [7-68(b)] shows that the imaginary part X will remain negative for all positive values of ωC but will be positive for negative values of ωC. The portion of the locus corresponding to negative ωC is shown dashed.

Notice that the locus of Fig. 7-30 gives us a complete picture of the impedance as ωC varies. Except for the extreme values $\omega C = 0$ and ∞, however, we do not know the points on the locus corresponding to particular values of ωC. Very often, this information is not required, in which case Fig. 7-30 will be adequate as it stands. If more quantitative information is desired, then we must label points on the locus with the corresponding values of ωC.

In discussing this example, we mentioned that either the frequency or the capacitance could be varied. Let us concentrate for a moment on the loci obtained when one element of a network is varied, all other elements and the frequency being held constant. We shall assume that there is only one excitation, a voltage source V_1, in the network. Let us write a set of loop equations and solve for the current I_1 in the input loop. The solution will be

$$\frac{I_1}{V_1} = \frac{\Delta_{11}}{\Delta} \tag{7-71}$$

where Δ is the determinant of the loop equations and Δ_{11} is its cofactor. Suppose a resistor R which appears only on the second loop is variable. The determinant Δ will have the form

$$\Delta = \begin{vmatrix} a_{11} & a_{12} & a_{13} & \cdots & a_{1n} \\ a_{21} & a_{22} & a_{23} & \cdots & a_{2n} \\ \cdots & \cdots & \cdots & \cdots & \cdots \\ a_{n1} & a_{n2} & a_{n3} & \cdots & a_{nn} \end{vmatrix} = \begin{vmatrix} a_{11} & a_{12} & a_{13} & \cdots & a_{1n} \\ a_{21} & a_{22}' + R & a_{23} & \cdots & a_{2n} \\ \cdots & \cdots & \cdots & \cdots & \cdots \\ a_{n1} & a_{n2} & a_{n3} & \cdots & a_{nn} \end{vmatrix}$$

The a's are the self-impedances and the mutual impedances of the loops. In the second form, the resistance R is shown explicitly, and a_{22}' is the self-impedance of loop 2 not including R.

Now let us expand Δ along the second row. The result will be

$$\begin{aligned} \Delta &= a_{21}\Delta_{21} + (a_{22}' + R)\Delta_{22} + a_{23}\Delta_{23} + \cdots + a_{2n}\Delta_{2n} \\ &= (a_{21}\Delta_{21} + a_{22}'\Delta_{22} + a_{23}\Delta_{23} + \cdots + a_{2n}\Delta_{2n}) + R\Delta_{22} \\ &= \Delta^o + \Delta_{22}R \end{aligned} \tag{7-72}$$

Note that the quantity within parentheses in the second line is the expansion of the determinant of the network when the resistance R is short-circuited. This determinant has been labeled Δ^o. It is not necessary to put the superscript o on Δ_{22}, since Δ_{22} will be the same whether R is shorted or not.

In a similar way, Δ_{11} can be expanded and written as

$$\Delta_{11} = \Delta_{11}{}^o + \Delta_{1122}R \tag{7-73}$$

where $\Delta_{11}{}^o$ is the cofactor of Δ^o and Δ_{1122} is the determinant obtained from the original determinant Δ if both the first row and first column and the second row and second column are removed. Again, the superscript is unnecessary on Δ_{1122}.

With these expressions for Δ and Δ_{11} substituted into Eq. (7-71), the input admittance of the network becomes

$$Y = \frac{I_1}{V_1} = \frac{\Delta_{11}{}^o + \Delta_{1122}R}{\Delta^o + \Delta_{22}R} \tag{7-74}$$

Note that R does not appear in any of the determinants in this expression. Hence the general form of the admittance is

$$Y = \frac{A + BR}{C + DR} \tag{7-75}$$

where A, B, C, and D are complex constants which depend on the frequency and on all the other network parameters except R. Although we assumed that R is in loop 2, this expression is valid no matter where R appears. If R is in loop k, then the cofactors appearing in Eq. (7-74) will be Δ_{kk} and Δ_{11kk}, but Eq. (7-75) will have the same appearance.

If we look upon this expression for Y as a function of R, we see that Y is the ratio of two linear polynomials in R. For this reason we call this expression a *bilinear* form. Note that the impedance, being the reciprocal of the admittance, also has the same form.

Now suppose that, instead of a resistance R, an inductance L or a capacitance C is the variable quantity. Then, instead of R in Eq. (7-74), there would be either $j\omega L$ or $1/j\omega C$. In either case the form of the admittance will still be bilinear, like Eq. (7-75) but with L or C appearing instead of R. As an alternative we can deal with X_L and X_C rather than L and C.

So much for impedances and admittances. Let us briefly consider transfer functions also. For example, returning to the discussion immediately preceding Eq. (7-71), rather than finding the current in loop 1, let us find the current I_k in loop k. The ratio of this current to the source voltage will be the transfer admittance Y_{k1}, and it will be expressed as

$$Y_{k1} = \frac{I_k}{V_1} = \frac{\Delta_{1k}}{\Delta} \tag{7-76}$$

Just as we previously expanded Δ and Δ_{11}, we can also expand Δ_{1k}. It will again be a linear function of the variable element. Hence the transfer admittance can also be written in the bilinear form given in Eq. (7-75).

All the other transfer functions—transfer impedance, voltage gain, and current gain—can also be expressed as the ratio of two determinants (do this) and, hence, can also be written as bilinear forms.

In conclusion, we can now make the following statement: *If only one element in a network is variable, then the expression for any of the network functions will be a bilinear function of the variable parameter.*

We shall continue to use Eq. (7-75) which expresses the admittance Y in terms of the variable parameter R. Keep in mind, however, that the same expression applies to the other network functions as well, and to the case of variable $L(X_L)$ or variable $C(X_C)$.

We now have the task of determining what type of curve is represented by the locus of a network function as given by the bilinear form in Eq. (7-75). For the example in Fig. 7-29 which we examined, we found the locus to be a circle. Other examples (which are suggested for you to do as homework problems) also lead to circles. Hence we are led to examine the possibility that the locus of Eq. (7-75) might always be a circle.

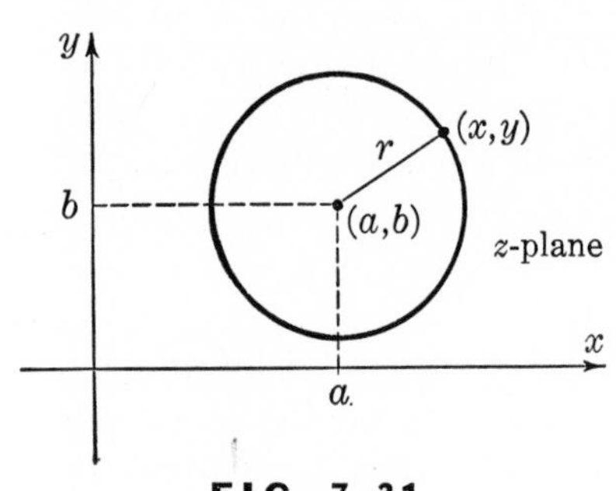

FIG. 7-31

A general circle

The general equation of the circle shown in Fig. 7-31 is

$$(x - a)^2 + (y - b)^2 = r^2 \quad (7\text{-}77)$$

In this expression x and y are the coordinates of points on the circle and a and b are the coordinates of the center; r is the radius. Now suppose we consider the plane in which the circle appears as a complex plane; if we let z represent the complex variable, then x is the real part of z, and y is the imaginary part of z. The center of the circle is then a point in the complex plane, having real part a and imaginary part b. The form of Eq. (7-77) is suggestive of the square of the magnitude of a complex number; that is, the sum of the squares of the real part and the imaginary part. If we let ρ be the complex number in question, then it appears that

$$\rho = (x - a) + j(y - b) = (x + jy) - (a + jb) \quad (7\text{-}78)$$

You can see that if we take the sum of the squares of $(x - a)$ and $(y - b)$, the result will indeed be the left side of Eq. (7-77).

The last form of writing Eq. (7-78) shows that ρ is the difference between two complex numbers: $(x + jy) = z$, which is a point on the circle, and $(a + jb)$, which is the center of the circle. If we designate the center as $z_0 = (a + jb)$, then the equation of the circle can be written in terms of complex quantities as

$$(z - z_0)(z - z_0)^* = \rho\rho^* = r^2 \quad (7\text{-}79)$$

since the magnitude square of a complex quantity is the product of the quantity and its complex conjugate.

If we can now show that the admittance Y satisfies an equation such as Eq. (7-79), we shall have proved that the locus of Y is a circle. In order to carry out this plan, let us use Eq. (7-75) for Y, and form the quantity $(Y - Y_0)(Y - Y_0)^*$, where Y_0 is still unknown. The result will be

$$(Y - Y_0)(Y - Y_0)^* = \frac{|A - CY_0|^2 + R[(B - DY_0)(A - CY_0)^* - (A - CY_0)(B - DY_0)^*] + R^2|B - DY_0|^2}{|C|^2 + R(CD^* + DC^*) + R^2|D|^2} \qquad (7\text{-}80)$$

(Don't throw up your hands at the complexity of this expression; we shall soon cut it down to size.)

In order for this expression to represent a circle, the quantity on the right must be a constant, namely, the square of the radius; that is, it must not depend on the varying parameter R. Both the numerator and the denominator of the right side are quadratic polynomials in R. The quantity will surely be independent of R if each coefficient in the numerator is equal to $|\rho|^2$ times the corresponding coefficient in the denominator, for then the numerator and denominator will cancel, leaving simply $|\rho|^2$ on the right. Thus the conditions that must be satisfied by the coefficients are

$$\begin{aligned} |A - CY_0|^2 &= |\rho|^2|C|^2 && (a) \\ |B - DY_0|^2 &= |\rho|^2|D|^2 && (b) \qquad (7\text{-}81) \\ (B - DY_0)(A - CY_0)^* &= |\rho|^2CD^* && (c) \end{aligned}$$

Each of the first two of these expressions states that the square of the magnitude of one complex number is equal to the square of the magnitude of another. From this we can conclude that the first complex number must be equal either to the second number itself or to its conjugate. Suppose we write

$$\begin{aligned} A - CY_0 &= \rho C^* && (a) \\ B - DY_0 &= \rho D^* && (b) \end{aligned} \qquad (7\text{-}82)$$

The first two of the expressions in Eq. (7-81) will be satisfied, and we find that the third one will also be satisfied.

Examining this last pair of equations reveals that there are two unknown quantities, the center Y_0 and the radius ρ. These equations can now be solved for the unknowns, yielding

$$\begin{aligned} Y_0 &= \frac{AD^* - BC^*}{CD^* - DC^*} && (a) \\ \rho &= \frac{BC - AD}{CD^* - DC^*} && (b) \end{aligned} \qquad (7\text{-}83)$$

Since we have been able to find a center and a radius such that Eq. (7-80) represents a circle, it follows that the locus of the admittance Y is a circle.

An alternative proof of this result can also be given. Starting from Eq. (7-75), express all complex quantities in rectangular form. For example, let the admittance Y be written $x + jy$. Now cross multiply the two sides of Eq. (7-75) and express each side in rectangular form. Now set the reals and imaginaries equal. This will yield two equations. Solve one of these for R and substitute into the other. The result will be an equation containing the variables x and y, together with known quantities. This equation can be put in the form of the circle in Eq. (7-77). You are urged to carry out the details.

In summary, we can now state the following theorem:

When only one element of a network is variable, the locus of any of the network functions is a circle.

In developing this theorem, we have dealt completely with the general case. In special cases one or more of the coefficients in Eq. (7-75) may take on special values, such as zero or one. We should examine the special forms taken on by the locus in certain such limiting cases. For example, suppose $D = 0$, $C = 1$, A real, and B imaginary. Then the real part of Y will be constant, only the imaginary part being variable. Hence the locus will be a straight line parallel to the imaginary axis. This seems to violate our previous conclusion about the locus being a circle. We can, however, look upon a straight line as a degenerate circle having an infinite radius and with a center at infinity. This limiting case is therefore included in the general one. Of course, the formulas for the center and the radius in Eq. (7-83) will not be useful in this case. The admittance of the parallel RC network in Fig. 7-29 has just such a locus.

Let us use the theorem we have just proved, together with the expressions for the center and the radius, to find the locus of the admit-

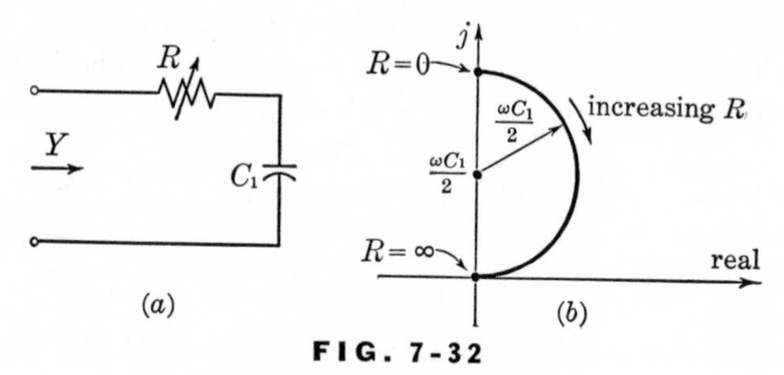

FIG. 7-32

Locus of series RC circuit

tance of the series RC circuit shown in Fig. 7-32(a). The variable parameter is the resistance R. The admittance is given by

$$Y = \frac{1}{R + \dfrac{1}{j\omega C_1}} \tag{7-84}$$

Comparing this with the general form in Eq. (7-75), we find the general coefficients to be

$$\begin{aligned} A &= D = 1 \\ B &= 0 \\ C &= 1/j\omega C_1 \end{aligned} \tag{7-85}$$

Substituting these into Eq. (7-83), we find the center and radius to be

$$\begin{aligned} Y_0 &= j\frac{\omega C_1}{2} \\ r &= |\rho| = \frac{\omega C_1}{2} \end{aligned} \tag{7-86}$$

The center of the circle is on the imaginary axis $\omega C_1/2$ units from the origin. The radius is $\omega C_1/2$. The locus is shown in Fig. 7-32(b). You should find the locus by the method used in the discussion of Fig. 7-29 as a check.

If the network under consideration is much more complicated than the examples we have considered, then calculation of the center and the radius by means of the formulas we have developed may not be very simple. It is not necessary, however, to find the complete locus in one huge calculation. The task can be broken down into a number of smaller steps.

As an illustration of the possibilities intimated in the preceding paragraph, consider the network shown in Fig. 7-33. It is assumed that the admittance Y_1 in parallel with the RC branch is fixed. The total

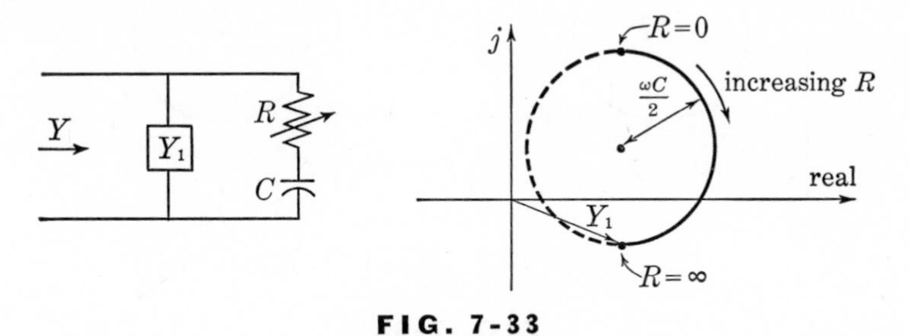

FIG. 7-33

Combination locus

admittance Y is the sum of the admittances of the two branches. Since Y_1 is a (complex) constant, the locus of Y is obtained by adding a constant to each point of the locus of the RC branch. Thus, with Y_1 represented by the directed line shown in Fig. 7-33(b), the locus of Y is simply the locus of the RC branch admittance, which is a circle, shifted by the complex constant Y_1, as shown in the figure.

In exploiting this procedure for the representation of network functions, it is often necessary to find the locus of a reciprocal quantity; that is, knowing the locus of admittance, it may be desired to obtain the locus of the corresponding impedance. We have already remarked that the loci of both impedances and admittances are circles when only one network element is varied, with straight lines admitted as special cases of circles. Thus, if the admittance locus is known, we need find just enough points on the reciprocal (impedance) locus to define the circle. Since three points suffice to define a circle, it is enough to find three corresponding points. All of these may lie on the circle or two may lie on the circle, the third point being the center.

First of all, suppose the Y circle passes through the origin. The reciprocal of zero is infinity. Hence one point on the Z locus will be at infinity. This implies that the Z locus will be a straight line.

Each point on the Z locus is related to its corresponding point on the Y locus in the following way: The magnitude is the reciprocal and the angle is the negative. Based on these facts, a geometrical construction of the reciprocal circle can be performed as follows: Figure 7-34 shows an admittance circle. Pass a line through the origin and the center P of the Y locus, and extend it to intersect the locus at the

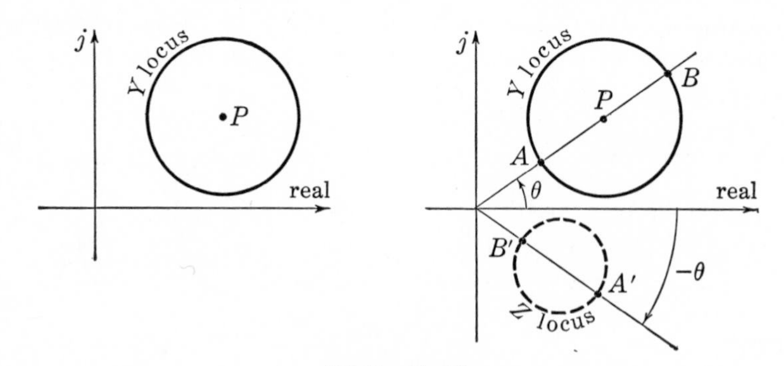

FIG. 7-34
Inversion of loci

points A and B. These points are closest and farthest from the origin, respectively. The line makes an angle θ with the real axis.

Now draw a line making an angle $-\theta$ with the real axis. The reciprocals of points A and B must lie on this line, say at points A' and B'. The distance from the origin to A' will be the greatest for all points on the reciprocal locus; similarly, the distance from B' to the origin will be the shortest. A' and B' must therefore be at opposite ends of a diameter. Hence the center of the reciprocal locus will be midway between A' and B'. The Z locus is now readily drawn; it is shown dotted in Fig. 7-34(*b*).

From the predominance of the words admittance and impedance in the preceding discussion, you may acquire the impression that transfer functions are excluded, even though the validity of the results for transfer functions has been certified. To remove such an impression we shall now consider an example of the locus of a transfer function.

Consider the network shown in Fig. 7-35(*a*). This is a bridge

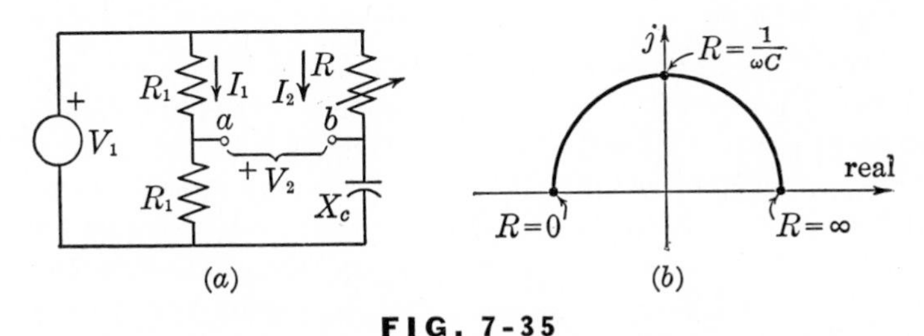

FIG. 7-35

Phase-shifting network and its locus

circuit whose output terminals are open circuited. (The load might be the grid circuit of a vacuum tube which has a high impedance, as we shall discover later.) It is desired to find the locus of the output voltage V_2. An analytic expression for V_2 is easily calculated. Thus

$$V_2 = R_1 I_1 - \frac{1}{j\omega C} I_2 = R_1 \left(\frac{V_1}{2R_1}\right) - \frac{1}{j\omega C}\left(\frac{V_1}{R + \dfrac{1}{j\omega C}}\right) \tag{7-87}$$

$$= \left(\frac{1}{2} - \frac{1}{j\omega CR + 1}\right) V_1$$

Finally,

$$G_{21} = \frac{V_2}{V_1} = \frac{-1 + j\omega CR}{2 + j2\omega CR} \tag{7-88}$$

When this expression is compared with the general bilinear function given in Eq. (7-75), we find the general coefficients to be

$$A = -1 \qquad C = 2$$
$$B = j\omega C \qquad D = j2\omega C$$

Using Eq. (7-83), we now find that the center of the circle lies at the origin and that the radius is 1/2. Thus the locus has the form shown in Fig. 7-35(*b*). Since R is always positive, only half the circle is pertinent. Note that the magnitude of the output voltage is constant at one half the magnitude of the input voltage; only its angle varies as R changes. For this reason this network is called a *phase shifter*.

It is of collateral interest in this example to express the transfer function in terms of the complex variable s. Thus Eq. (7-88) becomes

$$G_{21} = \frac{1}{2} \frac{s - \dfrac{1}{RC}}{s + \dfrac{1}{RC}} \tag{7-89}$$

There is one pole and one zero having the interesting property that the zero is the negative of the pole. This property actually defines a class of networks called *all-pass networks*. (This class itself is a subclass of the networks called *nonminimum-phase* networks.) If you go through the details of determining the magnitude of the voltage-gain function from the pole-zero diagram, you will find that the length of the line from the zero to any point on the $j\omega$ axis is the same as the length of the line from the pole, for all values of the variable resistance. The ratio of the two line lengths is, therefore, unity independent of R, and the magnitude of the function is simply equal to the constant multiplier, which is 1/2. This result agrees with the result obtained from the locus diagram, as it should.

Up to this point we have been discussing the loci of network functions when a network parameter is the variable. Let us now discuss the case when the frequency is variable and the network parameters are fixed. In this case the locus is no longer a simple curve, since the transfer functions are not simply bilinear functions of frequency (except when there is only one energy-storage element in the network). Hence, a simple analytical expression for the locus will not be available.

Some familiarity with variable-frequency loci can be obtained by considering some examples. Let us examine the admittance locus of

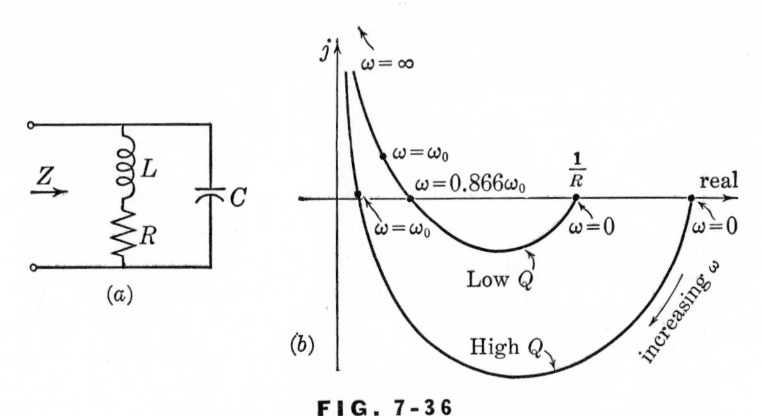

FIG. 7-36

Resonant circuit and its admittance locus

the parallel resonant circuit shown in Fig. 7-36 as the frequency is varied. The admittance is given by

$$Y = j\omega C + \frac{1}{R + j\omega L} = \frac{1 - \omega^2 LC + j\omega CR}{R + j\omega L} = \frac{1 + jQ_0\left(\dfrac{\omega}{\omega_0} - \dfrac{\omega_0}{\omega}\right)}{RQ_0^2\left(1 - j\dfrac{\omega_0}{Q_0\omega}\right)} \tag{7-90}$$

where

$$\omega_0^2 = \frac{1}{LC}$$

$$Q_0 = \frac{\omega_0 L}{R} = \frac{1}{\omega_0 CR} = \frac{1}{R}\sqrt{\frac{L}{C}}$$

are the familiar resonant frequency and the Q.

The locus is shown in Fig. 7-36(b) for two values of Q_0. Note that the point on the locus at which $\omega = \omega_0$ does not occur when Y is real. This point is closer to the real axis for the high-Q circuit, however, than it is for the low-Q case. The magnitude of Y at any point on the locus is the length of the line drawn from the origin to the point. The minimum magnitude can be found by means of a pair of dividers, by placing one end at the origin and swinging an arc of ever-increasing radius until the arc becomes tangent to the locus. For the high-Q locus in Fig. 7-36(b) the points of minimum $|Y|$, unity power factor (or real Y), and $\omega = \omega_0$ are so close together as to be almost indistinguishable. For the low-Q locus, however, the point of minimum $|Y|$ is seen to lie somewhere between the points of unity power factor and the point at which $\omega = \omega_0$. These results are in agreement with the discussion of the parallel tuned circuit in Sec. 5-6.

The complex loci take on a more symmetrical shape if the parts corresponding to negative values of ω are included. Thus the complete locus for the parallel tuned circuit will take the form shown in Fig. 7-37, the part corresponding to negative ω being dashed. This part of the locus is necessary for a complete mathematical description of the admittance function.

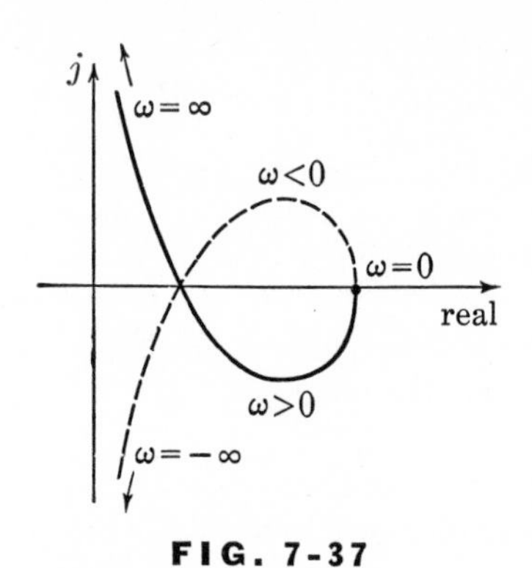

FIG. 7-37

Locus including negative value of ω

In the applications of the complex locus in feedback and control systems, very often it is not desired to have a detailed labeling of the complete locus. It is required only to have an idea of some general characteristics of the locus, such as the points corresponding to zero and infinite frequency, the magnitude of the function when the angle has certain values (such as 180, 135, or 225 deg), the angle of the function when the magnitude has certain values (such as 1), whether or not certain areas of the plane are enclosed by the locus, etc. Such general characteristics can often be determined without the need for detailed calculations. Several general theorems regarding the loci of network functions can be proved to aid in this determination. The considerations that enter into the proofs of these theorems, however, are beyond our scope, and we shall not pursue the subject any further here.

Problems

7-1 In the network of Fig. P7-1, find the branch voltage across the inductor: (*a*) by the node method of analysis; (*b*) by loop equations; (*c*) by the use of Thévenin's theorem at the appropriate point.

7-2 In the network of Fig. P7-2 find the impedance at the terminals of the current source: (*a*) by combining branches in series and in parallel; (*b*) by solving for V_1 in terms of I_1 using loop equations; (*c*) by solving for V_1 in terms of I_1 using node equations.

7-3 Find the current $i(t)$ and the output voltage $v_2(t)$ in the network of Fig. P4-14: (*a*) by the method of loop equations; (*b*) by the method of node equations.

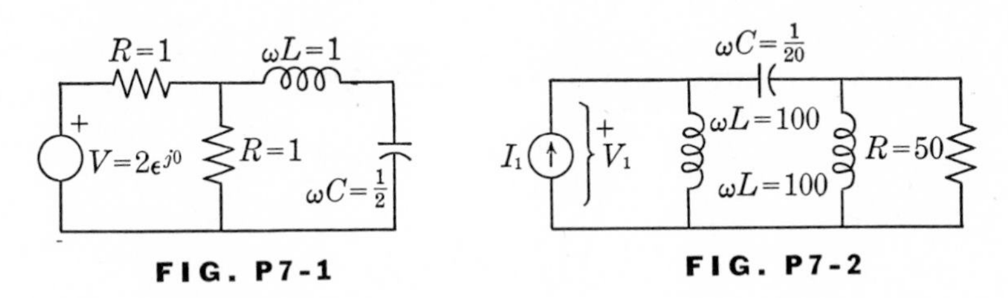

FIG. P7-1

FIG. P7-2

7-4 Solve Prob. 4-9 by (*a*) the method of loop equations and (*b*) the method of node equations.

7-5 In Figs. P4-10 and P4-12 find the transfer admittance I/V_g, and the input impedance by (*a*) using loop equations and (*b*) using node equations.

7-6 In the network of Fig. P7-6, find an expression for the voltage ratio V_2/V_1: (*a*) by the loop method of analysis; (*b*) by the node method of analysis; (*c*) by one or more applications of the Thévenin or Norton theorems; (*d*) by the ladder-network method.

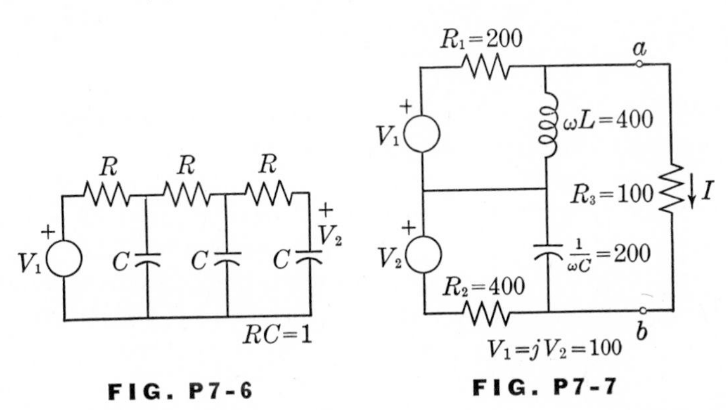

FIG. P7-6

FIG. P7-7

7-7 In the network of Fig. P7-7, find the current in R_3: (*a*) by the method of loop equations; (*b*) by the method of node equations; (*c*) by using Thévenin's theorem. (*d*) Find the value of R_3 that will make the power dissipated in it a maximum, and find the value of this maximum power.

7-8 In the network of Fig. P7-8, $RC = 2L/R$. Find an expression for the voltage-gain function V_2/V_1: (*a*) using loop equations; (*b*) using node equations; (*c*) using Norton's or Thévenin's theorem; (*d*) using the step-by-step method of ladder networks.

7-9 Find the Thévenin equivalent of the networks shown in Fig. P7-9 at the designated terminals: (*a*) utilizing the tee-pi or pi-tee transformations; (*b*) using loop equations or node equations.

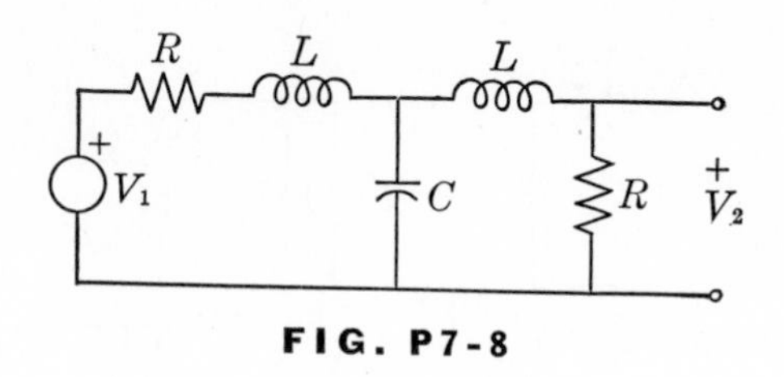

FIG. P7-8

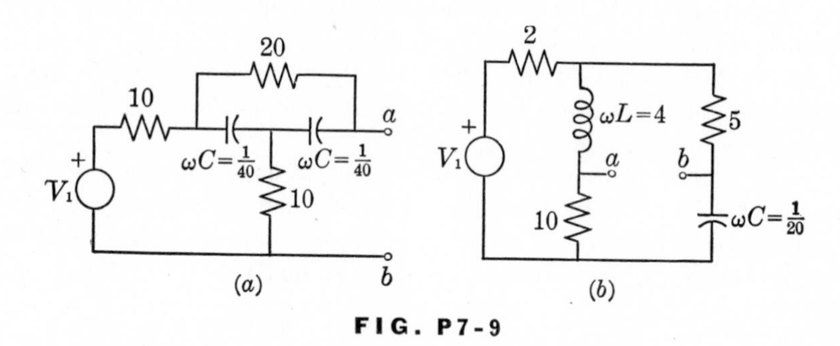

FIG. P7-9

7-10 In the network of Fig. P7-10, find the voltage across R when its value is adjusted to make the power dissipated in it a maximum.

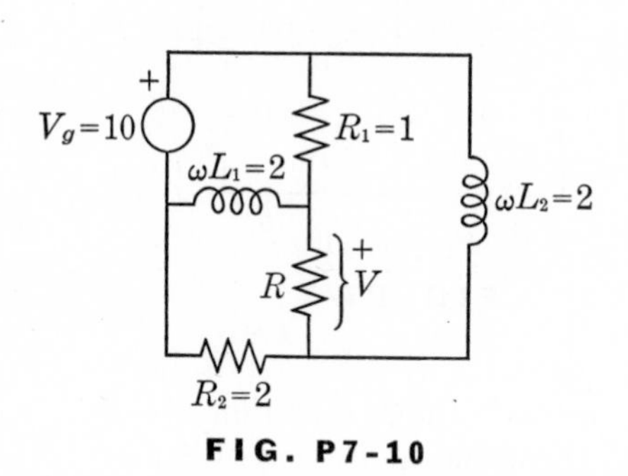

FIG. P7-10

7-11 It is desired to find the Thévenin equivalent of a physical device at a particular frequency by measurements at the terminals. The open-circuit voltage magnitude is found to be 20 volts; the short-circuit current magnitude is found to be 2.2 amp. Next a 10-ohm resistor is connected across the terminals, and the resulting voltage across it is found to be 11.1 volts. Finally, when a variable capacitor is connected at the terminals, it is found that the resulting current reaches a maximum at a particular value of C. Find the voltage and impedance of the Thévenin equivalent. What is the maximum power which this device can supply?

7-12 Apply the tee-pi transformation to the networks shown in Fig. P7-12. Are the equivalent pi's realizable at all frequencies, at one frequency, or at none?

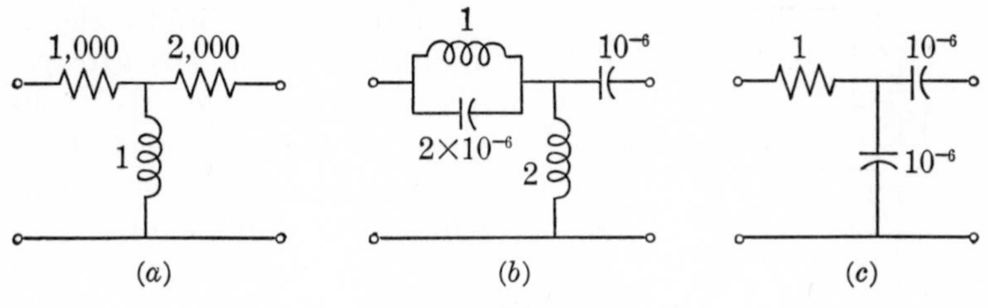

FIG. P7-12

7-13 Use the principle of superposition to find the designated steady-state response in the networks shown in Fig. P7-13. (*a*) $v_g = 2\cos 2t$; $i_g = 5\sin t$. Find v_C. (*b*) $i_1 = \cos 4t$; $i_2 = 2\sin 10t$. Find v_L.

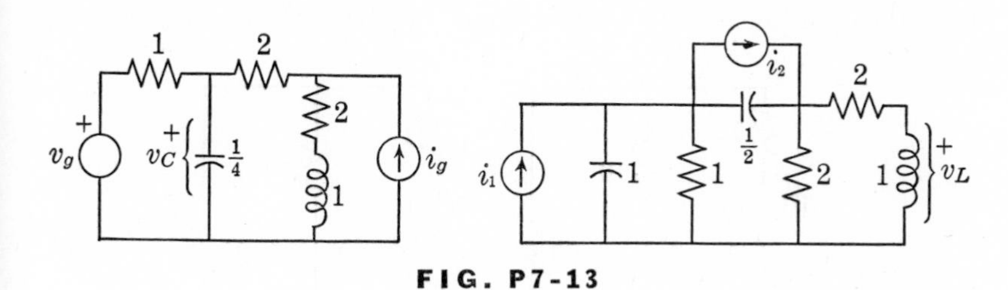

FIG. P7-13

7-14 In Fig. P7-14 find the value of ωC, in terms of the real and imaginary parts of Z, which will make the tee equivalent of the combined network consist of resistive branches only.

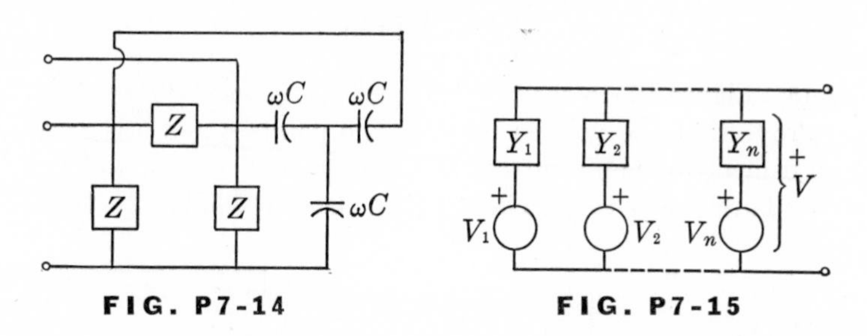

FIG. P7-14

FIG. P7-15

7-15 In Fig. P7-15 prove that

$$V = \frac{Y_1V_1 + Y_2V_2 + \cdots + Y_nV_n}{Y_1 + Y_2 + \cdots + Y_n}$$

This result is called *Millman's theorem.*

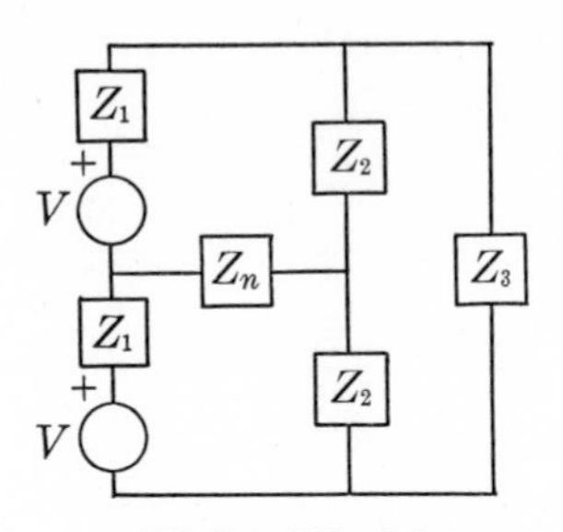

FIG. P7-16

7-16 Figure P7-16 shows a so-called *balanced three-wire power system.* Prove that the voltage across the branch marked Z_n, and so also the current through it, is zero. Show also that the voltage across Z_3 is twice that across Z_2.

7-17 Figure P7-17 shows a three-wire system in which the Z_n branch has been left open in one case and has been shorted in the other. When the system is balanced this will have no influence on the remaining currents. Now suppose R changes from the balanced value of 10 to $10 + x$. Use the compensation theorem to compute the current in the 20-ohm load in each case.

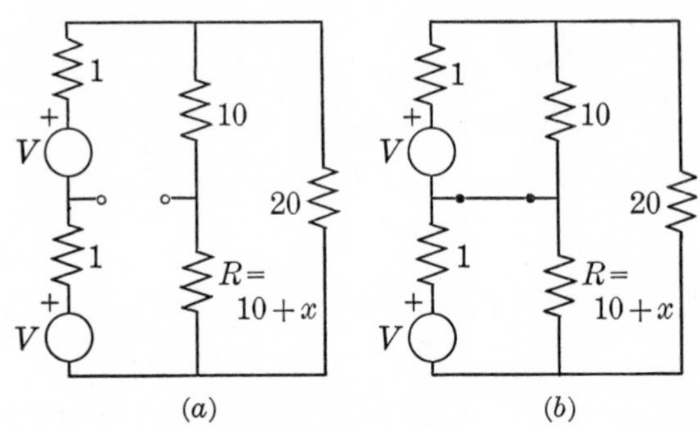

FIG. P7-17

7-18 Figure P7-18 shows a *balanced wye-connected three-phase system.* Prove that the voltage between the points a and b is zero. Do this by using Kirchhoff's current law at node a or node b, and again by applying Kirchhoff's voltage law around the two windows. Repeat by

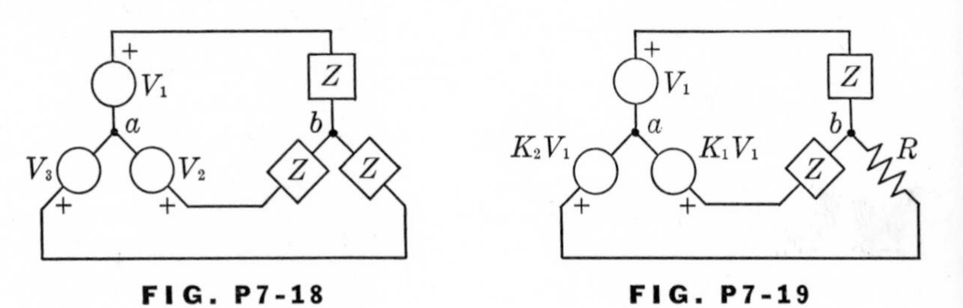

FIG. P7-18

FIG. P7-19

using Millman's theorem (Prob. 7-15) to find V_{ba} directly. Compare the required effort.

$$V_2 = V_1\epsilon^{j120^\circ}$$

$$V_3 = V_1\epsilon^{j240^\circ}$$

7-19 In Fig. P7-19 the three-phase system is unbalanced. Find the current in R, using the simplest method you know.

7-20 In the circuit of Fig. P7-20, the current measured by the ammeter is 10 amp. The meter itself, however, has a resistance of 1 ohm. Using the compensation theorem, find the value which the current will have in the absence of the meter. Find also the change in the current supplied by the source caused by the presence of the meter.

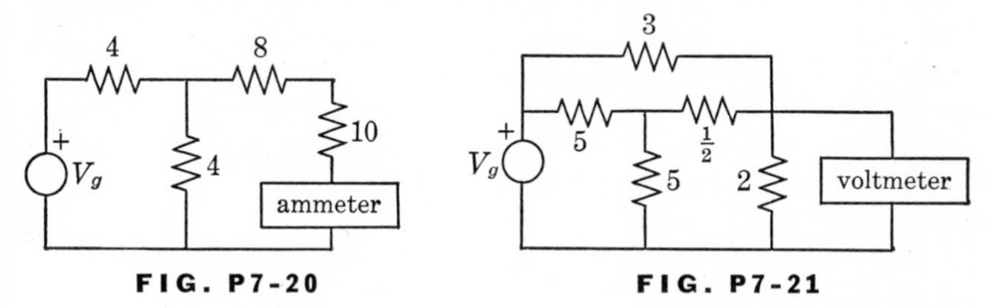

FIG. P7-20

FIG. P7-21

7-21 In Fig. P7-21 the voltage measured by the voltmeter is 100 volts. The meter itself, however, has a resistance of 1,000 ohms. Using the compensation theorem, find the value which the voltage will have in the absence of the meter. Find also the change in the current supplied by the source caused by the presence of the meter.

7-22 Find the admittance locus of the series RL circuit shown in Fig. P7-22 for: (*a*) variable R; (*b*) variable X.

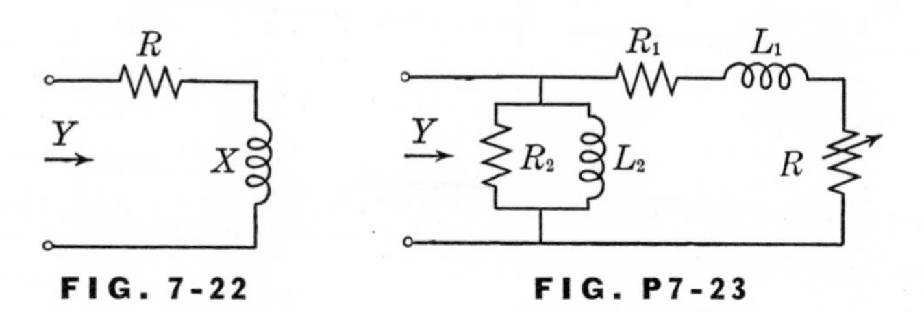

FIG. 7-22

FIG. P7-23

7-23 Find the admittance locus of the network shown in Fig. P7-23. This is an equivalent circuit of the induction motor in which R represents the mechanical load. Invert and find the impedance locus.

7-24 Using the result of Fig. P7-23, find the admittance locus of Fig. P7-24.

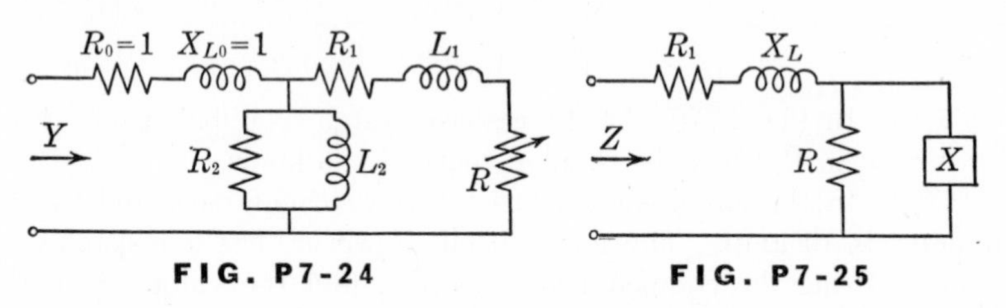

FIG. P7-24 FIG. P7-25

7-25 Find the impedance locus of the network shown in Fig. P7-25, assuming the variable quantity is the reactance X. On the locus indicate the points at which $X = kR$ where $k = \pm 1/2$, ± 1, and ± 2.

7-26 For the network in Fig. P7-26, find the impedance locus by starting from the admittance locus of the right-hand branch and, by inversions, working toward the input. Repeat by calculating the radius and center according to the formulas in the text.

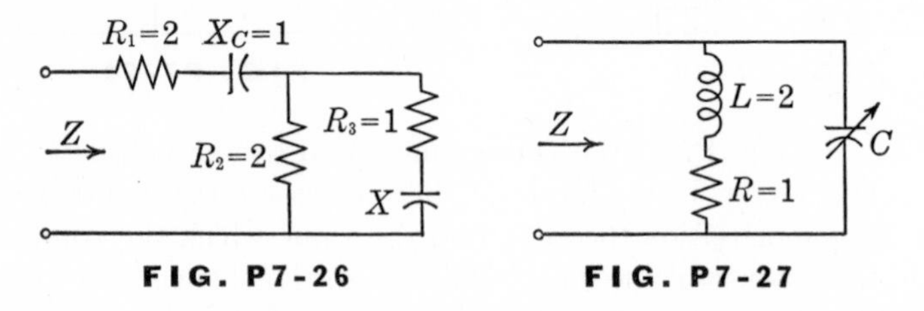

FIG. P7-26 FIG. P7-27

7-27 In the tuned circuit of Fig. P7-27, the capacitance is variable. Determine the impedance locus if the range of variation of the capacitance is $0 \leq C \leq 1$. Assume $\omega = 1$.

7-28 In the network of Fig. P7-28, the capacitance is variable. Sketch the locus of V_2/V_1 and from this find the value of X_C for which the angle of V_2/V_1 is a maximum.

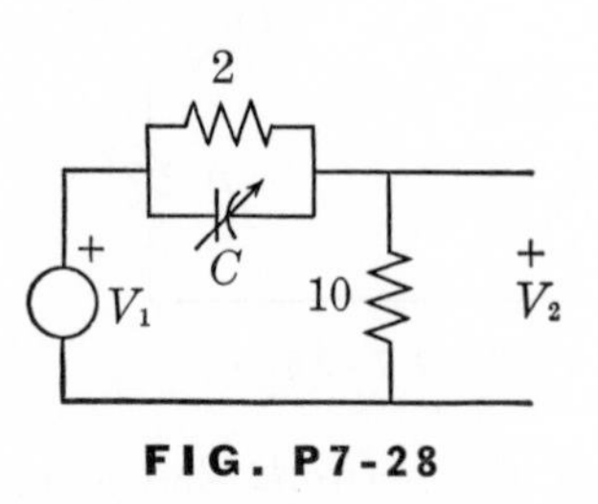

FIG. P7-28

7-29 The network of Fig. P7-29 is a model of an electronic device called a tunnel diode. The static *V-I* curve is shown in the figure. (It is clearly not linear.) The slope of the curve is the conductance G, which is seen to vary with the voltage. To determine the parameters of the diode, the real and imaginary parts of the input impedance are measured for several values of G, obtained by varying the voltage. At a frequency of 50 megacycles per second, the following three values of impedance are obtained for three values of voltage: $Z_1 = 5 + j0$, $Z_2 = 5 - j5$, $Z_3 = -1 - j4$. (*a*) Plot the locus of the impedance. (*b*) Assuming that all elements but G remain constant, determine the values of R, L, and C from the locus. (*c*) Find the values of G corresponding to the three measured points and show approximately the corresponding points on the V-I curve.

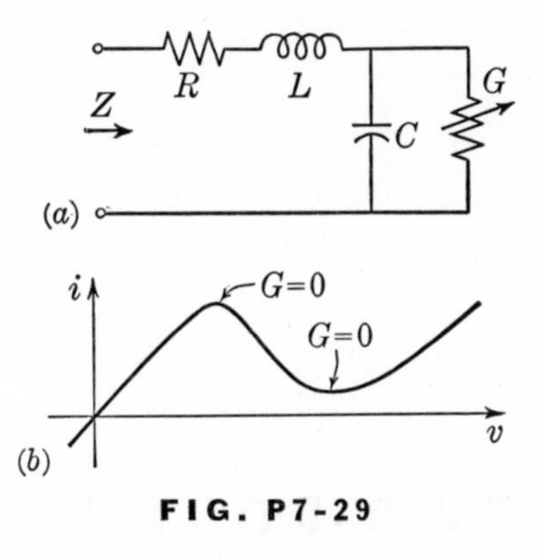

FIG. P7-29

TWO-PORT NETWORKS

8

IN our discussions of electrical networks so far, we have been concerned with the determination of currents and voltages, given the sources and the structure of the network. Our approach has been quite general, and we have not restricted ourselves as to the permissible structure of the network and the permissible locations of the sources. Very often, however, we are not interested in the internal construction of the network; we merely want to know the relationships among voltages and currents at the terminals of the network to which external connections can be made. Figure 8-1 shows the simplest network from this point of view; it has just one pair of terminals. These terminals constitute the only points of entry to or exit from the network. Borrowing an ordinary word from the English language, we call this a *one-port* network. Either the voltage or the current is the excitation, the other one being the response.

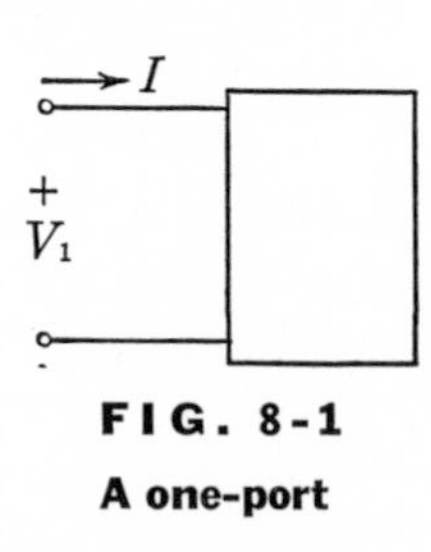

FIG. 8-1
A one-port

Other networks may have more pairs of terminals to which connections can be made. It is appropriate to call these networks *multiports.*

Figure 8-2(a) shows a network having two pairs of terminals, hence, a *two-port.* Although there is a total of four terminals, no connections are to be made from one terminal of a pair to a terminal of the other pair. The external signals carrying the power or information are fed

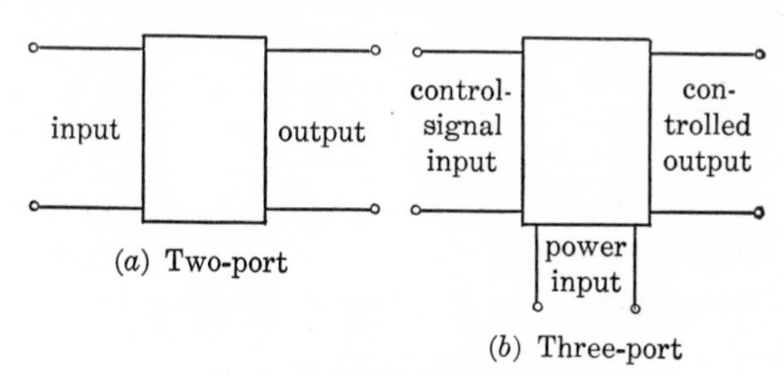

FIG. 8-2

A number of multiports

into one of the ports, which, for this reason, is called the *input* port. The network transmits the signals to the other port, which is called the *output* port. An example of such a two-port may be a filter, an amplifier, or a transmission line for carrying power or for carrying telephone signals.

In other systems different varieties of signals may enter the system at different ports. For example, Fig. 8-2(b) shows a *three-port.* The power is supplied at one of the ports, while a signal which is to control the power enters a second port. The desired signal appears at the output. A modulator can be described by just such a three-port. The so-called carrier enters at one port, the modulating signal at another. The modulated carrier comes out at the third port. The modulating process is not a linear one, however, so that this particular three-port must contain other elements besides the ones that now appear in our model.

The concepts of multiports which we shall develop are not all restricted to linear electrical networks, as illustrated in the preceding paragraph. They are not even restricted strictly to electrical networks. Other types of systems—mechanical, electromechanical, thermal, chemical—can all be described in terms of concepts which we shall discuss in this chapter. Of course, the variables used to describe these systems will not be currents or voltages.

In this chapter we shall restrict ourselves to the study of two-ports. The concepts we develop can be extended to higher-order multiports.

8.1 SHORT-CIRCUIT ADMITTANCE PARAMETERS

The discussion will be introduced in terms of the two-port shown in Fig. 8-3. As mentioned before, no connection is to be made between a terminal of one port and a terminal of the other. This will require that the instantaneous current which enters one terminal of a port must be equal to the instantaneous current which leaves the other terminal of the same port at any time. Clearly, the same condition must hold for the sinors. It is assumed that there are no sources inside the "black box."

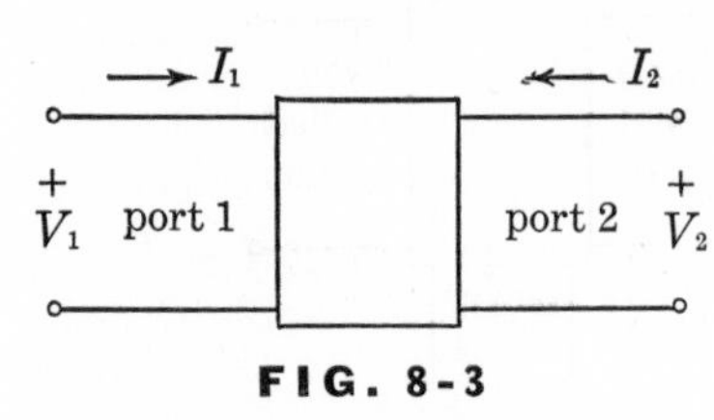

FIG. 8-3

Conventional references of two-port

Note the manner in which the reference of I_2 is chosen. For some purposes this is the more convenient reference. For other purposes, however, it is more convenient to choose the opposite reference. The reference we have chosen has become more or less standard.

In the case of the one-port network, there is a single voltage and a single current whose ratio is the impedance (or admittance), and this quantity completely characterizes the network. But in the case of the two-port, there are two voltages and two currents. How should we describe the behavior of the network at its terminals?

In all the work we have done to this point, we have found the response sinor (any response) to be linearly related to all source sinors, whether the sources are voltage or current. (For example, look at the solutions of the loop or node equations in Chap. 7.) Let us, therefore, assume that the voltages are the excitations and the currents are the responses and write

$$I_1 = y_{11}V_1 + y_{12}V_2 \qquad (a)$$

$$I_2 = y_{21}V_1 + y_{22}V_2 \qquad (b) \qquad \text{(8-1)}$$

At the moment, the y_{ik} coefficients simply represent proportionality factors relating the current sinors to the voltage sinors. Dimensionally, they should be admittances; this is the reason for our use of the symbol y. Again, it is more or less standard to use lower-case letters for these coefficients.

These equations show that, in the case of the two-port, four quantities are needed to give a complete characterization of the network. This is considerably more than the single quantity needed to characterize a one-port. Once these four quantities are known, the description of the V-I relationships at the terminals will be complete.

It is very simple to obtain an interpretation for these proportionality factors. Suppose the output port is short-circuited, as shown in Fig. 8-4(a), and a voltage source is applied at the input port. The

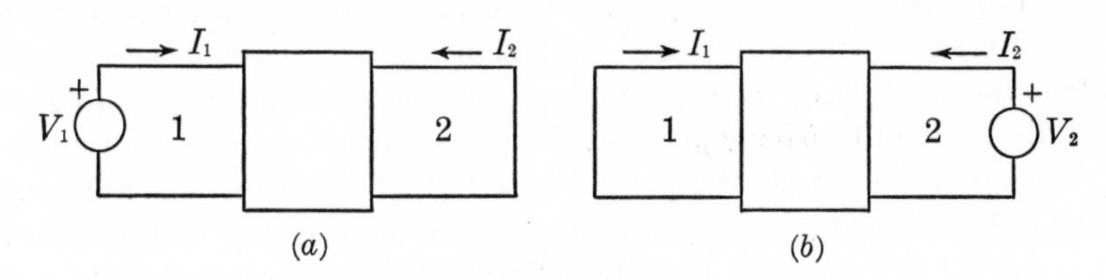

FIG. 8-4

Short-circuit parameters

short-circuit forces V_2 to be zero. Setting V_2 equal to zero in Eq. (8-1), we get

$$y_{11} = \left.\frac{I_1}{V_1}\right|_{V_2=0} \qquad (a)$$

$$y_{21} = \left.\frac{I_2}{V_1}\right|_{V_2=0} \qquad (b) \qquad (8\text{-}2)$$

Since V_1 is the only source in the network, the quantity y_{11} is, by definition, the admittance (the input admittance) at port 1 with port 2 short-circuited.

We observe that y_{21} is the ratio of a response-current sinor to a driving-voltage sinor. We call it the *transfer admittance* from port 1 to port 2, with port 2 short-circuited. (Sometimes it is also called the *forward short-circuit transfer admittance.*) Note the order of the subscripts, which is consistent with our definition of the transfer functions in Sec. 4-7.

Now suppose that port 1 is shorted, as shown in Fig. 8-4(b), and a voltage source is applied at port 2. The short circuit causes V_1 to be zero. Setting V_1 equal to zero in Eq. (8-1), we get

$$y_{22} = \left.\frac{I_2}{V_2}\right|_{V_1=0} \qquad (a)$$

$$y_{12} = \left.\frac{I_1}{V_2}\right|_{V_1=0} \qquad (b) \qquad (8\text{-}3)$$

In this case V_2 is the only source in the network. Hence, by definition, y_{22} is the input admittance at port 2 when port 1 is short-circuited. Since port 2 is normally the output port, this admittance is also called the short-circuit *output admittance.* The quantity y_{12} is again the ratio of a current-response sinor to a voltage-excitation sinor. We call it the *transfer admittance* from port 2 to port 1 when port 1 is short-circuited. (Sometimes it is also called the *backward short-circuit transfer admittance.*)

As a group, these y quantities are parameters which completely characterize a two-port. We call them the *short-circuit admittance parameters.* Two of them, y_{11} and y_{22}, are input admittances. (These are also called the *driving-point* admittances, since they refer to response functions which are at the point where the driving functions appear.) The other two are transfer admittances from one port to the other.

Let us now consider how the y parameters can be calculated for a given network. If the network is simple enough, the y parameters can be written down by inspection, using Eqs. (8-2) and (8-3). For example, consider the pi network shown in Fig. 8-5(a). If we short-circuit the output port, Y_2 and Y_3 will be in parallel. Hence the input admittance will simply be their sum. To find I_2, note that it is the current in branch Y_3 and that V_1 is the voltage across the branch. Owing to the references, the relationship will be $I_2 = -Y_3V_1$. Thus we find

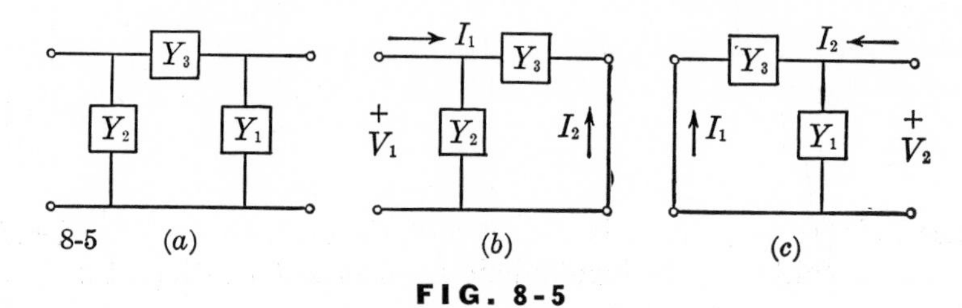

FIG. 8-5

Short-circuit parameters of a pi

$$y_{11} = Y_2 + Y_3 \qquad (a)$$
$$y_{21} = -Y_3 \qquad (b) \qquad (8\text{-}4)$$

Next we short-circuit the input port, as shown in Fig. 8-5(c). This time, Y_1 and Y_3 are in parallel, so that the admittance at port 2 is simply their sum. I_1 is the current in Y_3 while V_2 is the voltage across it; hence $I_1 = -Y_3V_2$, the sign being due to the references. As a result,

$$y_{22} = Y_1 + Y_3 \qquad (a)$$
$$y_{12} = -Y_3 \qquad (b) \qquad (8\text{-}5)$$

We notice that y_{12} and y_{21} are the same for the pi. We shall soon see that this result is not restricted to the pi.

For a more complicated network, the same procedure of short-circuiting the ports and computing the currents can be followed. As an alternative, look back at Fig. 8-3 and assume that voltage sources are applied at the two ports. If we write loop equations, they will take the general forms given in Eq. (7-16), except that all the source voltages but V_1 and V_2 will be zero. This assumes that the loops are chosen in such a way that V_1 appears only in loop 1 and V_2 appears only in loop 2. The solution for any of the loop currents is given in Eq. (7-19). In the present case we are interested only in I_1 and I_2. Hence we have

$$I_1 = \frac{\Delta_{11}}{\Delta} V_1 + \frac{\Delta_{21}}{\Delta} V_2 \qquad (a)$$

$$I_2 = \frac{\Delta_{12}}{\Delta} V_1 + \frac{\Delta_{22}}{\Delta} V_2 \qquad (a) \qquad (8\text{-}6)$$

These expressions are identical in form with those defining the y parameters in Eq. (8-1). By comparison, we see that

$$y_{11} = \frac{\Delta_{11}}{\Delta} \qquad y_{12} = \frac{\Delta_{21}}{\Delta}$$

$$y_{21} = \frac{\Delta_{12}}{\Delta} \qquad y_{22} = \frac{\Delta_{22}}{\Delta} \qquad (8\text{-}7)$$

where Δ is the determinant of the loop equations. These expressions are very important, not only for computing values of the y parameters but in determining various properties that the y's possess as a function of frequency, since the determinants contain the frequency as a variable. As an exercise, use these formulas to compute the y parameters of the pi network of Fig. 8-5.

From Eq. (8-7) we see that the expression for y_{12} involves the cofactor Δ_{21} while that for y_{21} involves the cofactor Δ_{12}. From these equations we find that

$$y_{12} = \frac{\Delta_{21}}{\Delta_{12}} y_{21} \qquad (8\text{-}8)$$

It would certainly be worthwhile to determine the relationship between the two cofactors Δ_{12} and Δ_{21}. We have a suspicion (strengthened by the fact that for the pi network at least $y_{12} = y_{21}$) that these two cofactors are equal. For this purpose let us examine the determinant of a set of loop equations. We have seen that the determinant has the form

$$\Delta = \begin{vmatrix} a_{11} & a_{12} & a_{13} & \cdots & a_{1n} \\ a_{21} & a_{22} & a_{23} & \cdots & a_{2n} \\ a_{31} & a_{32} & a_{33} & \cdots & a_{3n} \\ \cdots & \cdots & \cdots & \cdots & \cdots \\ a_{n1} & a_{n2} & a_{n3} & \cdots & a_{nn} \end{vmatrix} \tag{8-9}$$

The elements on the main diagonal are the self-impedances of the loops, whereas those off the main diagonal are $\pm$ the mutual impedances. The off-diagonal elements are equal in pairs; that is, $a_{ik} = a_{ki}$, since the mutual impedance common to two loops is the same, no matter which loop you decide to mention first. A determinant for which this condition is true is called a *symmetrical* determinant. The elements in any one row in such a determinant are equal to the corresponding elements in the corresponding column. Thus any row can be interchanged with the corresponding column, and the determinant will look the same.

Now let us turn to the cofactors. Δ_{12} is obtained by removing row 1 and column 2; Δ_{21} is obtained by removing row 2 and column 1. Thus

$$-\Delta_{12} = \begin{vmatrix} a_{21} & a_{23} & \cdots & a_{2n} \\ a_{31} & a_{33} & \cdots & a_{3n} \\ \cdots & \cdots & \cdots & \cdots \\ a_{n1} & a_{n3} & \cdots & a_{nn} \end{vmatrix} \qquad (a)$$

$$-\Delta_{21} = \begin{vmatrix} a_{12} & a_{13} & \cdots & a_{1n} \\ a_{32} & a_{33} & & a_{3n} \\ \cdots & \cdots & \cdots & \cdots \\ a_{n2} & a_{n3} & & a_{nn} \end{vmatrix} = \begin{vmatrix} a_{21} & a_{31} & \cdots & a_{n1} \\ a_{23} & a_{33} & & a_{3n} \\ \cdots & \cdots & \cdots & \cdots \\ a_{2n} & a_{n3} & & a_{nn} \end{vmatrix} \qquad (b) \tag{8-10}$$

Note that everything enclosed by the dashed lines in the two determinants is the same; Δ_{12} and Δ_{21} differ only in the first row and first column. On the right side of the last equation, we interchanged the orders of the subscripts on the elements of the first row and first column. This is permissible since $a_{ik} = a_{ki}$. In this form Δ_{21} differs from Δ_{12} only by having the first row and first column interchanged. But, according to the theory of determinants (see any algebra book), the interchange of a row with the corresponding column does not change the value of a determinant. Hence $\Delta_{12} = \Delta_{21}$. It follows that

$$y_{12} = y_{21} \tag{8-11}$$

This is quite an important result. Remember that y_{21} is the ratio of the response sinor I_2 to V_1 when the excitation is V_1. Similarly, y_{12}

is the ratio of the response sinor I_1 to V_2 when the excitation is V_2. Since these two are equal, it is implied that the current response in the two cases will be equal when the voltage excitations are the same. This result is known as the *theorem of reciprocity*. It is often stated in the following form:

Suppose a voltage source V_1 at one pair of terminals in a network produces a current I_2 somewhere else, assuming V_1 is the only source present. Now remove the source, open the branch in which I_2 appeared, and place the source at the terminals thus formed, making sure that the source-reference plus is at the tail of the reference of the former current I_2. If the original position of the voltage source is now short-circuited, a current will result which is the same as the current I_2, assuming that the reference of this current is chosen so that the tail of the arrow is at the former location of the reference plus of V_1.

Note carefully from the proof of this theorem that its validity is established only for networks having symmetrical determinants, so-called *bilateral* networks. At the present time our model of electric circuits contains only bilateral elements. Later on, additional elements will be introduced into the model which will not be bilateral. You are cautioned to keep in mind this restriction on the validity of the reciprocity theorem.

8.2 OPEN-CIRCUIT IMPEDANCE PARAMETERS

Let us now return to the starting point of the discussion of two-ports and look again at Fig. 8-3 and at Eq. (8-1). In writing these equations, we assumed that the voltages were the excitations and the currents were the responses. Let us now suppose that the currents are the excitations and that the voltages are unknown. Again we would expect the response sinors to be linearly related to the excitation sinors. Hence, for the voltage-response sinors, we can write

$$\begin{aligned} V_1 &= z_{11}I_1 + z_{12}I_2 \qquad (a) \\ V_2 &= z_{21}I_1 + z_{22}I_2 \qquad (b) \end{aligned} \qquad (8\text{-}12)$$

where, again, we can look upon the z_{ik} coefficients simply as proportionality factors relating the voltage sinors to the current sinors.

To obtain an interpretation of these z coefficients, suppose we apply a current source at port 1 and leave port 2 open-circuited, as shown in Fig. 8-6(a). The open circuit causes I_2 to be zero. Hence, setting $I_2 = 0$ in Eq. (8-12), we get

$$z_{11} = \left.\frac{V_1}{I_1}\right|_{I_2=0} \qquad (a)$$
$$z_{21} = \left.\frac{V_2}{I_1}\right|_{I_2=0} \qquad (b) \qquad (8\text{-}13)$$

Since I_1 is the only source in the network, the quantity z_{11} is, by definition, the driving-point impedance at the input when the output is open-circuited. The quantity z_{21} is the ratio of a voltage-response sinor at one port to a current-excitation sinor at another port. We call it the *open-circuit forward-transfer impedance* from port 1 to port 2.

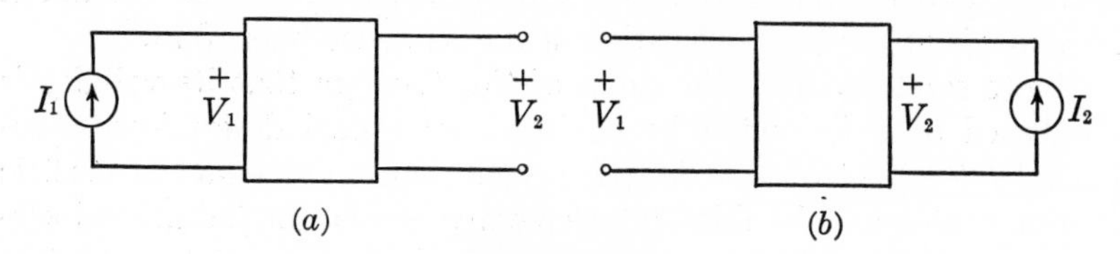

FIG. 8-6

Open-circuit parameters

Now let us apply a current source at port 2 and leave port 1 open-circuited, as in Fig. 8-6(b). The open circuit makes $I_1 = 0$, so that Eq. (8-12) yields

$$z_{22} = \left.\frac{V_2}{I_2}\right|_{I_1=0} \qquad (a)$$
$$z_{12} = \left.\frac{V_1}{I_2}\right|_{I_1=0} \qquad (b) \qquad (8\text{-}14)$$

In this case I_2 is the only source in the network. Hence z_{22} is the driving-point impedance at the output when the input is open-circuited; thus, the *open-circuit output impedance*.† The quantity z_{12} is the *open-circuit reverse-transfer impedance* from port 2 to port 1. Equations (8-13) and (8-14) can be looked upon as the definitions of the z parameters.

As a group, the z coefficients completely characterize a two-port; they are called the *open-circuit impedance parameters*. A word of

† There is some ambiguity in this terminology. Port 1 is called the input port and port 2 the output. Yet, in the present case, the only "input" quantity is at the output (namely, at port 2). The context usually makes clear whether the terms "input" or "output" refer to the location of a port or to a voltage or current.

caution here: z_{11} is not the reciprocal of y_{11}; z_{12} is not the reciprocal of y_{12}, etc. They are dual quantities, as you can see by comparing Eqs. (8-1) and (8-12).

Even though the z's and y's are not each others' reciprocals, they must be related in some manner, since we have claimed that either set of parameters constitutes a complete characterization of the two-port. The relationships are easily found. Suppose we solve Eq. (8-1) for the voltages in terms of the currents. The result will be

$$V_1 = \frac{\begin{vmatrix} I_1 & y_{12} \\ I_2 & y_{22} \end{vmatrix}}{\begin{vmatrix} y_{11} & y_{12} \\ y_{21} & y_{22} \end{vmatrix}} = \frac{y_{22}}{y_{11}y_{22} - y_{12}y_{21}} I_1 + \frac{-y_{12}}{y_{11}y_{22} - y_{12}y_{21}} I_2 \quad (a)$$

$$(8\text{-}15)$$

$$V_2 = \frac{\begin{vmatrix} y_{11} & I_1 \\ y_{21} & I_2 \end{vmatrix}}{\begin{vmatrix} y_{11} & y_{12} \\ y_{21} & y_{22} \end{vmatrix}} = \frac{-y_{21}}{y_{11}y_{22} - y_{12}y_{21}} I_1 + \frac{y_{22}}{y_{11}y_{22} - y_{12}y_{21}} I_2 \quad (b)$$

If we now compare this pair of equations with Eq. (8-12), we will find

$$z_{11} = \frac{y_{22}}{y_{11}y_{22} - y_{12}y_{21}}; \quad z_{12} = \frac{-y_{12}}{y_{11}y_{22} - y_{12}y_{21}}$$
$$z_{21} = \frac{-y_{21}}{y_{11}y_{22} - y_{12}y_{21}}; \quad z_{22} = \frac{y_{11}}{y_{11}y_{22} - y_{12}y_{21}} \qquad (8\text{-}16)$$

These are the desired relationships between the z's and the y's. Once the short-circuit admittances of a two-port are known, the open-circuit impedances can be calculated. The converse is also true. We can determine the appropriate relationships by proceeding in a dual manner. First we solve Eq. (8-12) for the currents, and then we compare coefficients with Eq. (8-1). The result should be just the duals of Eq. (8-16) with z's replaced by y's, and vice versa. Carry out the steps to reassure yourself.

A glance at the transfer impedances z_{12} and z_{21} in Eq. (8-16) shows that

$$z_{21} = \frac{y_{21}}{y_{12}} z_{12} \qquad (8\text{-}17)$$

Since $y_{12} = y_{21}$, we find the important result that $z_{12} = z_{21}$ for bilateral networks. This is an additional extension of the reciprocity theorem. Remembering that z_{21} and z_{12} are ratios of a voltage-response sinor to a current excitation, we can state this result in the following way.

Suppose a current source I_1 at one pair of terminals in a network produces a voltage V_2 at another pair of terminals. Now remove the current source from the first pair of terminals and place it at the second pair, with the tail of the current reference at the plus reference of the former voltage V_2. The voltage which now appears at the first pair of terminals is the same as the voltage V_2, assuming that the reference is chosen so that the plus sign is at the location of the tail of the former current I_1.

Let us now illustrate the use of the relationships among the z and y parameters by calculating the open-circuit impedances of the pi network of Fig. 8-5. The short-circuit admittances were given in Eqs. (8-4) and (8-5). Using these, we get

$$y_{11}y_{22} - y_{12}y_{21} = (Y_2 + Y_3)(Y_1 + Y_3) - Y_3^2 = Y_1Y_2 + Y_2Y_3 + Y_3Y_1$$

$$z_{11} = \frac{Y_1 + Y_3}{Y_1Y_2 + Y_2Y_3 + Y_3Y_1} \qquad (a)$$

$$z_{22} = \frac{Y_2 + Y_3}{Y_1Y_2 + Y_2Y_3 + Y_3Y_1} \qquad (b) \qquad (8\text{-}18)$$

$$z_{12} = z_{21} = \frac{Y_3}{Y_1Y_2 + Y_2Y_3 + Y_3Y_1} \qquad (c)$$

Whereas the short-circuit admittances of the pi network are simply related to the branch admittances, the same is not true of the open-circuit impedances.

In order to find the z parameters of a given network, it is not necessary first to find the y parameters and then convert by means of Eq. (8-16). The simplest procedure is to use the defining expressions for the z parameters given in Eqs. (8-13) and (8-14).

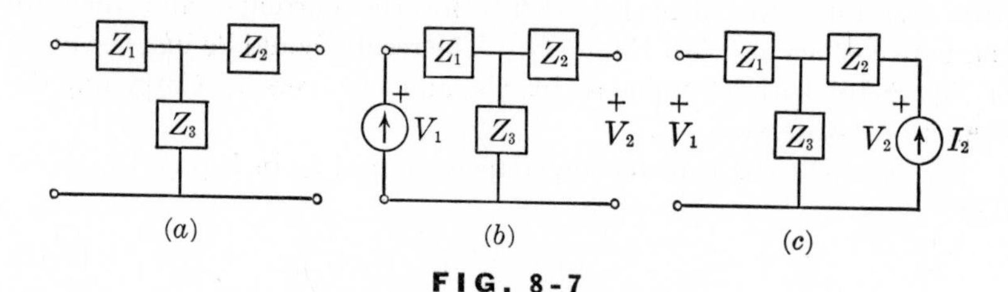

FIG. 8-7

Open-circuit parameters of a tee

As an example, consider the tee network shown in Fig. 8-7. To find the z parameters, let us first apply a current source I_1 at the input port, keeping the output port open, as shown in Fig. 8-7(b). This current flows through both Z_1 and Z_3. Hence

$$z_{11} = Z_1 + Z_3 \quad (a)$$
$$z_{21} = Z_3 \quad (b) \qquad (8\text{-}19)$$

Next we apply a current source I_2 at the output terminals, leaving the input open circuited, as shown in Fig. 8-7(c). This current flows through both Z_2 and Z_3. Hence

$$z_{22} = Z_2 + Z_3 \quad (a)$$
$$z_{12} = Z_3 = z_{21} \quad (b) \qquad (8\text{-}20)$$

We see that the z parameters of a tee network are related in a rather simple way with the branch impedances.

Whenever it is desired to refer to the y parameters or the z parameters of a network collectively, or to display quantitative relationships about them, such as the z parameters of the tee given in Eqs. (8-19) and (8-20), or the y parameters in terms of determinants given in Eq. (8-7), it is convenient to have some concise means of making such a display. Recall that the y and z parameters are coefficients of a set of equations. If we concentrate simply on the positions of the coefficients in these equations, the following layouts can be made:

$$\begin{bmatrix} y_{11} & y_{12} \\ y_{21} & y_{22} \end{bmatrix} \qquad \begin{bmatrix} z_{11} & z_{12} \\ z_{21} & z_{22} \end{bmatrix}$$

A rectangular array of quantities such as these is called a *matrix*. Each of the quantities in the array is called an *element* or an *entry* of the matrix. The elements form *rows* and *columns*, just like a determinant. A matrix, however, is *not* a determinant; it does not have a value, as a determinant does. It is simply a regularly laid out collection of things. The only significance of the brackets is to group together the things which are on display.

In the cases shown above, the matrix is *square*, since it has the same number of rows as columns. More generally, a matrix may have any number of rows and any number of columns. In particular, a matrix may have only one row, in which case it is called a *row matrix*, or only one column, in which case it is called a *column matrix*.

Each time it is desired to refer to a matrix, it is not necessary to lay out the entire array. Instead, a single symbol is used to characterize the matrix. For example,

$$[Y] = \begin{bmatrix} y_{11} & y_{12} \\ y_{21} & y_{22} \end{bmatrix} \qquad [Z] = \begin{bmatrix} z_{11} & z_{12} \\ z_{21} & z_{22} \end{bmatrix}$$

With this notation we can now rewrite Eq. (8-7), for example, in the following way

$$[Y] = \begin{bmatrix} \frac{\Delta_{11}}{\Delta} & \frac{\Delta_{21}}{\Delta} \\ \frac{\Delta_{12}}{\Delta} & \frac{\Delta_{22}}{\Delta} \end{bmatrix}$$

Similarly, the z parameters of the tee circuit can be written as

$$[Z] = \begin{bmatrix} Z_1 + Z_3 & Z_3 \\ Z_3 & Z_2 + Z_3 \end{bmatrix}$$

Linear algebraic equations lend themselves naturally to expression in terms of matrices, and an entire algebra of matrices has been developed which facilitates the solution of such equations. You will undoubtedly have occasion to study matrix theory in your subsequent career. For our purposes, we wish only to be able to write something occasionally in matrix form simply for the purpose of convenience. We shall not discuss matrix algebra and its application to the solution of algebraic equations.

We have seen that the external behavior of a two-port is completely specified in terms of four parameters. (Two of these are equal for a bilateral network; hence there are only three independent parameters in this case.) It is not necessary to know what the internal structure of the network is, as long as these parameters are known. In case there are two different two-ports having the same open-circuit or short-circuit parameters, their behaviors at the terminals will be identical. One of them can be replaced by the other without having any influence on the terminal voltages and currents. We say that these two-ports are *equivalent*. Of course, they are equivalent only as far as the terminals are concerned.

This idea leads to an alternative way of establishing the tee-to-pi transformation discussed in Chap. 7, for, if a tee and a pi are to be equivalent, their z parameters must be identical. (So also must their y parameters.) We have already calculated the z parameters of a pi in Eq. (8-18), and of a tee in Eqs. (8-19) and (8-20). The transformation is obtained by equating these parameters. You should go through the details and verify the previous results, being careful to note that numerical subscripts have been used here for the branches of both the tee and the pi.

This idea of the equivalence of two-ports, as far as their terminal behavior is concerned, leads us to inquire whether it is possible to replace any arbitrary two-port with a tee or a pi network which will be equivalent to it. We know that a bilateral two-port is characterized by three parameters. But a tee is also characterized by three quantities—its branch impedances. If we can solve for the three branch impedances of the tee in terms of the three parameters of the two-port (say the z parameters), the corresponding tee will be the equivalent of the two-port.

Equations (8-19) and (8-20) express the z parameters of the tee in terms of its branch impedances. It is necessary only to invert these and solve for the impedances. Z_3 is already given in Eq. [8-20(b)]. If this is inserted into Eqs. [8-19(a)] and [8-20(a)], the results will be

$$\begin{aligned} Z_1 &= z_{11} - z_{12} \\ Z_2 &= z_{22} - z_{12} \\ Z_3 &= z_{12} \end{aligned} \tag{8-21}$$

Thus the tee network shown in Fig. 8-8(a) will be the equivalent of any bilateral two-port.

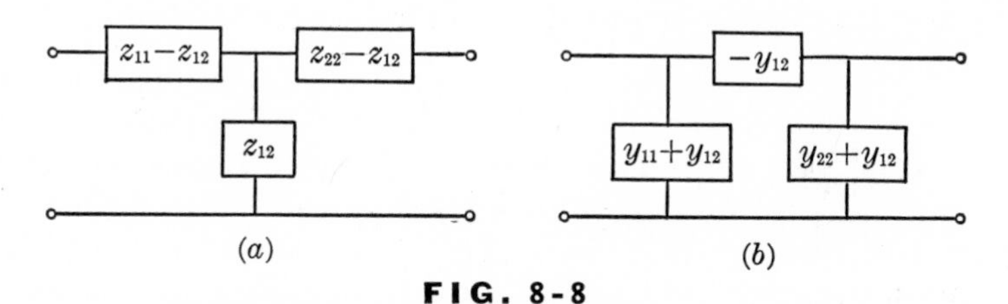

FIG. 8-8

Tee and pi equivalents of an arbitrary two-port

By going through a similar development, you can show that the pi network in Fig. 8-8(b) is equivalent to any bilateral two-port. (Carry out the required steps.)

Keep in mind that these "equivalent" networks are mathematical equivalents. It will not be possible to find actual tee or pi networks consisting of positive R, L, and C elements for an arbitrary network. For the purpose of computing responses at the terminals, however, this fact is immaterial.

Although the y parameters are defined with short-circuit terminations and the z parameters with open-circuit terminations, they represent the V-I relationships of the two-port, no matter what may be connected at the terminals. To illustrate, consider the situation depicted in Fig. 8-9. A two-port is terminated at the output port with an impedance Z. We wish to find the input impedance Z_1 and the

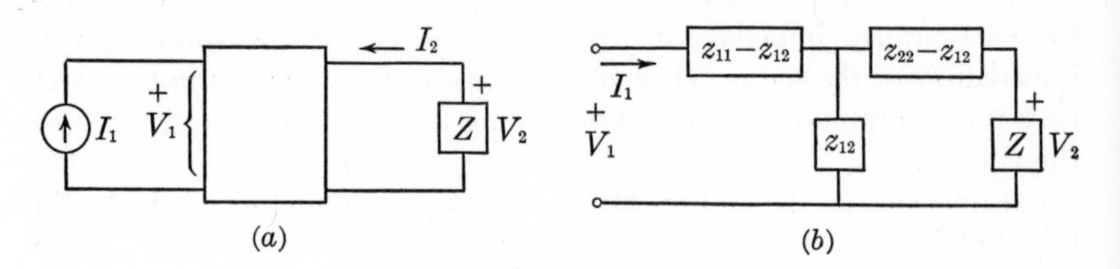

FIG. 8-9

Impedance-terminated two-port

transfer impedance $Z_{21} = V_2/I_1$. We assume that the input is excited by a current source. Because of the reference of I_2, the load impedance Z is the negative ratio of V_2 to I_2; $V_2 = -ZI_2$. Let us use this value for V_2 in Eq. [8-12(b)]. Thus

$$V_2 = z_{21}I_1 + z_{22}I_2 = -ZI_2$$

This equation can now be solved either for I_1 or for I_2, yielding

$$I_2 = \frac{-z_{21}}{Z + z_{22}} I_1 \qquad (a)$$

or (8-22)

$$I_1 = \frac{-(Z + z_{22})}{z_{21}} I_2 \qquad (b)$$

Substituting the second of these into Eq. [8-12(a)], we get

$$Z_1 = \frac{V_1}{I_1} = z_{11} - \frac{z_{12}z_{21}}{Z + z_{22}} = \frac{(z_{11}z_{22} - z_{12}z_{21}) + Zz_{11}}{Z + z_{22}} \qquad (8\text{-}23)$$

This is the desired input impedance. To find the transfer impedance, remember that $I_2 = -V_2/Z$. Using this in Eq. [8-21(a)] leads to

$$Z_{21} = \frac{V_2}{I_1} = \frac{z_{21}Z}{Z + z_{22}} \qquad (8\text{-}24)$$

This is the desired transfer impedance.

An alternative approach is to replace the two-port by a tee equivalent, as shown in Fig. 8-9(b). The calculations are now straightforward and are left for you to carry out.

Note that once the z parameters of the two-port and the load impedance Z are known, the input impedance and the transfer impedance can be found. It is curious to observe that the quantity z_{11} does not appear explicitly in the expression for the transfer impedance. An explanation of this circumstance can be obtained by considering the tee equivalent in Fig. 8-9. z_{11} appears in the input series branch only. Since the driving source is a current source, this branch will not influence the responses anywhere else but at the input terminals.

The expressions we have just derived are quite useful in many applications. Let us illustrate their utility in the problem of designing an attenuator. An *attenuator* is a resistive device used to reduce a voltage, usually by a known fraction. This may be required, for example, in the measurement of large voltages. Figure 8-10 shows a tee network which is to serve as an attenuator. It is placed between a load R_0 and a source. It is required that the presence of the attenuator have no influence on the impedance presented to the source. That is, at the input port the impedance should be R_0, the same as the load impedance.

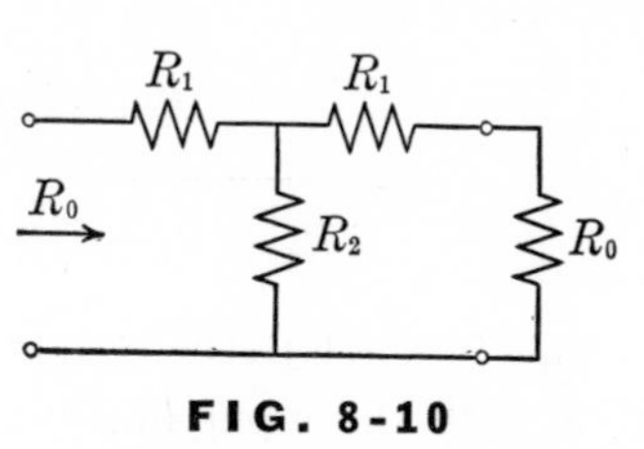

FIG. 8-10
A voltage attenuator

Of course, it is possible to compute the input impedance and the voltage-gain function directly from the diagram without regard for two-port theory. We shall use this problem, however, as a vehicle for illustrating the use of the expressions we have derived. In order to use Eqs. (8-23) and (8-24), we must have the z parameters of the tee. But these are given for the general tee in Eqs. (8-19) and (8-20). In the present case they will be

$$z_{11} = z_{22} = R_1 + R_2 \qquad (a)$$
$$z_{12} = z_{21} = R_2 \qquad (b) \qquad \text{(8-25)}$$

Substituting these expressions into Eq. (8-23) and imposing the condition that the input impedance should be R_0, we get

$$R_0 = (R_1 + R_2) - \frac{R_2^2}{R_0 + R_1 + R_2} \qquad \text{(8-26)}$$

It is now possible to find R_2 in terms of R_1, or vice versa. If we do the former, we get

$$R_2 = \frac{(R_0 - R_1)(R_0 + R_1)}{2R_1} \qquad \text{(8-27)}$$

This expression immediately shows that R_1 should be less than the load resistance if R_2 is to be a positive quantity.

Now let us turn to the determination of the voltage ratio. Since the input impedance is R_0, the input current will be $I_1 = V_1/R_0$. Hence, the voltage-transfer ratio will be $G_{21} = V_2/V_1 = V_2/R_0I_1 = Z_{21}/R_0$. Therefore, using Eq. (8-24) and the z parameters in Eq. (8-25), we get

$$G_{21} = \frac{V_2}{V_1} = \frac{Z_{21}}{R_0} = \frac{z_{21}}{R_0 + z_{22}} = \frac{R_2}{R_0 + R_1 + R_2} \tag{8-28}$$

In this expression we haven't yet used the restriction on R_2, given in Eq. (8-27), which is a result of the input impedance requirement. If we substitute Eq. (8-27) into the last expression, we get

$$G_{21} = \frac{\dfrac{(R_0 - R_1)(R_0 + R_1)}{2R_1}}{R_0 + R_1 + \dfrac{(R_0 - R_1)(R_0 + R_1)}{2R_1}} = \frac{R_0 - R_1}{R_0 + R_1} \tag{8-29}$$

This is a relatively simple expression. Since the network is resistive, both the input impedance and the voltage-gain functions are real. In a typical problem the desired voltage reduction (namely, G_{21}) will be specified. It is then a simple matter to solve this expression for R_1 and then to find R_2 from Eq. (8-27). The result of these steps will be

$$R_1 = \frac{1 - G_{21}}{1 + G_{21}} R_0 \qquad (a)$$

$$R_2 = \frac{2G_{21}}{1 - G_{21}^2} R_0 \qquad (b) \tag{8-30}$$

These expressions constitute the design equations. For any given load resistance R_0 and desired voltage ratio G_{21}, the required values of the attenuator resistances can be calculated.

8.3 THE CHAIN PARAMETERS

In describing the external behavior of a two-port up to this point, we have introduced two sets of parameters. The y parameters express the currents at both ends of the network in terms of the voltages at both ends. The z parameters reverse this description, giving the voltages in terms of the currents.

When a two-port is used to transmit signals from one port to the other, it is not likely that the voltages at the two ports, or the two currents, will be known. In such a case it is more appropriate to express the voltage and current at one port in terms of the voltage and current at the other. Since the relationships are linear, the appropriate expressions relating the input quantities to the output quantities will be

$$\begin{aligned} V_1 &= AV_2 - BI_2 \\ I_1 &= CV_2 - DI_2 \end{aligned} \tag{8-31}$$

Let us make some remarks about these equations. First of all, the coefficients are not all dimensionally the same. They are called the *ABCD* parameters, for obvious reasons.† They are also called the chain parameters since they are the natural way of describing the behavior of several two-ports connected one right after another in a chain. Second, the negative signs appear to be the result of mere whim. If you look at some other references, you may find that the same equations are written with a plus sign. The appearance of the negative sign is due to the reference direction of I_2 and has no other significance.

It is possible to interpret the *ABCD* parameters in terms of open-circuit and short-circuit terminations as we did for the z's and y's. Short-circuiting port 2 means $V_2 = 0$ and open-circuiting it means $I_2 = 0$. Under these conditions Eq. (8-31) yields

$$\begin{aligned} \frac{1}{A} &= \left.\frac{V_2}{V_1}\right|_{I_2=0}; \quad \text{open-circuit forward voltage gain} \\ -\frac{1}{B} &= \left.\frac{I_2}{V_1}\right|_{V_2=0} = y_{21}; \quad \text{short-circuit forward transfer admittance} \\ \frac{1}{C} &= \left.\frac{V_2}{I_1}\right|_{I_2=0} = z_{21}; \quad \text{open-circuit forward transfer impedance} \\ -\frac{1}{D} &= \left.\frac{I_2}{I_1}\right|_{V_2=0}; \quad \text{short-circuit forward current gain} \end{aligned} \tag{8-32}$$

Thus it is seen that the reciprocals of the *ABCD* parameters are all transfer functions of some kind, with transmission in the forward direction, from port 1 to port 2. B and C are seen to be the reciprocals of $-y_{21}$ and z_{21}, respectively. It is possible to express A and D also in

† These parameters were first introduced in the study of power transmission lines and were called *general circuit constants* at that time; "constants," because the frequency was fixed. Nowadays, since behavior with variable frequency is important, some people call them *general circuit functions*. There is, however, nothing more "general" about them than the other parameters under discussion in this section.

terms of z and y parameters. For example, we can set $I_2 = 0$ in Eq. (8-12), and then take the ratio. Similarly, we can set $V_2 = 0$ in Eq. (8-1), and then take the ratio. The results will be

$$\frac{1}{A} = \frac{z_{21}}{z_{11}} = \frac{-y_{21}}{y_{22}} \qquad (a)$$

$$\frac{1}{D} = \frac{-y_{21}}{y_{11}} = \frac{z_{21}}{z_{22}} \qquad (b) \tag{8-33}$$

The right-hand sides of these expressions are obtained by using Eq. (8-16) and their duals which relate the z and the y parameters.

There is still something troublesome about the appearance of the $ABCD$ system of equations in Eq. (8-31). There are a total of four parameters, and it does not appear that two of them are equal for bilateral networks, as is the case for $z_{12} = z_{21}$ and $y_{12} = y_{21}$. But note that the determinant of the $ABCD$ system of equations is $-(AD - BC)$. If we insert the values of A, B, C, and D in terms of the z and y parameters into this expression, we will find that

$$AD - BC = \frac{z_{12}}{z_{21}} = \frac{y_{12}}{y_{21}} = 1 \tag{8-34}$$

Thus, if three of the chain parameters are known, the fourth one is fixed by this expression. That is, for bilateral networks we corroborate the preceding evidence that only three parameters are sufficient to determine the behavior of the two-port.

Equation (8-31) expresses the voltage and current of port 2 in terms of those of port 1. These equations can be inverted to yield V_2 and I_2 in terms of V_1 and I_1. The details will be left for you to work out. The result will be

$$\begin{aligned} V_2 &= DV_1 - BI_1 \\ I_2 &= CV_1 - AI_1 \end{aligned} \tag{8-35}$$

These differ from Eq. (8-31) by having the parameters A and D interchanged. Using this set of equations, it is possible to give additional interpretations for the $ABCD$ parameters in terms of transfer functions under open-circuit or short-circuit terminal conditions. This time, transmission will be in the reverse direction. You can make up another tabulation like that of Eq. (8-32).

It was mentioned that the chain parameters offer a natural way of describing the behavior of two-ports which are connected in tandem in a chainlike manner. We shall illustrate this comment by calculating the *ABCD* parameters of two networks in cascade in terms of the parameters of the individual two-ports.

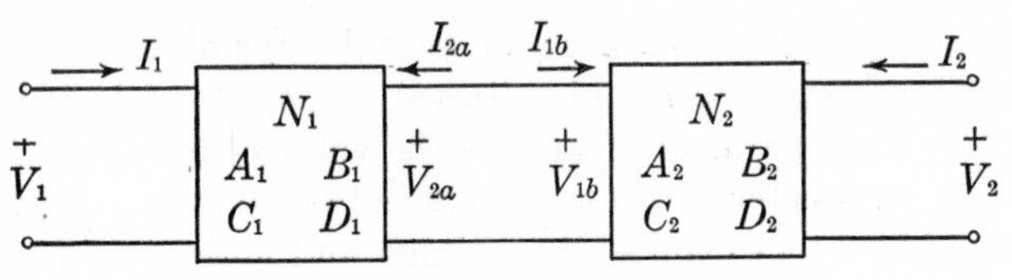

FIG. 8-11

Cascaded two-ports

Let the two networks be labeled N_1 and N_2, as illustrated in Fig. 8-11. The equations describing the behavior of the individual two-ports are

$$\begin{aligned} V_1 &= A_1V_{2a} - B_1I_{2a} \qquad & V_{1b} &= A_2V_2 - B_2I_2 \\ I_1 &= C_1V_{2a} - D_1I_{2a} \qquad & I_{1b} &= C_2V_2 - D_2I_2 \end{aligned} \tag{8-36}$$

From the diagram it is clear that

$$\begin{aligned} V_{2a} &= V_{1b} \\ -I_{2a} &= I_{1b} \end{aligned} \tag{8-37}$$

If we substitute these equations into the first of the previous two sets, and then insert the second of the two sets into the result, we will get

$$V_1 = A_1V_{1b} + B_1I_{1b} = A_1(A_2V_2 - B_2I_2) + B_1(C_2V_2 - D_2I_2) = AV_2 - BI_2$$

$$I_1 = C_1V_{1b} + D_1I_{1b} = C_1(A_2V_2 - B_2I_2) + D_1(C_2V_2 - D_2I_2) = CV_2 - DI_2$$

where

$$\begin{aligned} A &= A_1A_2 + B_1C_2 \\ B &= A_1B_2 + B_1D_2 \\ C &= C_1A_2 + D_1C_2 \\ D &= C_1B_2 + D_1D_2 \end{aligned} \tag{8-38}$$

These may look like complicated expressions, but the complication is only in the amount of algebraic manipulation involved; the calculation is straightforward. It should not be necessary to tell you not to commit these expressions to memory. We should remark parenthetically that the preceding development is greatly simplified if it is cast in the language of matrices.

8.4 THE HYBRID PARAMETERS

Up to this point we have found several ways of expressing two of the four variables (currents and voltages) of a two-port in terms of the other two. The z parameters express the voltages in terms of the currents; the y parameters express the currents in terms of the voltages; the chain parameters express the voltage and current at one port in terms of the voltage and current at the other. Purely from the standpoint of exhaustiveness, it is possible to mix up the variables and express the voltage at one port and the current at the other port in terms of the remaining two variables. That is, V_1 and I_2 can be expressed in terms of V_2 and I_1 (and vice versa). Thus there are six ways in which two of the four variables can be expressed in terms of the other two.

Let us write expressions relating V_1 and I_2 to V_2 and I_1. Since they must be linear, they will have the form

$$\begin{aligned} V_1 &= h_{11}I_1 + h_{12}V_2 && (a) \\ I_2 &= h_{21}I_1 + h_{22}V_2 && (b) \end{aligned} \tag{8-39}$$

Again we see that the coefficients (which we can call the h parameters) are not all dimensionally the same.

Since the variables are a mixed set, we refer to the h parameters as the *hybrid parameters*. If there were no advantage of some sort or other to be gained in describing a two-port by means of the hybrid parameters, we would not introduce them simply for the sake of completeness. As we shall discuss in a later chapter, a natural method of describing the behavior of certain devices, such as transistors, is in terms of the h parameters. This is reason enough to introduce them.

As before, interpretations of the hybrid parameters can be obtained under open-circuit or short-circuit conditions by setting $I_1 = 0$ or $V_2 = 0$ in Eq. (8-39). Thus

$$\begin{aligned} h_{11} &= \left.\frac{V_1}{I_1}\right|_{V_2=0} = \frac{1}{y_{11}}; && \text{short-circuit input impedance} \\ h_{12} &= \left.\frac{V_1}{V_2}\right|_{I_1=0} ; && \text{open-circuit reverse voltage gain} \\ h_{21} &= \left.\frac{I_2}{I_1}\right|_{V_2=0} = -\frac{1}{D}; && \text{short-circuit forward current gain} \\ h_{22} &= \left.\frac{I_2}{V_2}\right|_{I_1=0} = \frac{1}{z_{22}}; && \text{open-circuit output admittance} \end{aligned} \tag{8-40}$$

The right-hand sides of these expressions follow from a comparison with the y, z, or $ABCD$ systems of equations. Notice that no equivalent is given for h_{12}. For bilateral networks (which is the only variety we have considered so far) we find, from Eq. [8-35(a)], that $1/D$ can be interpreted as the open-circuit reverse voltage gain, which is h_{12}; hence $h_{12} = -h_{21}$. That is, the short-circuit forward current gain and the open-circuit reverse voltage gain simply differ by a factor of -1.

We shall not go through the algebraic work required to invert Eq. (8-39) and determine the relationships of I_1 and V_2 to V_1 and I_2. With the example of the other sets of equations before you, it will be worthwhile for you to carry out the necessary details and then find interpretations for the resulting parameters. These parameters are normally called the g parameters and are named g_{11}, g_{12}, g_{21}, and g_{22}.

In this chapter various sets of parameters were introduced in terms of which the terminal behavior of two-port networks can be described. Of necessity, a certain amount of algebraic manipulation was required. We tried to keep this to a minimum in order to avoid masking the basic results. The utility of some of the relationships that were developed may not be apparent at this point. As we go on to the following chapters, however, we shall find many instances where the ideas developed here will be useful.

We should point out that there are at least two other methods of description of the external behavior of two-ports. The pertinent parameters are the so-called *image parameters* and the *scattering parameters*. Clearly, these descriptions must be different from the ones that have been discussed. They cannot be simple linear relationships of two of the terminal variables in terms of the other two, because we exhausted all the possibilities. The scattering parameters are very useful in describing the behavior of microwave systems. The image parameters were used in early work in transmission networks and filters. We shall not discuss them in this book.

8.5 SYMMETRICAL TWO-PORTS

One special class of two-ports can be singled out for particular attention, both for its simplicity and for its utility. This is the class of symmetrical two-ports. We say that a *two-port is electrically symmetrical if reversing the ports will cause no change in the terminal voltages and currents.*

Using this definition, we can determine the consequences of symmetry on the two-port parameters by interchanging V_2 and I_2 with V_1 and I_1 in any of the systems of equations. (Do this.) The result will be

$$z_{11} = z_{22}; \quad y_{11} = y_{22}; \quad A = D \tag{8-41}$$

Some important symmetrical networks are the tee, pi, bridged tee, and lattice shown in Fig. 8-12. Notice that the bridged tee can reduce

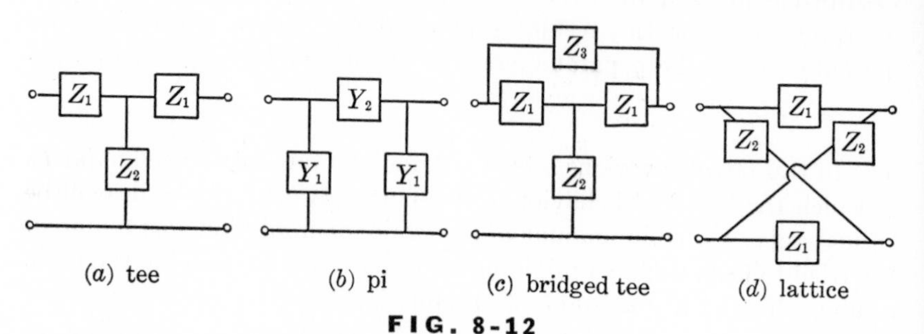

(*a*) tee (*b*) pi (*c*) bridged tee (*d*) lattice

FIG. 8-12

Some symmetrical two-ports

to either a tee or a pi. If Z_3 is infinite (open circuit), it becomes a tee; if Z_2 is zero (short circuit), it becomes a pi.

Note that the definition of symmetry refers to *electrical* symmetry. Another type of symmetry is *structural* symmetry. This type of symmetry exists if a vertical line through the network can be found so that the structure on one side is the same as the structure on the other. Each of the networks in Fig. 8-12 possesses structural symmetry. The attenuator pad in Fig. 8-10 is a symmetrical tee network.

It should be evident that a structurally symmetrical network will also be electrically symmetrical. Structural symmetry, however, is not necessary for electrical symmetry. Even though a network is not structurally symmetrical, it is possible for its electrical behavior to be the same if the ports are reverse. Figure 8-13 shows an example of a structurally unsymmetrical two-port which is electrically symmetrical. The open-circuit driving-point impedances are

$$z_{11} = z_{22} = \frac{8(s + 1)}{4s + 3} \tag{8-42}$$

(You are invited to corroborate this fact by computing the z's from the network.)

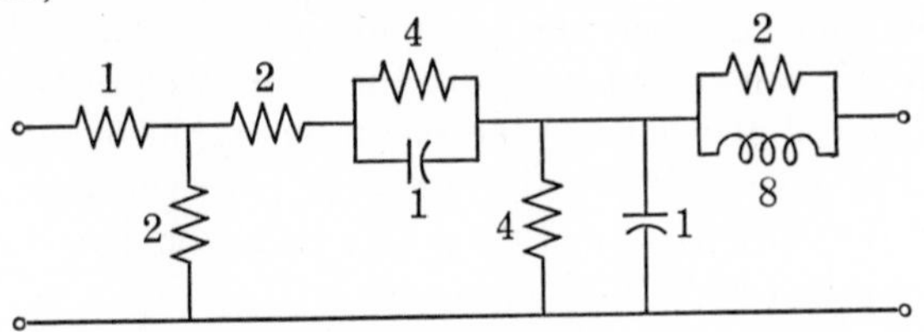

FIG. 8-13

Structurally unsymmetrical but electrically symmetrical two-port

Examination of the two-ports shown in Fig. 8-12 points up a basic distinctive feature of certain two-ports. In the case of the tee, the pi, and the bridged-tee, port 1 and port 2 have one terminal in common. We refer to such two-ports as *common-terminal*, or *unbalanced* two-ports. (Note that symmetry or lack of it has no influence on this property.) The lattice network does not have this property. In many complex systems it is desirable to have a common connection throughout the entire system, a so-called common ground. Unbalanced two-ports will satisfy this requirement.

All four of the two-ports shown in Fig. 8-12 are basic structures which have played dominant roles in the design of filters and equalizer networks.

8.6 SUMMARY

In this chapter we have introduced the concept of a two-port and have discussed a number of ways in which the terminal behavior of the two-port can be described. Two of the four variables which define the terminal behavior (two currents and two voltages) are expressed in terms of the other two by means of a pair of linear equations. The four coefficients of these equations constitute sets of parameters which completely describe the two-port behavior. For bilateral networks, which are the only kind we have discussed so far, only three of the four parameters are independent.

As you read this chapter you may have gotten the impression of an overpowering amount of algebraic manipulation. To some extent this impression is accurate. There is no doubt that a certain amount of pure algebra is involved in the development of this topic, by its very nature. In going through the manipulative aspects of the development, however, you should recognize these for what they are, and not become dismayed at what may appear to be an unending sequence of equation after equation. The hope is that, once you have become familiar with the basic pattern of these equations, the details will take on their proper perspective.

In this brief chapter perhaps the full power of the tool we have introduced has not become completely evident. We have omitted many topics, inclusion of which, perhaps, would have expanded your view. For example, we could have discussed methods of interconnecting two-ports, other than the cascade connection which was discussed in Sec. 8-3. Two two-ports can have each of their ports connected in series or in parallel; or one port in series and one in parallel. Just as the chain parameters constitute the natural set for describing the cascade connection, other parameters constitute the natural set for describing these other connections. You can readily appreciate, however, by glancing back at Eqs. (8-36) through (8-38), that the amount of algebra attendant on the discussion of these interconnections would not be insignificant. There is, therefore, a danger that the objective of providing breadth might become submerged in a sea of algebra.

The greatest utility of the concepts associated with two-ports is felt in the area of synthesis and design. Since this is beyond our level of attainment in this book, we shall be content with the present introductory exposure, with the hope that you may be stimulated enough to continue the study in further advanced books.

Problems

8-1 In Fig. P8-1 the structure of the network inside the box is not known. The short-circuit admittance parameters, however, are known. Find an expression for the transfer admittance $Y_{21} = I_2/V_1$ (*a*) first by using the equations defining the y parameters, Eq. (8-1), (*b*) then by using the pi equivalent of the network, and (*c*) finally, by applying Thévenin's theorem at the output terminals.

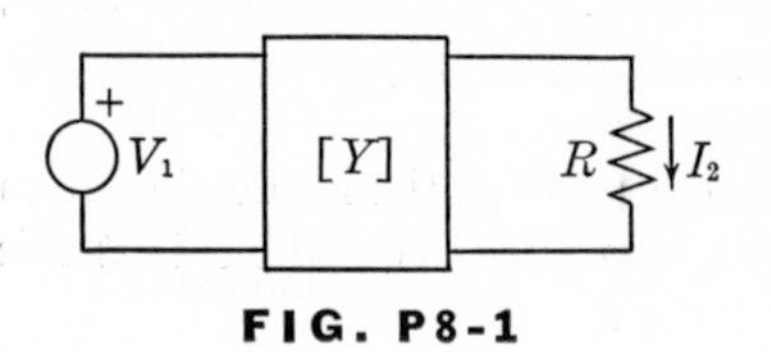

FIG. P8-1

8-2 Let $R = 10$ in Fig. P8-1. Use the results of the previous problem to find the output voltage across R for the networks described by the following parameters.

$$(a)\ [Y] = \begin{bmatrix} 2\epsilon^{j30°} & 5\epsilon^{-j135°} \\ 5\epsilon^{-j135°} & 2\epsilon^{j30°} \end{bmatrix} \qquad (c)\ [Y] = \begin{bmatrix} 3s+1 & -2s \\ -2s & 5s+2 \end{bmatrix}$$

$$(b)\ [Y] = \begin{bmatrix} \dfrac{10(s+1)}{s} & -10 \\ -10 & \dfrac{10(s+1)}{s} \end{bmatrix} \qquad (d)\ [Z] = \begin{bmatrix} \dfrac{10(s+1)}{s} & 10 \\ 10 & \dfrac{10(s+1)}{s} \end{bmatrix}$$

8-3 From Eq. (8-16) in the text, calculate the quantity $z_{11}z_{22} - z_{12}z_{21}$, which is the determinant of Eq. (8-12), and show that it is the reciprocal of the quantity $y_{11}y_{22} - y_{12}y_{21}$, which is the determinant of Eq. (8-1). Using this result, solve for the y parameters in terms of the z parameters. Do the same thing by first solving Eq. (8-12) for the currents in terms of the voltages and then comparing with Eq. (8-1).

8-4 The z parameters of the network inside the box of Fig. P8-4 are known. Find an expression for the transfer impedance $Z_{21} = V_2/I_1$ in terms of R_1, R_2, and the z parameters. Do this first by using the z

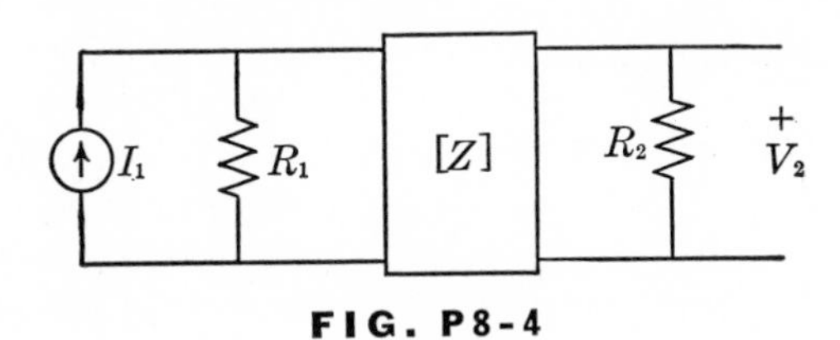

FIG. P8-4

system of equations. Repeat, using Thévenin's theorem and reciprocity.

8-5 Let the box in Fig. P8-4 contain the tee network shown in Fig. P8-5. (a) Find the z parameters of the tee and use them in the expression for the transfer impedance found in the previous problem. (b) Connecting R_1, R_2, and I_1 to the tee, compute the output voltage by another method, and compare results.

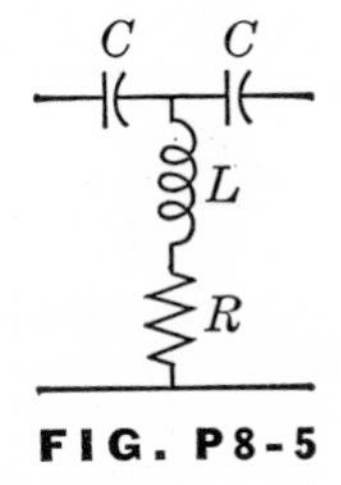

FIG. P8-5

8-6 Find the open-circuit impedance parameters and the short-circuit admittance parameters of the bridged-tee and lattice networks shown in Fig. 8-12 in the text.

8-7 In Fig. P8-7 the left-hand column contains networks whose element values are known. Each network in the right-hand column is to be equivalent at the ports to the corresponding one in the left. Find the branches of these networks. Do you expect to find a realizable two-port having a specified structure which is equivalent to any given two-port?

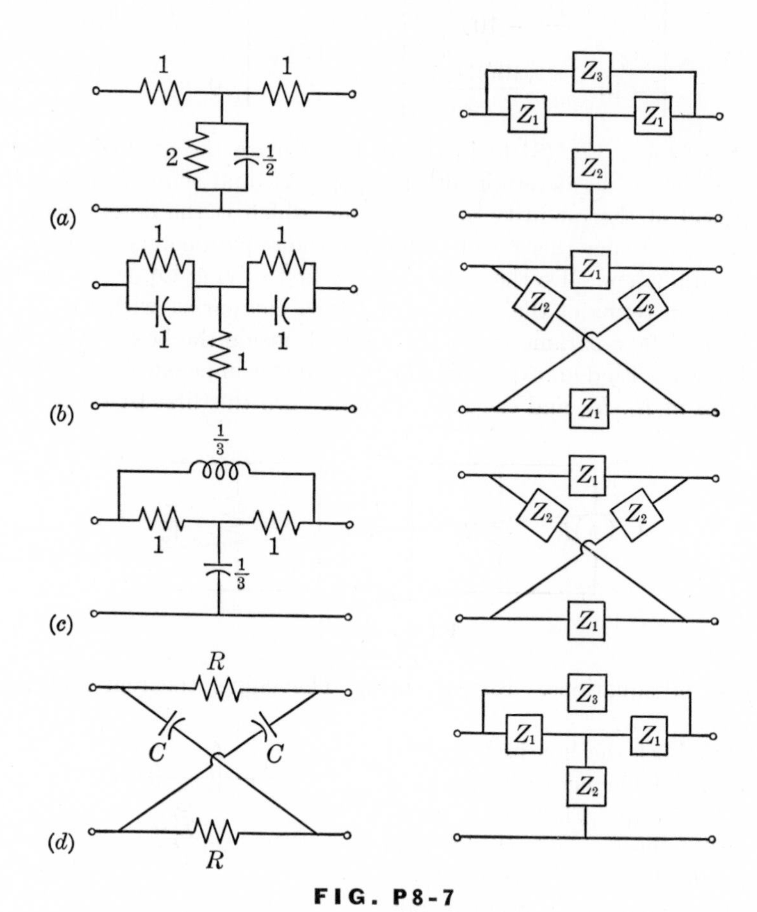

FIG. P8-7

8-8(*a*) By applying the basic definitions, find the *ABCD* parameters for the two-ports shown in Fig. P8-8. (*b*) Consider the tee to be a cascade connection of Figs. P8-8(*c*) and P8-8(*a*) and use the result in Eq. (8-38) to check your answer.

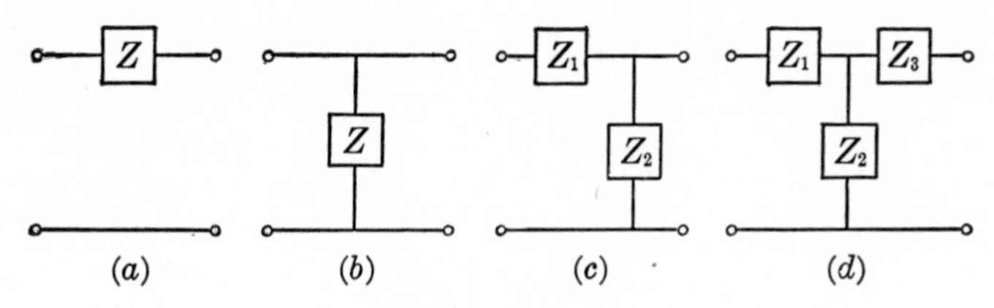

FIG. P8-8

8-9 It is desired to find expressions for each of the sets of two-port parameters in terms of the parameters of the other sets, such as the relationships among the z parameters and y parameters given in Eq. (8-16) in the text. Start from the y system and by placing the equations in the form of the other systems, identify the parameters of these systems in terms of the y's. Proceed in this manner to construct a table giving the parameters of each of the six sets of equations discussed in this chapter in terms of the other five.

8-10 A two-port is terminated by an impedance Z, as shown in Fig. 8-9 in the text. Use the $ABCD$ set of equations to find an expression for the input impedance in terms of Z and the chain parameters. Substitute the values of the z parameters in terms of the $ABCD$ parameters into Eq. (8-23) in the text in order to check your answer.

8-11 In Fig. P8-10 find the short-circuit driving-point admittance at port 1 as a function of s and show the locations of the poles and zeros in the complex plane. Repeat for the open-circuit impedance.

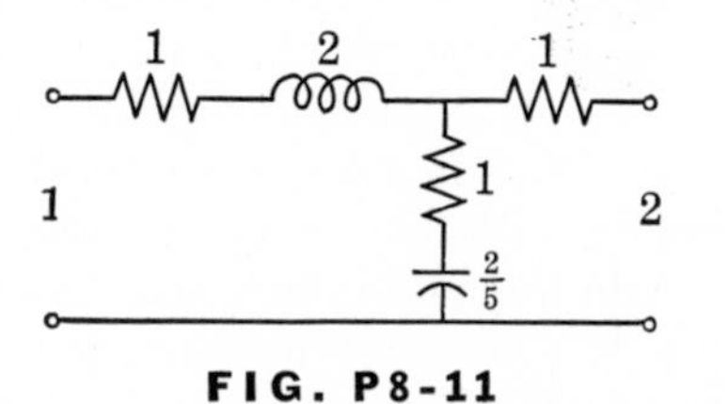

FIG. P8-11

8-12 It is desired to design an attenuator pad for a voltage reduction of 10 to 1 and a load resistance of 77 ohms. Find the element values.

8-13 A voltage attenuator working into a 50 ohm resistive load is to provide an attenuation of 40 db. Design a pad.

9 TRANSFORMERS

IT is now time to re-examine our model of electric circuits. Starting with a number of observations in the physical world, we have postulated a hypothetical model consisting of the three passive elements R, L, and C, and the two ideal sources, voltage source and current source. On this foundation we have built a fairly extensive theory concerning the variations of voltage, current, or power in the model.

In attempting to assess the degree of success achieved by the theory, it is necessary to return to the physical world for further observations; this time, observations of results which have been predicted by an analysis of the model. For example, in Chap. 3 we predicted the natural behavior of a series connection of R, L, and C. We saw how the response changes from an oscillatory one at low values of resistance, to a nonoscillatory decaying one at higher values of resistance. It is possible to verify these predictions by discharging a physical capacitor through a coil of wire and observing on an oscilloscope the waveshape of the current or the capacitor voltage.

In comparing experimental observations with theoretically predicted values, we might find that the agreement between predicted behavior and observed behavior is quite good. This is likely to be the case whenever the conditions under which the experimental observations are made correspond to those that were assumed when developing the model.

It is also possible to find that the observed behavior diverges considerably from the predicted behavior. This might happen, for example, if a coil of wire in a physical circuit is characterized simply by an inductance in the network model. If the coil is wound on a ferromagnetic core and the current is large, the flux-current relationship will be highly nonlinear; hence a linear model will be only a crude approximation. But even if no ferromagnetic material is present, the characterization of the coil solely by means of an inductance may not be adequate. For example, at a high enough frequency, the capacitance between the turns of the coil becomes (collectively) as important as the inductance. Hence the model representing the coil of wire must be modified, under these conditions, by the inclusion of other elements.

But even this type of modification will not be sufficient in some cases. That is, no matter how complicated a network structure is made up, consisting of our R, L, and C elements and ideal sources, the resulting behavior may not approximate that of certain electrical devices. Some such devices are transformers, electronic tubes, and semiconductor devices. We shall devote this chapter and the next one to the task of extending our model by postulating additional hypothetical elements in order to account for the behavior of such devices.

9.1 TRANSFORMER VOLTAGE-CURRENT RELATIONSHIPS

In Chap. 1 we discussed the observations first made by Faraday concerning the appearance of a voltage across the terminals of a coil of wire when it is in the presence of a changing magnetic field. There we gave explicit consideration to the case in which the magnetic field depends only on the current in the coil itself. (Refer to that discussion before going on.)

Let us now consider the situation shown in Fig. 9-1. Two coils are wound on a common core. They are said to be *magnetically* or *inductively coupled.* Each of the coils is connected to a circuit, at least one of which contains a source. Let us temporarily assume that the source voltages or currents have any arbitrary time-varying waveforms. According to Faraday's law, the voltages induced in the coils will be equal to the rate of change with time of the corresponding flux linkages. That is

$$v_1 = \frac{d\lambda_1}{dt} \qquad (a)$$
$$v_2 = \frac{d\lambda_2}{dt} \qquad (b) \qquad (9\text{-}1)$$

where λ_1 and λ_2 are the flux linkages of coil 1 and coil 2, respectively. The signs in these equations are valid only with the references of voltage and flux linkage shown in the figure.

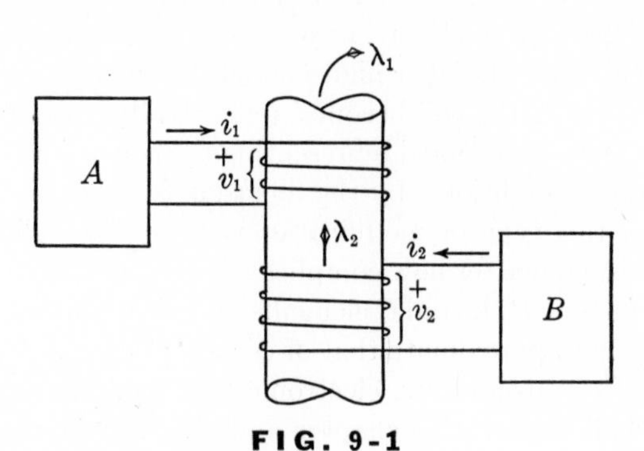

FIG. 9-1
Two coupled coils

The flux linkages depend on the currents i_1 and i_2. Quite generally, we can write

$$\lambda_1 = f_1(i_1, i_2) \qquad (a)$$
$$\lambda_2 = f_2(i_1, i_2) \qquad (b) \qquad (9\text{-}2)$$

where f_1 and f_2 are functions describing how the flux linkages vary with the currents. [Compare these with Eq. (1-6) in Chap. 1.] In actual physical cases these functions may not be very simple. (Because of the phenomenon of hysteresis, they are not even single-valued functions.)

We postulate a model, however, in which these functions are linear functions of the two currents. Thus, by postulate,

$$\lambda_1 = L_{11}i_1 + L_{12}i_2 \qquad (a)$$
$$\lambda_2 = L_{21}i_1 + L_{22}i_2 \qquad (b) \qquad (9\text{-}3)$$

where the quantities L_{ik} are real constants. [Compare these equations with Eq. (1-7) in Chap. 1.] L_{11} and L_{22} are called the *self-inductances* (or simply, the inductances), and L_{12} and L_{21} are called the *mutual inductances*. If we let first i_1, then i_2, go to zero, these equations show that L_{11} is the part of flux linkage λ_1 produced by current i_1, per unit value of i_1; and L_{22} is the part of flux linkage λ_2 produced by current i_2, per unit value of i_2. These parameters are positive quantities. Hence, in order to have a positive sign in front of the corresponding terms in Eq. (9-3), the references of current and flux linkage must be related in the manner shown in Fig. 9-1. This is usually remembered by thinking of grasping the coil in the right hand with the fingers in the reference direction of current; the direction of the thumb should then be the reference direction for flux (referred to as the *right-hand rule*).

L_{12} is the part of flux linkage λ_1 produced by current i_2, per unit value of i_2; and L_{21} is the part of flux linkage λ_2, produced by current i_1, per unit value of i_1. In contrast with the self-inductances, the mutual inductances may be either positive or negative. To appreciate this fact, suppose we choose the references of i_1 and λ_1 so as to yield a positive sign before the term $L_{11}i_1$, as shown in Fig. 9-1. Similarly, we choose the reference of i_2 and λ_2 so as to yield a positive sign before the term $L_{22}i_2$. There are two possibilities: either the references of λ_1 and λ_2 coincide, or they do not. If the references coincide, the fluxes are said to be *aiding;* then L_{12} and L_{21} will be positive. If the references do not coincide, the fluxes are said to be *opposing*; the mutual inductances will then be negative.

By making experimental observations on actual coils, we find that the two mutual inductances are very closely the same. On this basis we postulate that they are equal; $L_{12} = L_{21} = M$. The symbol M is commonly used for the mutual inductance.† The double subscripts on L_{11} and L_{22} are unnecessary in this simple case, so we shall use single subscripts L_1 and L_2.

If we substitute Eq. (9-3) into Eq. (9-1), making the changes in notation described in the last paragraph, the result will be

$$v_1 = L_1 \frac{di_1}{dt} + M \frac{di_2}{dt} \qquad (a)$$

$$v_2 = M \frac{di_1}{dt} + L_2 \frac{di_2}{dt} \qquad (b) \qquad (9\text{-}4)$$

These equations constitute the v-i relationships of a new component in our model. We call it a transformer and we represent it schematically as shown in Fig. 9-2. Whereas all the previous elements of our model have only two terminals, this one has four, grouped into two

† Since we are in the process of developing a model, we can postulate any relationship that we wish between L_{12} and L_{21}. Whether or not the resulting model adequately describes and predicts results in physical situations for which it is to be a model must be judged on the basis of experiment. It is also possible to prove the equality of L_{12} and L_{21} on the basis of field theory and an argument involving energy.

pairs. The two sides of the transformer are often called the *primary side*, or *branch*, and the *secondary side*, or *branch*. Each of the previous elements of our model is characterized by a single parameter; this one has three: L_1, L_2, and M.

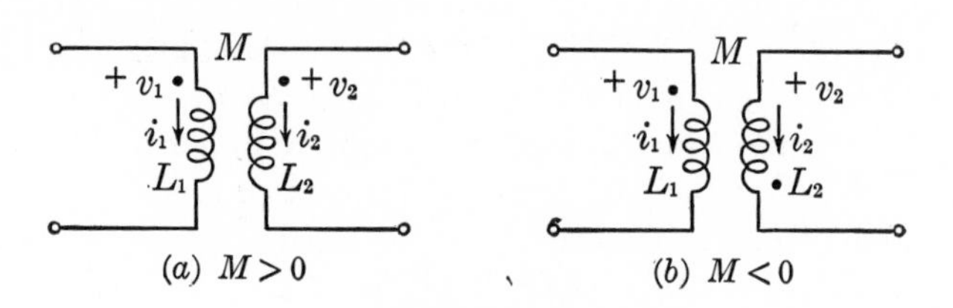

FIG. 9-2
A simple transformer

Let us now discuss the signs appearing in the v-i relationships of Eq. (9-4). Three sets of quantities are involved in arriving at these equations—the voltages, the currents, and the flux linkages. The reference of each voltage must be related to the reference of the corresponding flux linkage appropriately (as in Fig. 9-1) in order for the signs in Eq. (9-1) to be positive. Similarly, the reference of each current must be related to that of the corresponding flux linkage appropriately (as in Fig. 9-1), if the signs in Eq. (9-3) are to be positive. Since Eq. (9-4) is the result of combining Eqs. (9-1) and (9-3), the reference of each voltage must be related to the reference of the corresponding current in the manner shown in Fig. 9-2, if the signs in Eq. (9-4) are to be positive.

Note that the diagram in Fig. 9-2 is a schematic symbol. Even though this symbol which represents a transformer is suggestive of two coils of wire in close proximity, you should think of this symbol simply as a model. In the model it is not possible to show references for flux linkages. Hence we must use some other convention to indicate whether the fluxes are aiding or opposing. As a matter of fact, whether the fluxes are aiding or opposing is not an intrinsic property of the device but is dependent on the chosen references of the currents. Hence the convention we choose for indicating the sign of M must involve the current references in some way.

A common convention is to place a mark, such as a dot, at one terminal of each pair, as shown in Fig. 9-2. The positions of the dots relative to the current references must give the desired information about M.

Since the choice of branch-voltage and -current references can be made arbitrarily, we shall choose them in such a way as to simplify our work. Let us agree to place the plus sign of each *branch-voltage* reference at the tail of the reference arrow of the corresponding *branch current*. This means that the signs in Eq. (9-4) will always be positive when the v's and i's refer to branch variables. Now if the current-reference arrows are both directed toward a dot or both away from a dot, as in Fig. 9-2(a), this means that the two flux references in the corresponding physical pair of coupled coils coincide, and M is positive. If one current reference is directed toward a dot and one away, as in Fig. 9-2(b), then M is negative.

Note carefully that the voltages and currents in Eq. (9-4) are the branch variables. This fact must be borne in mind when we later discuss loop currents and node voltages.

In this procedure we have made no commitment as to the choice of voltage references relative to the dots. It is possible to guarantee that M will always be positive by making the convention that the branch-voltage plus signs are always taken at the dots. (In this case it will not even be necessary to show them.) Then, with our agreement about current references relative to voltage references, both branch currents will always be away from the dots and, hence, M will always be positive. In the following sections we shall illustrate both of these conventions.

As an illustration of these ideas, consider the situation shown in Fig. 9-3(a). With the current references chosen first, the flux references

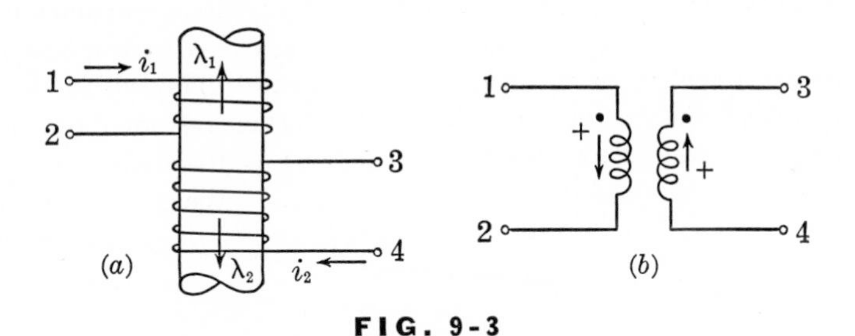

FIG. 9-3

Flux and current references

must be chosen as shown in order for L_1i_1 and L_2i_2 to carry positive signs in Eq. (9-3). Since the two references do not coincide, M is negative. The linear model of this circuit is shown in part (b) of the figure. Note that the voltage references are chosen with the plus signs at the tails of the corresponding arrows, in order that Eq. (9-4) may carry plus signs throughout.

9.2 COEFFICIENT OF COUPLING

According to Eq. (9-4), the mutual inductance is a measure of the voltage that is produced at one pair of terminals of a transformer owing to current variation in the other pair. The relative value of voltage induced in each of the two pairs of terminals by a current variation in one of them is expressed in terms of the *tightness*, or *closeness*, of coupling. To get a quantitative measure of the coupling, let us take the ratio of v_1 to v_2 first with i_1, then with i_2, equal to zero. From Eq. (9-4) we find

$$\left.\frac{v_1}{v_2}\right|_{i_1 \equiv 0} = \frac{M}{L_2} \qquad (a)$$

$$\left.\frac{v_1}{v_2}\right|_{i_2 \equiv 0} = \frac{L_1}{M} \qquad (b) \qquad \text{(9-5)}$$

The magnitude of either of these expressions may be less than unity or greater than unity; no law of nature will be violated in either case.

Let us now take the ratio of these two expressions; the result will be

$$\frac{\left.\dfrac{v_1}{v_2}\right|_{i_1 \equiv 0}}{\left.\dfrac{v_1}{v_2}\right|_{i_2 \equiv 0}} = \frac{M^2}{L_1 L_2} = k^2 \qquad \text{(9-6)}$$

We find that this ratio of the voltage ratios is a positive real constant k^2. The constant k is called the *coupling coefficient*. We shall now show that the coupling coefficient cannot exceed unity. This will be done on the basis that the energy stored in a transformer cannot be negative.

To start with, let us multiply the first of Eq. (9-4) by i_1, the second one by i_2, and then add the resulting equations. We get

$$v_1 i_1 = i_1 L_1 \frac{di_1}{dt} + i_1 M \frac{di_2}{dt} \qquad (a)$$

$$v_2 i_2 = i_2 M \frac{di_1}{dt} + i_2 L_2 \frac{di_2}{dt} \qquad (b) \qquad \text{(9-7)}$$

$$v_1 i_1 + v_2 i_2 = L_1 i_1 \frac{di_1}{dt} + M\left(i_1 \frac{di_2}{dt} + i_2 \frac{di_1}{dt}\right) + L_2 i_2 \frac{di_2}{dt}$$

$$= \frac{d}{dt}\left(\frac{L_1 i_1^2 + 2M i_1 i_2 + L_2 i_2^2}{2}\right) \qquad \text{(9-8)}$$

(You can verify that the second line of the last equation is correct by performing the indicated differentiation.) Note that the left-hand side represents the total power input to the transformer. Since power is the time derivative of energy, the quantity inside the parentheses represents the energy supplied to the transformer at any instant of time. This energy is stored in the transformer. (Recall that, for a simple inductor, the stored energy is $Li^2/2$, and compare.)

Let us extend the use of the symbol w_L to include the energy stored in a transformer. Thus, for the transformer,

$$w_L = \frac{1}{2}(L_1 i_1^2 + 2M i_1 i_2 + L_2 i_2^2) \tag{9-9}$$

By its very nature, the stored energy can never be negative, no matter what the values of the currents may be. If w_L can ever become negative, this will mean that the transformer is supplying energy to the external circuits.

Equation (9-9) is an example of a nonlinear function. The stored energy is a *quadratic* function of the currents, named so because the currents appear to the second power (or as a cross product, the sum of whose powers is 2). Such forms appear quite often in mathematics; they are given the name *quadratic forms*. In the present case the quadratic form in question is never negative, independent of the values of the variables. Quadratic forms having the property of never becoming negative, no matter what values the variables take on, are called *positive definite*. Conditions for a quadratic form to be positive definite can be discovered in terms of the mathematical properties of such forms. In the present case (which is relatively simple as quadratic forms go) we can find the conditions in the following way:

In the first place, if i_2 is zero, w_L becomes $L_1 i_1^2/2$. Since w_L is not to be negative, L_1 cannot be negative. This condition is certainly satisfied. If i_1 is zero, then w_L becomes $L_2 i_2^2/2$, which requires that L_2 be nonnegative, if w_L is to be nonnegative. This condition is also satisfied. Examination of Eq. (9-9) shows that the first and last terms are always positive, even if the currents are negative. The only way in which the entire expression can become negative is for the middle term to become negative. Presumably, the references are so chosen that M is positive. Hence the only way for the middle term to become negative is for i_1 and i_2 to be of opposite sign. Assuming this to be true, let us write

$$i_2 = -x i_1 \tag{9-10}$$

where x is a positive number, and substitute this into Eq. (9-9). If we now impose the condition that w_L be nonnegative, the result will be

$$L_1 - 2Mx + L_2 x^2 \geq 0 \tag{9-11}$$

The left side is a quadratic in x. The roots of this quadratic are found by the quadratic formula to be

$$x_{1,2} = \frac{M}{L_2} \pm \sqrt{\left(\frac{M}{L_2}\right)^2 - \frac{L_1}{L_2}} = \frac{1}{L_2}\left(M \pm \sqrt{M^2 - L_1L_2}\right) \tag{9-12}$$

The discriminant is $M^2 - L_1L_2$. If it is positive, the equation will have two real roots. A plot of the quadratic will then take the form shown in Fig. 9-4(a). There is a range of values of x between x_1 and x_2 for which the curve becomes negative. But this will violate the requirement that w_L should never be negative.

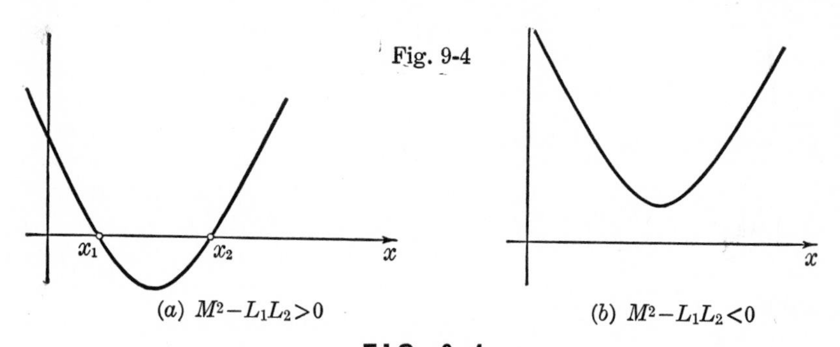

(a) $M^2 - L_1L_2 > 0$ (b) $M^2 - L_1L_2 < 0$

FIG. 9-4

Variations of the stored energy w_L

On the other hand, if the discriminant $M^2 - L_1L_2$ is negative, the equation will have complex roots, and a plot of the quadratic will take the form of the parabola shown in Fig. 9-4(b). The curve will always remain positive. In case the discriminant is zero, the equation will have a double real root, which means that the vertex of the parabola will lie directly on the x-axis, and this is a limiting permissible case.

In summary, then, in order for w_L to be positive for all possible values of the variables (that is, positive definite), we must require that

$$M^2 - L_1L_2 = L_1L_2(k^2 - 1) \leq 0$$

or

$$k^2 \leq 1 \tag{9-13}$$

This proves that the coupling coefficient can never be greater than unity.

An alternative method of proof consists in short-circuiting the secondary side, energizing the primary side, and computing the energy stored. Requiring that this energy never be negative again leads to the desired result. The details of this proof are left for you to supply.

If the coupling coefficient of a transformer has the limiting value $k = 1$, the transformer is said to be *perfect*, or *perfectly coupled.*

Having established that the coupling coefficient is no greater than 1, let us rewrite Eq. (9-6) as

$$M = k\sqrt{L_1L_2} \leq \sqrt{L_1L_2} \tag{9-14}$$

Thus the mutual inductance is no greater than the geometric mean of L_1 and L_2. It is certainly possible for the mutual inductance to be greater than either of the two self-inductances, but not greater than both of them simultaneously.

9.3 ANALYSIS OF TRANSFORMER NETWORKS

Up to this point we have made no restriction on the time dependence of voltages and currents. From here on, let us restrict ourselves to sinusoidal excitations and consider only the steady-state response. The voltages and currents are now represented by their sinors. The V-I relationships of the transformer given in Eq. (9-4) become

$$\begin{aligned} V_1 &= sL_1I_1 + sMI_2 && (a) \\ V_2 &= sMI_1 + sL_2I_2 && (b) \end{aligned} \tag{9-15}$$

We are now ready to consider solving network problems when the networks contain transformers in addition to the other elements of our model.

Consider the network shown in Fig. 9-5. If we write Kirchhoff's voltage law around the two loops, we get

$$\begin{aligned} Z_1I_1 + V_1 &= V_g && (a) \\ Z_2I_2 - V_2 &= 0 && (b) \end{aligned} \tag{9-16}$$

Note that I_1 and I_2 are loop currents; V_1 and V_2 are the branch voltages of the transformer branches with the references shown in the figure.

According to the discussion in Sec. 9-1, the branch-current references (which are not shown in the figure) are chosen with the tails at the voltage plus signs. Hence loop current I_2 has a reference opposite to that of the corresponding branch current. This means that in Eq.

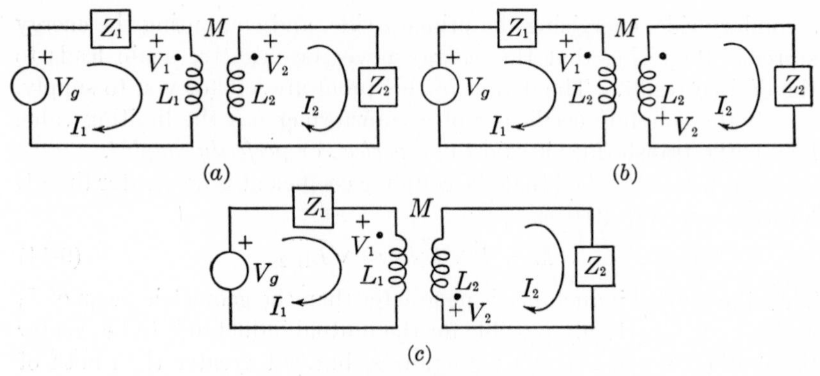

FIG. 9-5

Transformer network

(9-15) the sign of each term involving I_2 should be reversed if these equations are to be written in terms of loop currents. If this is done, and these equations are substituted into Eq. (9-16), the result will be

$$\begin{aligned} (Z_1 + sL_1)I_1 - sMI_2 &= V_g & (a) \\ -sMI_1 + (Z_2 + sL_2)I_2 &= 0 & (b) \end{aligned} \qquad (9\text{-}17)$$

Since the branch-current references (not the loop-current references) are both directed away from the dots, M is a positive quantity.

As an alternative approach, suppose we reverse the reference of V_2, as shown in Fig. 9-5(*b*), so that the secondary branch current becomes identical with loop current I_2 (because of our convention about branch-current reference relative to voltage reference). Instead of Eq. [9-16(*b*)], the equation of the second loop now becomes

$$Z_2I_2 + V_2 = 0 \qquad (9\text{-}18)$$

Since the branch currents and loop currents are identical, Eq. (9-15) is valid without change of sign. Hence the loop equations become

$$\begin{aligned} (Z_1 + sL_1)I_1 + sMI_2 &= V_g & (a) \\ sMI_1 + (Z_2 + sL_2)I_2 &= 0 & (b) \end{aligned} \qquad (9\text{-}19)$$

These equations differ from the previous ones in Eq. (9-17) by having positive signs in front of the mutual terms instead of negative signs. Observe from the figure, however, that the branch-current references are now directed oppositely relative to the dots, so that M is now negative. Hence the two sets of equations will give the same numerical results.

As a final variation, let us suppose one of the dots is reversed, as shown in Fig. 9-5(*c*). If we assume that the voltage-reference plus signs are at the dots, then the loop currents will be identical with the branch currents. Since all the references are the same as those of Fig. 9-5(*b*), Eq. (9-19) will again be valid. Now the branch-current references, however, are similarly directed relative to the dots, and so M is positive.

Sometimes it is convenient to regard M as an algebraic number which can take on either a positive or a negative value. In most of our work, however, there is no advantage in this procedure. Therefore, *we shall henceforth assume that M is always positive.* This means that we shall always choose the branch-voltage references with the plus signs at the dots, and the branch-current reference arrows with their tails at the corresponding voltage plus signs. We shall refer to this choice as the *standard references.* With this understanding, there will be no need to show these branch references. If both loop-current references coincide with the corresponding branch-current references, the signs in the V-I relationship in terms of loop currents will all be positive. If either one of the loop-current references is opposite to the corresponding branch-current reference, the sign of any term involving that particular current in the V-I relationships will be reversed.

To see the effect of this convention, glance back at Fig. 9-3. If M is to be positive, the fluxes must be aiding. Suppose the references of i_1 and λ_1 are chosen as shown. The reference of λ_2 is then fixed if λ_2 and λ_1 are to be aiding; it must be the opposite of that shown in the figure. This requires that the reference of i_2 also be reversed.

Let us now continue the discussion of the problem at hand, but without the series impedance Z_1 on the primary side. That is, we wish to consider the situation in which a transformer is inserted between a load Z_2 and a source, as shown in Fig. 9-6. Equation (9-17) applies, but with $Z_1 = 0$. The solution of these equations for the loop currents will be

$$I_1 = \frac{Z_2 + sL_2}{s^2(L_1L_2 - M^2) + sL_1Z_2} V_1 \qquad (a)$$

$$I_2 = \frac{sM}{s^2(L_1L_2 - M_2) + sL_1Z_2} V_1 \qquad (b) \qquad (9\text{-}20)$$

In terms of these currents, it is possible to find any of the network functions. Let us calculate the input impedance, the voltage-gain function, and the current-gain function. These will be

$$Z = \frac{V_1}{I_1} = \frac{s^2(L_1L_2 - M^2) + sL_1Z_2}{Z_2 + sL_2} = sL_1 - \frac{s^2M^2}{Z_2 + sL_2} \tag{9-21}$$

$$G_{21} = \frac{V_2}{V_1} = \frac{sMZ_2}{s^2(L_1L_2 - M^2) + sL_1Z_2} \tag{9-22}$$

$$\alpha_{21} = \frac{I_2}{I_1} = \frac{sM}{Z_2 + sL_2} \tag{9-23}$$

We shall examine each of these expressions in greater detail somewhat later.

The natural method of analysis of transformer networks seems to be the loop method. Let us, however, here contemplate carrying out a node analysis of the original network, which is given again in Fig. 9-7.

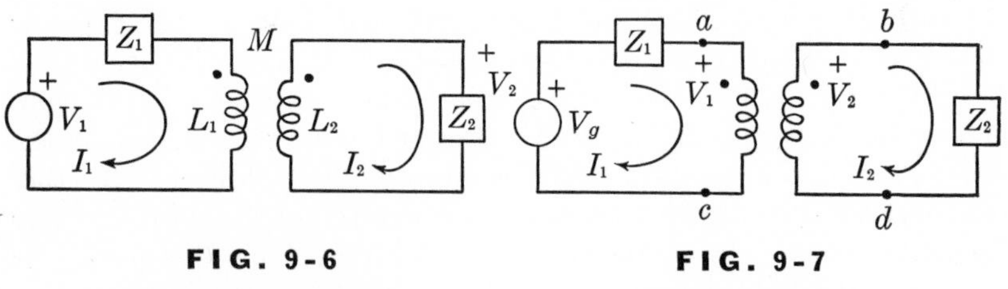

FIG. 9-6
Analysis of network

FIG. 9-7
Node analysis

The first step is to write Kirchhoff's current law at the nodes. There are four nodes, as shown. The current equation at node *c*, however, will be satisfied if it is satisfied at node *a*. Similarly, the current equation at node *d* will be satisfied if it is satisfied at node *b*. Hence there will be only two independent Kirchhoff current-law equations. But this disagrees with our previous result, first discussed in Chap. 2, which claims that there should be $N_v - 1$ independent equations.

This difficulty is easily overcome. Up until now, the networks we have considered have all been *connected;* that is they have been all in one piece, with a conductive connection from any one branch to any other branch. The network in Fig. 9-7 is in two parts which are *inductively* connected. We have found the number of independent current equations in this case to be 2, which is exactly $N_v - 2$.

If we call N_p the number of parts of the network, this number (the number of independent current equations) is $N_v - N_p$. This result is a general one, which we shall not prove but which is certainly reasonable. The number of independent Kirchhoff voltage-law equations should also be modified to read $N_m = N_b - (N_v - N_p)$ in case the network consists of N_p parts.

To write node equations, we must now choose one datum node in each part. The voltages at the nodes of each part are referred to the datum in that part. Thus, for Fig. 9-7, we can choose nodes c and d as datum nodes. Then $V_{ac} = V_1$ is the voltage of node a relative to that of node c; and $V_{bd} = V_2$ is the voltage of node b relative to that of node d. The voltage of one of the datum nodes relative to the other is completely arbitrary, since the two parts of the transformer are not connected. It is certainly possible, however, for this voltage to be zero. In this case we can assume that there is a conductive connection between nodes c and d. The number of nodes will be reduced by 1, but so also will the number of separate parts. Hence the number of independent equations will remain the same.

If we now write Kirchhoff's current law at nodes a and b, we get

$$I_1 = \frac{V_g - V_1}{Z_1} \qquad (a)$$
$$I_2 = -\frac{V_2}{Z_2} \qquad (b) \qquad (9\text{-}24)$$

In order to proceed, we need to substitute the V-I relationships of the transformer into this expression. This requires that we solve Eq. 9-15 for the currents in terms of the voltages. If we do this, we get

$$I_1 = \frac{L_2}{s(L_1L_2 - M^2)} V_1 + \frac{-M}{s(L_1L_2 - M^2)} V_2 \qquad (a)$$
$$I_2 = \frac{-M}{s(L_1L_2 - M^2)} V_1 + \frac{L_1}{s(L_1L_2 - M^2)} V_2 \qquad (b) \qquad (9\text{-}25)$$

These equations may be put in a slightly more compact form by substituting $M^2 = k^2L_1L_2$. The result will be

$$I_1 = \frac{1}{sL_1(1 - k^2)} V_1 - \frac{1}{sM(1/k^2 - 1)} V_2 \qquad (a)$$
$$I_2 = -\frac{1}{sM(1/k^2 - 1)} V_1 + \frac{1}{sL_2(1 - k^2)} V_2 \qquad (b) \qquad (9\text{-}26)$$

Either this set of equations or the set given in Eq. (9-25) constitutes the V-I relationship of the transformer with the currents expressed in terms of the voltages. Note that these equations are in terms of the branch variables with standard references.

To continue with the analysis, we now substitute these equations into Eq. (9-24), noting that loop current I_2 has a reference opposite to that of the standard branch current, and so the sign of I_2 in Eq. [9-26(b)] must be reversed. The result will be

$$\left(\frac{1}{sL_1(1-k^2)} + \frac{1}{Z_1}\right) V_1 - \frac{1}{sM(1/k^2 - 1)} V_2 = \frac{V_g}{Z_1}$$
$$-\frac{1}{sM(1/k^2 - 1)} V_1 + \left(\frac{1}{sL_2(1-k^2)} + \frac{1}{Z_2}\right) V_2 = 0 \qquad (9\text{-}27)$$

These are the node equations of the transformer network in Fig. 9-7. The solution is now straightforward, but we shall not pursue it any further.

Using the method of node equations when a transformer appears in a network is somewhat more difficult, computationally, than using the method of loop equations. For this reason, loop analysis is normally used in the analysis of transformer networks.

Although a transformer has been introduced as a device that has basically two pairs of terminals, it is possible to make connections between the terminals of one pair and those of the other. In particular, the branches can be connected in series or in parallel, with other elements connected externally. Such connections introduce no difficulty in obtaining a solution.

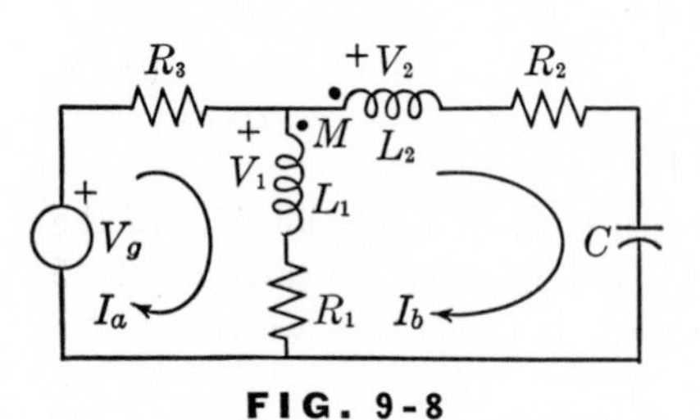

FIG. 9-8

Transformer network

As an example, consider the network shown in Fig. 9-8. Let us proceed to write loop equations, choosing the two windows as loops. In doing so, we shall substitute the V-I relationships into the voltage equations for all branches except the transformer branches. The result will be

$$R_3I_a + V_1 + R_1(I_a - I_b) = V_g \qquad (a)$$
$$V_2 + \left(R_2 + \frac{1}{sC}\right) I_b + R_1(I_b - I_a) - V_1 = 0 \qquad (b) \qquad (9\text{-}28)$$

The next step is to substitute the V-I relationships of the transformer from Eq. (9-15) and thus eliminate V_1 and V_2. Note carefully that I_1 and I_2 in Eq. (9-15) refer to the branch currents in the transformer. With the standard references, the branch currents in terms of the loop currents will be $I_1 = I_a - I_b$ and $I_2 = I_b$. Hence the V-I relationships in terms of loop currents will be

$$V_1 = sL_1(I_a - I_b) + sMI_b$$
$$V_2 = sM(I_a - I_b) + sL_2I_b \qquad (9\text{-}29)$$

Substituting these into Eq. (9-28) and collecting terms, we get

$$(R_1 + R_3 + sL_1)I_a - (R_1 + sL_1 - sM)I_b = V_g \qquad (a)$$

$$- (R_1 + sL_1 - sM)I_a + \left[R_1 + R_2 + s(L_1 + L_2 - 2M) + \frac{1}{sC_2}\right]I_b = 0 \qquad (b) \tag{9-30}$$

Notice how the mutual inductance enters into the second loop equation in the coefficient of I_b. It comes in once owing to V_2 and once owing to V_1.

The solution of these equations is now carried out in a straightforward manner.

9.4 PERFECT TRANSFORMERS AND IDEAL TRANSFORMERS

When discussing the coupling coefficient in Sec. 9-2, we found that the value of k was restricted to be no greater than unity. A transformer which has the limiting value of the coupling coefficient, namely $k = 1$, is called a *perfect transformer*, or a *unity-coupled* transformer. Let us now discuss the behavior of a perfect transformer.

The network of Fig. 9-9 is the one shown in Fig. 9-5 but with $Z_1 = 0$.

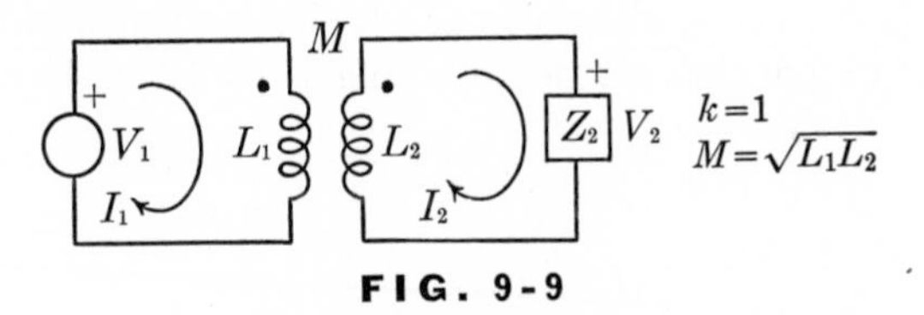

FIG. 9-9

Perfect transformer

We had previously calculated the input impedance, the voltage gain, and the current gain of this network. With $k = 1$ (or $L_1L_2 - M^2 = 0$), the voltage-gain function in Eq. (9-22) becomes

$$G_{21} = \frac{V_2}{V_1} = \sqrt{\frac{L_2}{L_1}} = \frac{1}{n} \tag{9-31}$$

In this case we find that the voltage gain is simply a constant, quite independent of the load impedance. We have labeled this constant $1/n$.

Next, let us turn to the current gain in Eq. (9-23). With $M = \sqrt{L_1L_2}$, this becomes

$$\alpha_{21} = \frac{I_2}{I_1} = \frac{s\sqrt{L_1L_2}}{sL_2 + Z_2} = \frac{1}{\sqrt{\frac{L_2}{L_1}} + \frac{Z_2}{s\sqrt{L_1L_2}}} = \frac{n}{1 + \frac{Z_2}{sL_2}} \tag{9-32}$$

The last step is obtained by inserting the value $n = \sqrt{L_1/L_2}$. We find that, unlike the voltage gain for a perfect transformer, the current gain depends on the load impedance Z_2. The current gain will also be relatively independent of Z_2 if the impedance of L_2 is much greater in magnitude than $|Z_2|$. For this to be true at all frequencies requires a large value of L_2, one approaching infinity. If L_2 approaches infinity, however, L_1 must do so also if the constant n, which is $\sqrt{L_1/L_2}$, is to remain finite and nonzero.

Finally, let us turn to the input impedance. With $L_1L_2 - M^2 = 0$, Eq. (9-21) becomes

$$Z = \frac{sL_1Z_2}{sL_2 + Z_2} = \frac{L_1}{L_2}\frac{sL_2Z_2}{sL_2 + Z_2} = n^2Z_2\frac{1}{1 + \frac{Z_2}{sL_2}} \tag{9-33}$$

We find that, aside from the factor $n^2 = L_1/L_2$, the impedance is that of the parallel combination of sL_2 and Z_2. If L_2 approaches infinity, this branch will be effectively open-circuited, and the impedance will approach n^2Z_2.

With the preceding discussion as a basis, let us define an *ideal transformer* as a two-port having the following *V-I* relationships:

$$V_2 = \frac{1}{n}V_1 \qquad (a)$$

$$I_2 = nI_1 \qquad (b) \tag{9-34}$$

where n is a real number called the *transformation ratio*, or the *turns ratio*. (Its value is arbitrary and not necessarily equal to $\sqrt{L_1/L_2}$.) When n has a magnitude greater than one, the ideal transformer is said to be *step-down;* when $|n|$ is less than one, it is said to be step-up. (The *up* and *down* refer to voltage; when the voltage is stepped up, the current will be stepped down, and vice versa, but the *V-I* product will remain the same.) If we compare the *V-I* relationships of the ideal transformer with the voltage gain and current gain for the prefect transformer, we see that an ideal transformer behaves like a perfect transformer whose primary and secondary inductances are both infinite, but with a finite ratio.

We represent the ideal transformer symbolically by the diagram in Fig. 9-10. Note the voltage and current references. Equation (9-34) applies for these references; if any one of the references is reversed, the corresponding sign in these equations will change. The diagram is like

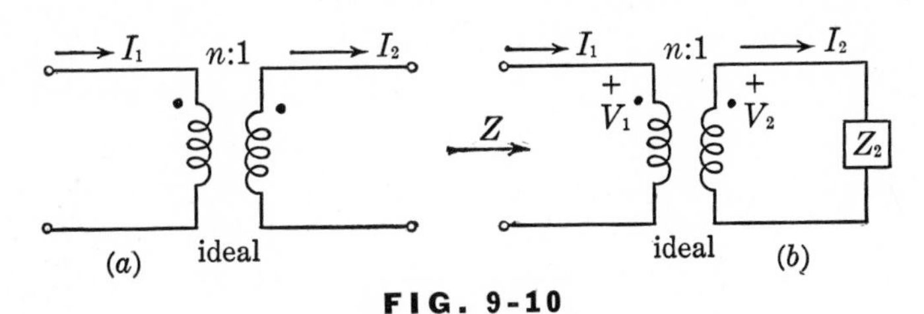

FIG. 9-10

An ideal transformer

that of an ordinary transformer, except that here there is a single parameter n, instead of the three inductances. The transformation ratio is shown as n:1.

Consider the ideal transformer with the load impedance Z_2 shown in Fig. 9-10(b). The input impedance Z is V_1/I_1. Taking the ratio of the defining equations of the ideal transformer and noting that $Z_2 = V_2/I_2$, we get

$$Z = \frac{V_1}{I_1} = n^2 \frac{V_2}{I_2} = n^2 Z_2 \tag{9-35}$$

This equation states that, when an impedance Z_2 is connected at one port of an ideal transformer, the impedance at the other port is Z_2 multiplied by the square of the turns ratio. Thus an ideal transformer acts as a transformer (changer) of impedances. This fact makes the ideal transformer useful for the purpose of matching, when it is desired to vary the load impedance of a network in order to obtain maximum power.

Note the distinction between a perfect transformer and an ideal transformer. The ideal transformer is defined in terms of the V-I relationships in Eq. (9-34) which involve a single parameter, n. On the other hand, a perfect transformer still needs two parameters to describe its behavior. (It needs only two instead of three, because, once two of the inductances are known, the third follows from the relationship $L_1L_2 = M^2$.) We shall subsequently have more to say about ideal transformers.

9.5 EQUIVALENT CIRCUITS AND TWO-PORTS

The transformer was introduced as a network element having two pairs of terminals. If no other elements are connected from the primary side to the secondary side, the transformer will constitute a two-port. Hence its terminal behavior can be described in terms of any of the sets of two-port parameters described in Chap. 8.

Consider the transformers shown in Fig. 9-11 as two-ports. The

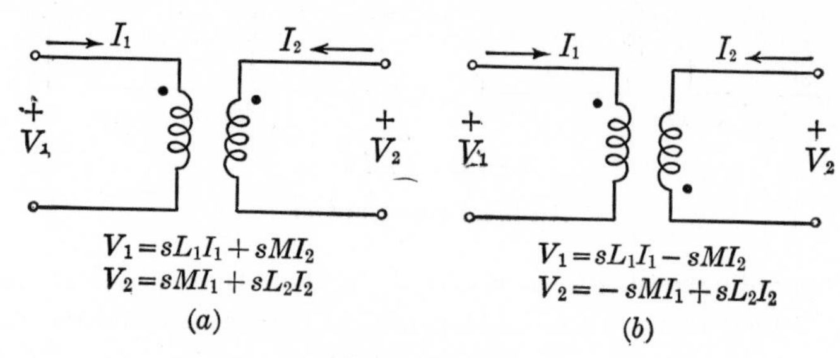

FIG. 9-11

Transformer as two-port

two possible arrangements of the dots are shown. Two sets of references must be reconciled. The references shown in the diagrams are the standard ones for two-ports. We agreed, however, that the branch references of the transformers were to be chosen in a special way. In the first figure the two-port references and the transformer branch references coincide. But in the second figure the secondary branch references are the opposites of the two-port references. These conditions lead to the V-I relationships shown in the figure, the net effect being a change in the sign before the mutual term in the second figure. These expressions are in the form of the z system of equations. By direct comparison, we find the z parameters of the transformer to be

	Fig. 9-11(a)	Fig. 9-11(b)
$z_{11} =$	sL_1	sL_1
$z_{22} =$	sL_2	sL_2
$z_{12} = z_{21} =$	sM	$-sM$

(9-36)

Very often, it is convenient to replace a transformer by a two-port which is equivalent to it. This can often lead to a simplification in the analysis. It can also lead to insights into the operation which might otherwise be obscured. In Chap. 8 we found that any two-port can be replaced by a tee equivalent. By referring to Fig. 8-8, we find the branch impedances of the tee to be

	Fig. 9-11(a)	Fig. 9-11(b)
$Z_3 = z_{21}$	sM	$-sM$
$Z_1 = z_{11} - z_{21}$	$s(L_1 - M)$	$s(L_1 + M)$
$Z_2 = z_{22} - z_{21}$	$s(L_2 - M)$	$s(L_2 + M)$

(9-37)

Thus the tee equivalents of the two cases take the forms shown in Fig. 9-12.

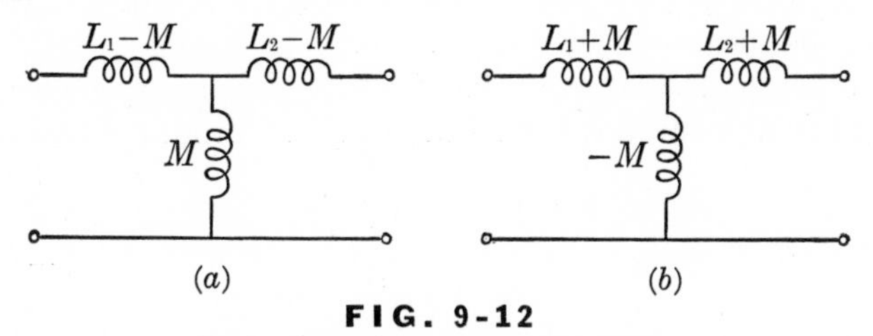

FIG. 9-12

Tee equivalents of transformer

Note that the series branch inductances in the first diagram involve a difference. Since it is possible for M to be greater than either L_1 or L_2 (but not both), one or the other of these two inductances may be negative. In the second diagram the shunt inductance will always be negative. In any of these cases, the tee equivalent of the transformer will not be physically realizable; nevertheless, it can still be used for the purpose of calculating the external behavior.

By utilizing the tee-to-pi transformation, it is possible to find a pi equivalent of the transformer also. Alternatively, the V-I relationship of the transformer given in Eq. (9-25), with the currents expressed in terms of the voltages, can be compared with the y system of equations of a two-port. The y parameters can then be written down directly. These y parameters can then be compared with the y parameters of a pi. The result of either of these procedures leads to the networks shown in Fig. 9-13. The calculations will be left for you to carry out.

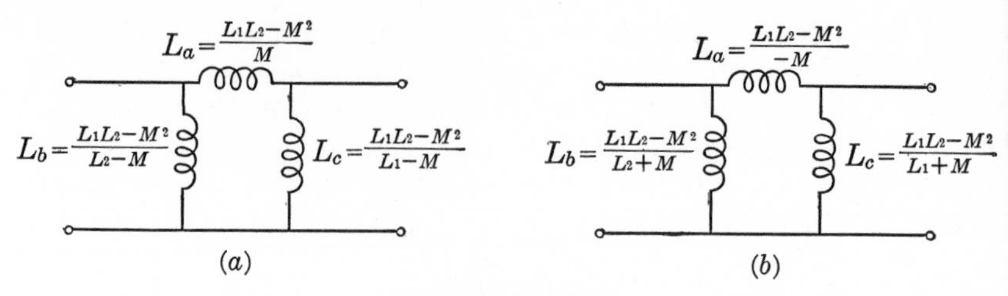

FIG. 9-13

Pi equivalents of transformer: (a) both dots at top; (b) one dot reversed

At this point let us make the observation that the tee and the pi equivalents are (conductively) connected networks; they are in one part. This property is different from that of a transformer which provides *isolation* between the primary and secondary sides. The thought may occur to you to connect an ideal transformer at either the output or the input of the tee or pi equivalent, thereby providing the desired isolation. This is a good idea, so let us follow it up.

For concreteness, let us assume that the dots are as shown in Fig. 9-11(a). Let us write the V-I relationships of an ideal transformer as

$$V_2 = \frac{1}{n} V_2' \qquad I_2 = nI_2' \tag{9-38}$$

where V_2' and I_2' are the primary voltage and current. If we quite formally insert these expressions into the V-I relationships of the transformer, the result will be

$$\begin{aligned} V_1 &= sL_1I_1 + s(nM)I_2' && (a) \\ V_2' &= s(nM)I_1 + s(n^2L_2)I_2' && (b) \end{aligned} \tag{9-39}$$

These are the V-I relationships of a two-port whose output current and voltage are I_2' and V_2', respectively. The z parameters of the two-port are seen to be $z_{11} = sL_1$, $z_{12} = snM$, and $z_{22} = sn^2L_2$. The tee network having these z parameters is easily found; it is shown in Fig. 9-14(a). The output variables of the tee, however, are the primed variables, whereas in the original transformer they were the unprimed ones. The primed and unprimed variables are related through Eq. (9-38), which represents an ideal transformer. Hence the complete network takes the form shown in Fig. 9-14(b). Note that the references of both I_2 and I_2' are reversed from those used in originally defining an ideal transformer. Hence the sign in the equation will remain unchanged.

The question arises as to what value should be assigned to the

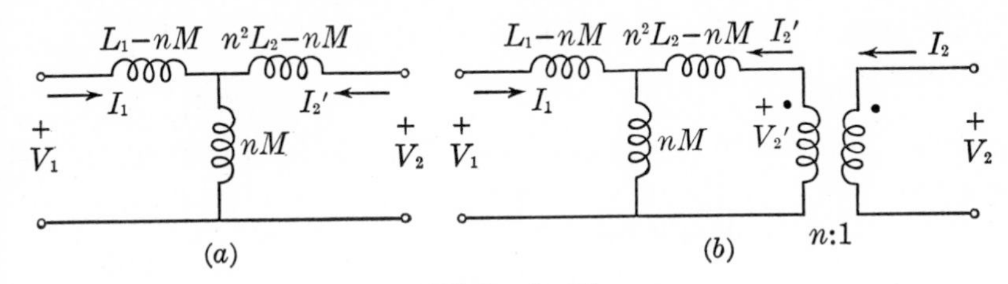

FIG. 9-14

Equivalent circuit of transformer

transformation ratio n. If you glance back at the preceding development, you will notice that no specific restrictions were placed on n; it may have any real value. Thus there are an infinity of equivalent networks of a transformer, all having the form of Fig. 9-14(b) with different values of the inductances.

When $n = 1$, the equivalent network reduces to the tee given in Fig. 9-12 but with a 1:1 ideal transformer at the output whose influence is merely to provide isolation. Another interesting value of n is $n = \sqrt{L_1/L_2} = kL/M = M/kL_2$. The corresponding network is shown in Fig. 9-15(a).

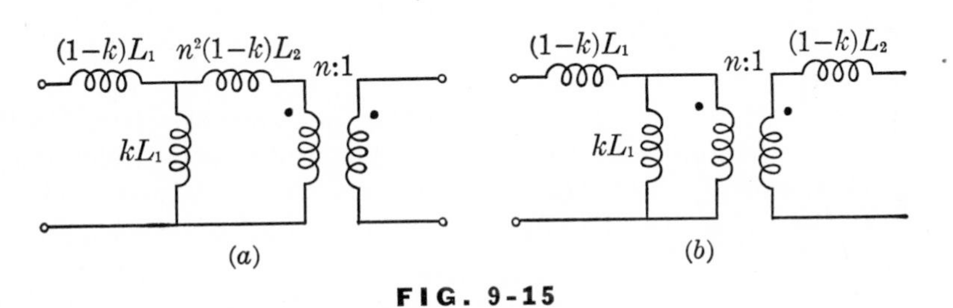

FIG. 9-15

Equivalent circuits of transformer: $n = \sqrt{\frac{L_1}{L_2}}$

If you remember that the impedance looking in the primary side of an ideal transformer is n^2 times the impedance in the secondary, then Fig. 9-15(a) can be redrawn as shown in Fig. 9-15(b).

In the special case of a perfect transformer, the coupling coefficient k is unity. Hence the series inductances in Fig. 9-15 will vanish. The equivalent circuit of a perfect transformer will then have the form shown in Fig. 9-16. The only difference between a perfect transformer and an ideal transformer is seen to be the appearance of L_1 across the primary terminals. The larger the value of L_1, the closer will a perfect transformer approach an ideal transformer, which corroborates our previous knowledge.

What was done starting at Eq. (9-38) can now be duplicated, but with the ideal transformer placed at the input terminals. There are no new ideas here, so we shall leave for you the details of developing these additional equivalent circuits.

Let us pause here briefly to make some general remarks. At the beginning of this chapter, we introduced a new element into our model which we called a transformer and which we defined in terms of a pair of V-I equations. In this section we have seen that this element itself can be represented as an interconnection of some inductances (one of the original elements), and something new, an ideal transformer. Hence it appears that we could have introduced the ideal transformer first as the new element in the model. Then a general transformer would follow as an interconnection of this new element with some old ones in the form of Fig. 9-14. The degree of abstraction required to leap from observations in the physical world to an ideal transformer defined by Eq. (9-34), however, is much greater than it is to go to the transformer defined by Eq. (9-4). The latter relationships constitute a simple extension of Faraday's law.

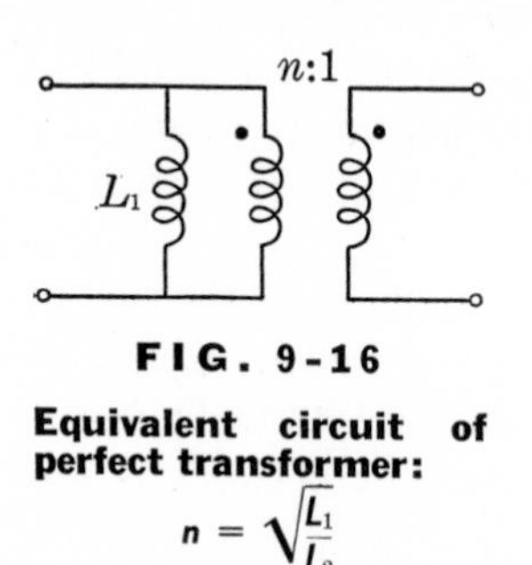

FIG. 9-16

Equivalent circuit of perfect transformer:

$$n = \sqrt{\frac{L_1}{L_2}}$$

Note that an ideal transformer is characterized by one parameter, n. A perfect transformer can be described in terms of two parameters, say n and L_1, as seen from Fig. 9-16. Finally, a general transformer requires three parameters to describe it. These may be either L_1, L_2, and M or L_1, k, and n.

When it is desired to determine the influence of a transformer in a network, it is often possible to get an approximate idea of this effect by assuming that the transformer is ideal. In order to gain an idea of the numerical error involved in such an approximation, let us turn back to the situation shown in Fig. 9-6. The load impedance is assumed to be real; $Z_2 = R_2$.

Suppose we take $L_1/L_2 = 1/4$. This means that the transformation ratio, as defined in Eq. (9-31), is $n = 1/2$. If the transformer is assumed to be ideal, the input impedance and the voltage gain will be

$$Z = \frac{L_1}{L_2} R_2 = \frac{R_2}{4}$$
$$G_{21} = \sqrt{\frac{L_2}{L_1}} = 2 \tag{9-40}$$

For our purposes, let us rewrite the expressions for the impedance and the voltage gain, given in Eqs. (9-21) and (9-22), in the following forms:

$$Z = R_2 \frac{L_1}{L_2} \frac{1 + (1 - k^2)\dfrac{sL_2}{R_2}}{1 + \dfrac{R_2}{sL_2}} \qquad (a)$$
$$G_{21} = \frac{V_2}{V_1} = \frac{k\sqrt{L_2/L_1}}{1 + (1 - k^2)\dfrac{sL_2}{R_2}} \qquad (b) \tag{9-41}$$

You are invited to go through the steps required to obtain these expressions, using the relationship $M^2 = k^2L_1L_2$. With $s = j\omega$, these expressions become

$$\frac{Z}{R_2/4} = \frac{1 + j(1 - k^2)\dfrac{\omega L_2}{R_2}}{1 - j\dfrac{R_2}{\omega L_2}} \qquad (a)$$
$$\frac{G_{21}}{2} = \frac{k}{1 + j(1 - k^2)\dfrac{\omega L_2}{R_2}} \qquad (b) \tag{9-42}$$

where each of them has been expressed as a fraction of its value for the ideal transformer. In each of these expressions, we find two parameters: the coupling coefficient k and the ratio of secondary reactance to load resistance $\omega L_2/R_2$ (or its reciprocal). With $k = 1$, the impedance will approach its ideal value as the ratio $\omega L_2/R_2$ gets larger. As mentioned before, for $k = 1$, the voltage gain attains its ideal value independently of the load.

With $k \neq 1$, however, increasing $\omega L_2/R_2$ indefinitely does not result in an improvement, either of the impedance or of the voltage gain. For example, taking $k = 0.98$ leads to the following values of the impedance:

$$\frac{Z}{R_2/4} = 0.999\epsilon^{j22.5°} \qquad \frac{\omega L_2}{R_2} = 5$$
$$\frac{Z}{R_2/4} = 1.070\epsilon^{j21.6°} \qquad \frac{\omega L_2}{R_2} = 10$$

We see that the magnitude of the impedance differs very little from its ideal value for $\omega L_2/R_2 = 5$, even though the angle is appreciably different. Increasing the $\omega L_2/R_2$ ratio causes a greater variation of the impedance, but, even in this case, the magnitude of the impedance differs from its ideal value only by about 7 per cent. Hence a rough idea of the behavior can be obtained by using the ideal value.

Before terminating this section, let us illustrate the use of an equivalent circuit in simplifying the analysis of a network containing a transformer. In the network shown in Fig. 9-17, it is desired to find the

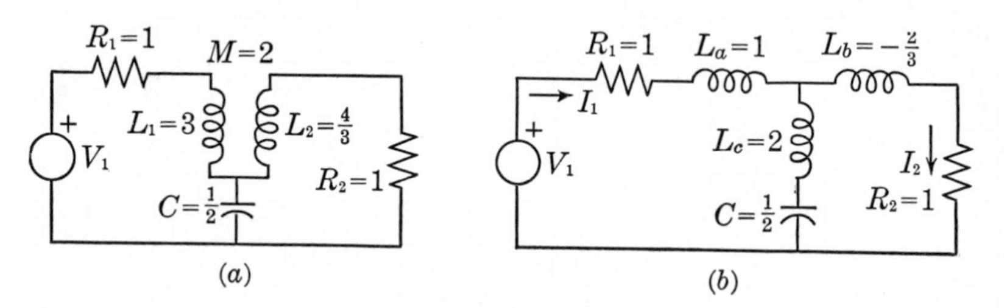

FIG. 9-17

Use of tee equivalent of a transformer

currents in the two resistors in response to a voltage source V_1. Of course, one approach is to use the method of loop equations. However, let us instead replace the transformer by a tee equivalent, as shown in Fig. 9-17(*b*). Note that one of the inductances has a negative value. To find I_1 it is enough to determine the input impedance of the network, which is easily done by series and parallel combinations of branches, as follows:

$$Z = 1 + s + \cfrac{1}{\cfrac{1}{2s + \cfrac{1}{s/2}} + \cfrac{1}{-\cfrac{2}{3}s + 1}} = \frac{13s^2 + 5s + 12}{4s^2 + 3s + 6}$$

$$I_1 = \frac{V_1}{Z} = \frac{4s^2 + 3s + 6}{13s^2 + 5s + 12} V_1$$

Finally, I_2 is obtained from the current-divider relationship as

$$I_2 = \frac{2s + \cfrac{2}{s}}{2s + \cfrac{2}{s} - \cfrac{2}{3}s + 1} I_1 = \frac{6(s^2 + 1)}{13s^2 + 5s + 12} V_1$$

9.6 CONTROLLED OR DEPENDENT SOURCES

Up to this point, we have determined several different circuits, all of which can be equivalent to a transformer. Let us now continue to look for an equivalent and again examine the *V-I* relationships, which are repeated here.

$$\begin{aligned} V_1 &= sL_1I_1 + sMI_2 \\ V_2 &= sMI_1 + sL_2I_2 \end{aligned} \tag{9-43}$$

Suppose we consider these to be the loop equations of a two-loop network. The first equation refers to the input port which has a voltage V_1 and a current I_1. It appears that I_1 flows through an inductor L_1, and that V_1 is equal to the voltage across this inductor, plus another voltage. Similarly, from the second equation it appears that I_2 flows through an inductor L_2, and that V_2 is equal to the voltage across this inductor, plus another voltage.

A possible network is shown in Fig. 9-18(*a*). This network is not yet complete. We still need to find branches to account for the voltages

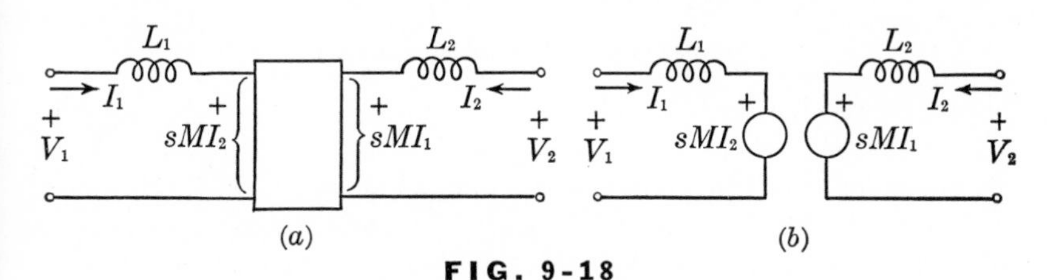

FIG. 9-18

Transformer equivalent

sMI_2 and sMI_1. One possibility is to assume that these quantities are the voltages of voltage sources; this assumption leads to the network shown in Fig. 9-18(*b*).

Let us examine these voltage sources. When we first defined a voltage source in Chap. 1, we claimed that the voltage was independent of the current at the terminals. Here we find that the value of each source voltage is dependent on the current in some other part of the network. Thus it is essentially different from the previously defined voltage source. We met such a source before when we discussed the compensation theorem in Sec. 7-7. There we replaced an impedance carrying a current I by a voltage source $V = ZI$. Although we did not call attention to this feature at the time, it is clear that the source voltage is dependent on the current.

Since we seem to have something whose character is different from anything we have previously dealt with, let us introduce a new hypothetical element into our model; in fact, two new elements. Let us define a *controlled, or dependent, voltage source* as a network element whose behavior is described by a voltage which is not independent but is a function of a voltage or current somewhere in the network in which it is connected. That is, a controlled voltage source may be *voltage controlled* or *current controlled.* The symbols representing these two possibilities are shown in Fig. 9-19(a) and (b). In general, the controlled

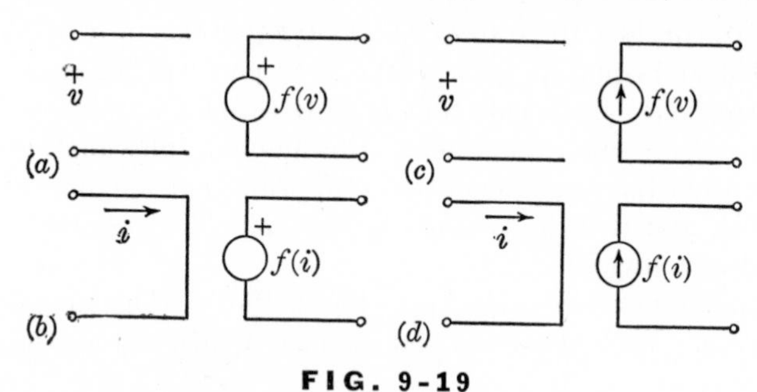

FIG. 9-19

Controlled or dependent sources

voltage source has four terminals. The terminals, however, may be connected so that there are effectively only three or only two terminals. The dependence of the source on the voltage or current is expressed in terms of the function f. We have used lower-case letters for the voltage and current, implying functions of time, in order to emphasize that we have a general situation here, not one involving sinors only.

In a similar way, we define a *dependent,* or *controlled, current source* as a network element whose behavior is described by a current which is a function of a voltage or current somewhere in the network in which it is connected. The controlled current source may be voltage controlled or current controlled. Figure 9-19(c) and (d) show the symbols representing these possibilities.

Having introduced these new elements into our model of electric networks, let us now return to the transformer equivalent network of Fig. 9-18. This network includes two controlled voltage sources, both of which are current controlled. The function $f(I)$ in this case is simply sM times the appropriate current.

This equivalent network can be used to replace the transformer in any network in which it appears. The presence of the controlled sources, however, prevents such procedures as combining branches in series and parallel, for example, as is possible with a tee equivalent. Nevertheless, in some situations this equivalent may prove to be of value. Furthermore, we shall discover in the next chapter, that controlled sources inevitably turn up in the models which are introduced to account for the behavior of certain electronic devices.

9.7 PHYSICAL TRANSFORMERS

After the initial introduction of the subject of this chapter, we have been dealing exclusively with models. This fact may have been obscured by our use of certain terms, such as turns ratio, which are suggestive of actual physical devices. Let us now turn back from the model to a consideration of the physical world.

In the physical world, mutual magnetic coupling may exist as an incidental effect whenever two or more circuits are in close proximity, or it may exist by design. Actual physical devices which are designed to produce in one circuit a response to an excitation applied in another circuit through the agency of mutual magnetic coupling are called transformers. It is unfortunate that the same name is used for a physical device and for a model which represents such a device. Whenever there is need to make the distinction clear, we shall say *physical transformer* to distinguish the physical device from the model.

A physical transformer consists of two or more coils of wire wound on a common core. The core may take one of many forms, such as a toroid or a three-legged rectangular frame. The model we have been discussing is for a two-winding transformer. Very often, ferromagnetic materials are used for the core whenever large values of inductance are required. The primary and secondary inductances will depend on the number of turns in the windings, the magnetic material of the core, and the geometry of the windings. If N_1 and N_2 are the number of turns in the primary and secondary, respectively, then the ratio N_1/N_2 can be called the turns ratio. (This is the origin of the name for the parameter n of an ideal transformer.) The inductance of a winding is proportional to the square of the number of turns. Thus, in a physical transformer, $L_1/L_2 = N_1^2/N_2^2$.

The size of the coupling coefficient depends on how well the flux paths linking the two windings coincide. The core material and the method of winding the coils have an influence on k. For high-permeability cores, the coupling coefficient can be made to approach unity. If the coils are wound by holding two wires together and winding them simultaneously, the result is called a *bifilar* winding. This also gives a value of k approaching unity. Thus perfect transformers can be approximated fairly well by actual physical transformers.

The windings, however, will inevitably have some resistance. Thus a model of a physical transformer should include resistance, as illustrated in Fig. 9-20(a). Furthermore, there will inevitably exist some capaci-

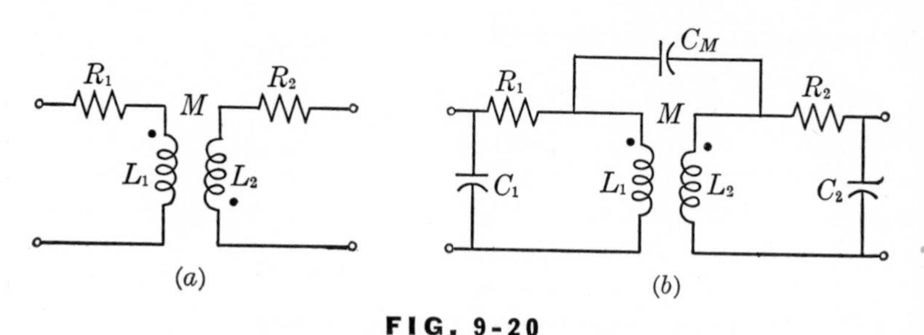

FIG. 9-20

Models of a physical transformer

tance between the turns of the windings. At low frequencies of operation, the effect of these capacitances may be negligible. In fact, the windings are often designed to minimize the capacitances. If the frequency of operation is high enough, however, the interturn and interwinding capacitances assume enough importance to be taken into consideration. Figure 9-20 might constitute an adequate model in such cases.

In any given problem, engineering judgment must be used to choose an appropriate model to fit the conditions of the problem. This judgment must be developed through a study of the behavior of the physical devices involved—in this case, transformers—under many conditions of operations. Although we are not here concerned with this phase of your engineering education, you must keep in mind that this is an important part of engineering.

Very often, magnetic coupling occurs in physical systems when it is not desired. A common example of this is the interference caused by power lines on telephone lines. In such cases steps are often taken to reduce the influence of external magnetic fields by means of *shields*, such as the metallic devices surrounding some of the parts in a television receiver. In order to predict the behavior of circuits which are influenced by the incidental coupling from other circuits, this coupling must be accounted for in the model representing the circuit.

Let us now discuss how the polarity marking (dot notation) which is used in analyzing transformers is related to the direction of winding of the coils in a physical transformer. For this purpose consider Fig. 9-21(*a*), which shows two windings on the two legs of a rectangular core.

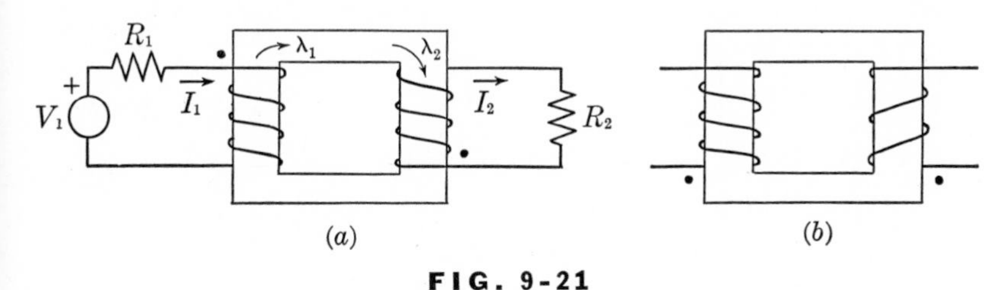

FIG. 9-21

Transformer windings and dots

Suppose we arbitrarily pick the reference for current I_1 as shown. If the relationship between flux linkage and current is to carry a positive sign, this will fix the reference for λ_1 to be upward in the left-hand leg of the core. If we want M to be positive, we must choose the reference of λ_2 to coincide with that of λ_1. Thus it must be downward in the right-hand leg. This, in turn, fixes the reference of I_2 to have the direction shown, if the relationship between λ_2 and I_2 is to carry a positive sign. With these references the dots must be placed at the upper terminal of one of the windings and at the lower terminal of the other, assuming the voltage references are the standard ones relative to the current references.

In summary, for the purpose of establishing the polarity marks of a physical transformer, we use the following procedure. A dot is arbitrarily placed at one terminal of one winding. A current reference directed toward this dot from outside the winding is assumed. This fixes the direction of the flux which would be produced by a positive current in this direction. Next we determine the direction of a current in the second winding which would produce a flux aiding the first one. We then place a dot at that terminal of the second winding toward which this current is directed from outside the winding. Use this scheme to verify the markings in Fig. 9-21(*b*).

Our study of transformers has included a consideration of two coupled coils only. Similar considerations prevail if more than two windings are involved. Figure 9-22(*a*) shows three coils wound on a common core. You can verify that the polarity markings will be as shown by going through the procedure which was just outlined. The marking implies that the three fluxes produced by positive currents, all directed toward the dots, will aid each other.

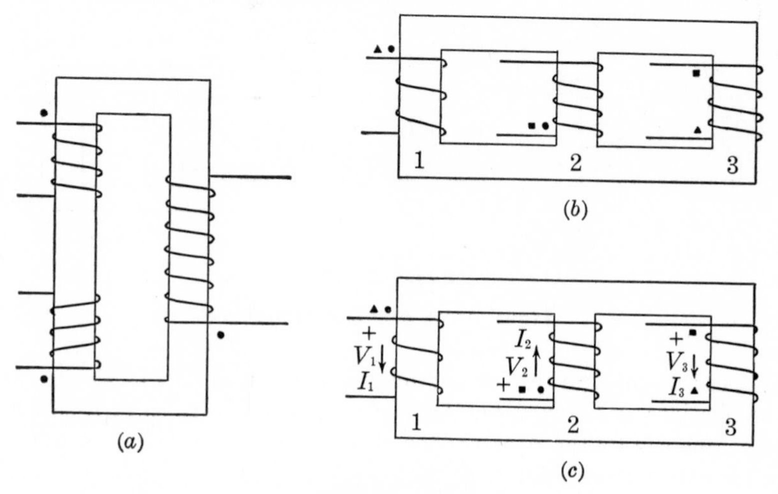

FIG. 9-22

Three winding transformers

This may not always be the case, however, when more than two windings are involved. That is, if we limit ourselves to just one set of dots (one dot per winding), it may be that there is no way of locating these dots so that positive currents into each of them will produce fluxes which are all mutually aiding. Such is the case in Fig. 9-22(*b*), as you can verify. The remedy is to assign a different set of polarity markings for each pair of windings. Dots with different shapes (triangles, squares, etc.) can be used for this purpose. The procedure which was outlined is carried out for each pair of windings. Thus, in Fig. 9-22(*b*), there will be three sets of dots. You should verify the polarity markings given in the figure.

Now let us consider the *V-I* relationships of the three-winding transformer. In the case of Fig. 9-22(*a*), the standard references can be chosen for the branch variables (voltage plus at dot, current tail at voltage plus) so that the signs in these relationships can be positive. In the case of Fig. 9-22(*b*), however, it is not possible to choose standard references for all the variables, since the two dots which are necessary for each winding do not always appear at the same terminal. The diagram is redrawn in Fig. 9-22(*c*), and branch references are chosen as shown. The windings are labeled 1, 2, and 3. Let us label the mutual inductances with double subscripts such as M_{12}, M_{13}, M_{23}. Then the V-I relationships in terms of branch variables will be

$$\begin{aligned} V_1 &= sL_1I_1 + sM_{12}I_2 - sM_{13}I_3 \\ V_2 &= sM_{12}I_1 + sL_2I_2 + sM_{23}I_3 \\ V_3 &= -sM_{13}I_1 + sM_{23}I_2 + sL_3I_3 \end{aligned} \tag{9-44}$$

Notice that the sign associated with the terms involving M_{13} is negative because I_1 and I_3 are oppositely directed relative to the corresponding dots (the triangular ones). The signs of the other mutual terms are positive since the other currents, in pairs, are similarly oriented relative to their dots.

When additional elements are connected at the terminals of the transformer, a loop analysis can be carried out in the normal manner, with Eq. (9-44) inserted as the V-I relationships of the transformer, keeping in mind that the currents are the branch currents.

9.8 SUMMARY

In this chapter we extended our model of electrical circuits. By observation in the physical world, it is found that the electrical behavior of certain circuits cannot be explained in terms of a model consisting of R, L, C, and independent sources. To account for the observations, we introduced a transformer as a new component. Unlike the previous elements, this one has four terminals. We characterized the transformer by three parameters, but only the mutual inductance was something new.

We found other networks equivalent to the transformer. The most general of these consists of a cascade connection of a tee network of inductors and a new device which we introduced, called an ideal transformer. This device is characterized by a single parameter, the transformation ratio n. From one point of view, then, we can consider that our model has been augmented by the addition of the ideal transformer.

We also introduced the concept of a controlled or dependent source.

This may be either a voltage source or a current source, and it may be dependent on either a current or a voltage. Although this concept was introduced, it was not exploited. We shall have much more to say about it in the next chapter.

Problems

9-1 In Fig. 9-22(a), number the windings from 1 to 3 and label the self- and mutual-inductances with appropriate subscripts. Let each winding, in turn, be excited by a sinusoidal source. Write expressions for the voltages of the other windings. Next, reverse the winding direction of the right-hand winding and repeat.

9-2 Write loop equations for the networks shown in Fig. P9-2 with particular emphasis on the signs of the terms containing the mutual inductances. Specify whether the mutual inductances are positive or negative.

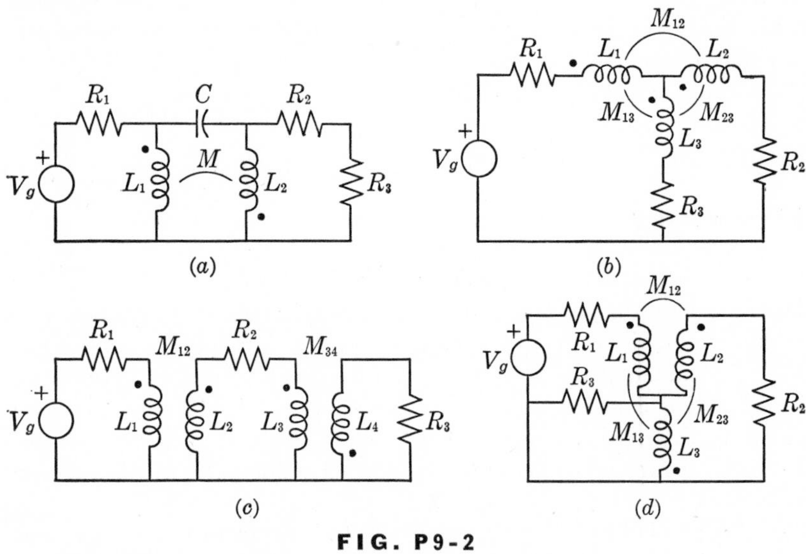

FIG. P9-2

9-3 From a consideration of Eq. (9-26), find an equivalent model of a transformer which involves voltage-controlled current sources.

9-4 Using the equivalent model shown in Fig. 9-18 in the text, compute the impedance Z when the two pairs of terminals of a transformer are connected in series, as shown in Fig. P9-4. What will this impedance become if one of the dots is reversed?

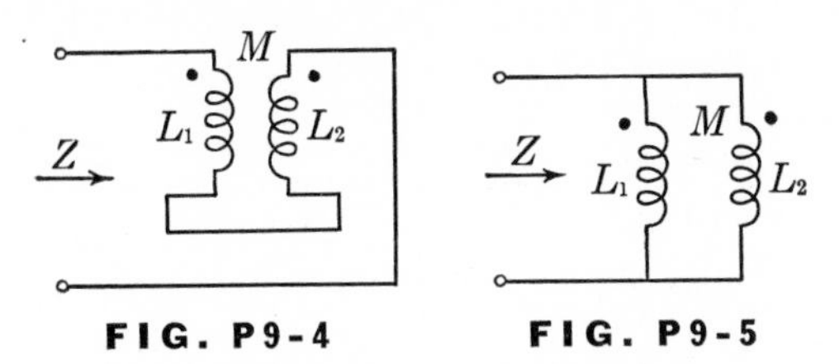

FIG. P9-4

FIG. P9-5

9-5 Using the equivalent network derived in Prob. 9-3, compute the equivalent impedance when the two pairs of terminals of a transformer are connected in parallel, as shown in Fig. P9-5. What will the impedance become if one of the dots is reversed?

9-6 In the networks of P9-6, find the input admittance and the voltage gain as functions of s. Redraw the networks, using a tee equivalent of the transformer, and again compute the input admittance by series and parallel combinations of branches.

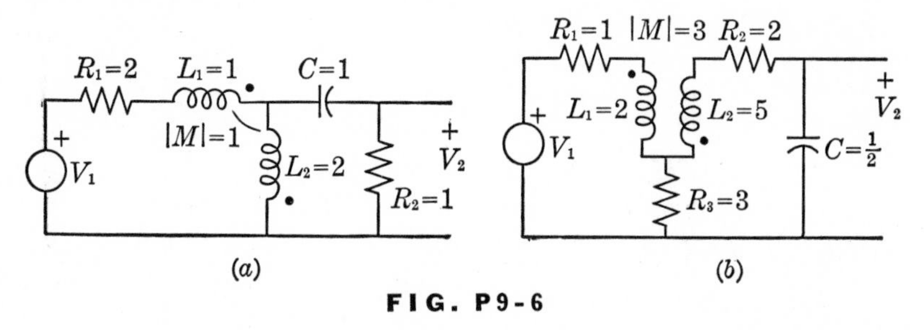

FIG. P9-6

9-7 In the network of Fig. P9-7, it is required that the input admittance have poles which lie on the real axis of the complex-frequency plane. Find the maximum permissible value of the transformation ratio n of the ideal transformer.

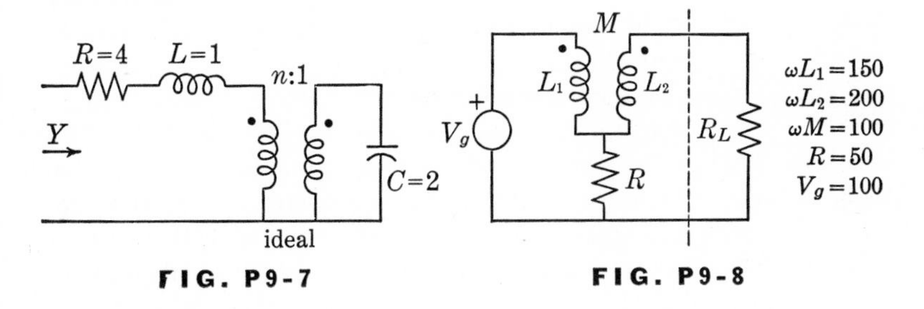

FIG. P9-7

FIG. P9-8

9-8 Find the Thévenin equivalent at the output terminals of the network shown in Fig. P9-8. Do this in literal form with the result given as functions of s. For the numerical values specified in the figure, find the value of R_L for which it will dissipate maximum power. Find this maximum power.

9-9 In the V-I relationships of a transformer, use the transformations $V_1 = nV_1'$ and $I_1 = I_1'/n$. From the resulting equations construct an equivalent model of a transformer and compare with Fig. 9-14 in the text.

9-10 For the special values (*a*) $n = L_1/M$ and (*b*) $n = M/L_2$, find the forms to which the circuit of Prob. 9-9 reduces. (Note the result for a perfect transformer, $k = 1$.) Repeat for the circuit of Fig. 9-14 in the text.

9-11 It is required to determine the inductances (self and mutual) of a physical transformer by measurements made at the terminals. Assume that the winding resistances can be neglected. The effect of the interwinding capacitances is to be accounted for by a capacitance C_x across the primary terminals. By utilizing the equivalent circuits found in Prob. 9-10, determine L_1, L_2, M, and C_x. For purposes of this experiment, there are available a variable-frequency sinusoidal generator, a voltmeter, and two capacitors C_1 and C_2.

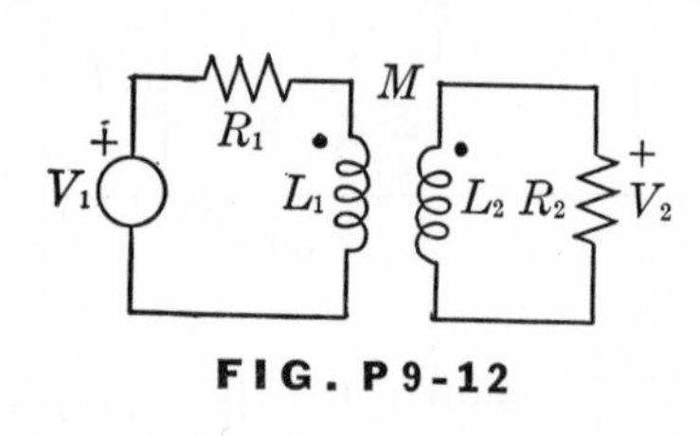

FIG. P9-12

9-12 In Fig. P9-12 the transformer has a coupling coefficient of 0.9995. The inductances are related by $R_2/L_2 = R_1/L_1$. Find the ratio of the two half-power frequencies of the voltage-gain function. If $R_2/L_2 = 50\pi$, find the two half-power frequencies.

9-13 Figure P9-13 shows the Thévenin equivalent of an amplifier which is to drive a loud-speaker. The loud-speaker is represented by a resistance $R = 16$. It is desired to have the maximum power delivered to the loud-speaker by using a transformer as shown. (*a*) Assume that the transformer is ideal and find the required turns ratio. (*b*) Assume that the transformer is not ideal but has a coupling coefficient of unity. Find the smallest permissible value that the primary reactance should have in order that the magnitude of the impedance load at the terminals of the amplifier should be at least 90 per cent of 900. If the amplifier is to operate over a frequency range of 200 to 3,000 radians per second, what is the smallest permissible value of the primary inductance? (*c*) With the value of inductance found in part (*b*), find the power delivered to the load at 200 radians per second and compare with the power delivered if the transformer were ideal.

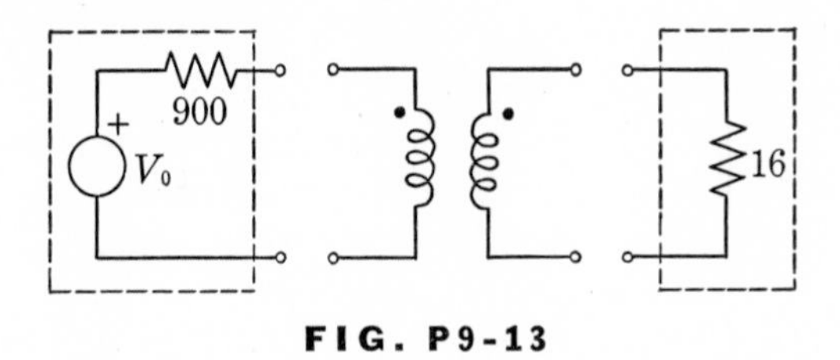

FIG. P9-13

9-14 In Fig. P9-8 find the open-circuit impedance parameters of the network, excluding the source and R_L, using the definitions of these parameters. Replace the transformer by a tee equivalent and repeat, using the z parameters of a tee.

9-15 Two identical transformers are connected as shown in Fig. P9-15. There is no coupling from one transformer to the other. Find the open-circuit impedance and short-circuit admittance parameters. From these, draw a tee equivalent and a pi equivalent.

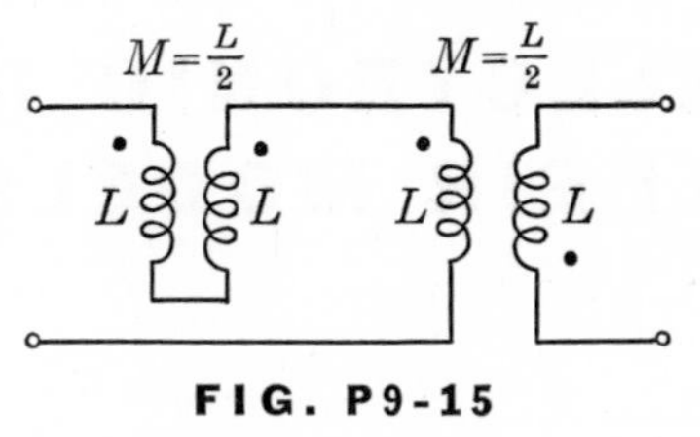

FIG. P9-15

9-16 Letting I_1 and I_2 be the primary and secondary current sinors of a transformer, prove that the average energy stored is

$$W_L = \frac{1}{2}(L_1I_1I_1^* + MI_1I_2^* + MI_1^*I_2 + L_2I_2I_2^*)$$

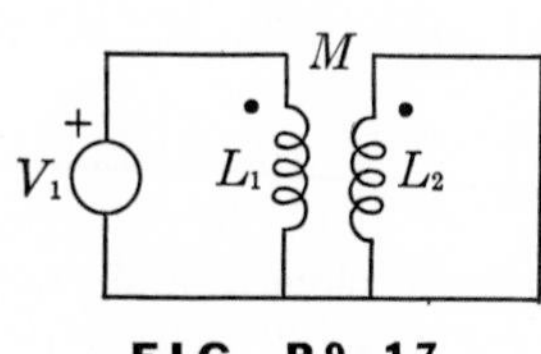

FIG. P9-17

9-17 The secondary of the transformer in Fig. P9-17 is shorted. (*a*) By computing the average energy stored, prove that the coupling coefficient is no greater than unity. (*b*) Replace the transformer by a tee equivalent and compute the equivalent inductance looking into the primary terminals. From this, show that $k \geq 1$.

9-18 Consider a transformer as a two-port. Find its *ABCD* matrix in terms of the primary inductance L_1, the transformation ratio n, and the coefficient of coupling k. Repeat if the transformer is perfect. Repeat if the transformer is ideal.

10 ELECTRONIC DEVICES AND THEIR MODELS

THE last chapter was devoted to the introduction of a new component into our model of electric circuits, the transformer. We are now ready to study the behavior of certain other physical devices called *electron tubes* and *transistors*.

The procedure which we have been employing in building a model of electrical circuits is now familiar. First we observe the relationship between the voltage and current of certain physical devices. From these empirical results, we postulate a "law" relating the variables. The law involves one parameter (such as R of a resistor) or more (such as L_1, L_2, M of a transformer) which characterize the model of the physical device, which we call the network elements. We shall again follow this approach. The physical devices which we shall now discuss, however, are relatively complicated, compared with, say, a pair of conducting plates which are placed side by side to form a physical capacitor. Hence we must devote somewhat more time to a discussion of the physical observations and the steps required in arriving at a model.

10.1 ELECTRON TUBES

A *thermionic vacuum tube* consists of two or more conducting electrodes inserted in an evacuated tube. One of the electrodes, called the *cathode*, is heated, and thereby emits electrons. By placing appropriate voltages at the electrodes, the electrons can be made to flow from the cathode to one or more of the other electrodes, thus constituting a current. We shall concern ourselves primarily with a vacuum tube having three electrodes, called a *triode*. A diagram representing a triode is shown in Fig. 10-1(a). In addition to the cathode, labeled K, there are two electrodes: the *plate* P (also called the *anode*) and the *grid* G.

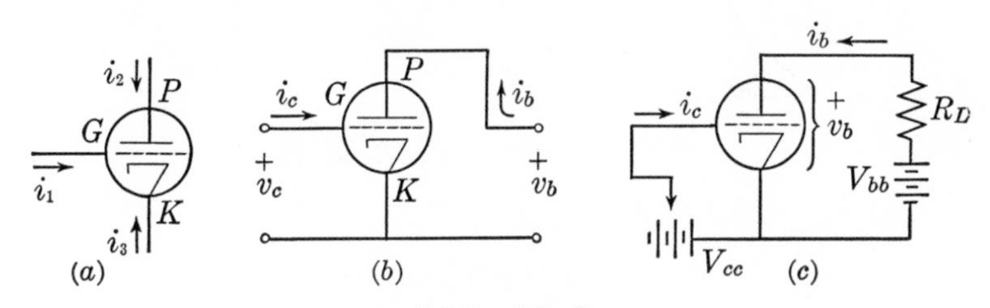

FIG. 10-1
A vacuum triode

In this three-terminal device there are three currents and three voltages. Application of Kirchhoff's current law to Fig. 10-1(a), however, leads to

$$i_1 + i_2 + i_3 = 0 \tag{10-1}$$

Thus, if two of the three currents are known, the third is fixed by this expression. Hence only two currents are independent. Similarly, the application of Kirchhoff's voltage law shows that only two of the three voltages are independent.

It is possible to consider the tube as a two-terminal-pair device, a two-port, by making one of the terminals common between the two ports. In Fig. 10-1(b) the cathode terminal is chosen to be the common terminal. The voltage of the plate relative to that of the cathode is labeled v_b and the voltage of the grid relative to that of the cathode is labeled v_c. Similar subscripts are used for the grid current and the plate current.

The relationships among the voltages and currents are found experimentally by an arrangement such as that shown in Fig. 10-1(c). If v_c is held constant, a curve of i_b against v_b can be obtained. This procedure can be repeated for other fixed values of v_c. The result is a family of curves of plate current i_b against plate voltage v_b with grid voltage v_c as a parameter, as shown in Fig. 10-2(a). This family of curves is called the *static plate characteristic*. The word static refers to the fact that the

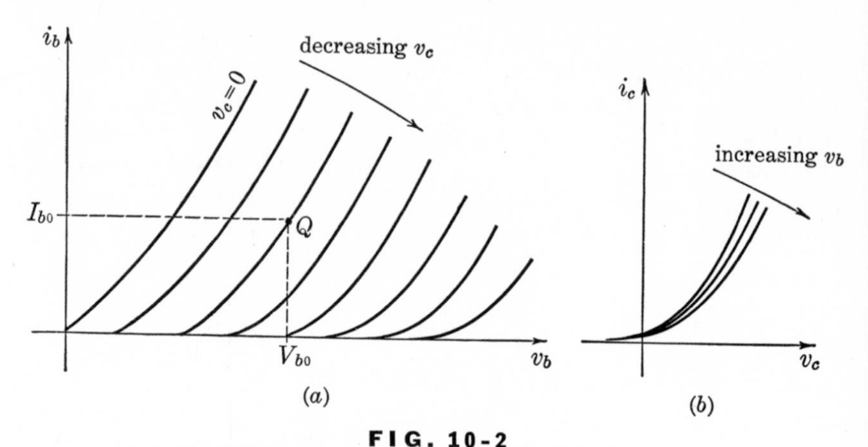

FIG. 10-2

Static plate and grid characteristics

measurements are made with direct voltages. Similarly, by holding the plate voltage constant at certain values and varying the grid voltage, curves of i_c versus v_c can be obtained. Such a family of curves, called the *grid characteristic*, is shown in Fig. 10-2(b). We see that grid current is almost zero for negative values of grid voltage and that it is almost independent of plate voltage when v_c is positive. We shall make the assumption that the grid current is identically zero for negative values of grid voltage. We shall further assume that the grid voltage is never driven positive, so we shall henceforth disregard the grid characteristic.

Instead of plotting the data in the manner shown in Fig. 10-2, it is also possible to plot i_b against v_c with v_b as parameter. This does not give any additional data but may be convenient for some purposes. For our purposes the plate characteristic is sufficient. In analytical form the plate characteristic can be expressed as

$$i_b = f(v_b, v_c) \tag{10-2}$$

where f is the function which is plotted in Fig. 10-2(a).

Suppose the batteries labeled V_{cc} (called the *grid bias*) and V_{bb} (called the *plate supply*) are adjusted so that the current has the value corresponding to the point Q in Fig. 10-2(a). This point is called the *operating point* or the *quiescent point*. Let the values of i_b, v_b, and v_c at this point be labeled I_{b0}, V_{b0}, and V_{c0}, respectively. These are the *quiescent values.* Now suppose that the plate voltage and the grid voltage take on small increments Δv_b and Δv_c, respectively. The corresponding value of i_b can be found by expressing the functional relationship in Eq. (10-2) in terms of a Taylor's series about the point Q.

In the case of a function of a single variable $y = f(x)$, the Taylor's series about the point (y_0, x_0) has the form

$$y = y_0 + y'(x_0)\Delta x + \frac{1}{2!} y''(x_0)(\Delta x)^2 + \cdots \tag{10-3}$$

where the primes indicate differentiation.

In the case at hand, i_b is a function of two variables, and the variations of both variables must be taken into account. The derivatives that appear in the Taylor's series must now be partial derivatives. (See any calculus book.) The series has the form

$$i_b = I_{b0} + \left.\frac{\partial i_b}{\partial v_b}\right|_Q \Delta v_b + \left.\frac{\partial i_b}{\partial v_c}\right|_Q \Delta v_c + \text{higher-order terms} \tag{10-4}$$

Note that the partial derivatives are evaluated at the operating point.

If the curves in Fig. 10-2 had been straight lines, then the higher-power terms in this expression would have vanished, because the coefficients of the terms are proportional to higher derivatives of the function. Even though the curves are not linear over their entire range, nevertheless, if the increments Δv_b and Δv_c are small, then the curves are approximately straight lines over this small range. In such a case the higher-power terms can be neglected, and Eq. (10-4) will become

$$\Delta i_b = \left.\frac{\partial i_b}{\partial v_b}\right|_Q \Delta v_b + \left.\frac{\partial i_b}{\partial v_c}\right|_Q \Delta v_c \tag{10-5}$$

where $i_b - I_{b0}$ has been renamed Δi_b.

The partial derivatives appearing in this expression are evaluated at the operating point, and are thus constant. Dimensionally, each of them is a conductance. By long-standing custom the following symbols are used for these quantities:

$$r_p \equiv \frac{\partial v_b}{\partial i_b} \qquad (a)$$

$$g_m \equiv \frac{\partial i_b}{\partial v_c} \qquad (b) \tag{10-6}$$

r_p is called the *plate resistance* and g_m is called the *transconductance* (or the *mutual conductance*). From the definition it is clear that r_p is the slope of the static characteristic at the operating point.

Let us now introduce some notation which will avoid using the Δ symbols in Eq. (10-5). Let

$$\begin{aligned} i_p &\equiv \Delta i_b && (a) \\ v_p &\equiv \Delta v_b && (b) \\ v_g &\equiv \Delta v_c && (c) \end{aligned} \tag{10-7}$$

That is, i_p, v_p, and v_g are *incremental* quantities measured from the quiescent values I_{b0}, V_{b0}, and V_{c0}, respectively. With these symbols Eq. (10-5) becomes

$$i_p = \frac{1}{r_p} v_p + g_m v_g \tag{10-8}$$

or

$$v_p = -\mu v_g + r_p i_p \tag{10-9}$$

where the symbol μ is used to replace the product $r_p g_m$. From the definitions of r_p and g_m, we find that

$$\mu \equiv \frac{\partial v_b}{\partial v_c} \tag{10-10}$$

This expression leads to the name *amplification factor* for μ.

Equation (10-8) or (10-9) constitutes the *V-I* relationship of the device we have been discussing. Note that the variables are not the total voltages and currents but simply the variations from the quiescent values. The quiescent values are fixed by the values of the polarizing batteries V_{bb} and V_{cc} and by the load resistance R_L in Fig. 10-1.

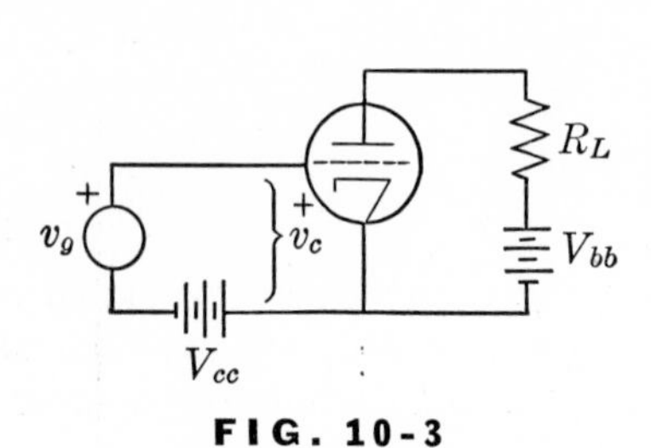

FIG. 10-3

Signal source in a tube circuit

Consider the circuit shown in Fig. 10-3. A voltage source v_g is placed in series with the bias battery V_{cc}. The total grid voltage is the sum of v_g and $-V_{cc}$, and, hence, v_g is the variation in the total grid voltage. We think of it as a *signal voltage*. Similarly, i_p and v_p are the *signal components* of the plate current and plate voltage. To complete the description we should also define i_g, the incremental grid current, but, according to our assumption, this will always be zero. Henceforth, we shall be dealing with the incremental variables almost always. When we say "plate current" or "grid voltage," etc., we shall mean the incremental values.

Now that we have the *V-I* relationship, let us try to find a network model consisting of interconnections of elements, which has this same equation as its V-I relationship. If we find such a model, it will be an "equivalent circuit" of the physical device. We can look upon Eq. (10-8) as expressing the balance of currents at a node. It appears that there are three branches connected at the node. One of these branches carries a current i_p; the second branch is a resistor r_p with a voltage v_p across it. The third branch carries a current $g_m v_g$ which is dependent on the voltage v_g. We must also keep in mind that, in addition to Eq. (10-8), we have the equation $i_g = 0$. This implies an open circuit between grid and cathode. With these considerations we see that the network shown in Fig. 10-4(a) has the V-I relationships in Eq. (10-8) and

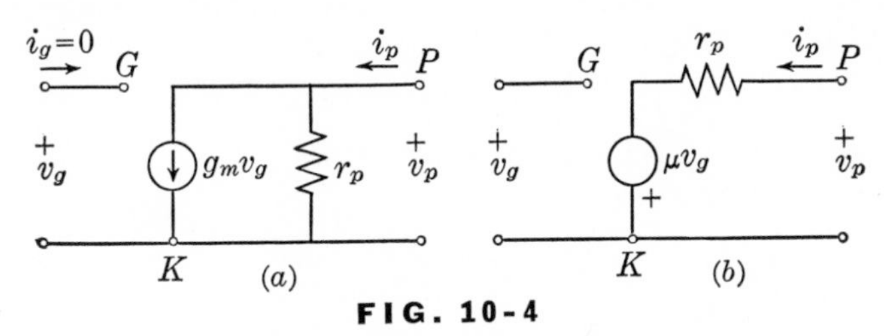

FIG. 10-4

Model of a vacuum triode

also satisfies $i_g = 0$. The current source which appears is a controlled source, controlled by the grid voltage v_g. The dependence of this current on the controlling voltage is relatively simple: it is directly proportional to it with a proportionality constant g_m. Note carefully its reference direction.

Let us now turn to Eq. (10-9), which is simply a rewriting of the previous equation. A network which has this equation as its V-I relationship is shown in Fig. 10-4(b). This time, we find a controlled voltage source, the controlling function again being a direct proportionality. Again note the reference polarity.

Let us now examine what we have accomplished. We started with experimental observations of the terminal voltages and currents of a physical device. These observations led us to the V-I relationships in Eq. (10-8) or (10-9). Note that these relationships are valid from two different points of view. In the first place, they would be correct if the curves in Fig. 10-2 were straight lines. Alternatively, we can assume that the incremental changes in the variables are small, and for small variations any curve is approximately straight. Based on this assumption, Eq. (10-8) or (10-9) is called a *small-signal* equation. Also, the networks in Fig. 10-4 are called the *small-signal equivalent* networks of a vacuum triode.

We have obtained a model of a vacuum triode which consists of a combination of a resistor and a controlled source. Note that this model is valid only under certain specific conditions. In the first place, operating conditions are to be adjusted so that the total grid voltage is not allowed to become positive. Second, the input signal should be small enough so that the departure of the plate characteristic from linearity is negligible.

There is another condition implicit in the procedure we followed which is not at once obvious. Since we have dealt with the static characteristic, we should not expect our conclusions to be valid if the input signal is a rapidly varying one. For rapid signal variations, other effects come into play which are not accounted for by the static characteristic. Thus there will inevitably be some capacitance between the electrodes, since these form a set of conductors in close proximity. For rapid signal variations—for example, high-frequency sinusoids—a more accurate model of a vacuum triode would take the form shown in Fig. 10-5. The capacitances are called the *interelectrode capacitances*. The subscripts used are standard. This network can be called the *high-frequency small-signal equivalent* as distinguished from the previous *low-frequency equivalent*.

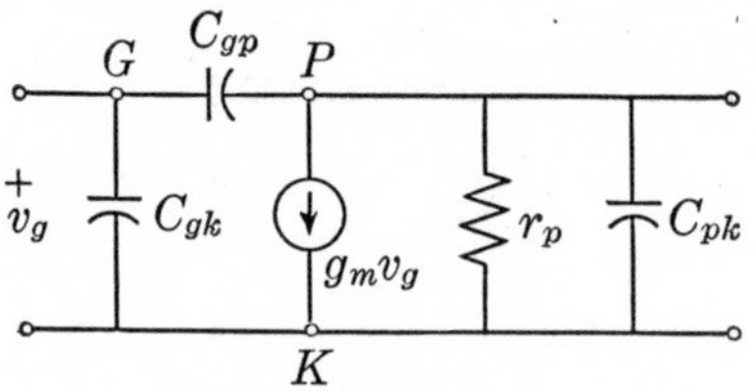

FIG. 10-5

More general model of a vacuum triode

10.2 IDEAL AMPLIFIERS

The plate resistance and the interelectrode capacitances, although important in the behavior of an electron tube, do not constitute its fundamental distinctive characteristic.

The feature which distinguishes the small-signal equivalent of the tube from other networks is the controlled source. Since we have already introduced controlled sources as elements in our model, it does not appear that anything new is added here. In the present case, however, the functional dependence of the source voltage or current on the controlling voltage has a special, albeit simple, form.

If we remove the plate resistance and the interelectrode capacitances, the model will contain only the controlled source. Since there is no current at the input, there will be no input power. Yet, if a load is connected between P and K, power will be absorbed by it. On the basis of this discussion, we shall make the following definition: An *ideal amplifier* is a two-port having the properties that (1) the ratio of the available average power output to power input is infinite and (2) either the output voltage or the output current is directly proportional to either the input voltage or the input current, the proportionality constant being independent of the load.

The definition is made in terms of available power output, because the actual power output depends on the load. Thus, if the load consists of a capacitor, the average power output will be zero, but, potentially, power will be available.

Figure 10-6 shows two types of ideal amplifiers. Both are voltage-controlled sources. In the first case the output voltage is K times the

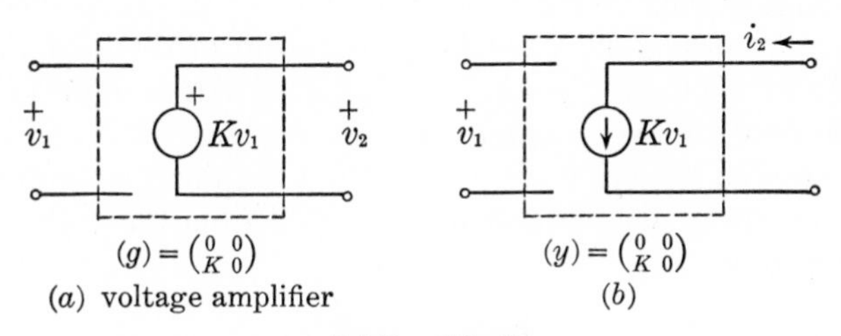

FIG. 10-6

Ideal amplifiers

input voltage, independent of the load. This is called an *ideal voltage amplifier*. In either case the input current, and hence the input power, is zero. Yet there is potentially a nonzero output power. Hence the ratio of output power to input power, called the *power amplification*, is infinite.

Two other types of ideal amplifiers are possible, as shown in Fig. (10-7). Both are current-controlled sources. In the first case the output current is K times the input current, independent of the load. It is called an *ideal current amplifier*.

The ideal amplifiers which have been introduced are two-ports. Hence their behavior can be described by one of the sets of parameters discussed in Chap. 8. Each of the ideal amplifiers is best described by a different set of parameters; the appropriate one is given in the corresponding figure. Note that in each case only one of the four parameters is nonzero, the one expressing the forward-transfer function: g_{21}, y_{21}, h_{21},

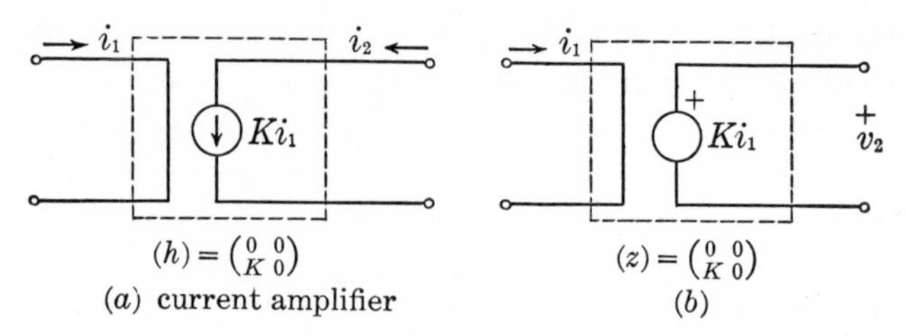

(a) current amplifier (b)

FIG. 10-7

More ideal amplifiers

or z_{21}. In each case we see that the forward-transfer function is not equal to the corresponding reverse-transfer function; that is, $z_{21} \neq z_{12}$, $y_{21} \neq y_{12}$, etc. This is a very significant observation. It means that an ideal amplifier is a nonbilateral device; that is, the reciprocity theorem is no longer valid when an ideal amplifier is included in a network.

An amplifier may be less than ideal in one of two ways: (1) the output voltage or current may not be independent of the load, or (2) the power amplification may be less than infinite. An example of the first case is afforded by the low-frequency small-signal model of a triode. If a resistance R_L is connected at the output terminals, as shown in Fig. 10-8(a), the output voltage can be computed from the voltage-divider relationship to be

$$v_p = \frac{-\mu R_L}{r_p + R_L} v_g \tag{10-11}$$

which is certainly dependent on the resistance R_L.

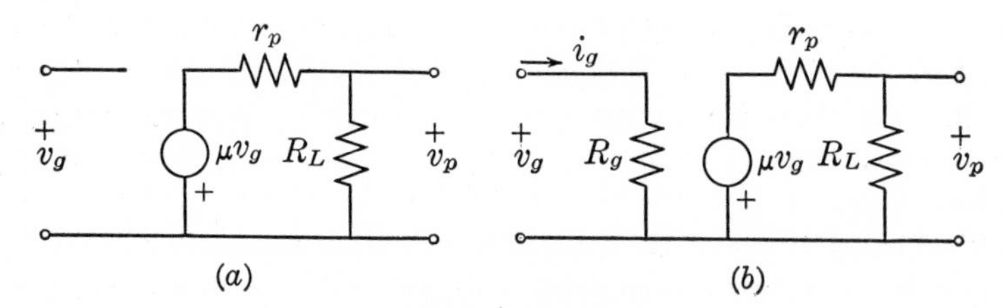

(a) (b)

FIG. 10-8

Nonideal amplifiers

In Fig. 10-8(b) a resistance R_g is placed across the input terminals, which means that the input current, and therefore the input power, will be nonzero. The amplifier will consequently be less than ideal.

The ideal amplifier now takes its place as another of our network elements. When the ideal amplifier is combined in various ways with other network elements, it is our hope that the resulting networks will be adequate models for an increasing number of electrical devices.

10.3 TRANSISTORS

Our study of the characteristics of electron tubes has led to the introduction of ideal amplifiers into our model of electrical circuits. Before we analyze the behavior of networks containing ideal amplifiers, let us continue our examination of the physical world by making observations on the external behavior of another device called a transistor.

A *transistor* is an electrical device consisting of several juxtaposed layers of semiconductor materials, such as germanium or silicon. Owing to the properties of the material, there are free-charge carriers which can be made to move from one layer to another, thus constituting a current. It is possible to control the flow of charge by externally applied voltages. We shall be concerned only with a three-layer transistor, a symbol of which is shown in Fig. 10-9(a). This symbol is not standard, and another variation is shown in Fig. 10-9(b). The three electrodes are called the *emitter* e, the *collector* c, and the *ba e* b.

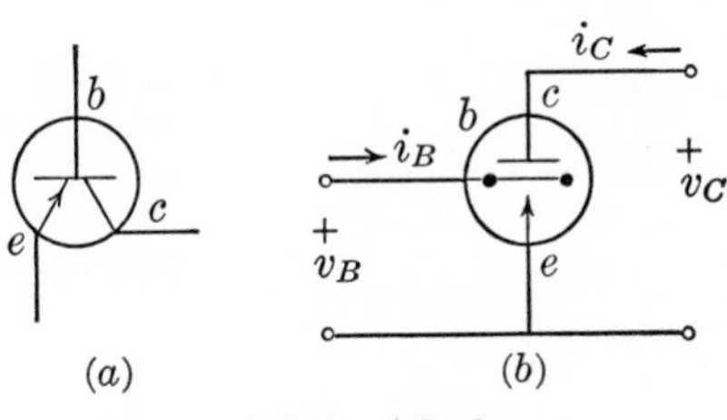

FIG. 10-9
A transistor

Since there are three terminals, there will be a total of three currents and three voltages. Only two currents and two voltages are independent, however, as in the case of the vacuum triode. When the transistor is considered as a two-port, one of the terminals is chosen to be common between input and output. The common terminal can be any one of them; in Fig. 10-9 the emitter is shown as the common terminal. The references and designations of the voltages and currents for the common-emitter connection are shown in the figure. Notice that capital letters are used as subscripts.

By means of external sources and circuits, the voltages and currents of a transistor may be varied and their interrelationships observed. The data can be presented in various forms. Of the four variables (two voltage and two current) two can be taken as independent variables and two as dependent. Plots of each of the dependent variables can be made against one of the independent variables, with the other independent variable as parameter.

Suppose we choose the two currents as independent variables. By direct measurement the curves plotted in Fig. 10-10 are obtained. In

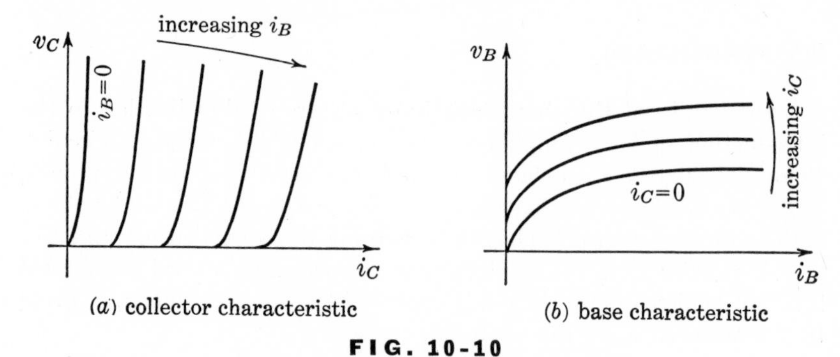

FIG. 10-10

Common-emitter static characteristics

the first set, curves of collector voltage are plotted against collector current, with the base current as parameter. This set of curves is referred to as the *collector characteristic*. In the second set, called the *base characteristic*, the base voltage is plotted against the base current with the collector current as a parameter. Two other families of curves can be plotted from the same data by interchanging the roles of the two currents as abscissa and parameter.

These families of curves can be represented analytically by the functions

$$V_B = f_1(i_C, i_B) \qquad (a)$$
$$V_C = f_2(i_C, i_B) \qquad (b) \qquad (10\text{-}12)$$

As we did in the case of the vacuum triode, we can expand these functions in Taylor's series about an operating point. If the curves were straight lines, only first-order terms would be present in the series. Alternatively, if the variations from the operating point are small, the higher-order terms can be neglected. In either case, we get

$$\Delta v_B = \frac{\partial v_B}{\partial i_B}\,\Delta i_B + \frac{\partial v_B}{\partial i_C}\,\Delta i_C \qquad (a)$$
$$\Delta v_C = \frac{\partial v_C}{\partial i_B}\,\Delta i_B + \frac{\partial v_C}{\partial i_C}\,\Delta i_C \qquad (b) \qquad (10\text{-}13)$$

In these expressions the partial derivatives are evaluated at the operating point. They represent the slopes of the curves in Fig. 10-10 and the slopes of the two other families of curves which were mentioned but not shown. Dimensionally, they are resistances. Let us define

$$r_{11} = \frac{\partial v_B}{\partial i_B} \qquad r_{12} = \frac{\partial v_B}{\partial i_C}$$
$$r_{21} = \frac{\partial v_C}{\partial i_B} \qquad r_{22} = \frac{\partial v_C}{\partial i_C} \tag{10-14}$$

In order to avoid the use of the incremental symbol Δ, let us also define the incremental variables

$$i_b = \Delta i_B \qquad v_b = \Delta v_B$$
$$i_c = \Delta i_C \qquad v_c = \Delta v_C \tag{10-15}$$

That is, the lower-case subscripts refer to the changes in the variables from the quiescent values. With this notation, Eq. (10-13) becomes

$$v_b = r_{11} i_b + r_{12} i_c \qquad (a)$$
$$v_c = r_{21} i_b + r_{22} i_c \qquad (b) \tag{10-16}$$

These equations constitute the *small-signal V-I* relationships of a transistor. In form, the equations are identical with the external equations of a two-port written in terms of the open-circuit impedance parameters, except that in the present case the z's are all real. In addition, the variables are not sinors but functions of time. These equations are to be compared with the small-signal equation of a triode given in Eq. (10-8) or (10-9). Note that the base current in the present case, which is analogous to the grid current in the triode, is not zero. In the case of the troide, we found two of the short-circuit parameters, y_{11} and y_{12}, to be zero. This is not the case here.

Let us now attempt to find a network consisting of elements in our model which has these equations as its *V-I* relationships. We might be tempted to look for a tee or a pi network. We know, however, that such a network is bilateral; that is, $z_{12} = z_{21}$. In the present case, by actually measuring the slopes of the characteristic curves, it is found that r_{12} is not equal to r_{21}. Hence, any "equivalent" network, consisting of bilateral elements only, is not possible.

One possible way out of the dilemma is to fall back on a model which includes two controlled sources, as we did in Fig. 9-18 for the transformer. Such a network is shown in Fig. 10-11. This network does, indeed, have Eq. (10-16) as its *V-I* relationship, and, at times, it can serve as a useful model of a transistor. Too great a burden, however, is placed on the dependent sources.

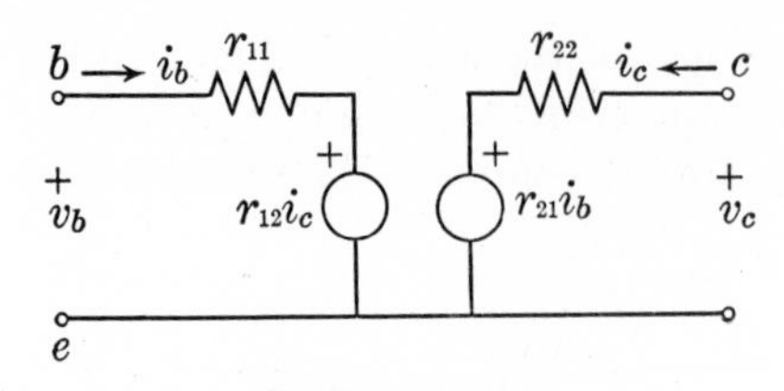

FIG. 10-11

Two-generator small-signal equivalent of a transistor

An alternative network can be obtained in the following way: Since the nonequality of r_{12} and r_{21} causes some difficulties, let us manipulate Eq. (10-16) in such a way that this difficulty is avoided. Possibly we can construct an equivalent in which the bilateral part is separated from the nonbilateral part. In the first part of Eq. (10-16), let us add and subtract the term $r_{12}i_b$. Also, in the second equation, let us add and subtract the term $r_{12}(i_b + i_c)$. After regrouping the terms, Eq. (10-16) will become

$$v_b = (r_{11} - r_{12})i_b + r_{12}(i_b + i_c) \qquad (a)$$
$$v_c = r_{12}(i_b + i_c) + (r_{22} - r_{12})i_c + (r_{21} - r_{12})i_b \qquad (b) \qquad (10\text{-}17)$$

(Show to yourself that these reduce to the previous ones.) Note that, if $r_{12} = r_{21}$, the last term in the second equation will vanish, and the resulting equations will be those of a tee network. When $r_{21} \neq r_{12}$, this last term can be represented as a dependent source in series with the output branch of the tee. Thus the network in Fig. 10-12(a) will have

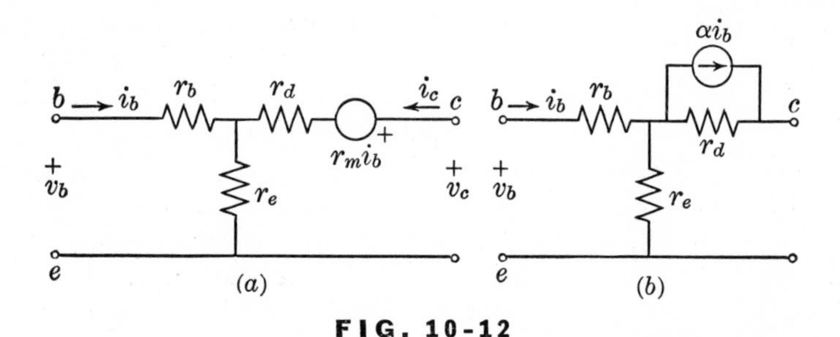

FIG. 10-12

Low-frequency small-signal equivalent of a transistor

these equations as its V-I relationships. The resistances shown in the diagram are defined as

$$\begin{aligned} r_b &= r_{11} - r_{12} \\ r_e &= r_{12} \\ r_d &= r_{22} - r_{12} \\ r_m &= r_{21} - r_{12} \end{aligned} \qquad (10\text{-}18)$$

If the controlled source in series with r_d is replaced by its current-source equivalent, the result will be as shown in Fig. 10-12(b). The parameter α is defined as

$$\alpha = \frac{r_m}{r_d} = \frac{r_{21} - r_{12}}{r_{22} - r_{12}} \tag{10-19}$$

The following typical values of the parameters of a transistor will give you an idea of the order of magnitude to be found for actual devices.

$$r_b = 400 \quad \text{ohms}$$

$$r_e = 25 \quad \text{ohms}$$

$$r_d = 6 \times 10^4 \text{ ohms}$$

$$r_m = 3 \times 10^6 \text{ ohms}$$

$$\alpha = 50$$

We have now succeeded in finding two alternative models (usually referred to as small-signal, low-frequency equivalent circuits) of a transistor. These are by no means the only possible models; many other models can be found. These variations arise in several ways. In the first place, the models in Figs. 10-11 and 10-12 themselves, or parts of them, can be converted into equivalent structures. For example, the tee of resistors in Fig. 10-12(a) can be converted to a pi. Second, the manipulations of Eq. (10-16) leading to Eq. (10-17) can be varied, thereby leading to alternative networks. In the third place, instead of choosing the currents as independent variables and expressing the voltages in terms of the currents, as in Eq. (10-12), the reverse can be done. This will lead to equations resembling the set defined by the y parameters in Chap. 8 which, in turn, will lead to additional networks.

But even these steps do not exhaust the possibilities. We can choose v_C and i_B as independent and express v_B and i_C in terms of these; or the reverse. This will lead to two other sets of equations in terms of the hybrid parameters. Finally, our development proceeded with the choice of the emitter terminal as common between the input and output ports. This is not necessary; either one of the other two can be chosen as common. These choices will lead to different models. Of course, some will have the same structure as the ones we have obtained, but the element values will be different.

Some of the various possible models which were suggested have no advantages to offer. Others are useful for various purposes, such as for design or for measurement of transistor parameters. It is not our purpose to undertake a complete study of transistor models, however, and so we shall not stress them.

As in the case of the vacuum tube, the transistor model which we have developed from the static characteristics should not be expected to be valid under all conditions of operation. For rapidly varying signals (high frequencies) the physical processes involved in the operation of the transistor introduce effects which must be represented by additional elements. The resulting models become quite extensive. Figure 10-13

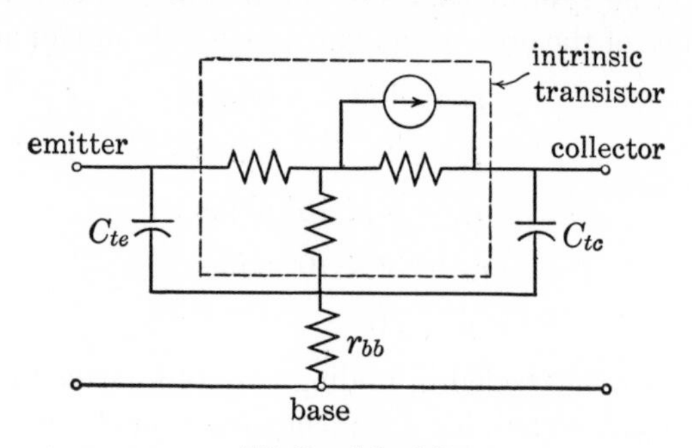

FIG. 10-13

High-frequency model of transistor

illustrates a model in which some of the more important additional elements are shown. The capacitances are called the emitter and collector *transition capacitances*, and r_{bb} is called the *base spreading resistance.* Note that this is a common-base connection. We shall not consider this topic any further here.

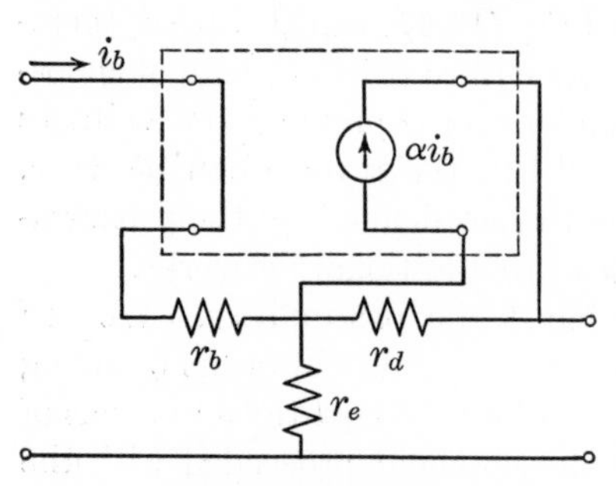

FIG. 10-14

Redrawing of transistor model

Glance back at the model of a transistor shown in Fig. 10-12(b). It consists of an interconnection of resistors and a controlled source. A redrawing of this network is shown in Fig. 10-14. This is an interconnection of a tee of resistors with an ideal current amplifier. Thus we see that a transistor can be characterized by an interconnection of elements of our model which have already been introduced; nothing new is required.

Let us now pause and look over our discussion of vacuum tubes and transistors with the purpose of obtaining a general view of electronic devices. Both devices have three terminals, and in both cases the starting point of our study consists of experimental observations of the external *v-i* relationships at the terminals. The changes of the terminal voltages and currents about the quiescent values are the *signals* which are important. In the immediate vicinity of the operating point, the curves can be assumed to be straight lines. The smaller the signals, the better this approximation. With the assumption of linear characteristics, the *v-i* relationships in terms of the signal variables become linear. The equations lead to the introduction of a new network element, the ideal amplifier. This element has two features which are different from the previous elements: (1) it is not bilateral; hence the reciprocity theorem will not be satisfied in any network in which it appears; and (2) it is active, rather than passive; the available output power is greater than the input power (by an infinite amount, by definition).

With this approach the way has been opened for the treatment of additional physical devices having any number of terminals, when and if they are invented. Their detailed characteristics may be quite different, just as those of the vacuum triode and the transistor are different from each other. Nevertheless, it seems that an interconnection of ideal amplifiers and other elements in our model will be sufficient to explain and predict the behavior of circuits incorporating such devices. If the present model proves insufficient for this task, we shall be obliged to augment it with additional hypothetical elements.

10.4 NETWORKS CONTAINING CONTROLLED SOURCES

We have now determined that certain vacuum tubes and transistors, when exposed to "small" signals, can be represented by linear models which, in addition to the usual elements, include ideal amplifiers, or controlled sources. In this section we shall examine our procedures for the analysis of networks when controlled sources are included among the types of elements present.

In this chapter we have been dealing with the voltage and current variables as arbitrary functions of time. Let us now assume that the signals are sinusoids, and let us restrict ourselves to the steady-state response. All the equations in the last sections which involved the signal variables can be rewritten in terms of sinors.

As a first illustration let us consider the network shown in Fig.

10-15. (This network is a model of a single-stage vacuum-tube amplifier.) It is desired to find the transfer-voltage ratio V_2/V_g. One ap-

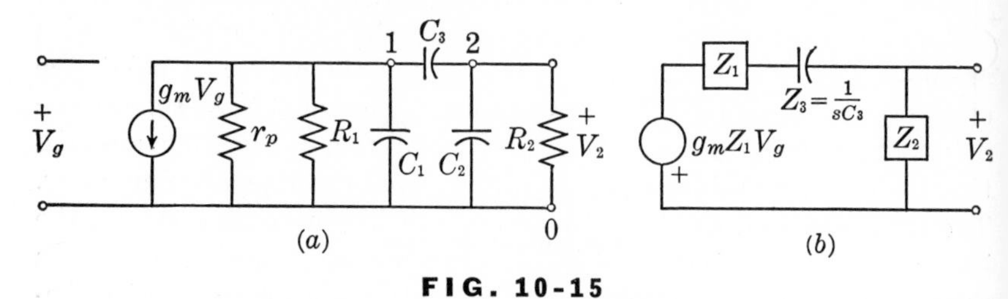

FIG. 10-15

Network with ideal amplifier

proach is illustrated in Fig. 10-15(b), where the controlled current source in parallel with the branch consisting of R_1, R_2, and C_1 in parallel is replaced by a voltage-source equivalent. In this figure

$$Z_1 = \frac{1}{sC_1 + \frac{1}{R_1} + \frac{1}{r_p}} \qquad (a)$$

$$Z_2 = \frac{1}{sC_2 + \frac{1}{R_2}} \qquad (b) \quad (10\text{-}20)$$

$$Z_3 = \frac{1}{sC_3} \qquad (c)$$

From the voltage-divider relationship the desired result is immediately written down as

$$\frac{V_2}{V_g} = \frac{-g_mZ_1Z_2}{Z_1 + Z_2 + Z_3} \qquad (10\text{-}21)$$

In this example we encountered nothing unusual in the solution of the problem owing to the presence of the controlled source.

As a second example, consider the network shown in Fig. 10-16. (This network is a model of a single-stage vacuum-tube amplifier including the grid-plate capacitance.) Again it is desired to find the transfer-voltage ratio V_2/V_g. As a preliminary step, let us replace the network to the left of nodes (1,0) by its Thévenin equivalent, as shown in Fig. 10-16(b).

We may now proceed by one of several alternative methods. One possibility is to replace the network to the left of the dashed lines in Fig. 10-16(b) by its Thévenin equivalent. Alternatively, the methods of loop equations or of node equations may be used. We shall outline

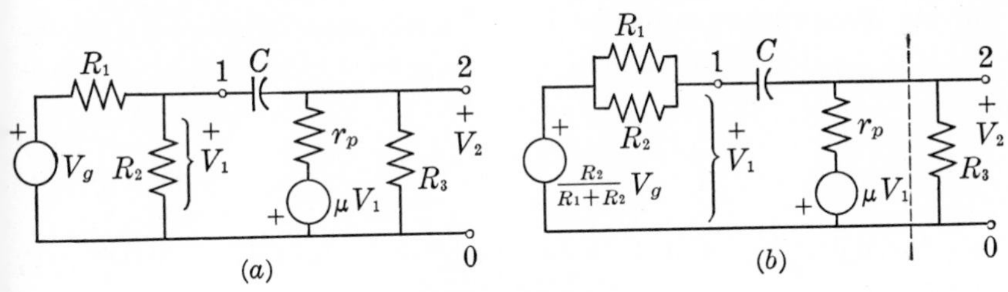

FIG. 10-16

Illustrative example

each of these procedures in order to point out the characteristic features that are introduced by the presence of controlled sources.

To start with, let us find the Thévenin equivalent of the network to the left of the dashed line. We shall do this by finding both the open-circuit voltage V_0 and the short-circuit current I_0. Look at Fig.

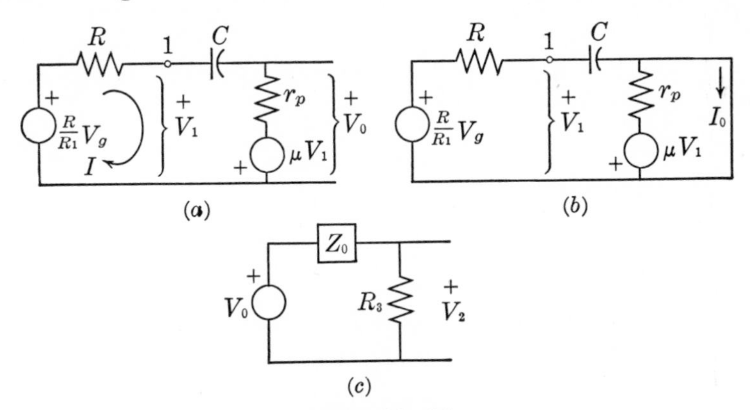

FIG. 10-17

Finding the Thévenin equivalent

10-17(a). (R had been used to represent the equivalent resistance of R_1 and R_2 in parallel.) The open-circuit voltage is given by

$$V_0 = r_p I - \mu V_1 \tag{10-22}$$

On the right side there are two unknowns, I and V_1. From Kirchhoff's voltage law we find

$$\frac{R}{R_1} V_g + \mu V_1 = \left(R + r_p + \frac{1}{sC}\right) I \qquad (a)$$

and

$$V_1 = \frac{R}{R_1} V_g - IR \qquad (b) \tag{10-23}$$

Solving these two equations for I and V_1, we get

$$I = \frac{(\mu + 1)R/R_1}{(\mu + 1)R + r_p + 1/sC} V_g \qquad (a)$$

$$V_1 = \frac{(r_p + 1/sC)R/R_1}{(\mu + 1)R + r_p + 1/sC} V_g \qquad (b) \qquad (10.24)$$

(Go through the details.) Substituting these into Eq. (10-22), we finally get

$$V_0 = \frac{(r_p - \mu/sC)R/R_1}{(\mu + 1)R + r_p + 1/sC} V_g \tag{10-25}$$

Let us turn next to the short-circuit current I_0; Fig. 10-17(b) shows the pertinent diagram. By applying Kirchhoff's current law at the upper node, we find that I_0 consists of the sum of two terms, as follows:

$$I_0 = \frac{RV_g/R_1}{R + 1/sC} - \frac{\mu V_1}{r_p} \tag{10-26}$$

Here V_1 is still unknown, but, because of the short circuit, V_1 is now equal to the voltage across the capacitor. From the voltage-divider relationship this voltage is

$$V_1 = \frac{1/sC}{R + 1/sC} \frac{R}{R_1} V_g \tag{10-27}$$

With this value of V_1, the short-circuit current becomes

$$I_0 = \frac{R}{R_1 r_p} V_g \frac{(r_p - \mu/sC)}{R + 1/sC} \tag{10-28}$$

With V_0 and I_0 determined, the Thévenin impedance becomes

$$Z_0 = \frac{V_0}{I_0} = \frac{r_p(R + 1/sC)}{(\mu + 1)R + r_p + 1/sC} \tag{10-29}$$

Figure 10-17(c) shows the resulting network. It is now a simple matter to find the desired voltage from this network. It is

$$\frac{V_2}{V_g} = \frac{R_3}{R_3 + Z_0} \frac{(r_p - \mu/sC)R/R_1}{(\mu + 1)R + r_p + 1/sC} \tag{10-30}$$

with Z_0 given in Eq. (10-29).

Let us return now to Fig. 10-16(b) and attempt to find the Thévenin impedance directly. If we remove both sources, including the controlled source, the resulting network will consist of r_p in parallel with the series connection of R and C. Clearly, this value of Z_0 will not agree with that found in Eq. (10-29), since the latter is dependent on μ. We seem to have encountered an obstacle.

A little reflection, however, will reveal that *controlled sources should not be removed when finding the Thévenin impedance.* In order to clarify this remark, refer to Sec. 7-4. In our development of Thévenin's theorem, we used the principle of superposition. On this basis, when finding the contribution to the response of the external source which we added, we removed all internal sources. (At that time we had not yet introduced controlled sources.) If controlled sources are present, however, these sources will influence the contribution of the external source to the response; hence they should not be removed.

Let us again return to Fig. 10-16(b) and, this time, remove only the independent source, retaining the controlled source. With a controlled source present, it is not possible to find the impedance by combining branches in series or in parallel. We must fall back on the definition of impedance as a ratio of voltage to current sinors. Hence we connect a voltage source at the terminals, solve for the resulting current, and then take the ratio. Alternatively, we can connect a current source at the terminals, solve for the voltage, then take the ratio.

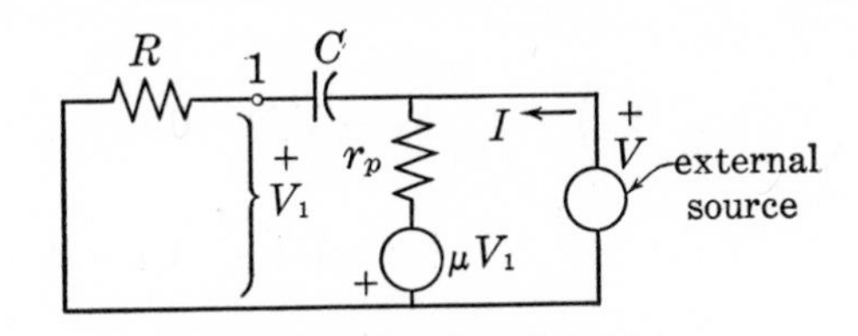

FIG. 10-18

Finding the Thévenin impedance

Let us follow the first plan, as shown in Fig. 10-18. By Kirchhoff's current law, I is the sum of the currents flowing in the other two branches. Thus

$$I = \frac{V}{R + 1/sC} + \frac{V + \mu V_1}{r_p} V \tag{10-31}$$

In this expression V_1 is still unknown. From the voltage-divider relationship we find it to be

$$V_1 = \frac{R}{R + 1/sC} V \tag{10-32}$$

Finally, with this value of V_1, the current becomes

$$I = \frac{(\mu + 1)R + r_p + 1/sC}{r_p(R + 1/sC)} V \tag{10-33}$$

Hence the Thévenin impedance will be

$$Z_0 = \frac{V}{I} = \frac{r_p(R + 1/sC)}{(\mu + 1)R + r_p + 1/sC} \tag{10-34}$$

This agrees with the previously found value in Eq. (10-29).

Perhaps the result which has been demonstrated here bears repeating. When applying the Thévenin or Norton theorems to networks containing controlled sources, "remove the sources" must be interpreted to mean, "remove the independent sources only; leave the dependent sources alone."

Let us now return to the original network in Fig. 10-16 and, this time, use the method of node equations to obtain a solution for the output voltage. Even if we use the network in the (*b*) part of the figure, we must still count the node labeled 1 as a separate node. We cannot combine R and C into a single branch, because, if we did, the voltage V_1 would be lost. But we should retain this voltage as one of the variables because the controlled source is dependent on V_1. Taking node 0 as datum, the node equations will be

$$\frac{1}{R}\left(V_1 - \frac{R}{R_1}V_g\right) + Cs(V_1 - V_2) = 0 \qquad (a)$$

$$Cs(V_2 - V_1) - \frac{1}{r_p}(V_2 + \mu V_1) + \frac{1}{R_3}V_2 = 0 \qquad (b) \tag{10-35}$$

After collecting terms, these equations become

$$\left(\frac{1}{R} + Cs\right)V_1 - CsV_2 = \frac{1}{R_1}V_g \qquad (a)$$

$$-\left(Cs + \frac{\mu}{r_p}\right)V_1 + \left(Cs + \frac{1}{r_p} + \frac{1}{R_3}\right)V_2 = 0 \qquad (b) \tag{10-36}$$

The solution of these equations is now straightforward. Although we shall not carry out the solution, let us examine the determinant of these equations. It is

$$\Delta = \begin{vmatrix} \left(\frac{1}{R} + Cs\right) & -Cs \\ -\left(Cs + \frac{\mu}{r_p}\right) & \left(Cs + \frac{1}{r_p} + \frac{1}{R_3}\right) \end{vmatrix} \tag{10-37}$$

We notice immediately that, unlike the determinant of bilateral networks, this determinant is unsymmetrical. This is in agreement with the fact that, for an ideal amplifier, $y_{12} \neq y_{21}$.

Finally, let us consider the method of loop equations to obtain a

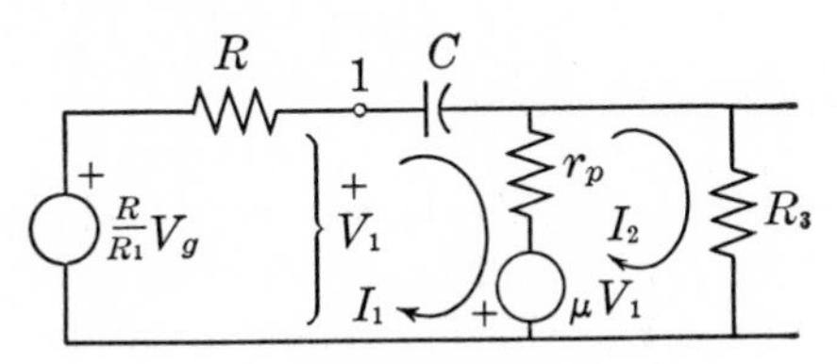

FIG. 10-19

Loop analysis of illustrative example

solution. Figure 10-16(*b*) is repeated in Fig. 10-19. There will be two independent loop equations. Choosing the windows as loops, we get

$$\left(R + \frac{1}{sC}\right) I_1 + r_p(I_1 - I_2) - \mu V_1 = \frac{R}{R_1} V_g$$
$$r_p(I_2 - I_1) + \mu V_1 + R_3 I_2 = 0 \tag{10-38}$$

Besides the two loop currents, there is a third unknown, V_1, which appears through the agency of the dependent source. Thus one additional equation is required. We already wrote such an expression in Eq. [10-23(*b*)] when finding the Thévenin equivalent. The current I in that equation is now I_1. Substituting this expression into Eq. (10-38), we get

$$\left[(\mu + 1)R + r_p + \frac{1}{sC}\right] I_1 - r_p I_2 = (\mu + 1)\frac{R}{R_1} V_g$$
$$-(r_p + \mu R)I_1 + (r_p + R)I_2 = -\mu \frac{R}{R_1} V_g \tag{10-39}$$

You can now carry out the solution in a straightforward manner.

Note again that the determinant of this set of equations is not symmetrical, in corroboration of the fact that the network is nonbilateral.

We have gone through this example in somewhat lengthy detail for the purpose of showing clearly the degree to which our analytic procedures are modified when controlled sources are present in a network. We have found that the methods of node equations and loop equations remain valid. Each controlled source, however, introduces one additional variable, namely, the voltage or current of the source. Sometimes this variable may itself be proportional to one of the other variables, in which case it is easily eliminated. (In the example under consideration, we found this to be the case when we used node analysis.) When this is not the case, an additional equation must be written for each dependent source expressing the dependence of the source voltage or current on the remaining variables. These equations must then be combined with the loop or node equations.

10.5 OPERATIONAL AMPLIFIERS

In a previous section of this chapter we introduced the concept of an ideal amplifier as a network element. We shall now discuss the characteristics of some networks which are made up of ideal amplifiers interconnected with other network elements.

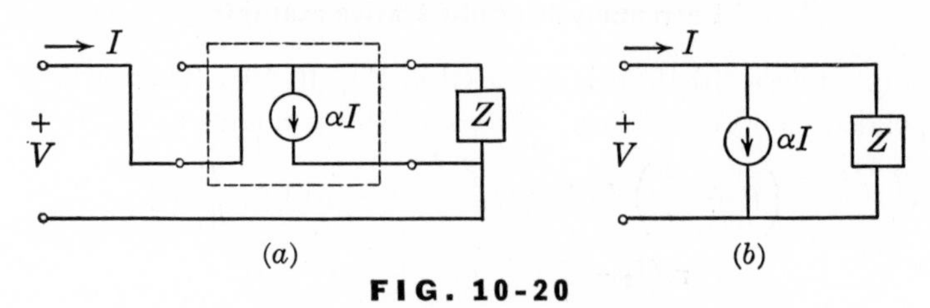

FIG. 10-20

Negative impedance converter

As a first example consider the network shown in Fig. 10-20(*a*). An impedance Z is connected to the output terminals of an ideal current amplifier in the manner shown. A redrawing is shown in part (*b*) of the figure. Kirchoff's current law shows the current in Z to be $(1 - \alpha)I$; hence the input voltage is $(1 - \alpha)ZI$. The input impedance is, therefore,

$$Z_1 = \frac{V}{I} = (1 - \alpha)Z \tag{10-40}$$

Nothing serious happens if α is a real number less than unity. If α is greater than 1, however, then the input impedance will be negative. For this reason the network is called a *negative impedance converter*. In particular, if Z is a resistance R, then at the input terminals the effect is that of a negative resistance. A negative resistance implies that power can be delivered by the network instead of being absorbed.

Although what we are discussing here is a model, it is possible to

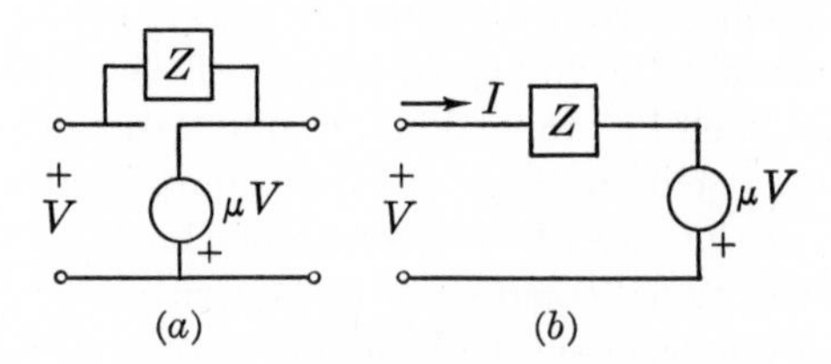

FIG. 10-21

Another impedance converter illustrating the Miller effect

build a physical circuit with electronic devices which behaves approximately as a negative impedance converter. This concept is quite useful in network synthesis.

As a second example, consider the network shown in Fig. 10-21(*a*). This time an impedance Z connects one of the input terminals of an ideal voltage amplifier to one of the output terminals, the other two terminals being connected directly. A redrawing is shown in Fig. 10-21(*b*). With a voltage V applied, the current at the terminals can be easily found. The input impedance is then found to be

$$Z_1 = \frac{V}{I} = \frac{Z}{\mu + 1} \tag{10-41}$$

Thus the input impedance is reduced by the factor $(\mu + 1)$. This network also has the property of converting an impedance into something else.

Again the network we are discussing is a model. Nevertheless, in certain vacuum-tube circuits this impedance-converting property assumes considerable importance. It is called the *Miller effect.*

We shall now turn to a particular network which, at the outset, does not appear to promise any far-reaching results. At the input terminal of the network of Fig. 10-21, we simply connect another series impedance, as shown in Fig. 10-22(*a*). In addition to being interesting, how-

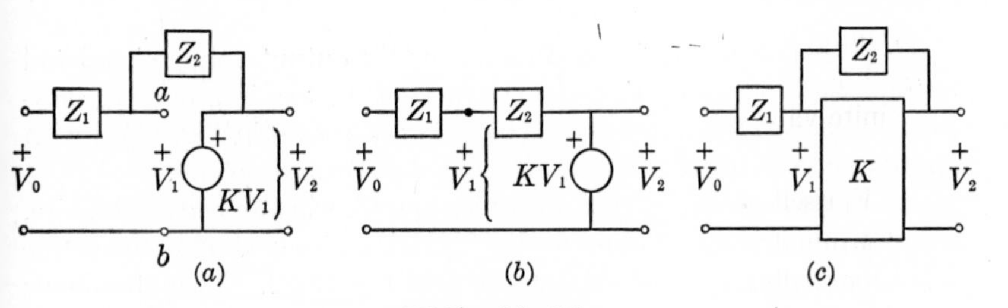

FIG. 10-22

An operational amplifier

ever, the physical counterpart of this network constitutes the major component in the electrical machine called an *analogue computer.* The network is redrawn in Fig. 10-22(*b*). It is again redrawn in Fig. 10-22(*c*) where the ideal voltage amplifier is drawn as a two-port. This diagram is fairly standard. Note that the only connection between the upper terminals of the two ports is through Z_2. Hence no current will flow into the upper terminal of the input port; the current in Z_2 will be the same as the current in Z_1.

Let us compute the output voltage V_2 in response to a voltage V_0. For this purpose we can write a voltage equation around the loop. Thus

$$V_0 = (Z_1 + Z_2)I_1 + KV_1 \tag{10-42}$$

Here there are two unknowns. We can also write another equation relating V_1 and I_1, as follows:

$$V_1 = I_1Z_2 + KV_1 \tag{10-43}$$

I_1 can be expressed in terms of V_1 from this equation and inserted into the preceding one, leaving only V_1 as unknown. If we now use the relationship $V_2 = KV_1$ and solve the resulting equation for V_2, we get

$$V_2 = \frac{V_0}{\dfrac{(1-K)}{K}\dfrac{Z_1 + Z_2}{Z_2} + 1} = \frac{-Z_2/Z_1}{1 - \dfrac{1}{K}(1 + Z_2/Z_1)} V_0 \tag{10-44}$$

This is the desired expression. The right-hand side was obtained by algebraic manipulations.

So far there is nothing very thrilling about this expression. Consider, however, the limiting form of the right-hand side as K gets larger and approaches infinity. The last term in the denominator becomes negligible compared with unity, and the expression reduces to

$$V_2 = -\frac{Z_2}{Z_1} V_0 \tag{10-45}$$

This is now a very interesting result; the output voltage is related to the input voltage by the negative ratio of the two impedances, *independent of the ideal amplifier*. The network shown in Fig. 10-22 with an infinite value of K is called an *ideal operational amplifier*.

If K is to be infinite, then V_2, which is KV_1, will also become infinite, unless V_1 itself is zero. We say that there is a *virtual short circuit* at the input terminals *a-b* of the ideal amplifier in Fig. 10-22. These terminals are not actually connected together, however, so that no current flows in the virtual short circuit.

Let us now examine the specific behavior of the operational amplifier when the impedances Z_1 and Z_2 are constituted of different elements. The simplest case will occur when both of these are resistors, as shown in Fig. 10-23(*a*). Then the output voltage sinor will be $V_2 = -aV_0$, where $a = R_2/R_1$. Thus the output voltage will be proportional to the negative of the input voltage. In particular, if the resistances are equal, V_2 will be $-V_1$. This particular network is called a *proportionor*.

Next let $Z_1 = 1/sC$ and $Z_2 = R$, as shown in Fig. 10-23(*b*). In this case Eq. (10-45) becomes

$$V_2 = -\frac{R}{1/Cs} V_0 = -RCsV_0 \tag{10-46}$$

Recall that our solutions in terms of sinors are based on the exponential waveform ϵ^{st}. Multiplication of the sinor by s implies differentiation of

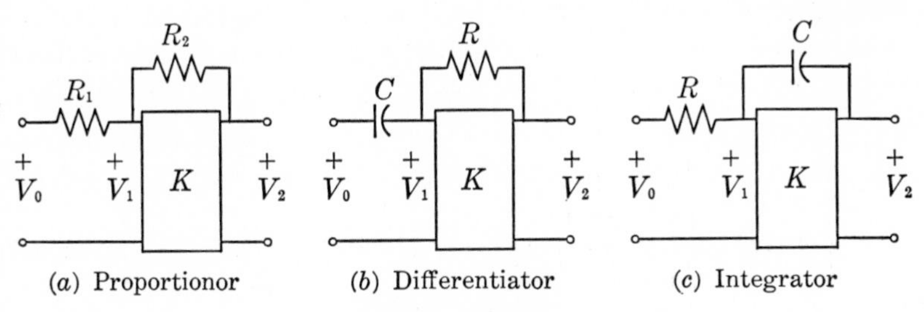

FIG. 10-23

Some specific operational amplifiers

the corresponding time functions. Hence the network in Fig. 10-23(*b*) is called a *differentiator*.

In order to gain additional familiarity with and insight into the behavior of this network, let us carry out a general analysis in terms of functions of time rather than sinors. That is to say, let us not assume that the waveforms are sinusoids. Furthermore, let us not yet assume that K is infinite. Using lower-case letters to represent functions of time as usual, we can write the following voltage equation around the input loop in Fig. 10-23(*b*).

$$v_0(t) = v_C(t) + v_1(t) \tag{10-47}$$

where $v_C(t)$ is the capacitor voltage (with an appropriate reference). The current through the resistor is

$$i = \frac{v_1 - v_2}{R} = v_1 \frac{(1 - K)}{R} \tag{10-48}$$

This is the same as the capacitor current, and so it will be given by $C\,dv_C/dt$. It is possible to solve for v_C in terms of the integral of the current. Let us instead, however, differentiate Eq. (10-47) and substitute i/C for dv_C/dt. If we then substitute Eq. (10-48) for i, the result will be

$$\frac{dv_0}{dt} = \frac{dv_C}{dt} + \frac{dv_1}{dt} = \frac{i}{C} + \frac{dv_1}{dt} = v_1 \frac{(1 - K)}{RC} + \frac{dv_1}{dt} \tag{10-49}$$

Finally, noting that $v_2 = Kv_1$, we get

$$v_2(t) = -RC\frac{dv_0}{dt} + \frac{1}{K}\left(v_2 + RC\frac{dv_2}{dt}\right) \qquad (10\text{-}50)$$

This is the desired relationship between the input and output voltages. It is a differential equation which must be solved to get an explicit expression relating v_2 to v_0. In the limit as K approaches infinity, however, the last term on the right will vanish, and the equation will reduce to

$$v_2(t) = -RC\frac{dv_0}{dt} \qquad (10\text{-}51)$$

Compare this expression with Eq. (10-46). Clearly, the latter will follow if we assume $v_1 = V_1\epsilon^{st}$ and $v_0 = V_0\epsilon^{st}$ in Eq. (10-51). In the ideal differentiator the output voltage is proportional to the derivative of the input voltage.

As the next possibility let us interchange the resistor and the capacitor, thus getting the network shown in Fig. 10-23(*c*). We shall again revert to the use of sinors. In the present case Eq. (10-45) reduces to

$$V_2 = -\frac{1}{RCs}V_1 \qquad (10\text{-}52)$$

Again thinking in terms of the exponential waveform ϵ^{st}, division of a sinor by s implies integration of the corresponding function of time. Hence the network is called an *integrator*. You are invited to carry out an analysis in the time domain assuming a finite value for K, as we did for the differentiator.

Finally, let us consider the network shown in Fig. 10-24(*a*). Here there are two inputs fed from two different sources. Since there is no current in the upper terminal of the input port of the ideal amplifier, the

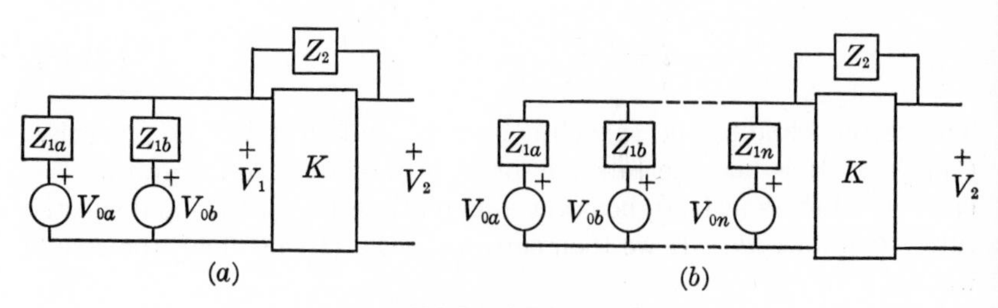

FIG. 10-24

Adder or summing amplifier

current in Z_2 will be the sum of the currents in Z_{1a} and Z_{1b}. Again assuming an infinite gain K, V_1 will be zero and V_2 will be equal to the negative voltage across Z_2. Thus

$$I = \frac{V_{0a}}{Z_{1a}} + \frac{V_{0b}}{Z_{1b}} \qquad (a)$$

$$V_2 = -Z_2 I = -\frac{Z_2}{Z_{1a}} V_{0a} - \frac{Z_2}{Z_{1b}} V_{0b} \qquad (b) \qquad (10\text{-}53)$$

We see that the output voltage consists of two terms, each of which has the same form as Eq. (10-45) for a single input.

In case all the impedances are resistors, the output voltage will consist of the negative sum of the two input voltages, each of which has had its scale changed by a constant factor. The amplifier is then called an *adder* or *summing amplifier*. For other, more general, impedances, each of the input voltages will have some mathematical operation performed on it before it is added to the other to form the output voltage.

It should be evident that the summing amplifier can be extended by connecting additional inputs, as shown in Fig. 10-24(*b*). The output voltage in this case will be

$$V_2 = -\frac{Z_2}{Z_{1a}} V_{0a} - \frac{Z_2}{Z_{1b}} V_{0b} - \cdots - \frac{Z_2}{Z_{1n}} V_{0n} \qquad (10\text{-}54)$$

Following this rather lengthy discussion, the reason for the name *operational amplifier* should now be evident. With different impedances Z_1 and Z_2, various mathematical operations are performed on the input voltage to yield the output voltage. It should also be clear that interconnecting several of these amplifiers, so that the output voltage of one is the input voltage of one or more others, will lead to a network which is capable of performing combinations of the operations of differention, integration, multiplication by a constant, addition, and sign change.

Let us now make an examination of the effects of the idealness which we have assumed. Consider Fig. 10-25(*a*). The tube with its load resistance replaces the ideal amplifier of Fig. 10-23(*c*). If the tube is replaced by its small-signal current-source model, Fig. 10-25(*b*) will result. Finally, letting the parallel combination of r_p and R_L be called R, and making a source conversion, we get Fig. 10-25(*c*). Notice that this diagram differs from the one in Fig. 10-22 by having a nonzero resistance R in series with the controlled source. The gain K has the value $-g_m R$.

Let us now compute the output voltage V_2. It will be given by

$$V_2 = RI + KV_1 \tag{10-55}$$

where we have written K for $-g_mR$. There are two unknowns on the right. An expression for I can be obtained from the V-I relationship of

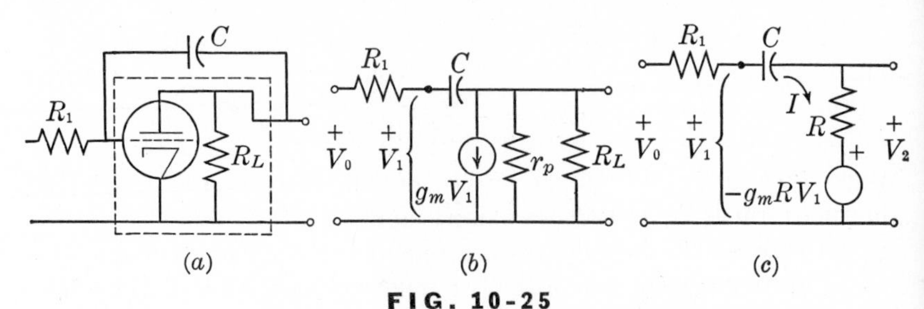

FIG. 10-25

Model of a physical operational amplifier

the resistor R_1. Another expression for I can be obtained from the V-I relationship of the branch consisting of the series connections of R and C. These expressions are

$$I = \frac{V_0 - V_1}{R_1} \quad (a)$$

$$I = \frac{V_1 - KV_1}{R + \dfrac{1}{sC}} \quad (b) \tag{10-56}$$

If these two expressions for I are equated, the result can be solved for V_1, yielding

$$V_1 = \frac{(R + 1/Cs)V_0}{R + R_1(1 - K) + 1/Cs} \tag{10-57}$$

This value of V_1 can now be inserted into Eq. [10-56(a)] to give

$$I = \frac{(1 - K)V_0}{R + R_1(1 - K) + 1/Cs} \tag{10-58}$$

Finally, the last two equations are substituted into Eq. (10-55). The result will be

$$\begin{aligned} V_2 &= \frac{R(1 - K)V_0}{R + R_1(1 - K) + 1/Cs} + \frac{K(R + 1/Cs)V_0}{R + R_1(1 - K) + 1/Cs} \\ &= \frac{(R + K/Cs)V_0}{R + R_1(1 - K) + 1/Cs} \\ &= -\frac{V}{R_1Cs}\left(\frac{1 + \dfrac{RCs}{K}}{1 - \dfrac{1}{K}\left(1 + \dfrac{1}{R_1Cs} + \dfrac{R}{R_1}\right)}\right) \end{aligned} \tag{10-59}$$

This is the desired expression. The last line was obtained by (1) dividing numerator and denominator by K, (2) factoring $-R_1$ from the denominator, and (3) factoring $1/Cs$ from the numerator. You are urged to verify each of the steps in this development.

This expression should be compared with the general one in Eq. (10-44) and the one in Eq. (10-45) for the ideal operational amplifier. In the present case $Z_2/Z_1 = 1/R_1Cs$. Note that if R is allowed to approach zero, the expression reduces to Eq. (10-44), which is heartening, since the network also reduces to that of Fig. 10-23(c). It further reduces to Eq. (10-45) if $K(= -g_mR)$ is allowed to approach infinity, which implies that g_m approaches infinity.

To get some idea of how great the numerical discrepancy is between the ideal case and the case under discussion, let us assume that, in a given situation, the amplifier gain K is -100, and that $R = R_1 = 1/\omega C$. For the ideal integrator, the output-voltage sinor will be jV_0. For the operational amplifier in Fig. 10-25 (which, incidentally, is called a Miller integrator) the output voltage given in Eq. (10-59) will become

$$V_2 = jV_0\left[\frac{1 + j\dfrac{1}{100}}{1 - \dfrac{1}{100}(2 - j1)}\right] = jV_0\frac{1 + j0.01}{0.98 + j0.01} \qquad (10\text{-}60)$$

$$= jV_0(1.02\underline{/0.1^\circ})$$

Thus the phase of the output voltage is only one tenth of a degree different from that of the ideal, while the amplitude is increased by about 2 per cent. It is clear, then, that a model in which the amplifier is assumed to be ideal leads to a tolerably good approximation.

We shall terminate our discussion of operational amplifiers at this point with a few remarks. We have seen that these amplifiers can perform the operations of differentiation, integration, and linear combination. It is reasonable to expect that interconnections of these components will perform mathematical operations which are combinations of these operations. An integrodifferential equation is just such a combination. The *analogue computer* is a machine which consists of interconnections of the physical counterparts of the models we have been discussing which (among other things) is capable of solving integrodifferential equations. It is a versatile tool of research. At some point in your education you will probably learn how to use one to solve engineering problems.

10.6 SUMMARY

In this chapter we took a giant step in extending our model of electrical circuits. We left the domain of "passive" devices and entered the domain of "active" ones. The two specific physical devices which we examined were the vacuum triode and the three-element transistor. These devices differ from the previous ones for which we found models in that they have three terminals instead of two (or four for the transformer). The procedure for developing a "small-signal" model which we followed, however, is universal; that is, if additional physical devices are discovered, essentially the same approach can be taken in arriving at a model, even if a different number of terminals is involved.

To obtain a complete appreciation of the "equivalent circuits" of these physical devices, and the modifications thereof that can be made under various operating conditions, it is necessary for you to learn a great deal concerning the physical bases for their behavior. Just as it is necessary for you to study the magnetic properties of transformer cores and to assess the influence of these properties on the external behavior, so also is it necessary for you to understand the mechanism of charge flow in the vacuum tube and transistor.

Whatever the physical laws may be that cause the physical device to behave in the observed manner, however, once a model has been established based on this behavior, the problem becomes one of network analysis. Techniques and procedures which can be brought to bear on the problem of analysis will be independent of the considerations that led to the particular model at hand. In this book we are concerned with methods of analysis of the model. Somewhere else you must study the behavior of physical devices.

Problems

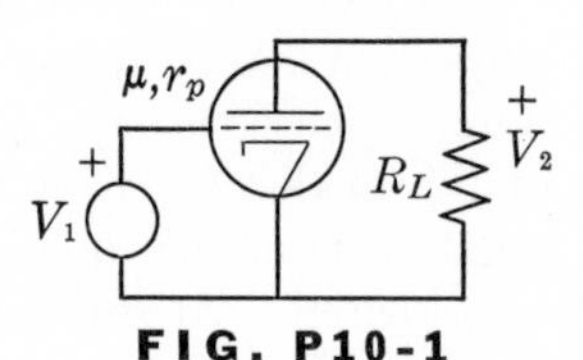

FIG. P10-1

10-1 Replace the tube in Fig. P10-1 by its small-signal equivalent and calculate the voltage-gain function. Sketch the variation of the magnitude of this function as R_L/r_p varies.

10-2 Find the short-circuit admittance parameters of the model of a triode shown in Fig. 10-5 in the text.

10-3 In the circuit of Fig. P10-3, assume that the tube can be replaced by its low-frequency small-signal equivalent. (*a*) Find the Thévenin equivalent to the left of terminals *a-b*; (*b*) Find the Thévenin equivalent at the terminals *c-b*; (*c*) If $v_1(t) = 2 \cos \omega t$, find the time function $v_k(t)$. Use the values $R_k = R_L = 2r_p$ and $\mu = 50$. (This circuit is called a phase inverter.) (*d*) Repeat the previous parts if the grid-cathode capacitance is to be taken into account. Let $C_{gk} = 1$.

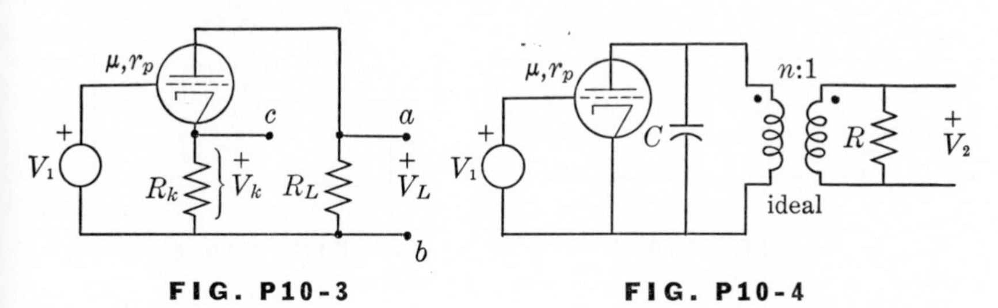

FIG. P10-3 FIG. P10-4

10-4 In the circuit of Fig. P10-4, assume that the tube can be replaced by its low-frequency small-signal equivalent. (*a*) Find an expression for the voltage gain V_2/V_1. (*b*) If $r_p = 1{,}000$ $\mu = 20$, $C = 10^{-9}$, and $R = 10$, find the value of the transformation ratio n for which the maximum power will be dissipated in R, when the frequency of the source approaches zero. (*c*) Find the frequency at which the power delivered to R is half of the maximum value. What is the angle of V_2 relative to that of V_1 at this frequency?

10-5 In the circuit of Fig. P10-5, the tubes are identical. Draw the low-frequency small-signal equivalent. Determine an expression for the output voltage V_2 in terms of V_1. Take $r_p = 2R/10$.

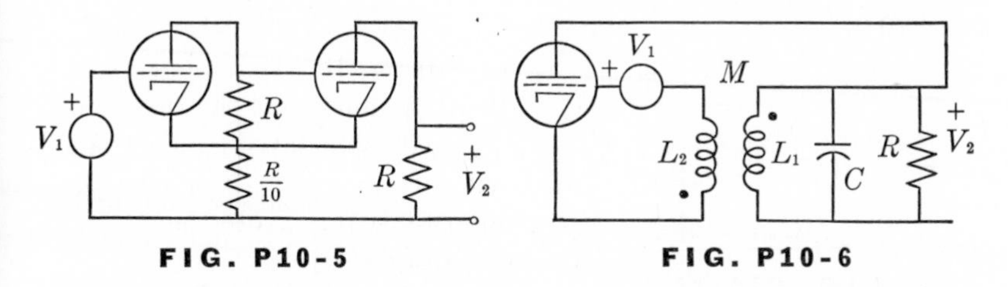

FIG. P10-5 FIG. P10-6

10-6 In the circuit of Fig. P10-6, assume that the tube can be replaced by its small-signal equivalent and that the transformer is perfect. (*a*) Draw an equivalent model replacing both the tube and the transformer with appropriate equivalents. (*b*) Calculate the voltage-gain function V_2/V_1. (*c*) Find the condition for which the poles of

V_2/V_1 will lie on the $j\omega$ axis and comment on the behavior of the circuit in this case.

10-7 Redraw the circuit of Fig. P10-7, using the small-signal equivalent of the tube. Find the voltage across R_L by first finding the Thévenin equivalent to the left of the dotted line.

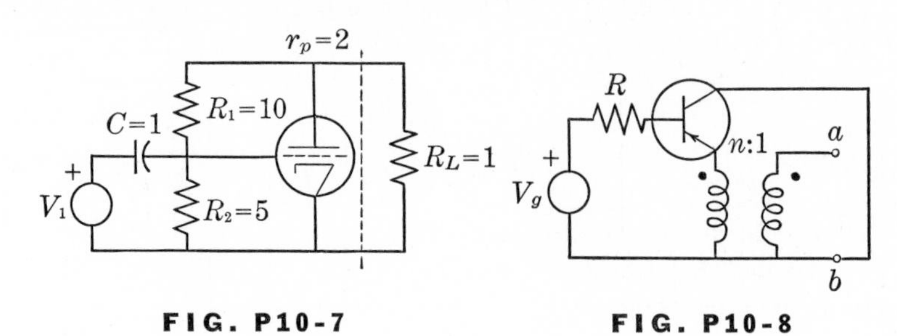

FIG. P10-7

FIG. P10-8

10-8 In the circuit of Fig. P10-8, assume that the small-signal equivalent of the transistor is valid and that the transformer is ideal. Find the Thévenin equivalent at the terminals a-b.

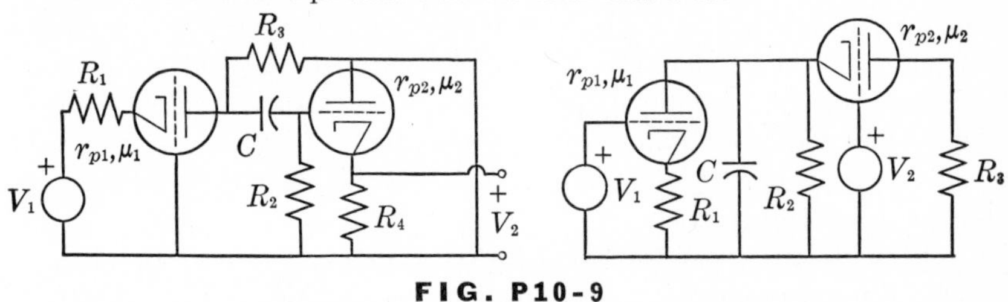

FIG. P10-9

10-9 Find the small-signal equivalents of the circuits shown in Fig. P10-9.

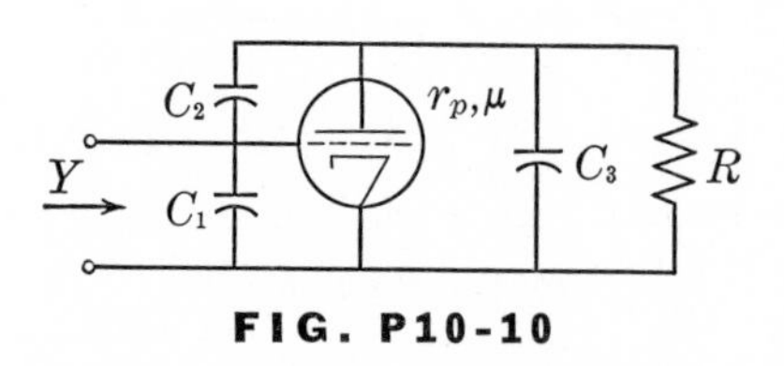

FIG. P10-10

10-10 In the circuit of Fig. 10-10, assume that the tube can be replaced by its low-frequency small-signal equivalent. Find the input admittance Y. Note the magnified effect of C_2.

10-11 Starting with the small-signal equations of a transistor in the common-emitter connection based on the z parameters, as given in the text, calculate the h parameters in terms of the z parameters. Find a small-signal equivalent circuit based on the h parameters.

10-12 Find the small-signal equivalents of the networks shown in Fig. P10-12 based on both the z parameters and the h parameters.

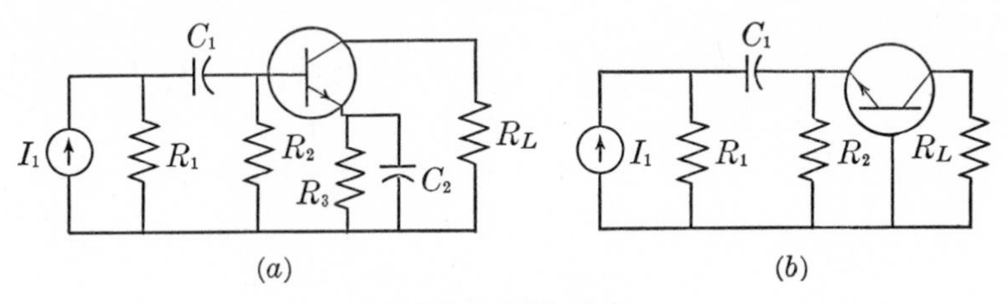

FIG. P10-12

10-13 In the network of Fig. P10-13, find the current-gain function I_2/I_1. Assume that the small-signal equivalent is valid and use the one based on the h parameters.

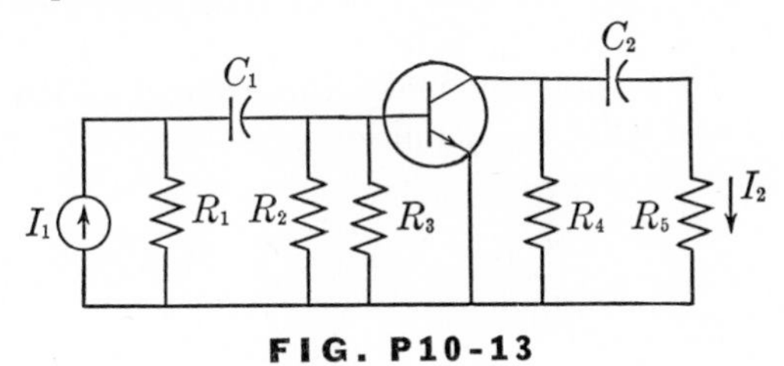

FIG. P10-13

10-14 In the circuit of Fig. P10-14, find the voltage-gain function V_2/V_1 when the capacitor is not present. Call it K. Next find the voltage gain as a function of s, expressed as a product of two factors one of which is K. Show the pole and zero on the complex plane. Finally, find the voltage ratio V_k/V_1. How do its poles and zeros differ from those of V_2/V_1? Sketch the frequency response for V_k/V_1, using ωCR_k for abscissa, and find the frequency range for which the value of $|V_k|$ is no less than 90 per cent of its zero-frequency value. Take $r_p = R = 10R_k$ and $\mu = 21$.

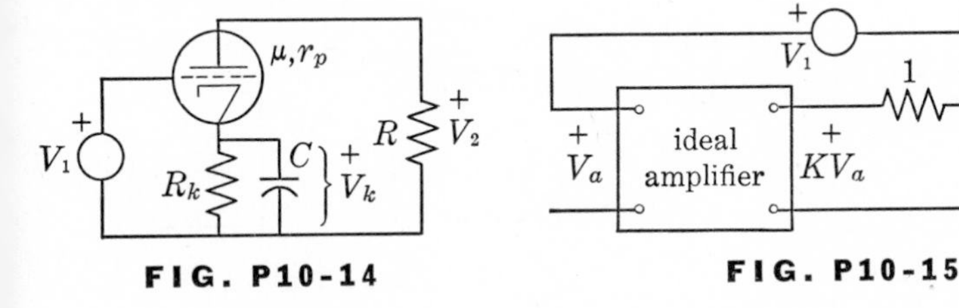

FIG. P10-14

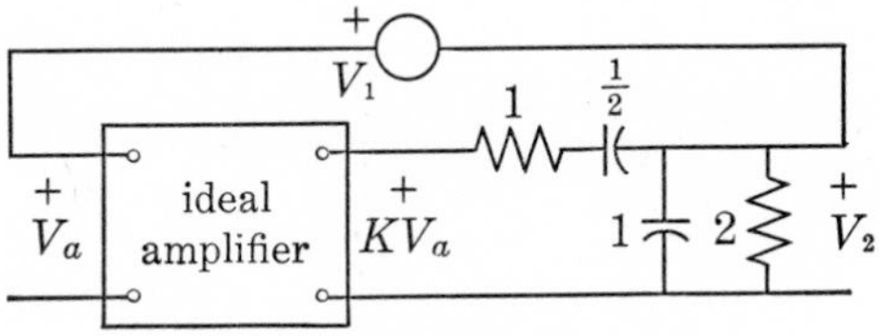

FIG. P10-15

10-15 The network in Fig. P10-15 includes an ideal amplifier. Find the transfer function V_2/V_1 and locate its poles in the complex plane, assuming (1) $K = 3$ and (2) $K = 6$.

10-16 The network in Fig. P10-16 is a model of an R-C-coupled amplifier stage. It is desired to find the variation of the magnitude and angle of $G_{21} = V_2/V_1$ with frequency. First convert the portion to the left of the dotted line to a Thévenin equivalent and find an expression

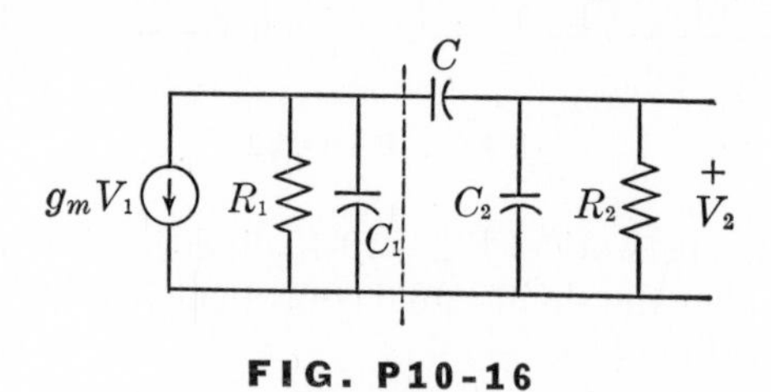

FIG. P10-16

for V_2/V_1. In this expression make the approximation $C_1 \ll C$ and $C_2 \ll C$. Make the following definitions:

$$R = \frac{R_1 R_2}{R_1 + R_2} \qquad \omega_1 = \frac{1}{C(R_1 + R_2)}$$

$$G_m = g_m R \qquad \omega_2 = \frac{1}{R(C_1 + C_2)}$$

Show that G_{21} can be written

$$G_{21} = \frac{-G_m}{1 + j\dfrac{\omega}{\omega_2} - j\dfrac{\omega_1}{\omega}}$$

Sketch the frequency-response curves under the assumption $\omega_2 \gg \omega_1$.

10-17 In Fig. P10-17 find an expression for the transfer admittance $Y_{21} = I_2/V_g$ as a function of s. Also find an expression for the input impedance $Z = V_g/I_g$.

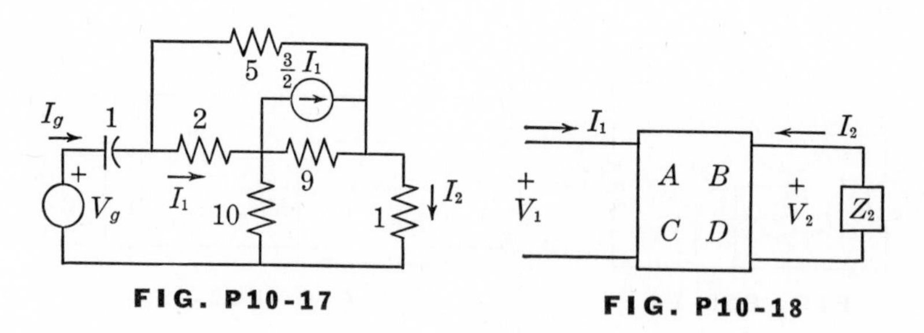

FIG. P10-17 FIG. P10-18

10-18 The two-port shown in Fig. P10-18 is characterized by the relationship $V_1 = -nV_2$, $I_1 = I_2/n$. (*a*) Write down its *ABCD* matrix. (*b*) Determine the input impedance when the two-port is terminated by an impedance Z_2. (*c*) How does this two-port differ from an ideal transformer? (*d*) Repeat if the terminal relationships are $V_1 = nV_2$, $I = -I_2/n$. (*e*) Compare the *VI* product at the input terminals with that at the output.

These two-ports are negative impedance converters, an example of which was discussed in Sec. 10-5.

10-19 Let the two-port in Fig. P10-17 be characterized by the equations $V_1 = -nI_2$, $I_1 = V_2/n$. (*a*) Write down its *ABCD* matrix. (*b*) Determine the input impedance when the two-port is terminated by an impedance Z_2. (*c*) This device is called a *gyrator*. How does it differ from an ideal transformer and a negative impedance converter? Is there dissipation of power within the device?

OPERATIONAL DIAGRAMS AND SIGNAL FLOW GRAPHS

11

IN all the preceding work in this book, we have been engaged in developing a network model in terms of which it is possible to explain and predict the response of an interconnection of electrical devices to one or more voltage or current excitations. This model was developed by making observations of the v-i relationships of many electrical devices. From these observations a set of laws was postulated relating voltages and currents of the branches. Each law applies to a hypothetical network element. In all cases—resistor, inductor, capacitor, ideal transformer, and ideal amplifier—there are just three mathematical operations involved relating the element response to the excitation: proportionality, differentiation, and integration.

To the present, we have represented each element in the model by means of a parameter (such as R) and by a distinctive graphical symbol, such as a broken, wiggly line for a resistor. Networks are made up of interconnections of these symbols. That is, in each v-i relationship, our interest has been directed at the physical constant, the parameter defined thereby. Thus, in writing $v = L\,di/dt$, the emphasis has been on L. It is possible, instead, to concentrate on the *mathematical operations* defined by the v-i relationships, and to symbolize each of these operations diagrammatically. A network may then be represented by combining these operational symbols into an operational diagram.

Just as it is possible to reduce a network structure by various transformations (series combinations, parallel combinations, tee-pi, etc.), it may be possible to find transformations to reduce the structure of the operational diagram also. This procedure will then constitute an alternative method of analysis. In this chapter we shall pursue the ideas just presented.

11.1 OPERATIONAL DIAGRAMS

Although the mathematical operations involved in the basic laws defining the network elements are only three in number, the procedure which we are about to develop is not restricted to these three operations alone. The response of any network, whether a single element or a complicated structure, is obtained by operating on the excitation in some manner. We represent this situation symbolically, as shown in Fig. 11-1(*a*). The arrows indicate a "flow." The excitation is the input

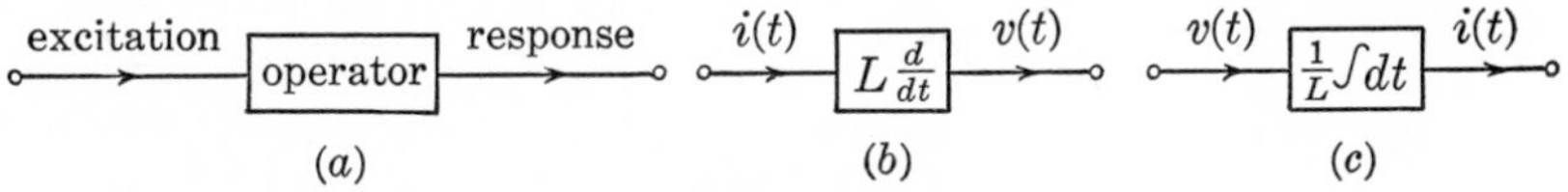

excitation → [operator] → response (*a*)

$i(t)$ → [$L\frac{d}{dt}$] → $v(t)$ (*b*)

$v(t)$ → [$\frac{1}{L}\int dt$] → $i(t)$ (*c*)

FIG. 11-1

Operational symbols

quantity; it is operated on by the operator, giving the output response. The special case of the inductor is represented by Fig. 11-1(*b*) or (*c*). Note that two different operational symbols, a differentiator and an integrator, are used for the inductor, depending on whether the excitation is the current or the voltage. This is a clear difference from the usual network-element representation where the symbol ⅏ represents the inductor independently of whether the voltage or the current is the excitation.

We shall be concerned here mainly with the sinusoidal steady-state response. In this case we deal with sinors, and the operations of differentiation and integration become replaced by multiplication and division by s, respectively. Some operational symbols for the network elements are shown in Fig. 11-2. Note that, even though the elements in question

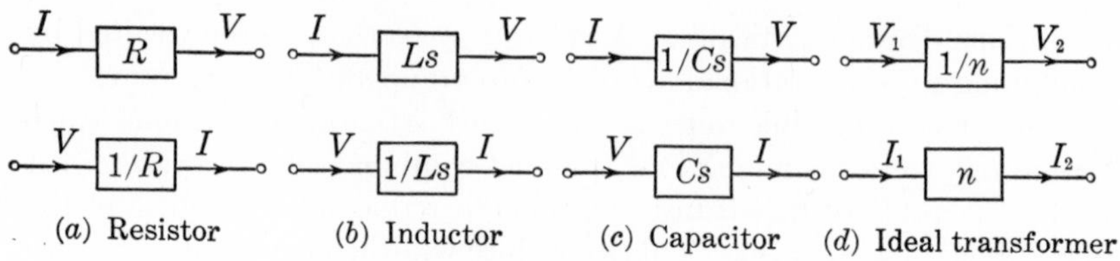
I
R
V
I
Ls
V
I
1/Cs
V
V1
1/n
V2
V
1/R
I
V
1/Ls
I
V
Cs
I
I1
n
I2
(a) Resistor
(b) Inductor
(c) Capacitor
(d) Ideal transformer

FIG. 11-2

Operational symbols of network elements

are bilateral, the flow of signals takes place only in the direction indicated by the arrows.

A different symbol is often given to the ideal voltage amplifier, as shown in Fig. 11-3. The reason for making this distinction is that the amplifier is not a bilateral device; that is, we would not draw an operational diagram with V_2 as the excitation and V_1 the response.

Having defined operational symbols for the network elements, we should also designate a symbol which represents summation or addition, in order to express Kirchhoff's laws operationally. Such a symbol is shown in Fig. 11-4; it is called a *summing point.* The signals entering

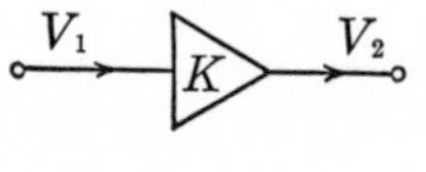
V1
K
V2

FIG. 11-3

Operational symbol of ideal amplifier

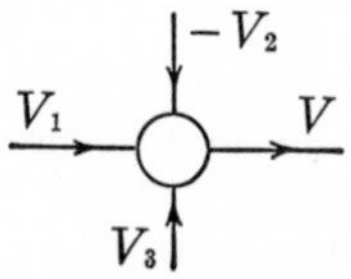
−V2
V1
V
V3

FIG. 11-4

Summing point

the point are indicated by arrows. Each signal which is to be subtracted is preceded by a minus sign; if there is no sign, the signal is to be added. The signal leaving the point is the algebraic sum of the signals entering it.

Any network, no matter how complicated its structure, can be represented by a single operational symbol, provided that there is one excitation and that a particular voltage or current is designated as a response. It is also possible to represent different parts of the over-all structure (down to as small a subdivision as the individual elements) by operational symbols and to interconnect these as required, to form an operational diagram of the network.

Let us illustrate the construction of an operational diagram of a

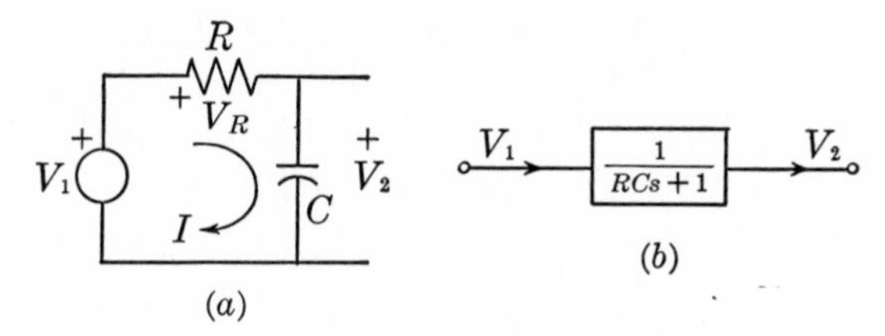

FIG. 11-5

Illustrative examples

network by means of the simple voltage divider shown in Fig. 11-5. From the voltage-divider expression, we can immediately write

$$V_2 = \frac{V_1/Cs}{R + 1/Cs} = \frac{1}{RCs + 1} V_1 \qquad (11\text{-}1)$$

Hence, with V_2 taken as the response, the operational diagram will take the form shown in Fig. 11-5(*b*). In an operational diagram the quantity which multiplies the input sinor to give the output sinor is called the *transmittance*. This may be an impedance or an admittance, as for the single elements, or a transfer function as in this case.

Let us now return to fundamental considerations and note that this network contains two branches, two nodes, and one loop. There will be a total of $2N_b = 4$ variables. The two branch currents can be made identical (by choice of references), however, and so let us take the input voltage V_1, the two branch voltages V_2 and V_R, and the current I as variables. The equations relating these variables are

$$\begin{aligned} V_R &= RI & &(a) \\ I &= CsV_2 & &(b) \qquad (11\text{-}2) \\ V_2 &= V_1 - V_R & &(c) \end{aligned}$$

The first two of these equations are represented by the operational symbols shown in Fig. 11-6(*a*). The last equation is Kirchhoff's voltage

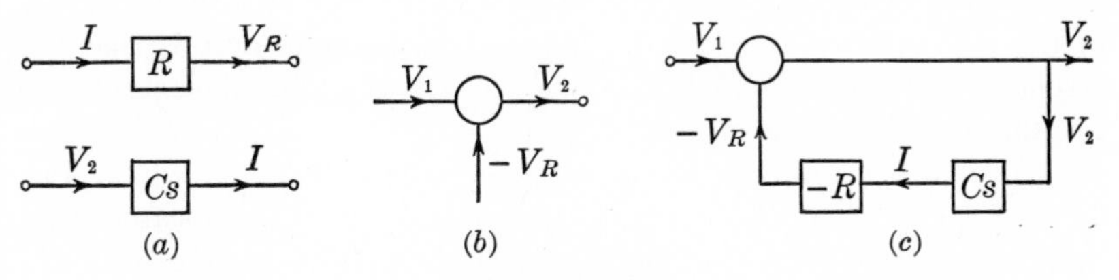

FIG. 11-6

Development of operational diagram of illustrative example

law and is represented by the summing point shown in Fig. 11-6(*b*). Combining these two parts leads to the operational diagram shown in Fig. 11-6(*c*). Such operational diagrams are also called *block diagrams*, because each operator is shown as a "block."

An alternative operational diagram can be obtained by rewriting Eq. (11-2) in the following way.

$$I = \frac{1}{R} V_R \qquad (a)$$

$$V_2 = \frac{1}{Cs} I \qquad (b) \qquad (11\text{-}3)$$

$$V_R = V_1 - V_2 \qquad (c)$$

The first two equations are represented by the blocks shown in Fig. 11-7(*a*); the last one is represented by the summing point in Fig. 11-7(*b*).

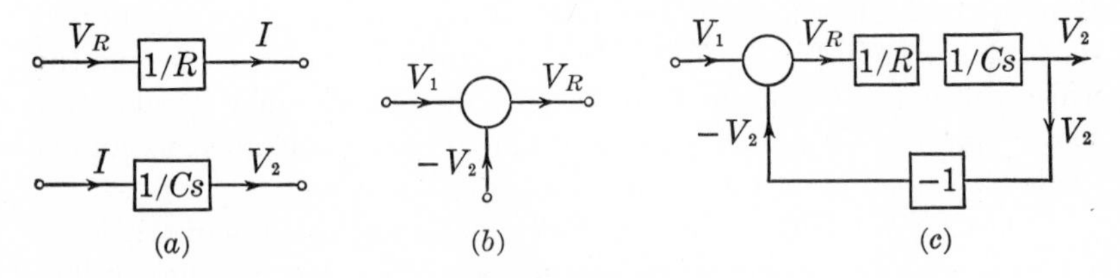

FIG. 11-7

Alternative operational diagram for illustrative example

The combination of these blocks results in the operational diagram shown in Fig. 11-7(*c*). Compare this diagram with the previous one in Fig. 11-6.

Two points should be noted here. We have succeeded in obtaining two different operational diagrams representing the same network [not counting the single-branch diagram of Fig. 11-5(*b*)]. This shows that there is no unique operational diagram for a given network. The specific diagram which is obtained will depend on the variables chosen for the analysis of the network, and on the manner in which the equations of the network are formulated. The second point to note is that the structure of the operational diagrams does not resemble the actual connection diagram of the elements.

As another example let us consider the network shown in Fig. 11-8(*a*). Here two *RC* networks are cascaded with an ideal voltage amplifier between them acting as an isolator. Considering each *RC* network as a unit, the equations representing this network will be

$$V_2 = \frac{1}{R_1C_1s + 1} V_1 \qquad (a)$$

$$V_3 = KV_2 \qquad (b) \qquad (11\text{-}4)$$

$$V_4 = \frac{1}{R_2C_2s + 1} V_3 \qquad (c)$$

The operational (block) diagram will take the form shown in Fig. 11-8(*b*).

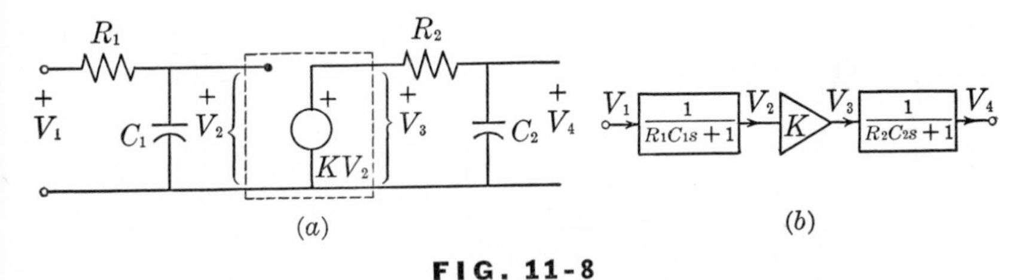

FIG. 11-8

Cascaded *RC* networks and the operational diagram

Alternatively, each *RC* network can be represented by one or the other of the diagrams shown in Fig. 11-6 or 11-7. These will then replace the left-hand and right-hand blocks in Fig. 11-8(*b*). Since our ultimate objective is to simplify the block diagram, however, there will be no advantage in this procedure.

Once an operational diagram of a network is obtained, the next task is to transform and reduce the structure of the diagram until only one block remains connecting the input and output variables (assuming a single input and a single output). The transmittance of this block will be the function expressing the ratio of the response to the excitation sinors. Before we undertake this task, let us make some modifications of our operational symbols for the purpose of still further simplifying operational diagrams.

11.2 SIGNAL FLOW GRAPHS

We first introduced a symbolic diagram in Fig. 11-1 to illustrate that a response is obtained by operating on an excitation in some manner. In this symbol there are three quantities of importance: the two signals, excitation and response, and the operation, which takes the form of a transmittance in the steady state. To designate the operation, it is not really necessary to show a rectangle, with the transmittance written inside. We can, instead, merely join the points, or nodes, which represent the excitation and response variables by directed lines, and write the transmittances alongside the lines. An operational diagram will now consist simply of directed lines joining various nodes.

In this system the operational diagrams first shown in Figs. 11-6(*c*) and 11-7(*c*) take the forms shown in Fig. 11-9. (The symbol T, with

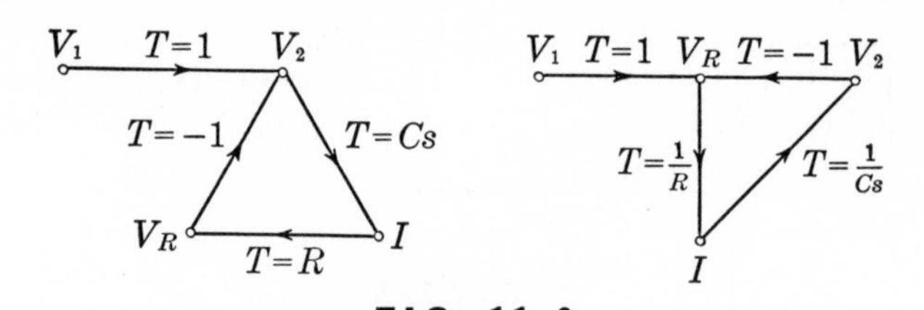

FIG. 11-9

Signal flow graphs of illustrative examples

appropriate subscripts, will be used to designate a transmittance.) These diagrams can be readily obtained from the equations given in Eqs. (11-2) and (11-3). First of all we place on the paper one node for each variable in the equations, and we label the nodes with the corresponding variable. In the present case the nodes are labeled V_1, V_2, V_R, and I. Then, taking each equation in turn, we draw directed lines from the nodes representing variables appearing on the right side to the node representing the particular variable appearing on the left for which the equation is written. Each branch thus formed is labeled in terms of the coefficient of the corresponding variable. The results are the diagrams in Fig. 11-9.

In this form the operational diagram is called a *signal flow graph.* It consists of a number of nodes, which represent the signal variables, connected by directed branches along which the signals "flow," in the direction of the arrow. Each node is like a telephone repeating station; it adds algebraically all the signals directed toward the node and transmits this sum along each branch directed away from the node. Thus the variable represented by a node is the sum of all signals entering the node. The signal leaving a node by a branch is multiplied by the transmittance of that branch in its passage along the branch.

A signal flow graph is thus a diagrammatic portrayal of the relationships existing among a set of variables. These relationships are also expressed by a set of equations. Thus a signal flow graph is equivalent to a set of equations. Note, however, that these equations are in a special form; each equation expresses one of the variables explicitly in terms of others. Given any set of (linear) equations, it is possible to draw a flow-graph representation of them, as we shall illustrate amply as we proceed. Conversely, given a signal flow graph, the corresponding equations can be written down directly.

Two specific groups of nodes in a signal flow graph deserve special classification. Any node which has only outgoing branches and no incoming ones is called a *source node*. This means that the corresponding variable is not explicitly written in terms of other variables, but other variables are expressed in terms of it. Thus it corresponds to an excitation, and so the name is an apt one.

All nodes which are not source nodes represent responses. If such a node has no outgoing branches, it is called a *sink node*. If we look at the graphs in Fig. 11-9, we notice that the node labeled V_1 is a source node having only one outgoing branch. Node V_2, however, which corresponds to the desired response, has both incoming and outgoing branches. Such a situation can always be remedied by adding an extra node through a transmittance of unity, as shown in Fig. 11-10. This corresponds to adding the trivial equation $V_2 = V_2$ to the set of equations describing the graph.

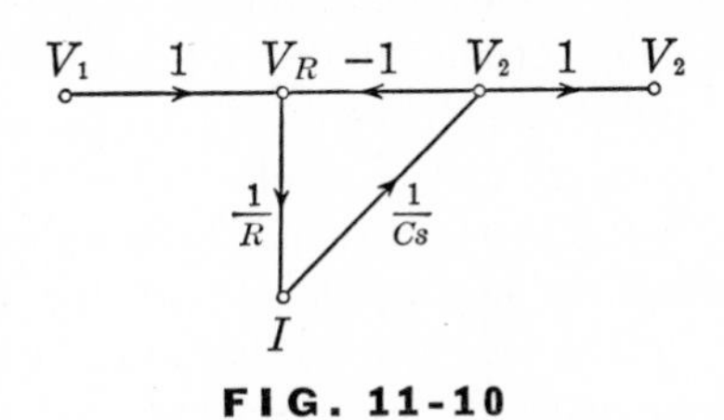

FIG. 11-10

Adding a trivial node

11.3 SETTING UP A SIGNAL FLOW GRAPH

Let us now investigate the fundamental considerations we face in drawing a signal flow graph from a given network. There are a number of different ways of writing equilibrium equations for a network. In each method there will be a different set of variables and a different number of equations, thus leading to different flow graphs.

Remember that, in a network containing no controlled sources, there are a total of $2N_b$ equations which completely specify the network performance. Of these, N_b are the branch $V = I$ relationships, $N_v - 1$ are the independent KCL equations, and $N_m = N_b - (N_v - 1)$ are the independent KVL equations. One method of solution consists of solving all $2N_b$ equations simultaneously. Many of these equations, however, can be eliminated by direct substitutions. Among the other methods of solution we have discussed are the method of node equations, which requires solving $N_v - 1$ simultaneous equations, and the method of loop equations, which requires solving $N_m = N_b - (N_v - 1)$ simultaneous equations.

When controlled sources are present, the number of equations is increased by the number of such sources. For each controlled source an equation must be written expressing its voltage or current in terms of the variables on which it depends. Otherwise, controlled sources introduce no additional complications.

The first task in setting up a signal flow graph is to choose a set of variables in terms of which the network performance is to be found. One possible approach is to choose the loop currents as variables. Then the flow graph will be a representation of the loop equations. Reducing the flow graph then amounts to solving the loop equations. A similar statement can be made about the choice of node voltages as variables. The full power of signal flow graphs, however, will not be felt if we simply use them as a graphical portrayal of loop or node equations.

The only criteria that must be satisfied by the set of equations which we eventually use in setting up the flow graph are that they be linearly independent and that there be enough of them to describe the network performance. The second criterion can be very easily satisfied. As we shall shortly see, it is very easy to eliminate redundant variables when reducing a flow graph. Hence, when in doubt regarding the need for an additional variable or an additional equation, it is wisest to include it, even if it should prove to be redundant.

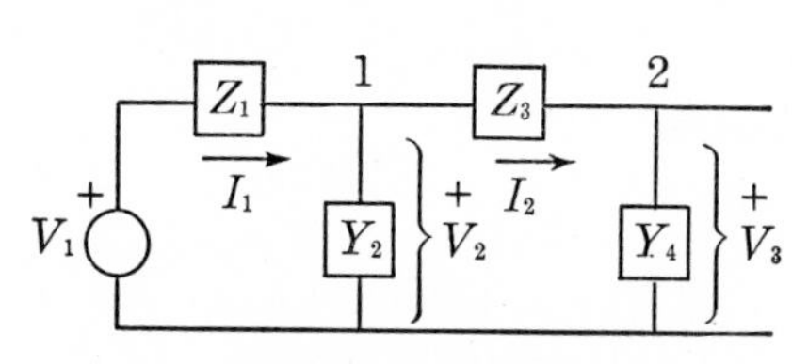

FIG. 11-11

Illustrative example

Let us illustrate the choice of variables and the method of writing the equations by means of an example. Figure 11-11 shows a four-branch ladder network whose series branches are labeled in terms of impedance and whose shunt branches are labeled in terms of admittance. As variables, let us choose the shunt branch voltages, as well as the source voltage, and the series branch currents. In terms of these, let us write Kirchhoff's voltage law around the two meshes and Kirchhoff's current law at the nodes labeled 1 and 2. We get

$$
\begin{aligned}
Z_1I_1 + V_2 - V_1 &= 0 & \text{(a)} \\
Z_3I_2 + V_3 - V_2 &= 0 & \text{(b)} \\
-I_1 + Y_2V_2 + I_2 &= 0 & \text{(c)} \\
-I_2 + Y_4V_3 &= 0 & \text{(d)}
\end{aligned}
\qquad (11\text{-}5)
$$

These equations form an independent set, since no one of them can be obtained as a linear combination of the others. Furthermore, all other branch variables can be found once the variables in Eq. (11-5) are known.

As the next step let us rearrange these equations in a cause-and-effect manner, so that each of them expresses one of the variables explicitly in terms of the remaining ones, as follows:

$$
\begin{aligned}
V_2 &= V_1 - Z_1I_1 & (a) \\
I_1 &= Y_2V_2 + I_2 & (b) \\
I_2 &= Y_4V_3 & (c) \\
V_3 &= V_2 - Z_3I_2 & (d)
\end{aligned}
\qquad (11\text{-}6)
$$

Notice that the second equation of the preceding set has been written last here. Each successive equation expresses one of the variables from the right side of the preceding equation explicitly in terms of other variables.

Starting from the first equation, the signal flow graph can be built up as shown in Fig. 11-12. Each variable is first represented by a node.

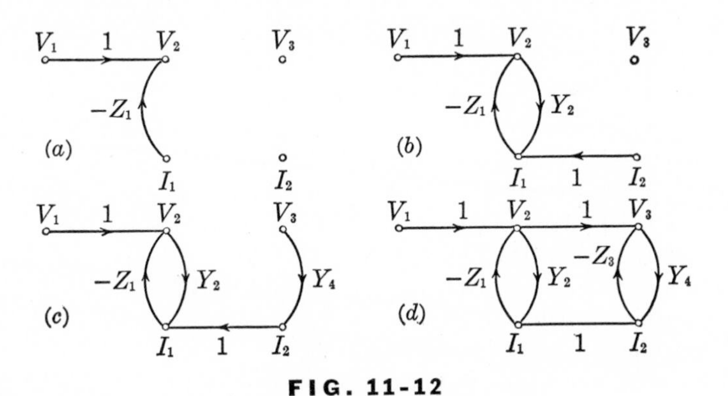

FIG. 11-12

Development of signal flow graph

The first equation shows that there is transmission from V_1 and I_1 to V_2, as shown in Fig. 11-12(a). The second equation shows that there is transmission from V_2 and I_2 to I_1, and the graph now takes the form shown in Fig. 11-12(*b*). Now we add a branch from V_3 to I_2, as required by the third equation. [See Fig. 11-12(*c*).] Finally, we add branches from V_2 to V_3 and from I_2 to V_3. The completed graph is shown in Fig. 11-12(*d*). Notice that V_1 is a source node but that there are no sink nodes.

At this point it is pertinent to ask whether there is any procedure for choosing variables which will guarantee a suitable signal flow graph, yet will not require an extremely large number of variables. In order to answer this question, we shall resort to considerations of the topological properties of networks, since the question of dependence or independence of variables and the appropriate number of independent equations is answered in terms of the network topology. You should look over Sec. 2-8 before proceeding.

Recall that a tree is a set of branches in a network connecting all the nodes without forming any closed loops. For a given tree, a link is any branch not on the tree. Recall also that connecting a link to the tree will form a closed path. Kirchhoff's voltage law then permits expressing the link voltages in terms of the tree-branch voltages. Thus a knowledge of the tree-branch voltages alone is sufficient to determine all the branch voltages. Analogously, recall that each tree branch, together with links, forms a cut set. Kirchhoff's current law then permits expressing the tree-branch currents in terms of the link currents. Thus a knowledge of the link currents alone is sufficient to determine all the branch currents.

In our discussion of topology in Chap. 2, we agreed to handle sources in a particular way for the purpose of counting nodes and branches. We agreed to count a voltage source in series with something as a single branch. Thus the node at the junction of the voltage source and the "something" was assumed not to exist. For our present purpose, let us partially rescind this agreement. That is, for the purpose of constructing a tree, and for this purpose only, let us count a voltage source as a separate branch, and let us count as a node the junction between it and anything connected in series with it.

We shall approach the task of choosing a set of variables with the preceding thoughts in mind. Suppose we choose a tree containing all the voltage sources in the network but none of the current sources. The current sources will all be in the corresponding link set. It is certainly possible to choose a tree so that it contains all the voltage sources, for, if not, a voltage source will be a link, and link voltages are determined in terms of tree-branch voltages. This means that the source voltage will not be independently specifiable, as it must be. A similar argument applies for choosing all current sources as links.

Now we write a set of Kirchhoff voltage-law equations around all the fundamental loops. (Recall that the fundamental loops are the ones obtained by adding the links one at a time to the tree branches.) In these equations we keep the tree-branch voltages as variables. For the links we insert the branch V-I relationships, however, thus changing the variables to the link currents.

Next we write a set of Kirchhoff current-law equations for all the fundamental cut sets. (Recall that the fundamental cut sets are those containing one tree branch only, together with links.) In these equations we keep the link currents as variables. For the tree branches, however, we insert the branch V-I relationships, thus changing the variables to the tree-branch voltages.

There are N_b equations in these two sets, and they are all independent. There are also N_b unknowns, the link currents and the tree-branch voltages. We write each of these equations in a form which expresses one of the variables explicitly in terms of the others present in the equation. This procedure constitutes a systematic method for setting up a signal flow graph.

In Fig. 11-11 the shunt branches (including the voltage source) were chosen as tree branches and the series branches as links. You should verify that Eq. (11-6) is in the form described.

Let us illustrate the procedure we have described by means of the network shown in Fig. 11-13. This will give us a chance to make some

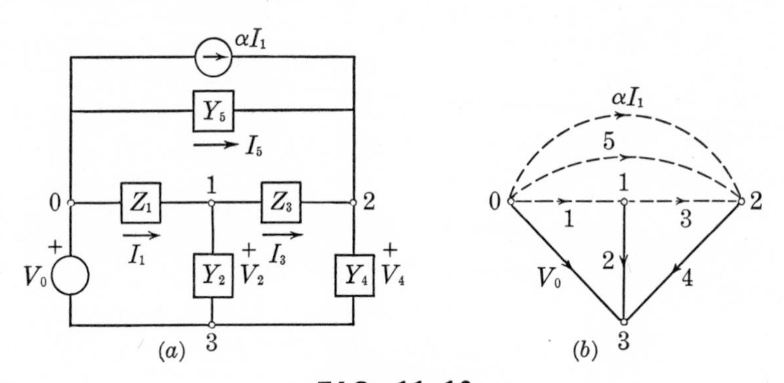

FIG. 11-13

Illustrative example

remarks concerning the details of the procedure. Note that a dependent current source is also included.

First of all, since we shall be dealing with branch variables, it will be necessary to choose references for them. We shall assume that standard references are always chosen; that is, each branch-voltage reference is at the tail of the corresponding branch-current arrow. Then it will be necessary to show only one of the references for each branch.

Figure 11-13(*b*) shows the topological graph of the network, the solid lines being the tree branches and the dashed lines the links. The arrows on the lines indicate the branch references. The fundamental loops are the ones formed by $V_0Z_1Y_2$, $Y_2Z_3Y_4$, and $V_0Y_5Y_4$. The appropriate equations will be

$$\begin{aligned} Z_1I_1 + V_2 - V_0 &= 0 && (a) \\ Z_3I_3 + V_4 - V_2 &= 0 && (b) \qquad (11\text{-}7) \\ Z_5I_5 + V_4 - V_0 &= 0 && (c) \end{aligned}$$

Note that the loop formed by the current source together with tree branches is not considered. This is consistent with our knowledge that there are only three independent *KVL* equations; it also agrees with our previous practice.

Now let us turn to the fundamental cut-set equations. These are identical with the *KCL* equations at nodes 0, 1, and 2. Note, however, that although we chose node 0 as a node, this was done only for the purpose of putting the voltage source in the tree. We do not write a current equation at this node. (This is analogous to not writing a voltage equation for a loop containing a current source.) The current equations at the other two nodes are

$$\begin{aligned} -I_1 + Y_2V_2 + I_3 \qquad\qquad &= 0 && (a) \\ -I_3 + Y_4V_4 - \alpha I_1 - I_5 &= 0 && (b) \end{aligned} \qquad (11\text{-}8)$$

We can combine these two sets of equations and at the same time write them in the appropriate form, as follows:

$$\begin{aligned} V_2 &= V_0 - Z_1I_1 && (a) \\ I_1 &= Y_2V_2 + I_3 && (b) \\ I_3 &= Y_4V_4 - \alpha I_1 - I_5 && (c) \qquad (11\text{-}9) \\ I_5 &= Y_5V_0 - Y_5V_4 && (d) \\ V_4 &= V_2 - Z_3I_3 && (e) \end{aligned}$$

The corresponding signal flow graph is built up in Fig. 11-14. Each equation is represented separately in the first five parts. The completed

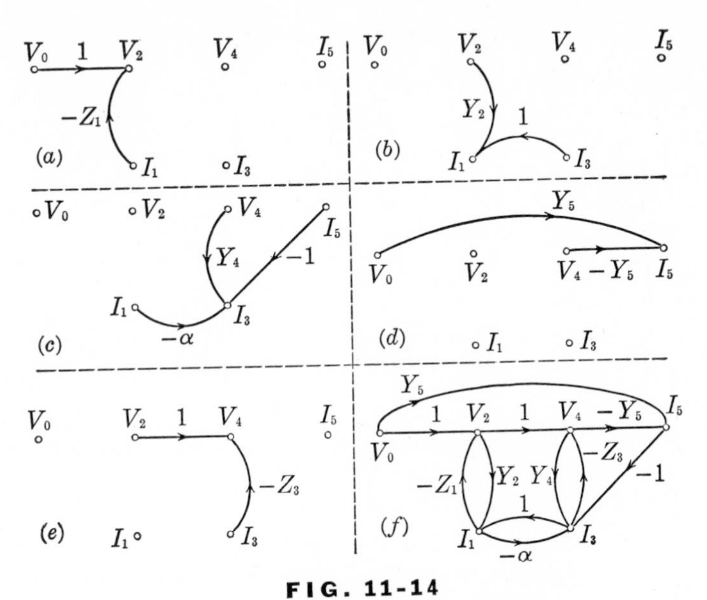

FIG. 11-14

Signal flow graph representing Fig. 11-13

graph is shown in part (*f*). You are urged to trace through the steps.

Note that only one equation is written explicitly for any one variable. It would be a mistake to write more equations for the same variable. (Why?)

Another point to note is that the appearance of the dependent source caused us no concern whatsoever. We could have designated by I_g the current of the source, and then written the equation $I_g = \alpha I_1$. This would have introduced another variable, and hence another node in the signal flow graph. Instead of αI_1 in Eq. [11-9(*c*)], we would have had I_g. Since the source current depends on only one other variable, however, it is easy to express it immediately in terms of this variable, rather than introducing an additional node which must eventually be eliminated from the graph.

Let us now inquire whether, after choosing the tree as we have done, it is really essential that the voltage equations we write should be those for the fundamental loops, and the current equations should be those for the fundamental cut sets. Suppose, for example, that we choose the windows in Fig. 11-13(*a*) for writing voltage equations. The only change from our previous choice will be to replace the loop consisting of $V_0Y_5Y_4$ by that consisting of $Y_5Z_3Z_1$. Thus Eq. [11-9(*d*)], which is the rewritten form of Eq. [11-7(*c*)], will be replaced by

$$I_5 = Y_5Z_1I_1 + Y_5Z_3I_3 \tag{11-10}$$

Everything else will stay the same. The only change in the signal flow graph will be to replace Fig. 11-14(*e*) by a partial graph with transmission from nodes I_1 and I_3 to I_5. (Make the appropriate change.)

An analogous discussion applies to the writing of current equations. There is no need to write the equations for the fundamental cut sets; it is enough to write current equations at the nodes ($N_v - 1$ of them) of the network. The only requirement is that there be $N_v - 1$ independent current equations and $N_b - N_v - 1$ independent voltage equations.

We can summarize this discussion of a systematic way of choosing variables and writing equations for a signal flow graph of a network as follows:

1. Choose a tree containing all the voltage sources but none of the current sources.
2. Write $N_m = N_b - (N_v - 1)$ independent voltage equations around appropriate loops, expressing the link voltages in terms of the link currents but keeping the tree-branch voltages as variables.
3. Write current equations at $N_v - 1$ nodes of the network, expressing the tree-branch currents in terms of the voltages but keeping the link currents as variables.
4. Rewrite the combined set of equations so that each equation is solved explicitly for one of the variables in terms of others. The variables are the tree-branch voltages and the link currents, a total of N_b.

11.4 REDUCTION OF A SIGNAL FLOW GRAPH

Up to this point we have concentrated on the task of setting up a signal flow graph to represent a given network. Our eventual purpose is to find responses to given excitations, so our next task should be a discussion of how to compute a response from a flow graph.

To start with, let us define some terms that describe the topological structure of a signal flow graph. If we start at a node and traverse a number of successive branches in the direction of the arrows, we shall trace out a *path*. Two types of path can be distinguished: an *open path*, or a *forward path*, along which the same node is not encountered more than once, and a *closed path*, which ends at the same node at which it starts. For example, in Fig. 11-12(d) the path $I_2I_1V_2V_3$ is an open path; the path $V_2I_1V_2$ is closed.

A *feedback loop* is defined as a set of branches which form a closed path. Any branch which is on a feedback loop is called a *feedback branch;* all others are called *cascade branches.* In Fig. 11-12(d) there are three feedback loops: $V_2V_3I_2I_1V_2$, $V_2I_1V_2$, and $V_3I_2V_3$. Thus all the branches except V_1V_2 are feedback branches. In the same way as for branches, nodes can be classified as *feedback nodes* or *cascade nodes,* depending on whether or not they lie on a feedback loop. Clearly, source nodes and sink nodes are cascade nodes. If a signal flow graph contains cascade branches only, it is called a *cascade graph;* otherwise it is a *feedback graph.* A cascade graph represents a relatively trivial set of equations, one which can be solved successively for one variable at a time. The flow graph resulting from Fig. 11-8 is a cascade graph.

We are now ready to consider procedures for the systematic reduction of signal flow graphs. Two simple configurations, together with the corresponding equations, are shown in Fig. 11-15. We have shown

$$V_2 = T_1V_1$$
$$V_3 = T_2V_2 = T_2(T_1V_1) = (T_1T_1)V_2$$

V_1 T_1 V_2 T_2 V_3 V_1 T_1T_2 V_3

$$V_2 = T_1V_1 + T_2V_1$$
$$= (T_1 + T_2)V_1$$

T_1 V_1 V_2 T_2 V_1 T_1+T_2 V_2

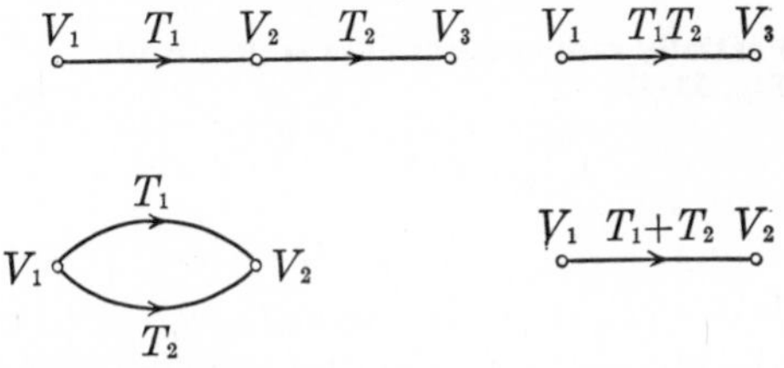

FIG. 11-15

Cascade and parallel transformations

the variables to be voltage, but they can be anything. The first of these shows a *cascade transformation.* An intermediate node is eliminated and the over-all transmittance is the product of the cascade transmittances. The second one is a *parallel transformation* which replaces two branches connecting the same two nodes by a single branch having an over-all transmittance which is the sum of the parallel transmittances. (Note that the two arrows should be pointing in the same direction.) In the signal flow graphs of Figs. 11-12 and 11-14, neither of these configurations shows up.

As the next step let us consider what it means to eliminate a node of the signal flow graph. In terms of the equations, it means that the variable corresponding to this node is to be eliminated. Let us carry out the discussion in terms of the flow graph of Fig. 11-12 and the corresponding equations in Eq. (11-6).

Suppose we wish to eliminate node V_2 from the graph, or variable V_2 from the equations. This can be done by substituting Eq. [11-6(*a*)] into Eqs. [11-6(*b*)] and [11-6(*d*)]. The equations now become

$$\begin{aligned} I_1 &= Y_2V_1 - Y_2Z_1I_1 + I_2 && (a) \\ I_2 &= Y_4V_3 && (b) \quad (11\text{-}11) \\ V_3 &= V_1 - Z_1I_1 - Z_3I_2 && (c) \end{aligned}$$

The signal flow graph corresponding to these equations is shown in Fig. 11-16. (Carry out the necessary steps.) In the equations we find something new. The first equation is an expression for I_1, but I_1 also appears on the right side. This means that in the graph there will be a transmission from node I_1 to itself. This is shown in the figure by the branch having a transmittance $-Y_2Z_1$ which closes on itself. Such a branch is called a *self-loop*.

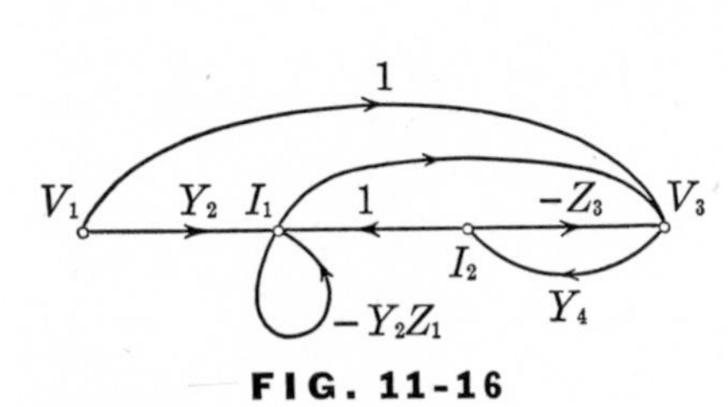

FIG. 11-16

Partially reduced signal flow graph of Fig. 11-12

As far as the remaining nodes are concerned, the graph in Fig. 11-16 is equivalent to the original one. Let us define a *path transmittance* as the product of transmittances of all branches which lie on a path. Observe that all paths from one node to another that did not pass through the eliminated node V_2 are still intact in the new graph. In the original graph, however, there is a path from I_1 to V_3 through V_2 having a path transmittance $(-Z)(1) = -Z_1$. Since the new graph is to be equivalent to the original one, we should expect that such a path with the same transmittance should exist between I_1 and V_3; in Fig. 11-16 we see that it does.

Similarly, in the original graph there is a path from I_2 to V_3 through V_2 and I_1 with a path transmittance $(1)(-Z)(1) = -Z_1$. In the new graph this path through I_1 with the appropriate path transmittance is preserved. Finally, the path from V_1 to V_3 through V_2 having a transmittance $(1)\ (1) = 1$ now becomes a direct path having the same path transmittance. The self-loop in the new graph arises from the feedback loop from I_1 to itself through V_2, having a path transmittance $-Z_1Y_2$.

This discussion serves to indicate a procedure whereby a node may be eliminated from a signal flow graph without going through the intermediate step of rearranging the equations. In the new graph it is only necessary to ensure that path transmittances between any two of the remaining nodes are preserved. But this is assured if path transmittances between nodes *adjacent* to the eliminated node are preserved. (Two nodes are adjacent if they have a direct connection without going through an intervening node.) For example, in eliminating node V_2, if the transmission from I_1 to V_3 is preserved, then certainly the transmission from I_2 to V_3 (through I_1) will be preserved.

Let us apply this criterion to the elimination of node V_3 in the graph of Fig. 11-16. We first draw the parts of the graph that were not connected to node V_3, as in Fig. 11-17(a). Now we check on the path

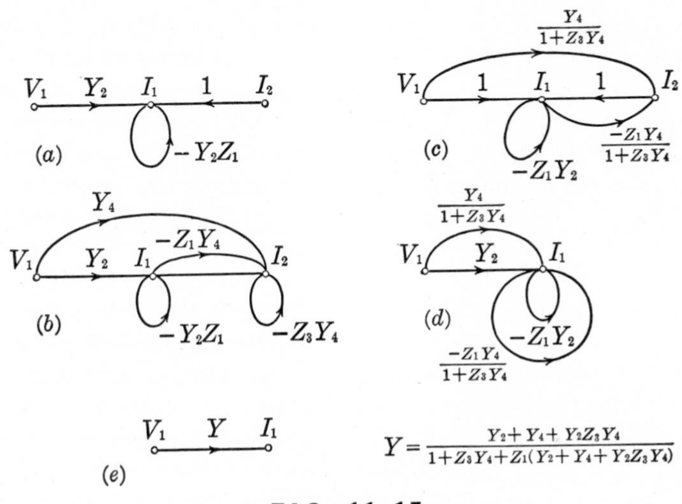

FIG. 11-17

Further reduction of signal flow graph

transmittances through node V_3 between each pair of nodes adjacent to V_3. In Fig. 11-16 there is a path from I_1 to I_2 through V_3 with a transmittance $-Z_1Y_4$; there is also a path from V_1 to I_2 with transmittance Y_4. Finally, there is a path between I_2 and itself through V_3, with transmittance $-Z_3Y_4$. This leads to a self-loop at I_2. The final graph is shown in Fig. 11-17(b).

As the next step we should eliminate either node I_1 or node I_2, leaving the one which is to represent the desired response. Here we run into some difficulty. The equations represented by the flow graph in Fig. 11-17(b) are

$$\begin{aligned} I_1 &= Y_2V_1 - Z_1Y_2I_1 + I_2 && (a) \\ I_2 &= Y_4V_1 - Z_1Y_4I_1 - Z_3Y_4I_2 && (b) \end{aligned} \qquad (11\text{-}12)$$

If we simply substitute the second equation for I_2 into the first one, the variable I_2 will still appear in the result and, hence, will not be eliminated. This holds true also if we try to eliminate I_1.

Suppose, however, that we first transpose the term involving I_2 from the right side of Eq. [11-12(b)] and then solve for I_2. The resulting equation will be

$$I_2 = \frac{Y_4}{1 + Z_3Y_4} V_1 - \frac{Z_1Y_4}{1 + Z_3Y_4} I_1 \qquad (11\text{-}13)$$

The signal flow graph representing this equation and Eq. [11-12(a)] is shown in Fig. 11-17(c). The price paid for the elimination of the self-loop at I_2 has been to divide the transmittance of each branch directed toward I_2 by $1 + Z_3Y_4$, which is 1 minus the transmittance of the self-loop. This is actually a general result which applies for the elimination of any self-loop, as you can verify by writing a general equation in the form $V_i = \sum T_{ij}V_j$, transposing the term involving V_1 from the right, then solving for V_1.

Since there is no longer a self-loop at I_2, this node can now be eliminated, leading to the graph shown in Fig. 11-17(d). (Verify this.) In this graph there are two self-loops at node I_1, and they are in parallel. Similarly, the two branches between V_1 and I_1 are in parallel. By applying the parallel transformation, these pairs of branches can be combined. Finally, the self-loop is removed by dividing the remaining branch transmittance by 1 minus the self-loop transmittance. The resulting graph is shown in Fig. 11-17(e). Since the response we have found is the input current, the transmission is the input admittance $Y = I_1/V_1$. It is given by

$$Y = \frac{Y_2 + \dfrac{Y_4}{1 + Z_3Y_4}}{1 + Z_1Y_2 + \dfrac{Z_1Y_4}{1 + Z_3Y_4}} = \frac{Y_2 + Y_4 + Z_3Y_2Y_4}{1 + Z_1Y_2 + Z_1Y_4 + Z_3Y_4 + Z_1Z_3Y_2Y_4} \qquad (11\text{-}14)$$

The task of verifying that this is indeed the input admittance of the network in Fig. 11-11 is left to you.

In summary, the procedure for systematically reducing a given signal flow graph is the following:

1. Any node not containing a self-loop can be removed, together with all its incoming and outgoing branches, by adding additional branches between nodes adjacent to the removed node, each one having a transmittance equal to the corresponding path transmittance through the removed node. The elimination of a cascade node having only one incoming and one outgoing branch (as shown in Fig. 11-15) is a special case.
2. A self-loop at a node can be eliminated by dividing the transmittances of all other branches directed toward the node by 1 minus the self-loop transmittance.
3. Parallel branches can be combined by adding their transmittances.

Successive application of these steps will reduce the signal flow graph to a cascade form with branches directed from the sources (the excitations) to the sinks (the responses).

The ideas which have been developed in this chapter constitute an alternative method of analysis. This method concentrates on the operational implications of the fundamental laws of network theory. Given a network which is to be analyzed, we first draw a signal flow graph representing the network. Such a graph is not unique but depends on the specific variables which are chosen for the analysis.

Once the signal flow graph is obtained, the next task is to compute any desired transfer function. One systematic step-by-step reduction process has been discussed.

Let us conclude this section by giving one additional example of the use of signal flow graphs in solving network problems. A vacuum-tube circuit, together with its low-frequency small-signal model, is shown in Fig. 11-18. The two tubes are assumed to be identical.

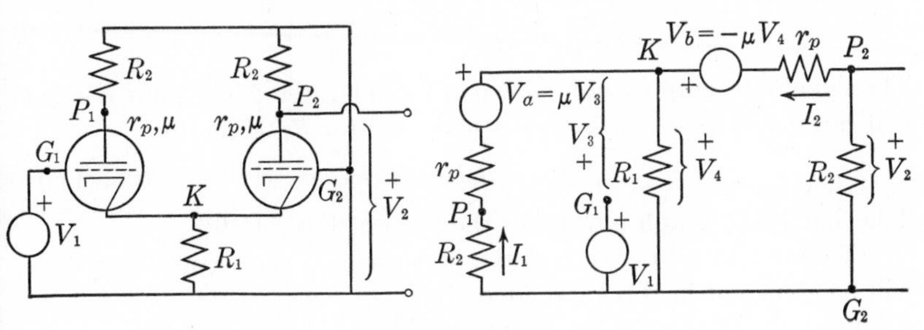

FIG. 11-18

Vacuum-tube circuit and its small-signal model

Suppose we choose as a tree the three voltage sources and the resistors R_1 and R_2. This means that we shall have as variables: V_1, V_a, V_b, V_2, V_4, I_1, and I_2. (We are not counting P_1 as a node but combining r_p and R_2 into a single branch.) But V_b is proportional to V_4. Hence, in addition to the source node V_1, there will be five nodes in the signal flow graph, and, consequently, five equations will be required. There are two independent *KVL* equations around the loops and two independent *KCL* equations (at nodes P_2 and K). The remaining equation expresses the dependence of the controlled source voltage V_a on the other variables. Thus the equations can be written in a straightforward way. We shall not, however, complete this formulation of the problem. Instead, we shall turn to a slightly different choice of variables which appears more natural to the problem at hand.

Let us choose as variables the input and output voltages V_1 and V_2, the grid-cathode voltage V_3 of the first tube, the voltage V_4 across the cathode resistor R_1, and the two plate currents I_1 and I_2. This set differs from the preceding one only in that V_3 replaces V_a. But V_a is proportional to V_3, so the set of variables is essentially the same. The point of the discussion here is that we are not now choosing the variables on the basis of the topology of the network, but on the basis of physically important quantities in the circuit.

Instead of writing the equations all at once, let us draw the signal flow graph in steps, writing the corresponding equations one by one. The grid-cathode voltage of the first tube is

$$V_3 = V_1 - V_4 \tag{11-15}$$

The corresponding entry in the signal flow graph is shown in Fig. 11-19(a).

Owing to the grid-cathode voltage, there will be a current in the first tube, given by

$$I_1 = \frac{\mu}{r_p + R_2} V_3 - \frac{1}{r_p + R_2} V_4 \tag{11-16}$$

The signal flow graph is now shown in Fig. 11-19(b). The plate currents will cause a voltage across the cathode resistor R_1, given by

$$V_4 = R_1 I_1 + R_1 I_2 \tag{11-17}$$

The signal flow graph now takes the form of Fig. 11-19(c).

With a voltage between grid and cathode of the second tube, there will be a current in this tube, given by

$$I_2 = -\frac{\mu}{r_p + R_2} V_4 - \frac{1}{r_p + R_2} V_4 \tag{11-18}$$

This adds two parallel branches to the signal flow graph, as shown in Fig. 11-19(*d*). Finally, the output voltage is given by

$$V_2 = -R_2 I_2 \tag{11-19}$$

The final graph is shown in Fig. 11-19(*e*).

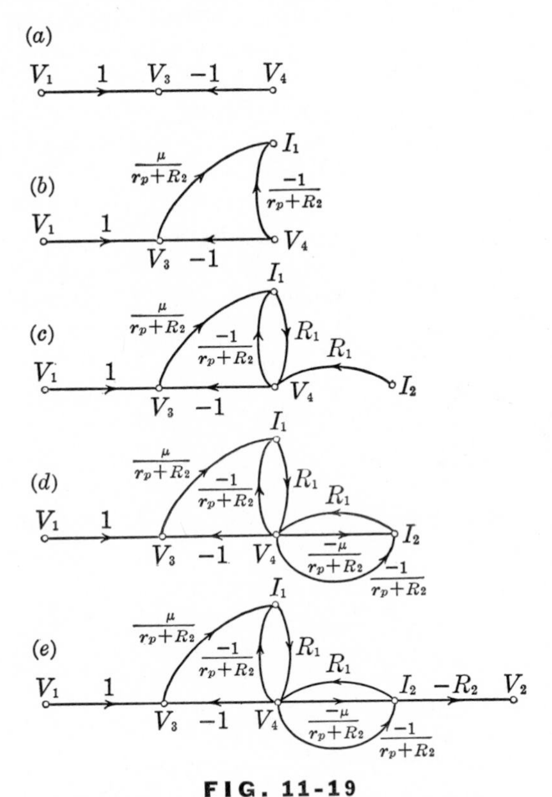

FIG. 11-19

Development of a signal flow graph

In developing this graph, we wrote the equations in an order that is presumed to indicate the sequence of cause and effect: The grid signal causes the plate current, which causes a cathode voltage, etc. These ideas are intuitive, however, and the actual "signal flow" is immaterial. As long as the equations are independent and the set contains an adequate number, that is all that is required. If intuitive ideas about signal flow will lead to such equations, so much to the good.

Now that we have the signal flow graph, let us proceed to reduce it. Note that there are a total of three feedback loops: $V_3I_1V_4V_3$, $I_1V_4I_1$, and $V_4I_2V_4$. No two branches are in cascade, so that the cascade transformation cannot be performed. There are two branches in parallel between V_4 and I_2, however, and these may be combined into a single branch.

In following the procedure which we have developed for eliminating nodes, let us inquire whether it is possible to eliminate more than one node at a time. This should be possible if it could be guaranteed that all path transmittances between the remaining nodes are unmodified. Let us illustrate this by eliminating nodes I_1 and V_3 together. The remaining graph is shown in Fig. 11-20(a). In the original graph there

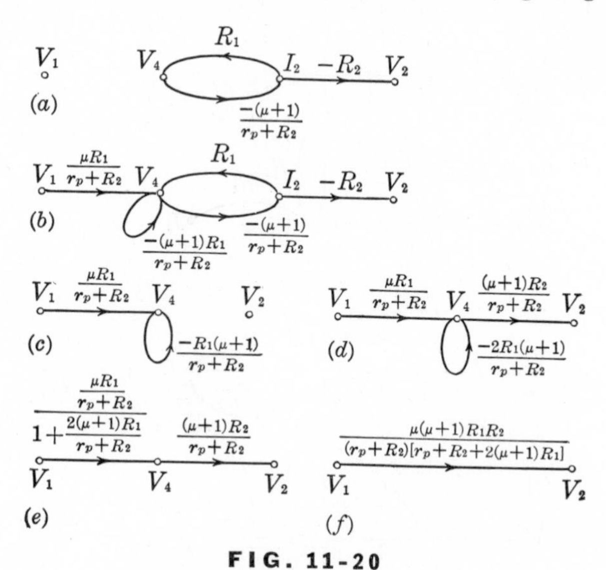

FIG. 11-20

Reduction of signal flow graph

is a path from V_1 to V_4 through V_3 and I_1, having a path transmittance $\mu R_1/(r_p + R_2)$. Hence we must now place a branch having this transmittance from V_1 to V_4. The only other paths in the original graph passing through V_3 or I_1 are the ones from V_4 to V_3 to I_1 back to V_4, having a path transmittance $-\mu R_1/(r_p + R_2)$, and from V_4 to I_1 back to V_4, having a path transmittance $-R_1/(r_p + R_2)$. Hence, in the new graph, we must place two self-loops at V_4 having these transmittances. Since these loops are in parallel, they can be combined into a single self-loop. The resulting graph is shown in Fig. 11-20(b).

Let us now make an observation concerning this simultaneous removal of two nodes. Look at the original signal flow graph in Fig. 11-19 and suppose that node V_3 alone is removed. This will not cause a self-loop to appear at I_1. Similarly, suppose that node I_1 alone is removed; this time, no self-loop will be formed at node V_3. Suppose a self-loop had been formed at one or the other of these nodes when the other was removed; then the simultaneous elimination would not have been possible, since, before this node could be eliminated, the self-loop would have to be removed, and we have made no provision for this step in the simultaneous-elimination process. The answer to our original question is now clear. Any number of nodes can be eliminated simultaneously as long as there is no feedback loop from any one of these nodes back to itself through any one of the others.

To continue the graph reduction, as the next step let us eliminate node I_2; Fig. 11-20(*c*) shows the preliminary step. The only paths through I_2 in Fig. 11-20(*b*) are from V_4 to V_2 and from V_4 back to itself. Hence we should insert in Fig. 11-20(*c*) a branch from V_2 to V_4, and a self-loop at V_4. This self-loop is in parallel with the one already there, and hence can be combined with it. The resulting graph is shown in Fig. 11-20(*d*).

There is now only one node remaining besides the source and sink, but this node has a self-loop. The self-loop can be removed if all incoming transmittances are divided by 1 minus the self-loop transmittance. In the present case there is only one incoming branch, and the resulting graph is shown in Fig. 11-20(*e*).

Finally, the cascade transformation leads to the result given in Fig. 11-20(*f*). You are invited to verify this result by using standard methods of analysis on the original network in Fig. 11-18.

Judging from the details with which we have gone through this example, you might get the impression that this procedure is long and drawn out. Actually, each of the steps is relatively simple and can be performed by observation. It takes much longer to discuss the procedure than actually to carry it out. One step which can be omitted, for example, is the writing of the equations. Since each equation is relatively simple, it can be visualized directly from the network diagram, and the signal flow graph can be drawn directly.

11.5 PERSPECTIVE

Let us now pause and take a critical look at the subject which was developed in this chapter. What we have is essentially a new method of solving a set of linear algebraic equations. The over-all method of analysis consists of two major parts: (1) setting up a signal flow graph for a given network, and (2), once the graph is at hand, calculating the graph transmittance for any one of the variables.

A step-by-step reduction procedure was introduced for gradually converting a graph to a single branch from source to sink. Since a node in the graph represents a variable in the corresponding set of equations, eliminating a node in the graph corresponds merely to eliminating a variable from the equations. Hence it might appear to be simply a matter of taste whether variables are systematically eliminated from the equations or nodes are eliminated from the graph. The graph reduction, however, has the advantage that more than one node can be eliminated at a time, subject to the restriction discussed in the last section. This is by no means a major advantage.

It should be remembered that a signal flow graph is simply a set of directed branches connected at nodes. On this basis we should expect that its topological properties play an important role in the relationships among the variables represented by the nodes. This is indeed the case. We are not yet equipped, however, to make a detailed study of these topological properties. Nevertheless, such studies have been successfully made, with the result that formulas have been developed for the over-all graph transmittance. The important thing about these formulas is that, once the graph is known, the over-all transmittance can be written down almost by inspection.

We shall here present one of these formulas without supplying a proof. It was initially developed by Samuel J. Mason in 1955. The formula involves a few concepts which we should first define. We have already defined the concepts of forward path, feedback loop, and path transmittance. We define a *loop transmittance* as the product of the transmittances of all branches on a loop. We say that two paths, open or closed, are *nontouching* if they have neither any branches nor any nodes in common. Letting T represent the over-all graph transmittance and T_k represent the transmittance of the kth forward path from the source to the sink, the formula is

$$T = \frac{1}{\Delta} \sum T_k \Delta_k \tag{11-20}$$

where Δ is the determinant of the graph and Δ_k is the determinant of that part of the graph which does not touch the kth forward path. Δ is given by

$$\Delta = 1 - \sum_j P_{j1} + \sum_j P_{j2} - \sum_j P_{j3} + \cdots \tag{11-21}$$

The first summation includes the loop gains of all the feedback loops. In the second summation we add up the products of loop gains, taken two at a time, of all loops which are nontouching. In the third summation we add up the product of loop gains, taken three at a time, of all loops which are nontouching, etc.

Since some of the concepts are new to you, this expression may seem quite complicated. In reality, it is marvelously simple. Let us illustrate its use by means of the signal flow graph given in Fig. 11-19 for which we calculated the graph transmittance in Fig. 11-20 by the step-by-step reduction process.

Counting the two parallel branches from V_4 to I_2 as a single branch, we see that there are a total of three feedback loops and only one forward path that goes from V_1 to V_3 to I_1 to V_4 to I_2 to V_2. All three of the feedback loops are touching, so everything beyond the first summation in Eq. (11-21) is zero. The loop gains of the three feedback loops are

$$\begin{aligned} &\text{Loop } V_3I_1V_4V_3\text{:} && \left(\frac{\mu}{r_p + R_2}\right)(R_1)(-1) = -\frac{\mu R_1}{r_p + R_2} \\ &\text{Loop } I_1V_4I_1\text{:} && R_1\left(\frac{-1}{r_p + R_2}\right) = -\frac{R_1}{r_p + R_2} \\ &\text{Loop } V_4I_2V_4\text{:} && \left(\frac{-(\mu + 1)}{r_p + R_2}\right)(R_1) = \frac{-(\mu + 1)R_1}{r_p + R_2} \end{aligned} \tag{11-22}$$

Hence Eq. (11-21) becomes

$$\Delta = 1 + \frac{2(\mu + 1)R_1}{r_p + R_2} \tag{11-23}$$

The path transmittance of the forward path is

$$T_1 = (1)\left(\frac{\mu}{r_p + R_2}\right)(R_1)\left(-\frac{(\mu + 1)}{r_p + R_2}\right)(-R_2) = \frac{\mu(\mu + 1)R_1R_2}{(r_p + R_2)^2} \tag{11-24}$$

Since all the loops touch the forward path, the quantity Δ_1 is simply 1. Hence, finally, the graph transmittance becomes

$$T = \frac{T_1}{\Delta} = \frac{\mu(\mu + 1)R_1R_2}{(r_p + R_2)[r_p + R_2 + 2(\mu + 1)R_1]} \tag{11-25}$$

which agrees with the value found in Fig. 11-20.

Note that it was not really necessary to write down the individual loop transmittances as in Eq. (11-22), since these are easily obtained by inspection. It is clear that the amount of computation involved, at least in this example, is a small fraction of that required by the node-elimination procedure.

The formula given in Eq. (11-20) is not the only one which has been developed for a graph transmittance. Other expressions were developed by C. L. Coates in 1959 and by Charles Desoer in 1960. As a matter of fact, the subject is still not closed, and perhaps some of you will someday contribute to our understanding of it.

Problems

11-1 Demonstrate the equivalence of the flow graphs shown in Fig. P11-1.

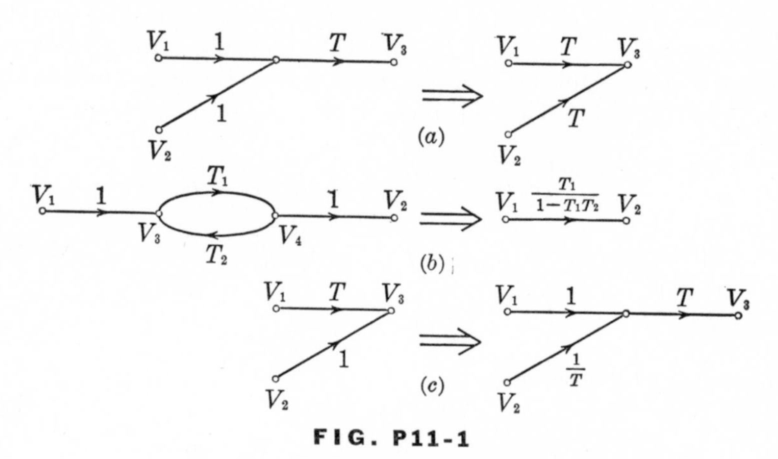

FIG. P11-1

11-2 In the flow graphs shown in Fig. P11-2, eliminate each node which is numbered, obtaining a new flow graph which is equivalent to the original one as far as the remaining nodes are concerned.

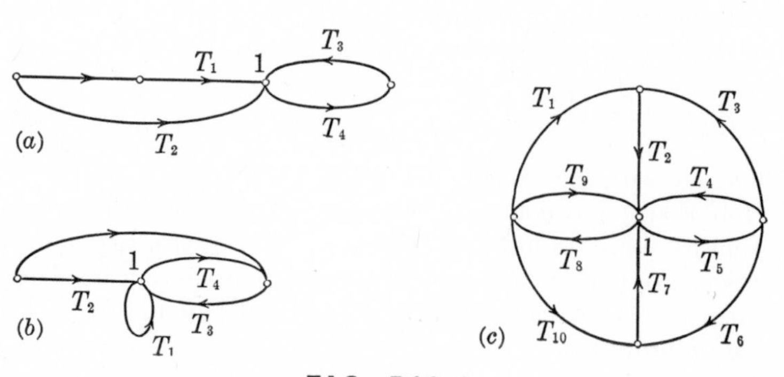

FIG. P11-2

11-3 Draw signal flow graphs representing the following sets of equations. Note the source nodes, the sink nodes, and the self-loops.

(a) $V_2 = V_1 - 3I_1$

$I_1 = 4I_2$

$I_2 = 5V_3 + 2V_2$

$V_3 = 2I_1 - 3I_2$

(b) $V_1 = 6V_2 - 3I_1 + 2I_2$

$I_1 = 2V_1 - 5I_2$

$I_2 = 3V_3 + 4I_1 - 2I_2$

$V_3 = V_1 - 3I_1$

(c) $I_1 = Y_1V_1 + Y_2V_2 + T_1I_2$

$I_2 = T_2I_1 + T_3I_2 + Y_3V_3$

$V_3 = T_4V_2 + Z_5I_1$

(d) $I_1 = 3V_1 - 2sV_2$

$I_2 = \dfrac{1}{s+1} I_1 + sV_1$

$V_2 = \dfrac{3}{s} I_1 - (2s + 1)I_2$

11-4 Choose one of the variables as a response in each of the graphs in the previous problem, and reduce the graphs to a single transmission from source node to sink, thus finding the transfer function. Find the transfer function again by solving the equations algebraically.

11-5 Write the equations represented by the signal flow graphs in Fig. P11-5.

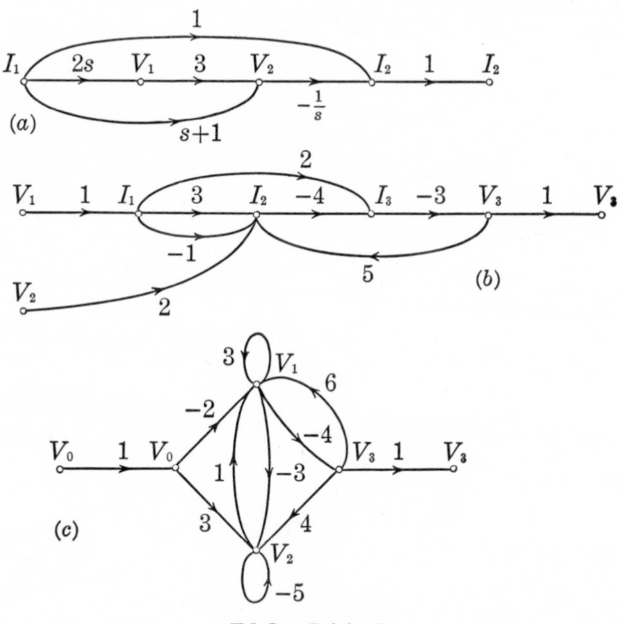

FIG. P11-5

11-6 Reduce each of the graphs in Problem 11-5 until only the source nodes and the sink nodes remain. Verify your results by solving the corresponding equations for the variables of the sink nodes.

11-7 Draw the signal flow graphs representing the networks in Problems 10-3 through 10-14. Solve the corresponding problem by reducing the graph.

11-8 Use the graph transmittance formula to find the appropriate transfer functions in problems 11-4, 11-5, and 11-7.

11-9 Figure P11-9 shows the diagrams of two amplifiers, the *cathode-coupled* amplifier and the *cascode* amplifier. Replace the tubes by their small-signal models. Find expressions for the indicated output voltage using signal flow graphs.

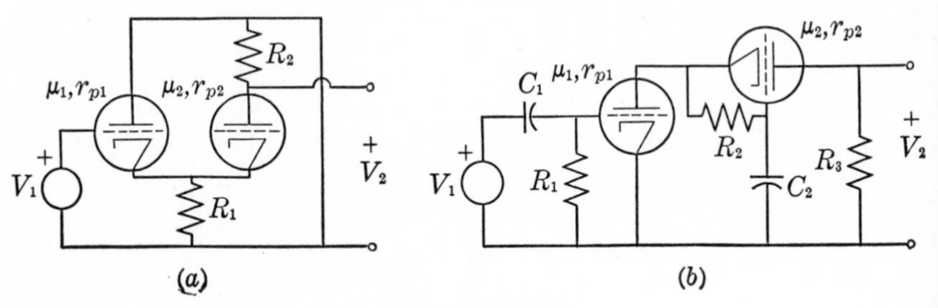

FIG. P11-9

NONSINUSOIDAL PERIODIC EXCITATIONS

12

AT several points early in this book, we made the comment that there are two aspects to network theory. One of these, and the one with which we have been concerned for the most part until the present, has to do with the consequences of interconnecting electrical elements to form a topological structure. The second aspect has to do with the signals which are the excitations and responses. To a large extent the present chapter will be devoted to this second aspect of the problem. The study of the properties of signals can be called *signal analysis*.

In the last eight chapters our attention was concentrated on the analysis of networks under the excitation of sinusoidal functions of time. Our interest was in determining the steady-state response. Remember that this study is but one phase of our over-all objective, which is to determine the complete response of a network for any excitation.

Before we attempt this ultimate problem, however, we shall momentarily remain in the realm of the steady state, choosing for excitations those functions which are periodic in time. Periodic functions have waveforms which recur periodically as time goes on. Hence we would intuitively expect the forced response of a network to such functions to recur periodically also, so that it is appropriate to refer to this as the steady-state response.

The sinusoidal function of time occupied our attention to a large extent in the preceding chapters. One reason for our interest in the sinusoid is that many of the waveforms actually used in electrical applications are sinusoidal, or approximately so. The greatest importance of the sinusoid, however, lies in the fact that any waveform whatsoever (subject to certain restrictions) can be represented as a combination of sinusoids.

The consequences of this fact are of fundamental importance in many branches of science and engineering. In electrical engineering this fact is of tremendous significance; it has even influenced our language and mode of thinking, so that we would hardly know how to express ourselves in certain areas without the concepts it supplies. We shall now concern ourselves with the exploitation of this fact. We shall not, however, devote much time to a detailed development of all the facets of the mathematical background. Instead, we shall outline the basic results, if only to introduce the pertinent notation. You are expected to consult other references for the main mathematical development.†

12.1 FOURIER SERIES

Infinite series consisting of trigonometric functions had been used in the solution of certain problems by such mathematicians as Euler and Lagrange as early as the late eighteenth century (1780). It was the French scientist Fourier, however, who, in 1822, recognized certain universal aspects of these series in the representation of periodic functions.

A periodic function is defined (see Chap. 1) as a function whose form repeats itself in uniform intervals of time $T = 2\pi/\omega_0$ called the period of the function. Analytically, this is expressed as

$$f(t + T) = f(t) \tag{12-1}$$

† For a rigorous mathematical treatment, see H. S. Carslaw, *Fourier Series and Integrals*, Cambridge University Press, New York, 1930. Other references are David K. Cheng, *Analysis of Linear Systems*, Chap. 5, Addison-Wesley Publishing Company, Inc., Reading, Mass., 1959; E. A. Guillemin, *Mathematics of Circuit Analysis*, Chap. 7, John Wiley & Sons, Inc., New York, 1949; B. J. Ley, S. G. Lutz, and C. F. Rehburg, *Linear Circuit Analysis*, Chap. 6, McGraw-Hill Book Company, Inc., New York, 1959.

Figure 1-24 in Chap. 1 showed some examples of periodic functions.

Fourier showed that a periodic function, subject to some restrictions which we shall soon mention, can be represented by means of an infinite series of trigonometric terms, as follows:

$$\begin{aligned} f(t) &= a_0 + a_1 \cos \omega_0 t + a_2 \cos 2\omega_0 t + \cdots + a_n \cos n\omega_0 t + \cdots \\ &\quad + b_1 \sin \omega_0 t + b_2 \sin 2\omega_0 t + \cdots + b_n \sin n\omega_0 t \\ &= \sum_{n=0}^{\infty} a_n \cos n\omega_0 t + b_n \sin n\omega_0 t \end{aligned} \tag{12-2}$$

In his honor, this series is called a Fourier series. The quantity $\omega_0 = 2\pi/T$ is the angular frequency of the periodic function. It is called the *fundamental* frequency. Each trigonometric function in the series has a frequency which is an integral multiple of the fundamental. These frequencies are called the *harmonics*.

The conditions which a periodic function must satisfy if it is to have a Fourier series are the following:

1. It may have discontinuities, but there must be only a finite number of them in one period.
2. It may be oscillatory, but there must be a finite number of maxima and minima in one period.
3. It must have a finite average value; furthermore, this should not be the result of cancellation of positive and negative areas. That is, the integral $\int |f(t)|\, dt$ over one period must be finite. This condition is stated concisely by saying that the function must be absolutely integrable over one period.

These conditions are called the *Dirichlet conditions*. It is clear that we would be hard pressed to find an engineering function which did *not* satisfy them.

If the periodic function $f(t)$ is given, the fundamental frequency ω_0 will be known; hence it remains only to find the coefficients of the series. From our knowledge that the average value of a sinusoid is zero, we note that, if both sides of Eq. (12-2) are integrated over one period, and if the series can be integrated term by term, only the constant term will contribute a nonzero value. Hence

$$\int_0^T f(t)\, dt = \int_0^T a_0\, dt$$

or

$$a_0 = \frac{1}{T} \int_0^T f(t)\, dt = \frac{1}{2\pi} \int_0^{2\pi} f(t)\, d(\omega_0 t) \tag{12-3}$$

The last form is obtained by inserting $T = 2\pi/\omega_0$ in the first part, and then changing the variable of integration from t to $\omega_0 t$. (The limits are also changed accordingly.)† But the result of integrating a periodic function over one period and then dividing by the period yields simply the average value of the function. Thus the constant term a_0 is the average value of $f(t)$; we call it the d-c component.

To find the remaining coefficients, we might think of performing some operation on Eq. (12-2) so that all the terms except one will vanish, just as in the determination of a_0. This thought is actually productive. Suppose we multiply both sides of the equation by $\cos k\omega_0 t$ and integrate over one period, again assuming that the series can be integrated term by term. (k is any integer.) The result will be

$$\int_0^T f(t) \cos k\omega_0 t \, dt = \sum_{n=0}^{\infty} a_n \int_0^T \cos n\omega_0 t \cos k\omega_0 t \, dt + b_n \int_0^T \sin n\omega_0 t \cos k\omega_0 t \, dt \qquad (12\text{-}4)$$

When n is any integer different from k, all the integrals on the right will be zero. This is easily verified by expanding the integrands by means of the identities

$$\begin{aligned} 2 \cos n\omega_0 t \cos k\omega_0 t &= \cos (n + k)\omega_0 t + \cos (n - k)\omega_0 t \\ 2 \sin n\omega_0 t \cos k\omega_0 t &= \sin (n + k)\omega_0 t + \sin (n - k)\omega_0 t \end{aligned} \qquad (12\text{-}5)$$

and noting that the integral of a sinusoid over one period is zero.

When $k = n$, the term $\cos k\omega_0 t \cos n\omega_0 t$ becomes $\cos^2 n\omega_0 t$, and the integral of this over one period is simply $T/2 = \pi/\omega_0$. (Verify this.) The coefficient of this term is a_n. Hence the result of this operation will be

$$a_n = \frac{2}{T} \int_0^T f(t) \cos n\omega_0 t \, dt = \frac{1}{\pi} \int_0^{2\pi} f(t) \cos n\omega_0 t \, d(\omega_0 t) \qquad (12\text{-}6)$$

This formula gives the coefficients of all the cosine terms.

In a completely similar manner, the b_n coefficients can be found by first multiplying Eq. (12-2) by $\sin k\omega_0 t$ and then integrating term by term, assuming that this is permissible. The integrals will again all vanish except when $n = k$. The result will give an expression for b_n.

† If you consult other references on Fourier series, you will find a number of different systems of notation. In mathematics texts the variable x is usually used for $\omega_0 t$. The mathematics of Fourier series is probably best developed in terms of such a dimensionless variable. Our purpose, however, is not to give the neatest possible mathematical development. We wish, concurrently, to use variables suggestive of the engineering interpretations.

The formulas for all the coefficients will be collected here for easy reference

$$a_0 = \frac{1}{T}\int_0^T f(t)\,dt = \frac{1}{2\pi}\int_0^{2\pi} f(t)\,d(\omega_0 t) \qquad (a)$$

$$a_n = \frac{2}{T}\int_0^T f(t)\cos n\omega_0 t\,dt = \frac{1}{\pi}\int_0^{2\pi} f(t)\cos n\omega_0 t\,d(\omega_0 t) \qquad (b) \qquad (12\text{-}7)$$

$$b_n = \frac{2}{T}\int_0^T f(t)\sin n\omega_0 t\,dt = \frac{1}{\pi}\int_0^{2\pi} f(t)\sin n\omega_0 t\,d(\omega_0 t) \qquad (c)$$

You are urged to carry through the details of the steps which were just suggested and verify the vanishing of all the integrals which were claimed to vanish. This property of the trigonometric sine and cosine functions (that is, the vanishing of the integral over one period of a product of two sinusoids whose frequencies are related as above) is called the *orthogonality* property.

Note that all the integrations cover one period ranging from 0 to T (or 0 to 2π). It is not necessary to cover this particular range as long as one period is covered; that is, the same results will be obtained if the range of integration is from t_1 to $T + t_1$, where t_1 is an arbitrary value of time.

There are still a few points that should be clarified in the preceding development. First of all, the procedure for calculating the *Fourier coefficients*, as the a's and b's are called, does not constitute a proof that the series exists and converges to the specified periodic function $f(t)$. This requires a fairly extensive development which, needless to say, we shall not indulge in here. Assuming that the series representation of $f(t)$ exists, however, then Eq. (12-7) will give the correct values of the coefficients.

Second, in order to arrive at these formulas, we assumed that the series could be integrated term by term. It is a mathematical theorem that term-by-term integration of a series is possible only if the series converges uniformly within the interval of integration. (See any calculus book.) As part of the proof of the Fourier theorem, it is proved that the series does indeed converge uniformly (except at points of discontinuity), so that the term-by-term integration is permissible. At a point of discontinuity, the Fourier series converges to the mean of the two values of the function on either side of the discontinuity.

For reference purposes let us state the preceding discussion as the following theorem which we shall call the *Fourier series theorem.*

If a periodic function $f(t)$ satisfies the Dirichlet conditions, then it can be represented by the series given in Eq. (12-2) with the coefficients specified by Eq. (12-7). The series converges to the function (and converges uniformly) at all continuous points; at a point of discontinuity t_1, the series converges to the mean of the values on each side of the discontinuity, that is, to $[f(t_1-) + f(t_1+)]/2$.

Although the series was written in Eq. (12-2) with both cosine and sine terms, it is possible to combine the sines and cosines by means of trigonometric identities, and to write the series either in terms of sines or of cosines. One possibility is

$$f(t) = a_0 + d_1 \cos(\omega_0 t + \phi_1) + d_2 \cos(2\omega_0 t + \phi_2) + d_3 \cos(3\omega_0 t + \phi_3) + \cdots = a_0 + \sum_{n=1}^{\infty} d_n \cos(n\omega_0 t + \phi_n) \tag{12-8}$$

where

$$d_n = \sqrt{a_n^2 + b_n^2} \qquad (a)$$

$$\phi_n = -\tan^{-1}\frac{b_n}{a_n} \qquad (b) \qquad (12\text{-}9)$$

We see that the function $f(t)$ is made up of a d-c component and an a-c component which consists of a *fundamental* and its *harmonics*. d_n and ϕ_n are called the *harmonic amplitudes* and *phase angles*, respectively. Note that d_n is the peak value of the harmonic, not its effective value.

A still different form can be obtained by expressing the cosine function in terms of exponentials. Thus

$$d_n \cos(n\omega_0 t + \phi_n) = \frac{d_n}{2}\left(\epsilon^{j(n\omega_0 t + \phi_n)} + \epsilon^{-j(n\omega_0 t + \phi_n)}\right)$$

$$= c_n \epsilon^{jn\omega_0 t} + c_n^* \epsilon^{-jn\omega_0 t} \tag{12-10}$$

where c_n is a complex quantity given by

$$c_n = \frac{d_n}{2}\epsilon^{j\phi_n} = \frac{1}{2}\sqrt{a_n^2 + b_n^2}\,(\cos\phi_n + j\sin\phi_n) = \frac{1}{2}(a_n - jb_n) \tag{12-11}$$

The right-hand side in terms of a_n and b_n was obtained by inserting Eq. (12-9).

With Eq. (12-10) inserted into Eq. (12-8), the series becomes

$$f(t) = c_0 + \sum_{n=1}^{\infty} c_n \epsilon^{jn\omega_0 t} + \sum_{n=1}^{\infty} c_n^* \epsilon^{-jn\omega_0 t} \tag{12-12}$$

(The constant term a_0 has been relabeled c_0 just for uniformity.) Let us, for convenience, use the notation $c_{-n} = c_n{}^*$, and in the second series on the right let us replace n by $-n$. The summation limits will then go from -1 to $-\infty$. The entire right-hand side can then be rewritten as a single summation as follows:

$$f(t) = \sum_{j=-\infty}^{\infty} c_n \epsilon^{jn\omega_0 t} \tag{12-13}$$

This expression is called the *exponential form* of the Fourier series. Note that the summation extends from $-\infty$ to $+\infty$. A formula for the c_n coefficients can be found by inserting those of a_n and b_n into Eq. (12-11). Alternatively, both sides of Eq. (12-13) can be multiplied by $\epsilon^{-jk\omega_0 t}$ and integrated over one period. In either case the result will be

$$c_n = \frac{1}{T} \int_0^T f(t) \epsilon^{-jn\omega_0 t}\, dt = \frac{1}{2\pi} \int_0^{2\pi} f(t) \epsilon^{-jn\omega_0 t}\, d(\omega_0 t) \tag{12-14}$$

Note that c_n is a complex quantity which contains the information of both the harmonic amplitude and the phase; but the magnitude of c_n, $|c_n|$, is one-half the nth-harmonic amplitude.

Since a periodic function goes through its periodic variations for all time, the specific point which is chosen as the origin of time should not be of great significance. Let us now determine what the effect will be on the Fourier coefficients if the origin of time is shifted by an amount t_1. That is, suppose the Fourier series for the function shown in Fig. 12-1 has been written, following which the origin is moved to

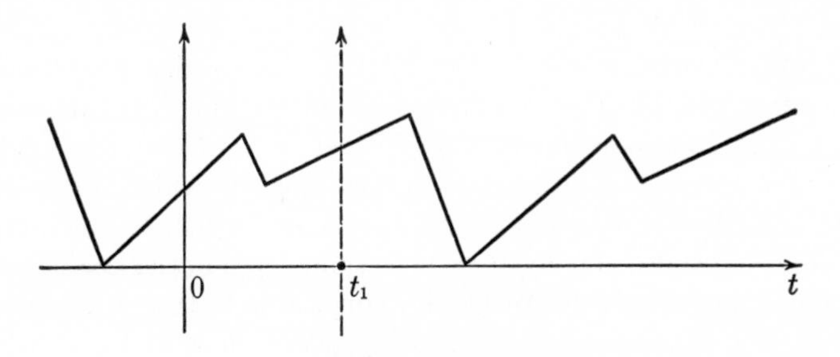

FIG. 12-1
Shifting the time origin

the point t_1. Any point t relative to the old origin will now have the value $t - t_1$. Hence the effect on the Fourier series is simply to replace t by $t - t_1$. If this is done in the exponential form of the series given in Eq. (12-13), the result will be

$$f(t - t_1) = \sum_{n=-\infty}^{\infty} c_n \epsilon^{jn\omega_0(t-t_1)} = \sum_{n=-\infty}^{\infty} (c_n \epsilon^{-jn\omega_0 t_1}) \epsilon^{jn\omega_0 t} \tag{12-15}$$

We see that the coefficients are modified. Since the magnitude of $\epsilon^{jn\omega_0 t_1}$ is unity, however, it is only the phase which is modified. The phase of each harmonic is shifted by an amount proportional to the frequency of that harmonic. This is the only influence in the shifting of the time origin.

The preceding result can often be exploited in finding the real form of the Fourier series of a given function. Thus, if by shifting the time origin it is possible to make the phase of each harmonic become an even or odd multiple of $\pi/2$, only cosine or sine terms will appear. This is a computational simplification.

12.2 THE FREQUENCY SPECTRUM

In discussing the properties of a function of time, a signal, we are accustomed to emphasizing certain aspects of its shape, such as its amplitude, its duration, etc. These aspects describe the signal in the *time domain.* We have now found that a periodic signal can be described in an alternative manner, by means of a Fourier series, as a superposition of sinusoids whose frequencies are *harmonically* related. (That is, all the frequencies are integral multiples of a lowest one, the fundamental.)

In other words, a signal is completely described if a statement is made, either analytically or graphically, of its functional dependence on time. It is also completely described in terms of the harmonic amplitudes and harmonic phases of the sinusoids which make up the signal. All the information about the signal which is conveyed by specifying $f(t)$ is also conveyed by specifying the Fourier coefficients c_n. By glancing at Eq. (12-14) or (12-7), you will note that c_n (that is, the harmonic amplitudes and phases) is dependent on ω_0 and n. Just as a graphical portrayal of the signal can be obtained by plotting $f(t)$ against t, so also a graphical portrayal can be obtained by plotting the harmonic amplitudes and phases against $n\omega_0$. Since n is an integer, however, the abscissa does not take on all-continuous values, but only discrete values.

Let us now illustrate these ideas by means of some examples. Consider first the pulse train shown in Fig. 12-2. The pulse amplitude

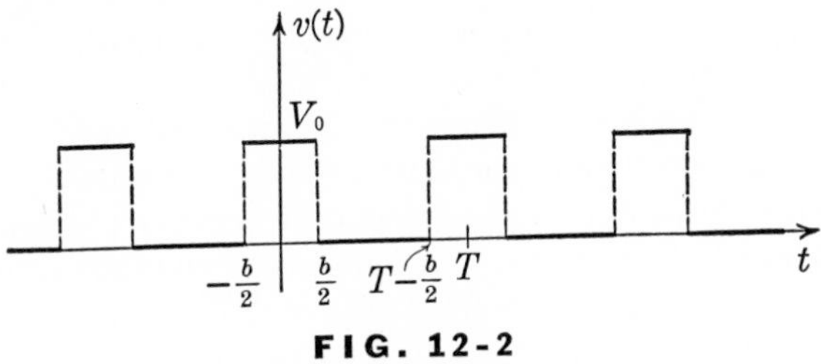

FIG. 12-2

Pulse train

is V_0, the pulse width is b, and the pulse repetition rate is $1/T$. (The radian frequency is $\omega_0 = 2\pi/T$.) The origin of time is chosen at the center of a pulse. Analytically, the function can be expressed as

$$\begin{aligned} v(t) &= V_0 \qquad \frac{-b}{2} < t < \frac{b}{2} \\ v(t) &= 0 \qquad \frac{b}{2} < t < T - \frac{b}{2} \\ v(t + T) &= v(t) \end{aligned} \tag{12-16}$$

Let us first calculate the coefficients of the exponential form, using Eq. (12-14). Instead of integrating from 0 to T, it is more convenient to integrate from $-b/2$ to $T - b/2$; this range still covers one period. Over part of this range the integrand will be zero, so the only contribution will come from the interval $-b/2$ to $+b/2$. Thus

$$c_n = \frac{1}{T} \int_{-b/2}^{b/2} V_0 \epsilon^{-jn\omega_0 t}\, dt = \frac{2V_0}{n\omega_0 T} \sin \frac{n\omega_0 b}{2} \tag{12-17}$$

As n takes on all values from $-\infty$ to ∞, this expression gives the corresponding values of c_n. Note that c_n and c_{-n} are the same for the same value of n in this example. The series representing the pulse train becomes

$$\begin{aligned} v(t) &= \frac{2V_0}{\omega_0 T} \left(\frac{\omega_0 b}{2} + \sin \frac{\omega_0 b}{2}\, \epsilon^{j\omega_0 t} + \frac{1}{2} \sin \omega_0 b\, \epsilon^{j2\omega_0 t} + \frac{1}{3} \sin \frac{3\omega_0 b}{2}\, \epsilon^{j3\omega_0 t} + \cdots \right. \\ &\qquad \left. + \sin \frac{\omega_0 b}{2}\, \epsilon^{-j\omega_0 t} + \frac{1}{2} \sin \omega_0 b\, \epsilon^{-j2\omega_0 t} + \frac{1}{3} \sin \frac{3\omega_0 b}{2}\, \epsilon^{-j3\omega_0 t} + \cdots \right) \\ &= \frac{V_0}{\omega_0 T} \left(\omega_0 b + \frac{4}{n} \sum_{n=1}^{\infty} \sin \frac{n\omega_0 b}{2} \cos n\omega_0 t \right) \end{aligned} \tag{12-18}$$

Combining the exponential leads to the trigonometric form given in the last line.

The constant term which is bV_0/T can also be found by calculating the average value of the function directly. The area of one pulse is bV_0. This is distributed over one period, thus giving an average value bV_0/T, in agreement with the value found above.

Note that the c_n coefficients, as given in Eq. (12-17), are purely real in this example. This means that the harmonic phases are all zero or 180 deg, which conclusion is borne out by the series in Eq. (12-18). Equation (12-17) can be rewritten in the following form:

$$c_n = \frac{bV_0}{T} \frac{\sin \dfrac{n\omega_0 b}{2}}{\dfrac{n\omega_0 b}{2}} \tag{12-19}$$

If we think of $n\omega_0 b/2$ as a continuous variable x, then the right-hand side has the form of $(\sin x)/x$. This function arises very often in signal analysis and is called the *sampling function;* that is,

$$Sa(x) \equiv \frac{\sin x}{x} \tag{12-20}$$

This function is shown plotted in Fig. 12-3. Note that $(\sin x)/x$ approaches 1 when x goes to zero.

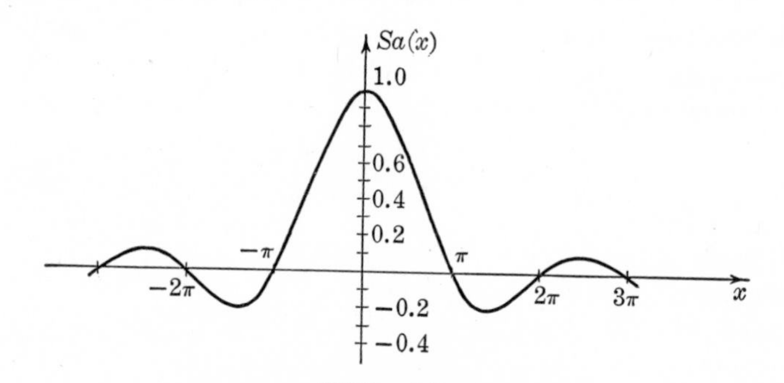

FIG. 12-3

The sampling function (sin x)/x

Let us now consider the graphical portrayal of the harmonic amplitudes and phases. Remember from Eq. (12-8) that d_n is the nth harmonic amplitude, and from Eq. (12-11) that the magnitude of c_n is just half of d_n. We have a choice, then, of portraying d_n or $|c_n|$. For definiteness, let us speak in terms of $|c_n|$ and still refer to it as the harmonic amplitude.

In order to display a graph, it will be necessary to give a numerical value to $\omega_0 b$ or, alternatively, to the ratio of the pulse width b to the period T. Figure 12-4 shows a plot of $|c_n|$ for the value $T/b = 5$.

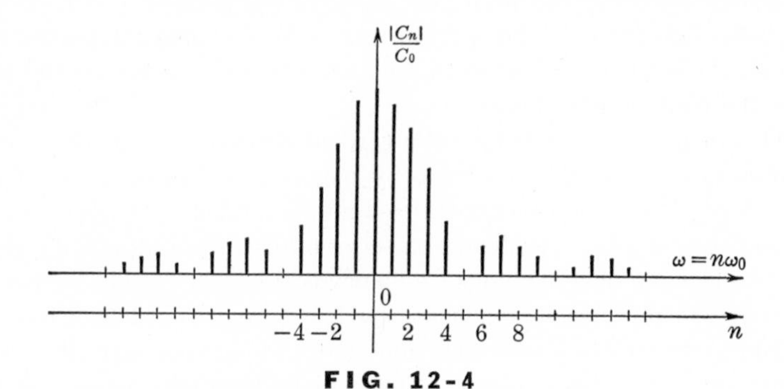

FIG. 12-4

Amplitude spectrum of pulse train

The graph is normalized by plotting $|c_n|/c_0$, instead of $|c_n|$, where $c_0 = bV_0/T = V_0/5$. The abscissa can be taken either as the frequency, which takes on the discrete values ω_0, $2\omega_0$, etc., or as the index n, which takes on integral values. Note that every fifth harmonic amplitude is zero.

The harmonic phase can also be portrayed in the same way. In the present case the phase is zero whenever c_n is positive, and π radians whenever c_n is negative. That is, $\phi_n = 0$ for $n = 0$ to 5, 10 to 15, 20 to 25, etc., and $\phi_n = \pi$ for $n = 5$ to 10, 15 to 20, etc.

These plots of $|c_n|$ and ϕ_n as a function of n are collectively called *frequency spectra;* individually, they are called the *amplitude spectrum* and the *phase spectrum*, respectively. For obvious reasons, they are also given the names *line spectra* or *discrete spectra.*

Consider the result of reducing the value of $\omega_0 b$ by reducing the frequency ω_0 or increasing the period. What would happen to the line spectrum in Figure 12-4? For example, suppose that the period is increased by a factor of four over what it was, thereby making $T/b = 20$. The argument of $\sin n\omega_0 b/2$ becomes $n\pi/20$, which becomes π for $n = 20$. That is, the amplitude of the twentieth harmonic will now be zero instead of the fifth harmonic. But since the fundamental frequency is now four times less than before, the actual frequency $\omega = n\omega_0$ at which the spectral amplitude is zero remains the same. This means that, keeping the scale in Fig.

12-4 the same, there will be four times as many lines in the spectrum, and no other effect. Of course $c_0 = bV_0/T$ will now be four times smaller, but, since the ordinate in Fig. 12-4 is normalized, this will have no influence. You can intuitively appreciate that, if the period is increased further and made to approach infinity, the lines in the spectrum will get denser and denser until, in the limit, no lines can be distinguished, only a continuous spectrum. The waveform is now no longer a pulse train but a single pulse (since we must wait an infinite time before the next one in the train comes along).

This intuitive argument can be verified rigorously, and the concept of representing a periodic function by means of a superposition of frequency components can be generalized and extended to nonperiodic, or transient, waveforms. Instead of a series of discrete components, there is now an integral of continuous frequency components. The frequency spectrum will now be a *continuous spectrum*, as opposed to a discrete spectrum. In this book we shall not undertake a thorough discussion and exposition of this subject. This will undoubtedly be one of the topics you will wish to pursue in more advanced studies.

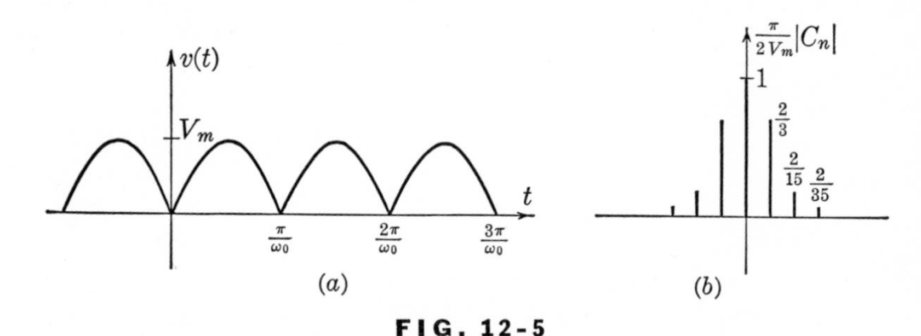

FIG. 12-5

Full-wave rectified sinusoid and its amplitude spectrum

Let us return now to our consideration of frequency spectra with another example. Consider the full-wave rectified sine function shown in Fig. 12-5. The frequency of the sinusoid is ω_0. Analytically, this function can be expressed as

$$v(t) = V_m \, |\sin \omega_0 t| \tag{12-21}$$

In evaluating the c_n Fourier coefficients using Eq. (12-14), it is necessary to break the range of integration into two parts: from $t = 0$ to π/ω_0, the function $v(t)$ will be $V_m \sin \omega_0 t$; from $t = \pi/\omega_0$ to $2\pi/\omega_0$, it will be $-V_m \sin \omega_0 t$, since $\omega_0 t$ is negative there. Thus the coefficients will be

$$c_n = \frac{V_m}{T}\int_0^{\pi/\omega_0} \sin \omega_0 t \epsilon^{-jn\omega_0 t}\,dt - \frac{V_m}{T}\int_{\pi/\omega_0}^{2\pi/\omega_0} \sin \omega_0 t \epsilon^{-jn\omega_0 t}\,dt \qquad (12\text{-}22)$$

The most expeditious way of evaluating these integrals is to express the sine in terms of exponentials. You should carry out the integrations. The resulting expression will be

$$c_n = -\frac{1 + (-1)^n}{\pi(n^2 - 1)} V_m \qquad \text{all } n \qquad (a)$$

or (12-23)

$$c_n = -\frac{2}{\pi(n^2 - 1)} V_m \qquad n \text{ even} \qquad (b)$$

We see that all the odd harmonics vanish, leaving only even harmonics. If we write the series and group the exponentials, we get the following trigonometric form:

$$v(t) = \frac{2V_m}{\pi}\left(1 - \frac{2}{3}\cos 2\omega_0 t - \frac{2}{15}\cos 4\omega_0 t - \frac{2}{35}\cos 6\omega_0 t - \cdots\right) \qquad (12\text{-}24)$$

The amplitude spectrum is shown in Fig. 12-5(*b*). Notice that, in comparison with the pulse train, the harmonic amplitudes of the rectified sinusoid decrease more rapidly as the order of the harmonic increases. This is an illustration of a general property of Fourier coefficients which we shall now briefly mention.

If the waveform of a function has rapid variations, a discontinuity being the most violent, it should intuitively be expected that a large number of harmonic components will be required to build up such a rapid variation. This intuitive expectation can be rigorously demonstrated. In the case of the pulse train, we found the harmonic amplitudes to decrease as $1/n$ (neglecting the variation of $\sin n\omega_0 b/2$). The waveform here contains discontinuities. In the present case the rectified sinusoid is not discontinuous, but its derivative is discontinuous at the beginning, at the end, and at the center of the period. We found the harmonic amplitudes to go down approximately as $1/n^2$ as n increases. It is a property of the Fourier coefficients that they decrease faster with increasing n the "smoother" the waveform of the function represented by the series. The "smoothness" of the waveform is described by the number of times it can be differentiated.

As a final example let us discuss an amplitude-modulated wave. An analytical expression for such a wave can be written

$$v(t) = f(t) \cos \omega_0 t \qquad (12\text{-}25)$$

Suppose $f(t)$ is a periodic function with fundamental frequency ω_1.

Then it can be represented as a Fourier series. Let us write it in the following form:

$$f(t) = \sum_{n=0}^{\infty} d_n \cos (n\omega_1 t + \phi_n) \tag{12-26}$$

Then the modulated wave $v(t)$ can be written

$$\begin{aligned} v(t) &= \sum_{n=0}^{\infty} d_n \cos (n\omega_1 t + \phi_n) \cos \omega_0 t \\ &= \sum_{n=0}^{\infty} \frac{d_n}{2} \{\cos [(\omega_0 + n\omega_1)t + \phi_n] + \cos [(\omega_0 - n\omega_1)t - \phi_n]\} \end{aligned} \tag{12-27}$$

The last form is obtained by using the appropriate trigonometric identity. (You should write this expression out at length to get a feeling for it.)

There are a few points to notice about this result. The harmonic amplitudes of the modulated wave $v(t)$ are half the corresponding amplitudes of the modulating wave $f(t)$. More important, the frequency spectrum of the modulated wave is a replica of that of the modulating wave, but moved up the frequency axis so that the spectrum is centered about the *carrier* frequency ω_0. The amplitude spectra of $f(t)$ and $v(t)$ are shown in Fig. 12-6. Notice that we have here plotted d_n instead of $|c_n|$; the difference is only a scale factor of 2 and the fact that c_n takes on values for negative n but d_n does not.

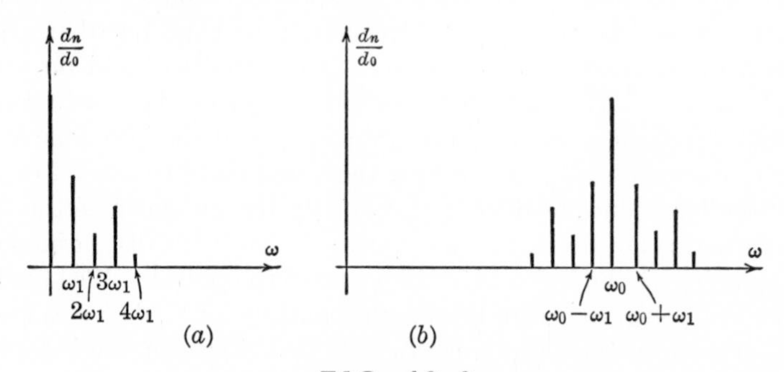

FIG. 12-6

Spectrum of amplitude-modulated wave

In Fig. 12-6(*b*) the harmonics of the modulating wave are clustered on either side of the carrier frequency. These two frequency intervals are called the *sidebands*—the *upper sideband* and the *lower sideband*.

In discussing Fourier series up to this point, we have assumed that there are an infinite number of frequency components. Generally, an infinite number will indeed be necessary to represent an arbitrary periodic function. Not all periodic functions, however, contain an infinite number of frequency components. For example, the function

$$f(t) = \cos^2 \omega_0 t = \frac{1}{2} + \frac{1}{2} \cos 2\omega_0 t$$

is a periodic function (sketch it) which has only a d-c component and a second harmonic. In fact, the function represented by the sum of any finite number of terms of a Fourier series will be a periodic function.

Any function whose frequency components are limited to a finite frequency interval, such as $\cos^2 \omega_0 t$, is called a *band-limited* function. In engineering work we often approximate a given function by the first few terms of its Fourier series. Thus the function we deal with is band limited.

There are many other topics useful in the evaluation of Fourier series which we could well discuss at this point. Some of these are: (1) the even and odd symmetry properties of certain waveforms, recognition of which would simplify computations; (2) the numerical evaluation of Fourier coefficients when signals are specified graphically; (3) differentiation and integration of series; etc. As already mentioned, however, this chapter is not intended to be a self-contained development of Fourier series; rather, our intention is to discuss the conceptual framework of Fourier analysis and to exhibit a few results useful in network analysis when periodic functions are applied to a network.

12.3 STEADY-STATE RESPONSE TO PERIODIC WAVEFORMS

So far in the chapter we have been concerned with analysis of signals; networks have not made an appearance. We are now ready to discuss the steady-state response of networks when the excitations are periodic functions.

Let us introduce the subject by means of an example. Suppose that the saw-tooth voltage waveform shown in Fig. 12-7 is applied to the series RC circuit. It is required to find the steady-state current. From the Fourier-series theorem we know that this waveform can be represented by a Fourier series, since the steady-state current will be periodic. Applying the formulas for the calculation of the coefficients, we find the Fourier series representing the sawtooth to be

$$v(t) = \frac{4}{\pi}\left(-\sin \pi t + \frac{1}{2}\sin 2\pi t - \frac{1}{3}\sin 3\pi t + \cdots\right) \qquad (a)$$

$$a_n = 0 \qquad (b) \qquad (12\text{-}28)$$

$$b_n = \frac{4}{n\pi}(-1)^n \qquad (c)$$

(You should carry out the details.)

Each of the frequency components can be considered to represent the voltage of a source, the actual source in the diagram being the series connection of the individual sources. Since the network is linear, the response to the sum of these excitations is the sum of the responses to the individual terms. As a matter of fact, just such a situation was discussed in Chap. 7, when we introduced the superposition principle.

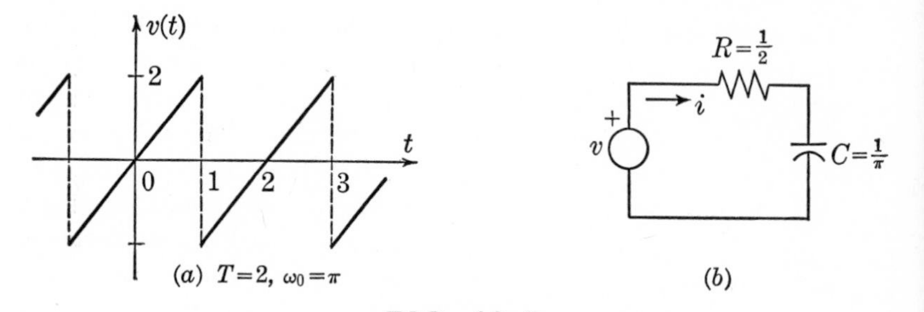

FIG. 12-7

Sawtooth voltage applied to series *RC* circuit.

Since each of the component terms is a sinusoid, the steady-state response to each term can be found by means of sinors. It is necessary only to be cautioned that each sinusoid has a different frequency, and this must be taken into account when computing the impedance. Furthermore, the b_n coefficients are the amplitudes—not the rms values—of the sinusoids.

Remembering the relationship between the c_n coefficients of the exponential form and the coefficients of the trigonometric form, we observe that c_n is proportional to the sinor representing the nth harmonic. (In fact, $c_n = d_n\epsilon^{j\phi_n}/2 = \sqrt{2}$ rms value $\epsilon^{j\phi_n}/2 = \text{sinor}/\sqrt{2}$.) Hence it is worthwhile dealing with the c_n coefficients. Using Eq. (12-28) in Eq. (12-11), we find

$$c_n = -j\frac{b_n}{2} = \frac{b_n}{2}\epsilon^{-j90°} = \frac{2}{n\pi}(-1)^n\epsilon^{-j90°} \qquad (12\text{-}29)$$

Let us now proceed to find the steady-state response to each sinusoid. The current sinor at each harmonic frequency will be the voltage sinor at that frequency multiplied by the admittance of the network evaluated at that frequency. The admittance of the RC circuit is

$$Y(jn\omega_0) = \frac{1}{R - j\dfrac{1}{n\omega_0 C}} = \frac{1}{\dfrac{1}{2} - j\dfrac{1}{n}} \tag{12-30}$$

For each harmonic, the appropriate value of n must be substituted. Thus

$$\begin{aligned}
&\text{Fundamental:} && Y(j\omega_0) = \frac{1}{0.5 - j0.1} = 0.895\epsilon^{j63.4°}\\
&\text{Second harmonic:} && Y(j2\omega_0) = \frac{1}{0.5 - j0.5} = 1.414\epsilon^{j45°}\\
&\text{Third harmonic:} && Y(j3\omega_0) = \frac{1}{0.5 - j0.333} = 1.66\epsilon^{j33.6°}\\
&\text{Fourth harmonic:} && Y(j4\omega_0) = \frac{1}{0.5 - j0.25} = 1.79\epsilon^{j26.5°}
\end{aligned} \tag{12-31}$$

Etc.

The current sinors (actually the c_n coefficients for the current) can now be found by multiplying the corresponding quantities in Eqs. (12-29) and (12-31). From physical considerations we know that there will be no d-c current in the circuit. This is borne out by the fact that Y in Eq. (12-30) becomes zero when n is zero. For the current, the c_n's are

$$\begin{aligned}
&\text{Fundamental:} && \frac{2}{\pi}\epsilon^{-j90°}(0.895\epsilon^{j63.4°}) = \frac{2}{\pi}(0.895\epsilon^{-j26.6°})\\
&\text{Second harmonic:} && \frac{1}{\pi}\epsilon^{-j90°}(1.414\epsilon^{j45°}) = \frac{2}{\pi}(0.707\epsilon^{-j45°})\\
&\text{Third harmonic:} && \frac{2}{3\pi}\epsilon^{-j90°}(1.66\epsilon^{j33.6°}) = \frac{2}{\pi}(0.553\epsilon^{-j56.4°})\\
&\text{Fourth harmonic:} && \frac{1}{2\pi}\epsilon^{-j90°}(1.79\epsilon^{j26.5°}) = \frac{2}{\pi}(0.448\epsilon^{-j63.5°})
\end{aligned} \tag{12-32}$$

Etc.

The trigonometric series for the current can now be written as

$$\begin{aligned}
i(t) = \frac{4}{\pi}[0.895 \cos(\pi t - 26.6°) + 0.707 \cos(2\pi t - 45°)\\
+ 0.553 \cos(3\pi t - 56.4°) + 0.448 \cos(4\pi t - 63.5°) + \cdots]
\end{aligned} \tag{12-33}$$

To find the steady-state current exactly requires that we calculate all harmonic components. An acceptable approximation, however, will be obtained if only a relatively few terms are retained, the number of terms depending on how fast the series converges and on how close an approximation is required.

As another example let us consider the power-supply filter network shown in Fig. 12-8. The box on the left can be considered a voltage

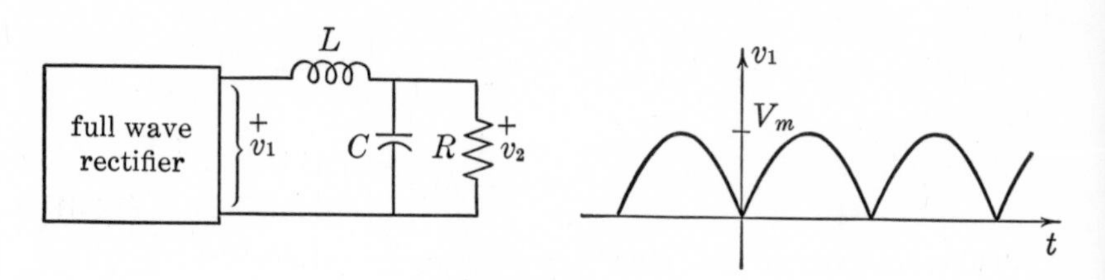

FIG. 12-8

Rectifier filter network

source having a full-wave rectified voltage waveform. It is desired to find the steady-state output voltage.

The Fourier series for the rectified sinusoid has already been calculated. [See Eqs. (12-22) and (12-23).] It remains to find the transfer-voltage ratio. This is easily found from the voltage-divider relationship to be

$$G_{21}(j\omega) = \frac{V_2}{V_1} = \frac{\dfrac{R}{j\omega RC + 1}}{\dfrac{R}{j\omega RC + 1} + j\omega L} = \frac{1}{1 - \omega^2 LC + j\omega L/R} \tag{12-34}$$

The frequency takes on only discrete values which are integral multiples of the fundamental frequency, $\omega = n\omega_0$. Hence G_{21} becomes

$$G_{21}(jn\omega_0) = \frac{1}{1 - n^2\omega_0{}^2 LC + jn\omega_0 L/R} \tag{12-35}$$

If we multiply the c_n coefficients of the exponential form of the series for the input voltage by $G_{21}(jn\omega_0)$ for corresponding values of n, the result will be the coefficients of the exponential form of the series for the output voltage. The result is tabulated for the numerical values given in Fig. 12-7 as follows

n	c_n	G_{21}	$G_{21}c_n$
0	$2/\pi$	1	$\frac{2}{\pi}$
2	$-2/3\pi$	$-0.0394\epsilon^{j14.6°}$	$\frac{2}{\pi}(0.0131)\epsilon^{j14.6°}$
4	$-2/15\pi$	$-0.0098\epsilon^{j7.2°}$	$\frac{2}{\pi}(0.00065)\epsilon^{j7.2°}$

Remembering that the magnitude of c_n is half the harmonic amplitude, the series for the output voltage becomes

$$v_2(t) = \frac{2}{\pi} V_m[1 + 0.0262 \cos (2\omega_0 t + 14.6°) + 0.0013 \cos (4\omega_0 t + 7.2°) + \cdots] \quad (12\text{-}36)$$

Although only two of the harmonic components (called the *ripple*) are calculated, it is apparent that the next term is quite negligible. In fact, the amplitude of the largest ripple component (the second harmonic) is only about 2.6 per cent of the d-c value, whereas in the input waveform this percentage is about 67 per cent. The reason this circuit is called a filter is now obvious.

Having discussed these examples, the general procedure for finding the steady-state response of a network to a periodic excitation can now be stated. First we find the Fourier spectrum of the input waveform, calculating as many terms as are needed for an acceptable approximation. The most appropriate coefficients to calculate are the c_n coefficients of the exponential form, because these are proportional to the sinors of the corresponding harmonics.

Next we determine the network function, whether driving point or transfer, which is appropriate for the problem. In one of the examples, this was an admittance; in the other, it was a voltage-transfer function. This function is evaluated at the appropriate harmonic frequencies. Finally, the product of each c_n with the network function evaluated at the corresponding harmonic frequency gives the corresponding coefficient in the Fourier series of the response.

Observe that in the output waveform no frequency components can exist which are not present in the input waveform. This is a direct result of the linearity of the network. The effect of the network is simply to modify the harmonic amplitudes and phases. By contrast, in a nonlinear network the output waveform will, in general, contain frequency components which are not only absent in the input wave but are not even harmonics of the input frequency.

A note of pessimism should be injected here. Although the process we have described is straightforward, and the numerical calculations are not extremely tedious, yet, when the output series has been calculated, the waveshape of the output is not, in general, clear from the analytical expression for the series. If the actual form of the output wave is important, it is necessary to carry out a laborious point-by-point plot.

In many cases, however, the exact form of the wave is not important. For example, in the case of the filter problem, it is only required to know the approximate amplitude of the ripple, which is simply the second-harmonic amplitude.

These comments accentuate and amplify some of the introductory remarks of this chapter in which it was stated that the *conceptual* framework of Fourier analysis (also called *frequency analysis*), as opposed to the quantitative relationships it supplies, provides a means of thought whose importance perhaps surpasses the importance of the quantitative results.

Let us further illustrate this comment by considering the network shown in Fig. 12-9(a). A sinusoidal source v_1 with a frequency ω_0 is available, but a sinusoid of three times the frequency is desired. Using

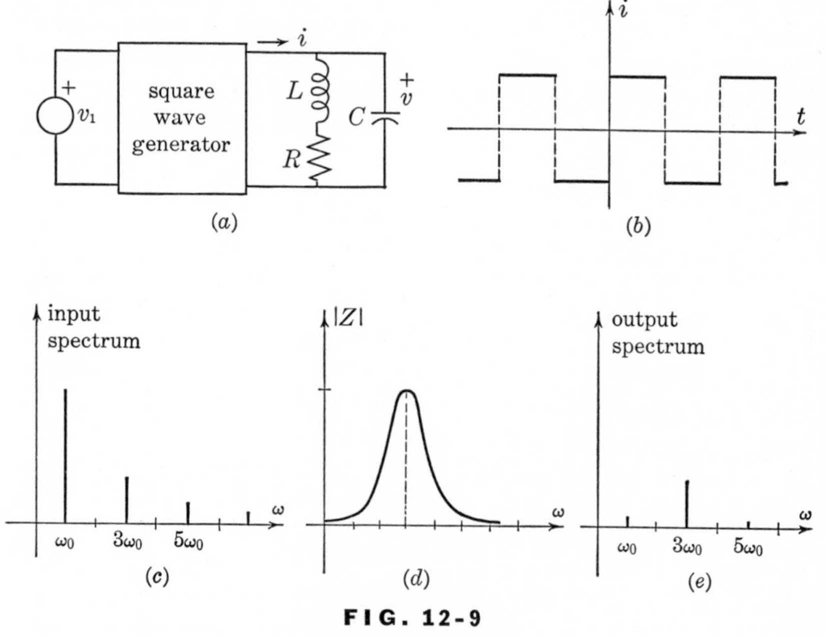

FIG. 12-9

Frequency multiplier

v_1, it is possible to generate a square wave by means of the box shown in the figure. (This network is obviously nonlinear, since the square wave will contain harmonics not present in the source v_1.)

For our purposes, let us assume that the current i supplied by the box to the parallel tuned circuit is a square wave. Its amplitude spectrum is shown in Fig. 12-9(c). The parallel resonant circuit is assumed to be high Q and tuned to the third-harmonic frequency. The resonance curve is shown in Fig. 12-9(d). The amplitude spectrum of the output is simply the product of these two graphs.

Because the response of the tuned circuit is low at the fundamental frequency, the amplitude of this frequency component in the output will be low, and the higher the Q the lower it will be. The same will be true of the fifth and higher harmonic amplitudes. The third harmonic amplitude, however, will remain relatively high, since the peak of the amplitude response of the tuned circuit occurs at that frequency. A possible output spectrum is shown in Fig. 12-9(e). Although a sine wave has not been obtained, since the output waveform contains other frequency components, these so-called distortion components are small. The over-all network is called a *frequency doubler*.

The major point to be made here is that, although quantitative results are involved in this example, this aspect is of less significance than the conceptual, qualitative analysis in terms of frequency spectra. This type of analysis can be applied in many specific engineering situations, and it is all a consequence of the concepts of Fourier analysis.

Very often, it may be desired to obtain approximate relationships concerning the response which can be expected from a network. In such a case the frequency response of the network can be idealized to simplify computation. As an illustration note that the essential features of the impedance plot in Fig. 12-9(d) are the relatively large values near $\omega = 3\omega_0$ and the relatively low values for all frequencies outside a range centered at $3\omega_0$. These features are represented approximately by the idealized band-pass amplitude response shown in Fig. 12-10. If the square wave had been applied to a network having this idealized response (there is no such network), then the output would have been a pure sinusoid with a frequency $3\omega_0$, without any of the other frequency components ap-

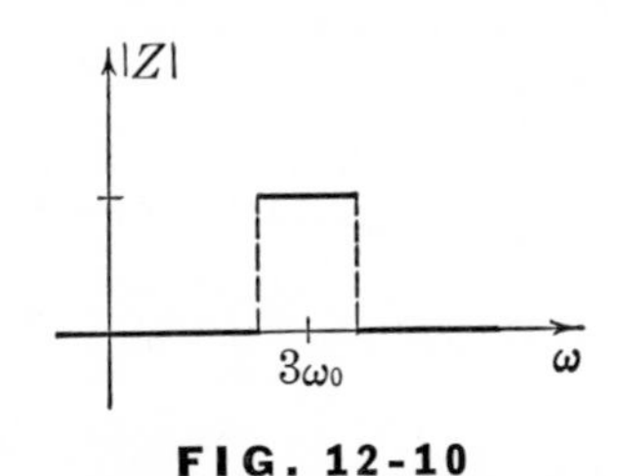

FIG. 12-10

Idealized band-pass response

pearing in Fig. 12-9(e). Thus a first-order approximation to the correct result would be obtained in terms of idealized frequency response.

Another question, similar to the preceding one, which may be answered by means of Fourier concepts is the following: In a signal-transmission network the fidelity with which the signal is transmitted is an important property. The only modifications of the signal which can be tolerated are a possible reduction in its amplitude and a possible delay in time. That is, if the signal going in is $f(t)$, the signal coming out should be $Kf(t - t_1)$; any other modification would distort the signal.

From the discussion in Sec. 12-1 of the influence of a shift in the time origin on the Fourier coefficients, we recall that the function $Kf(t - t_1)$ will have Fourier coefficients which are $K\epsilon^{-jn\omega_0 t_1}$ times the coefficients of the function $f(t)$. This means that the transfer function of the network should be $K\epsilon^{-jn\omega_0 t_1}$. That is, the amplitude response should be constant for all frequencies, and the phase response should be proportional to frequency. A network having these characteristics (*flat amplitude, linear phase*) is called a *distortionless transmission network*. It is impossible for a physical network to have such a transfer function. (In Fig. 7-35, however, we did see one example of a network having a flat amplitude response, an all-pass network, though not a linear phase.) An important engineering problem is the design of an electric circuit whose behavior approximates that of the ideal distortionless network.

12.4 POWER AND ENERGY

One of the problems of importance in network theory, no matter what the waveform of the excitation, is the determination of the power transmitted through a network, the power dissipated in it, and the energy stored in it. Recall that, at each instant of time, the power dissipated in a network and the energy stored in an inductor and capacitor are

$$p(t) = v(t)i(t) \qquad (a)$$

$$w_L(t) = \frac{1}{2} Li^2(t) \qquad (b) \qquad (12\text{-}37)$$

$$w_C(t) = \frac{1}{2} Cv^2(t) \qquad (c)$$

In each case we have the square of a voltage or current, or a product of the two.

Of greater interest than the instantaneous variations of these quantities is their average value over a period of the function. For *sinusoidal* voltages and currents, these average values are

$$P = V_{\text{rms}} I_{\text{rms}} \cos \theta \qquad (a)$$

$$W_L = \frac{1}{2} L I_{\text{rms}}^2 \qquad (b) \qquad (12\text{-}38)$$

$$W_C = \frac{1}{2} C V_{\text{rms}}^2 \qquad (c)$$

We shall now inquire into the values of these quantities when the waveforms are periodic but nonsinusoidal. Let the voltage and current be expanded in Fourier series as follows:

$$\begin{aligned} v(t) &= V_0 + V_{m1} \cos (\omega_0 t + \alpha_1) + V_{m2} \cos (2\omega_0 t + \alpha_2) + \cdots \\ i(t) &= I_0 + I_{m1} \cos (\omega_0 t + \beta_1) + I_{m2} \cos (2\omega_0 t + \beta_2) + \cdots \end{aligned} \qquad (12\text{-}39)$$

The notation we have used here is more in keeping with the physical significance of the function at hand, rather than the general notation which we used before for the Fourier series of an arbitrary function. Thus V_0 and I_0 represent the d-c components; V_{mk} and I_{mk} represent the amplitudes of the kth harmonics.

The instantaneous power is the product of v and i. When the two series in Eq. (12-39) are multiplied, the result will be a fairly extensive expression. Two types of terms, however, can be distinguished: (1) products involving the same harmonic frequency, and (2) products involving different frequencies. To find the average power these terms must be integrated over one period. But, according to the orthogonality property of sinusoids, which we have already noted, the integral of each cross-product term will disappear. The integral of the term involving the product of the two kth harmonics will be

$$\begin{aligned} \int_0^T v_k i_k \, dt &= \int_0^T V_{mk} I_{mk} \cos (k\omega_0 t + \alpha_k) \cos (k\omega_0 t + \beta_k) \, dt \\ &= V_{mk} I_{mk} \int_0^T \cos (k\omega_0 t + \beta_k + \theta_k) \cos (k\omega_0 t + \beta_k) \, dt \end{aligned}$$

In the last step we have defined $\theta_k = \alpha_k - \beta_k$ as the phase difference between the kth harmonic voltage and current. Now, expanding the first cosine by the appropriate identity and integrating, we get

$$\int_0^T v_k i_k \, dt = T \frac{V_{mk} I_{mk}}{2} \cos \theta_k \qquad (12\text{-}40)$$

With these considerations we can now express the average power as

$$P = \frac{1}{T}\int_0^T vi\,dt$$
$$= V_0I_0 + \frac{1}{2}V_{m1}I_{m1}\cos\theta_1 + \frac{1}{2}V_{m2}I_{m2}\cos\theta_2 + \cdots \qquad (12\text{-}41)$$
$$= P_0 + P_1 + P_2 + P_3 + \cdots$$

where

$$P_k = \frac{1}{2}V_{mk}I_{mk}\cos\theta_k \qquad (12\text{-}42)$$

is the average power of the kth harmonic. (You should fill in all the omitted steps.)

This is a very interesting result. It states that the average power input to a network when the signals are periodic is the sum of the average powers of the individual harmonics. This theorem (and its generalization to random signals) is called *Parseval's theorem.*

The average power can be computed in either of two ways. If the waveforms of the voltage and current are known as functions of time, a time integration is performed. Or, if the frequency spectra are known, the power can be computed as a summation of the harmonic powers. Of course, these two procedures are completely equivalent, since there is a one-to-one correspondence between a waveform and its frequency spectrum. Nevertheless, determination of the average power from the frequency spectra permits the acquisition of additional information which cannot be obtained by the alternative method.

Suppose, for example, that the network to which the power is applied is frequency selective, such as the tuned circuit previously considered. That is, certain frequency components are "passed" to the output while others are "stopped." It is of importance to know what fraction of the power is contained in the frequency components which fall within the pass band of the network. For this purpose it is necessary to carry out the summation in Eq. (12-41) only over the appropriate frequency components. It is thus possible to speak of the power contained within a particular frequency band.

Another particularly useful relationship can be derived from Eq. (12-41) by considering the network to be simply a resistor. Then $v = Ri$, and, when this expression is inserted, we get $V_{mk} = RI_{mk}$ and $\theta_k = 0$, since the voltage and current are in phase. Thus we find

$$P_R = R\left(I_0{}^2 + \frac{1}{2}I_{m1}{}^2 + \frac{1}{2}I_{m2}{}^2 + \cdots\right) \tag{12-43}$$

Remembering the definition of the rms value of a periodic waveform, we can now express the rms value of the current as

$$I_{\text{rms}} = \sqrt{I_0{}^2 + I^2_{\text{rms }1} + I^2_{\text{rms }2} + \cdots} \tag{12-44}$$

where we have used the fact that for a sinusoid $2I_m{}^2 = I^2_{\text{rms}}$.

In words, this result states that *the rms value of a periodic function is equal to the square root of the sum of squares of the rms values of the harmonic components, including the d-c component.*

What has been done in terms of the average power input to a network can also be repeated for the average energy stored in an inductor or a capacitor. To start, we substitute the appropriate Fourier series in Eqs. [12-37(b)] and [12-37(c)]. The calculations will again involve the product of two Fourier series, except that now both of them will represent either the voltage or the current. The same considerations of orthogonality will be pertinent. Instead of both V_m and I_m in Eqs. (12-40) and (12-41), we shall have one or the other. Furthermore, the angles θ_k will be 0. Thus, corresponding to Eq. (12-40), we have

$$\int_0^T i_k{}^2\,dt = T\frac{I_{mk}{}^2}{2} \qquad \text{or} \qquad \int_0^T v_k{}^2\,dt = T\frac{V_{mk}{}^2}{2} \tag{12-45}$$

Finally, adding terms, the result will be

$$\begin{aligned} W_L &= \frac{1}{2}L(I^2_{\text{rms }1} + I^2_{\text{rms }2} + \cdots) = \frac{1}{2}LI^2_{\text{rms}} & (a) \\ W_C &= \frac{1}{2}C(V^2_{\text{rms }1} + V^2_{\text{rms }2} + \cdots) = \frac{1}{2}CV^2_{\text{rms}} & (b) \end{aligned} \tag{12-46}$$

Thus we find that the same relationships that express the average energies for sinusoidal waveforms apply to the nonsinusoidal periodic case also, in terms of the rms values of voltage and current.

Problems

12-1 Find the Fourier series representations of the periodic waveforms shown in Fig. P12-1. Whenever convenient, shift the vertical axis, or the horizontal, to simplify computation of the coefficients. Compare rapidity of convergence. Plot the frequency spectra.

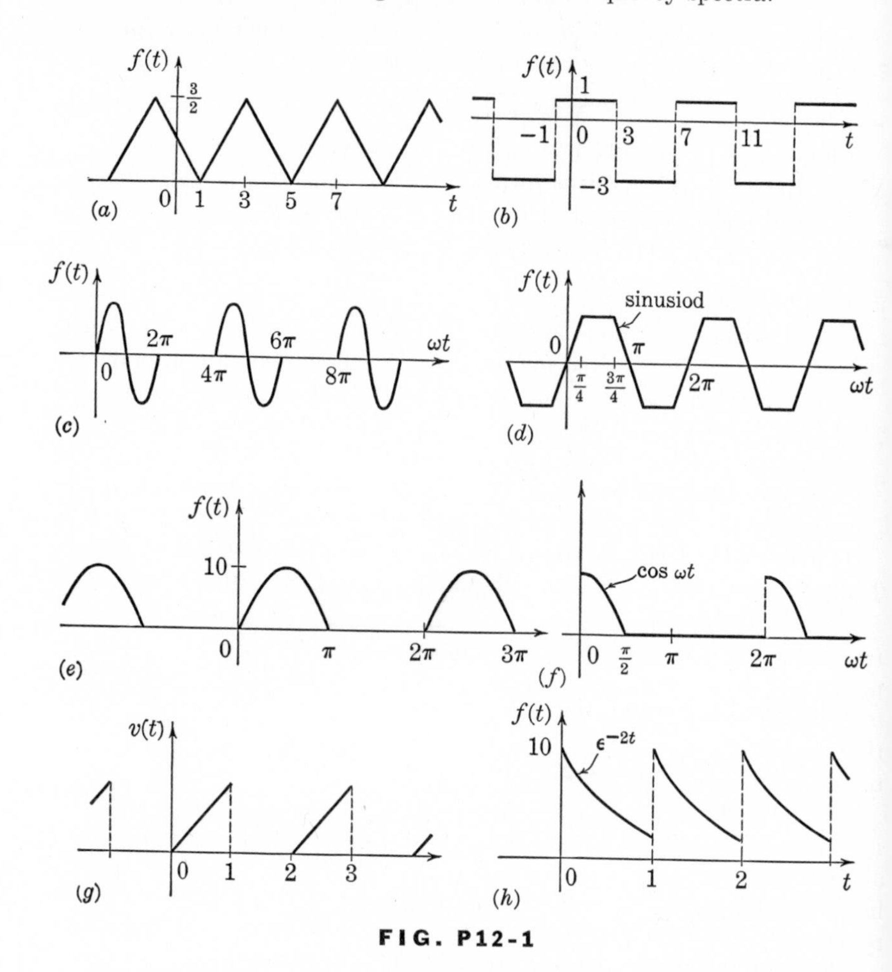

FIG. P12-1

12-2 Find the Fourier series representing a full-wave rectified sinusoid by using the series for the half-wave rectified sine shown in Fig. P12-1(e), in conjunction with a shift of the time origin.

12-3 Find the Fourier series representations of the two functions shown in Fig. P12-3. Differentiate, term by term, the first series and compare with the second one. Evaluate the second series at the points of discontinuity.

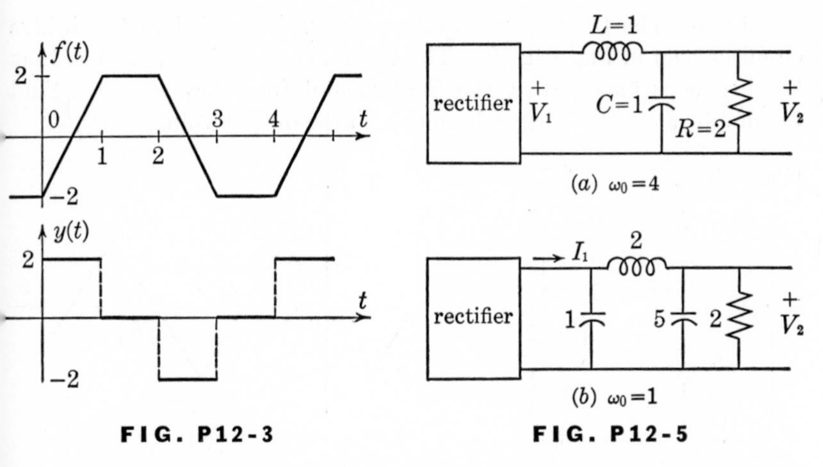

FIG. P12-3 **FIG. P12-5**

12-4 Find the Fourier-series representation of the following functions. (Use the binomial theorem to find the exponential form, as one alternative.)

(*a*) $f(t) = \cos^3 t$ (*c*) $f(t) = \cos^n \omega t$
(*b*) $f(t) = \sin^4 2t$ (*d*) $f(t) = 2\cos^2 \omega t \cos 10\omega t$

The last function is an amplitude-modulated wave.

12-5 Let the input to the filter networks shown in Fig. P12-5 be a full-wave rectified sinusoid. (*a*) Compute the ratios of second-harmonic and fourth-harmonic amplitudes to the d-c component in the filter output. (*b*) Repeat if the input is half-wave rectified. In part (*a*) the rectifier is assumed to be a voltage source; in part (*b*) it is assumed to be a current source.

12-6 A voltage source having one of the waveforms in Fig. P12-1 is applied to a series RL circuit: $R = 1$, $L = 2$. (*a*) Find the harmonic amplitudes and phases of the voltage across the inductor for the first few harmonics. Plot the waveform. (*b*) Find the power dissipated in the resistor by using the appropriate Fourier series.

12-7 Let the waveforms in Fig. P12-6 represent the current in a resistor. Find the power dissipated using the Fourier series. Also find the power by direct calculation and compare.

12-8 The current waveform in a circuit whose applied voltage is $10 \cos \omega t$ is the waveform shown in Fig. P12-1(*d*). (Can this circuit be linear?) Find the power supplied by the source.

12-9 Repeat the previous problem if the source is a d-c voltage of 100 and the current has the waveform of Fig. P12-1(*e*).

12-10 The triangular waveform shown is applied to the bridge circuit shown in Fig. P12-10. The variable branch is adjusted so that the bridge is balanced at the fundamental frequency $\omega_1 = 1$. Find the amplitude of the largest harmonic in the output voltage $v_2(t)$.

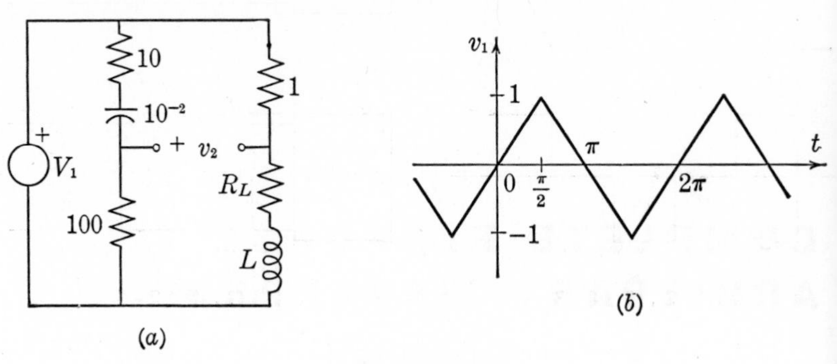

FIG. P12-10

COMPLETE RESPONSE TO ARBITRARY WAVEFORMS

13

WITH this chapter we have reached a turning point in our studies. It is now time to stand back and assess whatever progress we have achieved toward fulfilling our originally stated objective. This objective was the determination of the complete response of an arbitrary network, made up of various elements of the model which we have introduced, to excitations of arbitrary waveform.

In the first place, for a network containing resistors only, we found the pertinent equations describing the behavior of the network to be algebraic equations independent of the excitation waveforms. This fact is not changed if the network contains ideal amplifiers (and even ideal transformers) as well. For this case we have been able to find the complete response by solving the algebraic equations.†

When a network contains storage elements also, however, the equilibrium relationships are integrodifferential equations. The complete solution of such equations contains two components—the forced component and the natural component. In Chap. 3 we discussed a

† A simple extension permits the application of the same methods to networks containing any one kind of element—only inductors, or only capacitors. Again, ideal amplifiers and ideal transformers can also be included.

procedure, often called the *classical method*, for solving differential equations by finding the two components separately and then combining them. We carried out the solution of first-order and second-order networks by this procedure with certain simple "elementary" excitation functions such as polynomials in t, sinusoids, and exponentials.

The classical method of solution becomes increasingly difficult as the complexity of the network increases. But this is not its major drawback. The method for finding the forced response becomes computationally very difficult when the excitation is a function which cannot be simply expressed as a combination of the "elementary" functions. Remember that the forced response is assumed to have the same waveform as the excitation. But if the analytical form of the excitation function is different over different intervals of time, the solution for each such interval must be found separately. The procedure of assuming the forced response to have the same waveform as the excitation must be applied once for each interval of time. The natural component of the response must then be combined with the forced component in each interval.

Remember that there are as many constants in the natural response as there are natural frequencies. In the first interval these constants must be evaluated so that the solution fits the state of the network at the initial time, the initial conditions. With the complete solution in the first interval determined, the value of the response at the beginning of the second interval (which is the end of the first interval) must be found. The constants in the natural response of the second interval must now be computed so that the complete response fits the previously determined value at the beginning of the interval. This process is then repeated until all the intervals are exhausted. Clearly, this entire process leaves something to be desired.

Compare the amount of labor involved in this procedure with the relative ease of the sinor method of solution of the sinusoidal steady state. The relative simplicity of the sinor method is a direct consequence of the fact that the differential equations are transformed to ordinary algebraic equations, even though the variables and the coefficients in the equations are complex. If you meditate about this circumstance, you will undoubtedly begin to wonder whether a similar result cannot, perhaps, be accomplished by some sort of transformation even when the excitation is an arbitrary one and when we are interested in the complete response, not just the steady state. In this chapter we shall pursue this thought and exploit its consequences.

Historically, it was the English engineer Oliver Heaviside who first used transform methods (he called them operational methods) in the solution of network problems. The methods which he introduced were a product of his keen insight and inspiration. Since he was mainly interested in the solution of practical problems, he was indifferent to rigorous mathematical justification of his techniques. This led to strong antagonisms between Heaviside and the mathematicians of his day, and to bitter criticisms of his heuristic methods. Since then, his methods have been improved, and mathematical justification has been supplied.

Oliver Heaviside, like Michael Faraday before him, was one of those rare individuals who possess an intuitive insight into the workings of nature. With bold leaps of the imagination, they introduce new concepts and new ideas which seem strange and unorthodox to their contemporaries, and so not easily acceptable. When these ideas are refined, however, and made familiar, often by lesser men, they become the warp and woof of our knowledge.

13.1 THE LAPLACE TRANSFORM

It will be very helpful to recall here the basic features of the calculation of the steady-state response of a network to a sinusoidal excitation. The first step is to replace the sinusoidal time functions representing the variables by their sinors. We think of this step as a transformation. The integrodifferential equations resulting from the application of the fundamental laws are transformed into algebraic equations with the sinors as variables, since the operations of integration and differentiation become replaced by the operations of division by $j\omega$ (or s) and multiplication by $j\omega$, respectively. These equations are then solved for the sinors. Finally, the sinors are replaced by the sinusoidal functions of time which they represent. This last step can be thought of as the inverse of the first step, the inverse transformation.

It is our desire to seek a similar sequence of steps which will permit the calculation of a network response when the excitation is any given function of time. Furthermore, we are interested in the *complete* response, which means that the initial conditions must somehow be introduced.

In Chap. 12 we discussed the concept of a frequency spectrum and the fact that a periodic function can be represented by a series of sinusoids. We also mentioned briefly that the same idea can be extended to include functions which are not periodic, in which case the spectrum becomes a continuous one. For a periodic function $f(t)$ the following expressions were developed in the preceding chapter.

$$c_n = \frac{1}{T}\int_0^T f(t)\epsilon^{-jn\omega_0 t}\,dt \qquad (a)$$
$$f(t) = \sum_{n=-\infty}^{\infty} c_n\epsilon^{jn\omega_0 t} \qquad (b) \qquad (13\text{-}1)$$

where c_n is the general coefficient of the exponential form of the Fourier series representing $f(t)$.

This pair of equations has a mutual relationship; one of them converts a function of time $f(t)$ to a quantity c_n which is a function of $\omega = n\omega_0$. The other reconverts the c_n's back to a function of time. We can think of them as a pair of *transforms*, the first one being a *direct* transform and the second an *inverse* transform.

Without concentrating on the details, let us list the basic features of the expression in Eq. [13-1(a)]. First of all, the right side contains an integral; although the limits here are specific ones, basically we have a lower one and an upper one, say a and b. Next, the integrand is a product of two functions, one of which is $f(t)$; although the second function is a specific one here, its basic feature is that it is a function of both t and another variable n (or $\omega = n\omega_0$). Thus the relationship can be expressed in the following form:

$$F(x) = \int_a^b f(t)K(x,t)\,dt \qquad (13\text{-}2)$$

Such expressions arise quite often in many branches of science and engineering. We call such an expression an *integral transform.* The function $K(x,t)$ is called the *kernel* of the transform. After integration, the resulting function $F(x)$ is no longer a function of t; $F(x)$ is called the *transform* of $f(t)$. Many mathematical properties of integral transforms can be studied independently of the specific form of the kernel, and independently of the specific physical interpretation. Of course, we shall not enter into such studies. Using the Fourier interpretation, however, $F(x)$ is a kind of "spectrum" function of $f(t)$.

Consider the possibility of using an integral transform to solve the problem we have posed ourselves. If $f(t)$ is the variation with time of a network variable, using the transform of Eq. (13-2) certainly eliminates the time t, but it introduces another variable x. We do not yet know, however, to what kind of equations the differential equations in t are converted under such a transformation. It is conceivable that this may even depend on the kernel and on the limits of integration. Let us, therefore, decide on these.

In Eq. [13-1(a)] the integration is over one period. Nonperiodic functions, however, do not recur in any finite time interval. Another way of thinking of this is to say that nonperiodic functions recur only after an infinite time. Hence, choosing the upper limit at ∞ seems reasonable. As for the lower limit, the origin of time is arbitrary. That particular time following which it is desired to find a response, such as an instant of switching, is usually chosen as $t = 0$. For this reason it seems reasonable to choose 0 as the lower limit. Finally, as for the kernel, we have already indicated that for nonperiodic functions the Fourier spectrum loses its discreteness. Thus the nth harmonic frequency $n\omega_0$ of the Fourier series becomes just any frequency ω. We could, therefore, take the kernel to be $\epsilon^{-j\omega t}$.

This, however, is unsatisfactory for the following reason: Suppose $f(t)$ is constant, like a battery voltage. The integral transform with a kernel $\epsilon^{-j\omega t}$ would then become

$$\int_0^\infty (1)\epsilon^{-j\omega t}\,dt = \frac{1}{j\omega}(1 - \epsilon^{-j\omega\infty}) \tag{13-3}$$

By $\epsilon^{-j\omega\infty}$ we mean the limit of $\epsilon^{-j\omega t}$ as $t \to \infty$. But this limit does not exist; hence the integral transform of a constant, with this kernel, does not exist. A similar conclusion is reached if $f(t)$ is a sinusoid. Since the constant and the sinusoid (direct and alternating current) are two of the most important functions of interest, it behooves us to choose another kernel.

The situation is remedied if we can keep the exponential function but make its exponent real, say ϵ^{-xt}. Then, if $f(t)$ is a constant, substitution of the upper limit of the integral leads to a definite, finite result. This thought leads us to choose

$$F(x) = \int_0^\infty f(t)\epsilon^{-xt}\,dt \tag{13-4}$$

as our integral transform. In this development, however, we swung from a purely imaginary exponent $-j\omega t$ to a purely real exponent $-xt$. This is not really essential. It is enough to have a *complex* exponent; its real part can then act as a convergence factor to alleviate the difficulty engendered by the upper limit. And we have just the right quantity for this purpose, namely, the complex frequency variable $s = \sigma + j\omega$. Using s instead of x in Eq. (13-4), we get

$$F(s) = \int_0^\infty f(t)\epsilon^{-st}\,dt \tag{13-5}$$

This is still an integral transform, of course, but one with a specific kernel and specific limits. The integral defines a function of the variable s which is called the Laplace transform of $f(t)$ after the French mathematician who first introduced it in the eighteenth century. Sometimes it is convenient to use a symbolic notation for the purpose of referring to the transform. We write

$$F(s) = \mathcal{L}[f(t)] \tag{13-6}$$

where the symbol on the right is a script $\mathcal{L}$; we read this equation as "$F(s)$ is the Laplace transform of $f(t)$."

Although, as we said, s can be a complex quantity as long as its real part is great enough, when dealing with the integral in Eq. (13-5), which we call the *Laplace integral*, it is perhaps simpler to look upon s as a real quantity. Questions of convergence are then more easily handled.

13.2 EVALUATION OF SOME TRANSFORMS

Before we begin to discuss the use of the Laplace transform in the solution of network problems, let us spend some time familiarizing ourselves with some of the manipulations involved in evaluating transforms, and acquainting ourselves with the transforms of some simple functions.

As a first example take the exponential function ϵ^{xt}, where x may be real, positive or negative, or complex.

$$\mathcal{L}(\epsilon^{xt}) = \int_0^\infty \epsilon^{xt}\epsilon^{-st}\,dt = \left.\frac{\epsilon^{-(s-x)t}}{s-x}\right|_0^\infty \tag{13-7}$$

If the upper limit is to lead to a finite result, the real part of s, which is σ, should be greater than x. (Or if x is complex, σ should be greater than the real part of x.) Hence we get

$$\mathcal{L}(\epsilon^{xt}) = \frac{1}{s-x}; \quad Re\,(s) > Re\,(x) \tag{13-8}$$

The exponential function is a transcendental function, whereas $1/(s - x)$ is an algebraic function. Thus the Laplace transform has converted a function which is "high up" in the hierarchy of functions to one which is "lower down." This is a fairly general property of the Laplace transform, as you will observe.

With the transform of the exponential at hand, the transform of several other functions related to the exponential can be obtained simply. Consider the unit step function shown in Fig. 13-1(*a*). This can

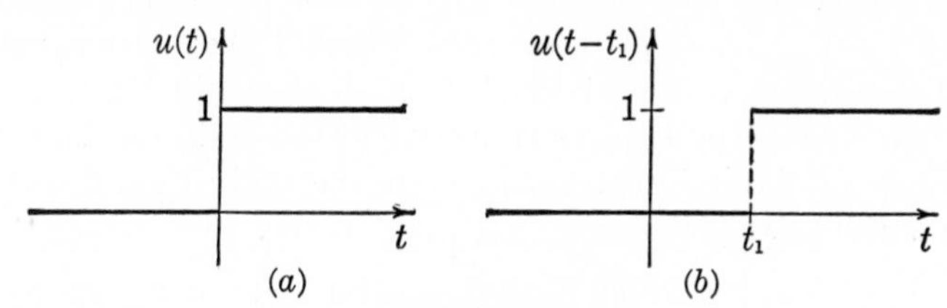

FIG. 13-1
Unit step function

be regarded as an exponential function ϵ^{xt}, defined for positive values of t, in which x goes to zero. Symbolically, we can write

$$\mathcal{L}[u(t)] = \mathcal{L}[\lim_{x\to 0} \epsilon^{xt}] \tag{13-9}$$

If we simply let x go to 0 in Eq. (13-8), we can write

$$\lim_{x\to 0} \mathcal{L}(\epsilon^{xt}) = \lim_{x\to 0} \frac{1}{s - x} = \frac{1}{s} \tag{13-10}$$

The question now arises whether finding the Laplace transform of the exponential and *then* taking the limit, as in this expression, leads to the same result as taking the limit (as x goes to zero) *first* and then taking the Laplace transform, as in Eq. (13-9). We were confronted with a similar type of question in the last chapter when we wished to know whether we could integrate the series in Eq. (12-4) term by term.† The answer is again the same and involves the uniform convergence of the integral in Eq. (13-7). That is, it is a theorem of mathematics that the two operations can be interchanged if the integral is uniformly convergent. In the present case this condition is known to be satisfied whenever the integral converges at all, namely for $Re\ (s) > Re\ (x)$. Hence the two operations can be interchanged. The conclusion is that

$$\mathcal{L}[u(t)] = \frac{1}{s} \tag{13-11}$$

Of course, the same result can be obtained by direct integration of the Laplace integral, which you are urged to do. We went through the

† This question can also be formulated in terms of interchanging two operations by asking whether the terms must be added first and *then* integrated or *first* integrated and then added.

preceding development in considerable detail because we shall later have occasion to discuss a similar case in which the interchange of two such operations is not mathematically justified.

The discontinuity of the unit step function shown in Fig. 13-1(a) occurs at $t = 0$. You will recall from Chap. 1 that a step function whose discontinuity occurs at time $t = t_1$, as shown in Fig. 13-1(b), is written $u(t - t_1)$. The Laplace transform of this displaced unit step is found easily by direct integration. Since the range from 0 to t_1 contributes nothing, we get

$$\mathcal{L}[u(t - t_1)] = \int_{t_1}^{\infty} (1)\epsilon^{-st}\, dt = -\frac{1}{s}\,\epsilon^{-st}\Big|_{t_1}^{\infty} = \frac{1}{s}\,\epsilon^{-t_1 s} \qquad (13\text{-}12)$$

Again, in order for the integral to converge, it is necessary that the real part of s be large enough; in this case, $\sigma > 0$ is sufficient.

We see that the transform of the unit step which is shifted along the axis t_1 units of time, is the same as that of the undisplaced step multiplied by $\epsilon^{-t_1 s}$. This result is of considerable significance and will be generalized soon.

Let us next find the Laplace transform of the trigonometric functions $\sin \omega t$ and $\cos \omega t$. Each of these can be written as a linear combination of two exponentials. Thus

$$\sin \omega t = \frac{\epsilon^{j\omega t} - \epsilon^{-j\omega t}}{2j} \qquad (a)$$

$$\cos \omega t = \frac{\epsilon^{j\omega t} + \epsilon^{-j\omega t}}{2} \qquad (b) \qquad (13\text{-}13)$$

Hence we can write

$$\mathcal{L}(\sin \omega t) = \mathcal{L}\left(\frac{\epsilon^{j\omega t} - \epsilon^{-j\omega t}}{2j}\right) = \mathcal{L}\left(\frac{\epsilon^{j\omega t}}{2j}\right) - \mathcal{L}\left(\frac{\epsilon^{-j\omega t}}{2j}\right)$$

$$= \frac{1}{2j}\left(\frac{1}{s - j\omega} - \frac{1}{s + j\omega}\right) = \frac{\omega}{s^2 + \omega^2} \qquad (13\text{-}14)$$

In this sequence of steps, we performed one that needs comment. We wrote the Laplace transform of a difference (or negative sum) of two functions as the difference of the Laplace transforms of the two functions. Thus the operation of taking the Laplace transform is commutative with the operations of addition or subtraction. In the second line we used Eq. (13-8) for the transform of an exponential, first with $j\omega$, then with $-j\omega$ replacing x.

In a similar way we can find

$$\mathcal{L}(\cos \omega t) = \frac{s}{s^2 + \omega^2} \qquad (13\text{-}15)$$

Note that, when computing the transforms of sin ωt and cos ωt, the "x" of the exponential is $\pm j\omega$, which is purely imaginary. Hence the condition $Re\ s > Re\ x$ given in Eq. (13-8) requires that $Re\ s > 0$.

Since hyperbolic functions can also be expressed as linear combinations of sinusoids, their Laplace transforms can also be found in this way. The details will be left to you.

Another simple function is $f(t) = t$. We can find its Laplace transform by direct integration, as follows:

$$\mathcal{L}(t) = \int_0^\infty t\epsilon^{-st}\,dt = -\frac{1}{s}\,t\epsilon^{-st}\Big|_0^\infty + \frac{1}{s}\int_0^\infty \epsilon^{-st}\,dt \tag{13-16}$$

The right-hand side is obtained by integration by parts. At the lower limit, the first term disappears. At the upper limit, one factor of the product $t\epsilon^{-st}$ becomes infinite, the other zero, assuming that s is real and positive or is complex with a positive real part. Application of L'Hôpital's rule, however, shows that the limit approached by $x\epsilon^{-x}$ is zero as x approaches infinity, as long as x is positive (if it is real). Thus the first term on the right disappears at both limits for the proper values of s. The remaining integration is routine and yields

$$\mathcal{L}[t] = \frac{1}{s^2} \tag{13-17}$$

We can find the Laplace transform of higher powers of t in the same way—integration by parts. You should carry out the details at least for t^2. For t^n the result is

$$\mathcal{L}(t^n) = \frac{n!}{s^{n+1}} \tag{13-18}$$

Let us now collect the Laplace transforms we have computed into a table, as shown in Table 13-1. The first column gives the function of time and the second column its Laplace transform.

Notice that all but one of the Laplace transforms which we have determined so far have been rational functions of the variable s; that is, ratios of two polynomials. You will recall that a rational function can be represented by specifying its poles (values of s at which the denominator vanishes) and its zeros (values of s at which the numerator vanishes) in the complex plane. The third column of the table shows these points by ×'s and ○'s respectively.

Having calculated the Laplace transform of a few simple, but important, functions, the question naturally arises whether, given any function of time, the Laplace transform can always be found. Of course a function $f(t)$ will have a Laplace transform if it can be inserted into

TABLE 13-1

$f(t)$	$F(s)$
ϵ^{at}	$\dfrac{1}{s-a}$
$u(t)$	$\dfrac{1}{s}$
$u(t-t_1)$	$\dfrac{1}{s}\epsilon^{-t_1 s}$
$\sin \omega t$	$\dfrac{\omega}{s^2+\omega^2}$
$\cos \omega t$	$\dfrac{s}{s^2+\omega^2}$
t	$\dfrac{1}{s^2}$
t^n	$\dfrac{n!}{s^{n+1}}$
$\sinh at$	$\dfrac{a}{s^2-a^2}$
$\cosh at$	$\dfrac{s}{s^2-a^2}$

the Laplace integral and the integration performed. We saw, in a few of the preceding examples, that the convergence of the integral for a given function definitely depends on the value of s. For different functions $f(t)$ the required range of s is different. A function may approach a constant as time increases (like the unit step), or it may oscillate (like the sine and cosine), or it may even increase with increasing time (like a real exponential), and it can still have a transform.

Is it possible that, for some functions, no values of s can be found for which the Laplace integral converges? Although it is difficult to find examples of such functions, nevertheless they exist. Remember that the role of the exponential ϵ^{-st} (at least the real part of s) is to act as a convergence factor. If a function increases fast enough with t, however, this convergence factor may not be effective.

A function which is not Laplace transformable is $f(t) = \epsilon^{t^2}$. Such functions, however, are of little engineering interest. Remember that the linear network model we have been using is an approximation which is best when the variables are not large—the "small-signal" case. If the variables become excessively large, then the linear model will no longer be a valid approximation. Certainly, if an applied voltage varies as ϵ^{t^2}, its value will become quite large as time goes on, thereby causing us to lose interest in it.

In more advanced work on Laplace-transform theory, it is shown that, if a function is "piecewise continuous" and of "exponential order," it is transformable; that is, the function is permitted to have a finite number of isolated discontinuities as long as, between each two points of discontinuity, it is continuous. Being of "exponential order" means that it should not rise any faster for large t than an exponential. Any such function can be multiplied by a real exponential $\epsilon^{-\sigma t}$ which, for some value of σ, causes the product $f(t)\epsilon^{-\sigma t}$ to approach zero as $t \rightarrow \infty$.

In the future, when we refer to solving a problem for "any arbitrary" function of time, we shall mean any transformable function, with the knowledge that any function we are likely to meet will be transformable.

Glance back at the time functions in Table 13-1; all except the step function $u(t)$ are not identically zero for negative values of t. Since the range of integration of the Laplace integral starts at $t = 0$, however, the Laplace transform is quite indifferent as to the behavior of the function for negative t. Thus all of the functions illustrated in Fig. 13-2

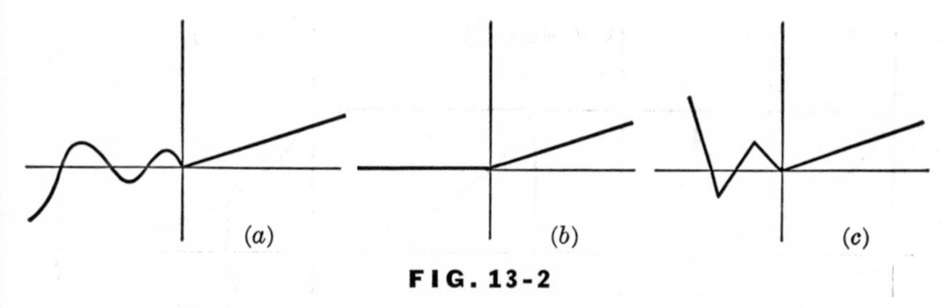

FIG. 13-2

Three time functions having the same transform

have the same Laplace transform, even though their behavior is radically different for negative t.

Since negative values of t are without interest, we shall henceforth assume that all functions we deal with are zero for negative t. To indicate this fact analytically, we multiply a function $f(t)$ by the unit step $u(t)$. Thus $f(t)u(t)$ is simply $f(t)$ for positive values of t, since $u(t)$ is, there, 1; and for negative values of t it is 0, since $u(t)$ is, there, 0.

This procedure of defining all functions of time to be zero for negative t introduces the possibility of a discontinuity at $t = 0$. Thus the function $\cos \omega t$ approaches the value 1 when t approaches 0 either from positive values or from negative values. Cosine $\omega t\, u(t)$ is identically zero for negative t, however, and jumps to the value 1 at $t = 0$. This is likely to cause some ambiguity in taking Laplace transforms, since the lower limit of the integral is zero. Does the range of integration start just below $t = 0$ or just above? Although an adequate procedure can be derived in either case, we shall assume that $t = 0$ is approached from the positive side. A common way of indicating this is to use the notation $(0+)$, but we shall simply use the simpler notation (0) with the understanding that it means $(0+)$.

13.3 THE REAL SHIFTING THEOREM

In Eq. (13-12) we found the Laplace transform of a unit step which is shifted in time so that it occurs at a time t_1 rather than at time $t = 0$. It is worthwhile to determine the transforms of other functions which are displaced in time also. An example of such a function is shown in Fig. 13-3.

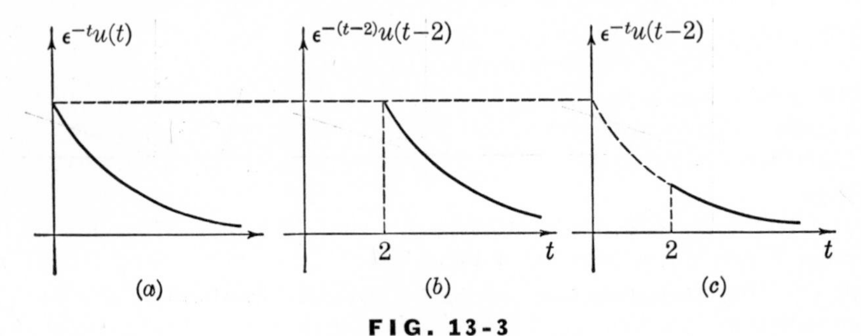

FIG. 13-3
Shifted exponential

In the first part is shown the exponential ϵ^{-t} which has been "cut off" for negative t by the multiplier $u(t)$. It is an easy matter to visualize sliding this function any distance along the horizontal axis, say two units, as shown in Fig. 13-3(b). Another point of view is to imagine that the vertical axis has been moved *to the left* two units. This amounts to a change of variable, the new variable being $(t - 2)$. Hence we express this function analytically by inserting $(t - 2)$ for t in the function $\epsilon^{-t}u(t)$. The result is $\epsilon^{-(t-2)}u(t - 2)$.

Observe that this is quite different from the function $\epsilon^{-t}u(t - 2)$ which is shown in Fig. 13-3(*c*). Multiplying the original exponential ϵ^{-t} by $u(t - 2)$ causes the product to vanish for all values of t up to $t = 2$, and to take on the values of ϵ^{-t} for $t > 2$. The exponential remains fixed, but we slide the unit step to the right, thereby annihilating everything in its path up to the occurrence of the step.

Let us now find the transform of a general shifted function, like that of Fig. 13-3(*b*), which is expressed as $f(t - t_1)u(t - t_1)$. From the definition of the Laplace transform, we have

$$\mathcal{L}[f(t - t_1)u(t - t_1)] = \int_0^\infty f(t - t_1)u(t - t_1)\epsilon^{-st}\,dt = \int_{t_1}^\infty f(t - t_1)\epsilon^{-st}\,dt$$

Since the integrand is zero up to $t = t_1$, there will be no contribution to the integral from 0 to t_1, thus permitting the last form. Now let $t - t_1 = x$; then $dt = dx$, and ϵ^{-st} becomes $\epsilon^{-s(t-t_1+t_1)} = \epsilon^{-st_1}\epsilon^{-sx}$; when $t = t_1$, $x = 0$. Hence

$$\mathcal{L}[f(t - t_1)u(t - t_1)] = \epsilon^{-st_1}\int_0^\infty f(x)\epsilon^{-sx}\,dx = \epsilon^{-st_1}F(s) \qquad (13\text{-}19)$$

Note that the t which appears in the usual way of writing the Laplace integral is simply a dummy variable of integration and can be replaced by any other symbol, like x in this expression.

The result expressed by Eq. (13-19) is called the *real shifting theorem* (*real* in contrast with *complex*, which will be discussed later). It is seen to be the generalization of Eq. (13-12), which can be included in the present result by taking $f(t) = 1$. The effect of shifting a time function t_1 units on the horizontal axis is simply a multiplication of its Laplace transform by ϵ^{-st_1}. This theorem is very useful in its applications.

To illustrate its use let us calculate the Laplace transform of the pulse shown in Fig. 13-4. The pulse has an exponential top; it starts at $t = 2$ and ends at $t = 3$. To write an analytical expression for this function, we can conceive of it as the difference between the two functions shown in Fig. 13-4(*b*). The first of these is an exponential shifted two units to the right. The second one is also an exponential shifted two units to the right, but its value is forced to be zero up to $t = 3$ through the agency of the multiplier $u(t - 3)$. Hence

$$\begin{aligned} f(t) &= 5\epsilon^{-(t-2)}u(t - 2) - 5\epsilon^{-(t-2)}u(t - 3) \\ &= 5\epsilon^{-(t-2)}[u(t - 2) - u(t - 3)] \qquad (13\text{-}20 \end{aligned}$$

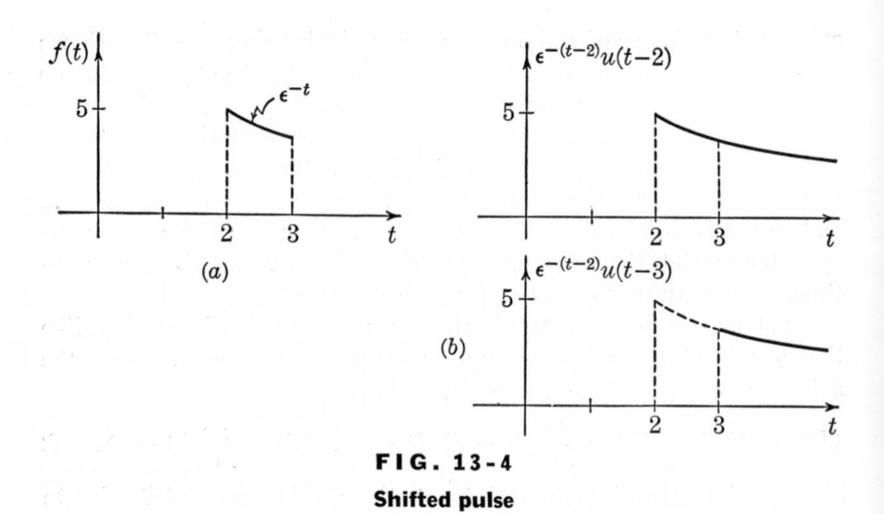

FIG. 13-4

Shifted pulse

The last line of this expression suggests another point of view. The quantity within the brackets is the difference between two step functions, one occurring at $t = 2$, the other at $t = 3$. Thus it is a square pulse, as shown in Fig. 13-5(a). When this pulse multiplies a function of time, it has the effect of "chopping off" the function for values of t less than $t = 2$ and greater than $t = 3$. It is like a gate which permits the entrance of a function only over its width.

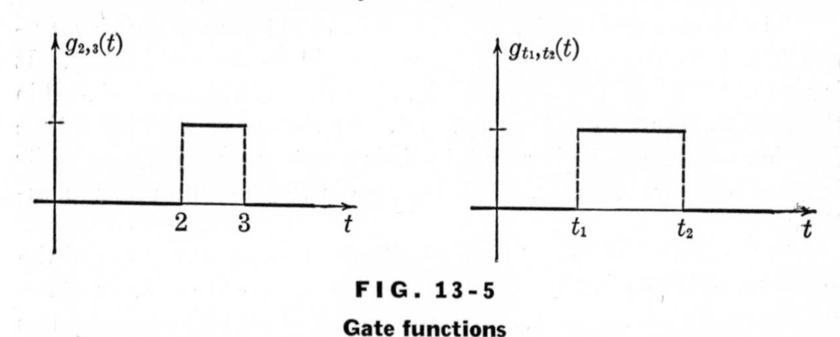

FIG. 13-5

Gate functions

More generally, we shall use the symbol $g_{t_1,t_2}(t)$ to designate a gate function, as shown in Fig. 13-5(b), which starts at t_1 and ends at t_2, its width being $t_2 - t_1$. In this notation the gate in Fig. 13-5(a) is $g_{2,3}(t)$.

Returning now to Eq. (13-20), we see that the shifting theorem can be applied to the first term, but the second term is not in the correct form. This can be corrected, however, by adding and subtracting 1 in the exponent. Thus

$$f(t) = 5\epsilon^{-(t-2)}u(t-2) - 5\epsilon^{-1}\epsilon^{-(t-3)}u(t-3) \qquad (13\text{-}21)$$

Since the transform of ϵ^{-t} is $1/(s+1)$, application of the shifting theorem leads to

$$F(s) = \mathfrak{L}[f(t)] = \frac{5}{s+1}\epsilon^{-2s} - \frac{5}{s+1}\epsilon^{-(3s+1)} \qquad (13\text{-}22)$$

It is seen that the transforms resulting from the application of the shifting theorem are not rational functions of s, since they contain exponentials.

13.4 APPLICATION OF THE LAPLACE TRANSFORM

Now that we have gained some familiarity with the manipulative aspect of the calculation of Laplace transforms, let us return to our major concern. We have yet to establish that use of the Laplace transformation will permit a simplification of the calculation of a network response.

In order to provide motivation for the properties that we shall now discuss, let us consider the series RLC circuit shown in Fig. 13-6. A voltage $v_1(t)$ is switched into the circuit at time $t = 0$, at which time the circuit is *inert* or *dead* or *relaxed;* that is, there is no current in the inductor and there is no voltage across the capacitor from some previous connection of the elements. It is desired to find an expression for the current as a function of time following the switching. Writing a loop equation, the following integrodifferential equation is obtained.

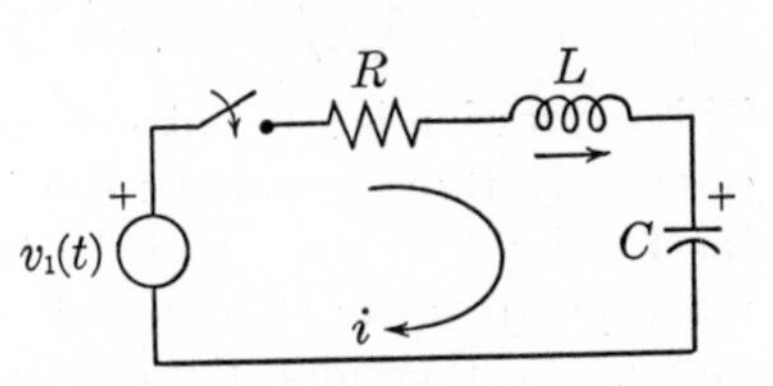

FIG. 13-6

Series *RLC* network with voltage excitation

$$L\frac{di}{dt} + Ri + \frac{1}{C}\int_0^t i(x)\,dx = v_1(t) \qquad (13\text{-}23)$$

Our plan is to use the Laplace transform with the hope that the equation will be converted to an algebraic equation. To pursue this plan, we multiply both sides of the equation by ϵ^{-st} and integrate from 0 to ∞. Symbolically, we have

$$\mathfrak{L}\left[L\frac{di}{dt} + Ri + \frac{1}{C}\int_0^t i(x)\,dx\right] = \mathfrak{L}[v_1(t)] \qquad (13\text{-}24)$$

On the left-hand side we are to take the Laplace transform of a sum of a number of terms. Is this the same as the sum of their transforms? The answer is affirmative, since integration and addition are commutative. In fact, we already used this property when finding the transform of a sine function in terms of that of an exponential. A similar question is whether the transform of a term like Ri is the same as R times the transform of i. The answer is again affirmative, since integration and multiplication by a constant are commutative. These two properties are quite evident, but, just for emphasis, let us state them in a general way as follows:

If $f_1(t)$ and $f_2(t)$ are two Laplace transformable functions having the Laplace transforms $F_1(s)$ and $F_2(s)$, then

$$\mathcal{L}[a_1 f_1(t) + a_2 f_2(t)] = a_1 F_1(s) + a_2 F_2(s) \tag{13-25}$$

where a_1 and a_2 are constants.

Notice that we are using lower-case letters to represent functions of t and capital letters to represent transforms. This is a fairly standard notation, and we shall adhere to it. The transform of a voltage $v(t)$ will be written $V(s)$; that of a current $i(t)$ will be written $I(s)$.

Returning now to Eq. (3-24) and using Eq. (3-25), we have

$$L\mathcal{L}\left(\frac{di}{dt}\right) + R\mathcal{L}(i) + \frac{1}{C}\mathcal{L}\left(\int_0^t i(x)\,dx\right) = V_1(s) \tag{13-26}$$

This expression is still not very useful, because on the left side we have essentially three variables: the transform of the current i, the transform of its derivative, and the transform of its integral. If we are to make progress, we should find relationships among these three transforms. Furthermore, these relationships should be algebraic if we are to make any headway. For example, it would gain us nothing if it were to turn out that the Laplace transform of the derivative of a function is equal to the derivative of the transform, because, if it were, then a differential equation would be converted into another differential equation, and we would be back where we started.

Finding the Laplace transform of the derivative and the integral of a function in terms of the transform of the function will be our next concern.

Transform of a Derivative

In this discussion we shall use a general notation without committing ourselves as to the physical variables represented by the symbols.

Let $f(t)$ and its derivative $f'(t) = df/dt$ be Laplace transformable functions, $F(s)$ being the transform of $f(t)$. The Laplace transform of the derivative is by definition

$$\mathcal{L}\left(\frac{df}{dt}\right) = \int_0^\infty \frac{df}{dt} \epsilon^{-st}\, dt \tag{13-27}$$

On the right side let us use integration by parts. Define

$$\begin{aligned} u &= \epsilon^{-st} & du &= -s\epsilon^{-st}\, dt \\ dv &= \frac{df}{dt}\, dt & v &= f(t) \end{aligned} \tag{13-28}$$

and insert in the formula

$$\int_0^\infty u\, dv = uv\Big|_0^\infty - \int_0^\infty v\, du \tag{13-29}$$

The result will be

$$\mathcal{L}\left(\frac{df}{dt}\right) = f(t)\epsilon^{-st}\Big|_0^\infty + s\int_0^\infty f(t)\epsilon^{-st}\, dt \tag{13-30}$$

The integral in the last term is $F(s)$ by definition. In the first term the lower limit causes no difficulty. Since $f(t)$ is assumed to be transformable, that means some value of s exists which will cause the product $f(t)\epsilon^{-st}$ to approach zero as $t \to \infty$. Hence, at the upper limit, the first term vanishes. The final result is

$$\mathcal{L}\left(\frac{df}{dt}\right) = sF(s) - f(0) \tag{13-31}$$

[Remember that $f(0)$ means $f(0+)$, the value of f just after $t = 0$.]

This is the desired expression, and our hopes are realized. Aside from the constant $f(0)$, which is the initial value of the function $f(t)$, differentiation in the time domain has been replaced by multiplication by s in the s domain.

By repeated application of this formula, the Laplace transform of the derivative of a function of any order can be found. Thus, for the second derivative, we get

$$\begin{aligned} \mathcal{L}[f''(t)] &= s\mathcal{L}[f'(t)] - f'(0) = s[sF(s) - f(0)] - f'(0) \\ &= s^2F(s) - sf(0) - f'(0) \end{aligned} \tag{13-32}$$

where $f'(0)$ is the value of the derivative at $t = 0$. You can carry out the details for higher-order derivatives.

Transform of an Integral

Next we turn to the definite integral of a function. Let us define

$$g(t) = \int_0^t f(x)\,dx \qquad (13\text{-}33)$$

If the function g is the integral of f, then f must be the derivative of g. Thus

$$f(t) = \frac{dg}{dt} \qquad (13\text{-}34)$$

But now we can use the preceding result about the derivative. Assuming that $g(t)$ and its derivative [that is, $f(t)$ and its definite integral] are transformable, and using Eq. (13-31), we get

$$\mathcal{L}[f(t)] = \mathcal{L}\left(\frac{dg}{dt}\right) = sG(s) - g(0) \qquad (13\text{-}35)$$

where $G(s) = \mathcal{L}[g(t)]$ is the transform of the definite integral we are seeking. To find $g(0)$, put $t = 0$ in Eq. (13-33), which leads to $g(0) = 0$. Hence

$$\mathcal{L}\left[\int_0 f(x)\,dx\right] = \frac{1}{s}F(s) \qquad (13\text{-}36)$$

This is again a gratifyingly simple result. Integration in the time domain has been converted to division by s in the s domain. As for differentiation, the result can be extended to higher-order integrals; we shall, however, omit any further consideration here.

We have now established that there is a linear algebraic relationship between the transform of a function and the transform of its derivative and of its integral. We should expect, therefore, that an integrodifferential equation will be converted to an algebraic equation. Let us return to the example of the series RLC circuit which we left at Eq. (13-26). Using Eqs. (13-31) and (13-36) and remembering that the initial current in L is zero, this expression becomes

$$LsI(s) + RI(s) + \frac{1}{Cs}I(s) = V_1(s)$$

or

$$I(s) = \frac{V_1(s)}{Ls + R + \dfrac{1}{Cs}} \qquad (13\text{-}37)$$

Observe that the integrodifferential equation has been converted to an algebraic equation in one variable, for which we can solve.

The form of this expression is quite familiar. If $V_1(s)$ and $I(s)$ had represented sinors, this expression would have been Ohm's law for sinors, since the denominator is simply the impedance of the circuit written as a function of s. This is a truly amazing and useful result, and we shall explore its consequences in the following sections.

There is still something of fundamental importance, however, to consider here. We start the solution of a problem by writing one or more integrodifferential equations which we then transform. Then we solve for the transform of the desired response, as in Eq. (13-37) for our example. But this is not yet the solution; it is the Laplace transform of the solution. We must find some way of getting the function of time from its transform. This operation is referred to as the *inverse transform* and is given the symbol $\mathcal{L}^{-1}$. Thus the expression

$$f(t) = \mathcal{L}^{-1}[F(s)] \tag{13-38}$$

is read "$f(t)$ is the inverse transform of $F(s)$." This expression is the "mate" of Eq. (13-6). (Write them down side by side.) The two together constitute a *transform pair*, just like Eq. (13-1).

Actually, there is an integral formula, called the *inversion integral*, which permits the calculation of a function $f(t)$, given its transform $F(s)$. Understanding this formula and evaluating it, however, require a knowledge of more mathematics then we have at our disposal here.

Fortunately, this difficulty can be overcome in a simple way. You will recall that, using the definition of a Laplace transform, we were able to calculate and tabulate the transforms of some specific functions of time. This same process can be carried out for many other functions. If we build up a table of such transform pairs, then, when we encounter a transform function, we enter the table and seek its corresponding function of time. The transform-pair table plays the same role as an integral table.

If this plan is to work, we should first be reassured that to any given function $F(s)$ there corresponds only one specific function of time $f(t)$. That is, how can we be sure that two different functions of t do not have the same Laplace transform? As a matter of fact, recalling that the limits of the Laplace integral go from 0 to ∞, if two functions are the same for positive values of t but different for negative values of t, their Laplace transforms will be the same, as we have already mentioned. Hence the table will give information only for positive values of t. We shall assume that all functions are zero for negative t. Aside from negative t, the problem of uniqueness still remains. It is not possible for us to prove the uniqueness property here. However, if you accept the statement that there exists a valid formula (the inversion integral) from which a function of time $f(t)$ can be calculated for a given $F(s)$, this formula must always lead to a unique result. Hence if, for a given transform function $F(s)$, we can find a corresponding $f(t)$ by any means whatsoever, this must be the only one.

One other point should be clarified. When developing the relationship between the transform of a derivative of a function and the transform of the function itself, we assumed that the function and its derivative were transformable. At the start of a problem, however, the response function is unknown; how do we know that it, and its derivative, and its integral are transformable, a circumstance which must be true if we are to use Eqs. (13-31) and (13-36)? A possible answer to this question is as follows: Even though we may have no specific knowledge of the transformability of the response function, we assume that it (along with any of its derivatives and integrals which appear in the equations) is transformable, and we carry out the solution. It is an easy matter, by inserting the result into the original equation and by testing to see whether it satisfies the original equation, to verify whether or not the response function which is obtained is a solution.

Let us terminate this discussion by considering a very simple example. Suppose an exponential pulse of voltage is applied to a series RL circuit and it is desired to find the resulting current, assuming that initially there was no current. Assume that the appropriate differential equation is

$$\frac{di}{dt} + 2i = 10\epsilon^{-3t} \tag{13-39}$$

The next step is to transform the equation.

$$\mathcal{L}\left(\frac{di}{dt}\right) + 2\mathcal{L}(i) = 10\mathcal{L}(\epsilon^{-3t}) \tag{13-40}$$

If $\mathcal{L}(i) = I(s)$, then the transform of the derivative is $sI(s)$, since the initial current $i(0)$ is 0. To find the transform of the exponential, we can either calculate it directly or look at Table 13-1. The result will be

$$sI(s) + 2I(s) = \frac{10}{s+3}$$

or

$$I(s) = \frac{10}{(s+2)(s+3)} \tag{13-41}$$

If we had a more extensive table than Table 13-1, we would be able to look up the inverse transform of this function. Lacking this, let us resort to ingenuity. In the first place, what form would we expect the complete response to have, based on our previous study in Chap. 3? In the forced response we would expect an exponential like the driving function. In the natural response we would also expect an exponential, one whose exponent contained the natural frequency, which is $s = -2$ in this case. Hence we would expect the form of the complete response to be

$$i(t) = K_1\epsilon^{-2t} + K_2\epsilon^{-3t} \tag{13-42}$$

Suppose we now find the Laplace transform of this expression, using Table 13-1. (You will soon have the transform of an exponential memorized.) It is

$$I(s) = \frac{K_1}{s+2} + \frac{K_2}{s+3} \tag{13-43}$$

Compare this with Eq. (13-41). If we can only put Eq. (13-41) in this form, then we can recognize each of the terms as the transform of an exponential and immediately write down the inverse transform. Actually, this form is simply a *partial fraction*. The constants K_1 and K_2 can be evaluated by equating Eqs. (13-41) and (13-43), clearing the fractions, and equating coefficients of like powers of s. Thus

$$K_1(s+3) + K_2(s+2) = 10$$

$$(K_1 + K_2) = 0 \qquad K_1 = 10$$

$$3K_1 + 2K_2 = 10 \qquad K_2 = -10$$

With these values Eq. (13-43) becomes

$$I(s) = \frac{10}{s+2} - \frac{10}{s+3} \tag{13-44}$$

If you do not yet recognize the transform of an exponential, you will have to glance at the table. In any case, the corresponding function of time will be

$$i(t) = 10\epsilon^{-2t} - 10\epsilon^{-3t} \tag{13-45}$$

You can verify that this function is a solution by inserting it into the original differential equation and by noting that, at $t = 0$, it gives the correct value of $i(0)$.

13.5 TRANSFORM NETWORKS

We have now established the basic groundwork for finding the complete response—forced and natural together—of a network when it is subjected to any (Laplace transformable) excitation. Let us summarize here the steps in the procedure.

1. We apply the fundamental laws (Kirchhoff's laws and the v-i relationships) to the network and obtain a set of equations which, in general, will contain derivatives and integrals of the unknowns.
2. We transform these equations, thereby converting the equations to algebraic ones. The variables are now Laplace transforms. In this step, initial conditions are introduced.
3. Next we solve the algebraic equations for the transform variables.
4. Finally, we use the inverse transformation to convert the transforms back to functions of time. Note that the complete solution is obtained, and the initial conditions are already incorporated.

Of course, we still know very little about the techniques for carrying out the last step. We shall correct this deficiency to some extent in the next section.

For the present, there is still some room for improving the over-all procedure in the first two steps. You will recall that the sinusoidal steady-state solution by means of sinors can be formulated in just such a sequence of steps. We found that it was not necessary, however, to perform the first two steps separately; that they could be combined into a single step by transforming the v-i relationships of the elements, and by writing Kirchhoff's laws directly in terms of the transformed variables (namely, the sinors). The same thing can be done here.

First of all, note that, by applying the Laplace transform to Kirchhoff's laws, we find that these laws are satisfied by the transforms of the variables as well as by the variables themselves. For example, when writing Kirchhoff's voltage law around a loop, it can be written in terms of the Laplace transforms of the voltages rather than the voltages themselves.

Next let us examine the v-i relationships of the elements. They are shown in the following listing, together with their transforms.

Resistor	*Inductor*	*Capacitor*
$v(t) = Ri(t)$	$v(t) = L\dfrac{di}{dt}$	$v(t) = \dfrac{1}{C}\displaystyle\int_0^t i(x)\,dx + v(0)$
$i(t) = \dfrac{1}{R}v(t)$	$i(t) = \dfrac{1}{L}\displaystyle\int_0^t v(x)\,dx + i(0)$	$i(t) = C\dfrac{dv}{dt}$
$V(s) = RI(s)$	$V(s) = LsI(s) - Li(0)$	$V(s) = \dfrac{1}{Cs}I(s) + \dfrac{v(0)}{s}$
$I(s) = \dfrac{1}{R}V(s)$	$I(s) = \dfrac{1}{Ls}V(s) + \dfrac{i(0)}{s}$	$I(s) = CsV(s) - Cv(0)$

There is no difficulty in interpreting the first column. In the case of the resistor, Ohm's law applies for the transformed variables as well as the variables themselves.

The V-I relationships of the inductor and capacitor are complicated slightly by the appearance of the initial current and initial voltage, respectively. In case these initial values are zero, the transformed V-I relationships are the same as Ohm's law for sinors. Thus

$$V(s) = Z(s)I(s) \qquad \text{or} \qquad I(s) = Y(s)V(s) \tag{13-46}$$

where $Z(s) = Ls$ or $1/Cs$, except that the variables are now Laplace transforms rather than sinors. We have emphasized that the impedance and admittance are each a function of s by writing $Z(s)$ and $Y(s)$.

Turning back to Chap. 4, you will recall that it was precisely because of these two facts—that Kirchhoff's laws hold in terms of the transformed variables (sinors, in that case) and that the V-I relationships take the form of Ohm's law—that we were able to develop all the concepts and procedures of the sinusoidal steady-state solution. You will recall that we generalized the concept of frequency by defining a complex-frequency variable s; and we interpreted the procedures of sinor analysis to consist of finding the forced response to an exponential excitation of the form ϵ^{st}.

Now, for an initially relaxed network, we find the same V-I relationships except that the variables are Laplace transforms instead of sinors. Instead of defining network functions, such as impedance, transfer-voltage ratio, etc., as ratios of sinors, we define them as ratios of Laplace transforms. When written in terms of s, both definitions lead to identical functions.

As a consequence of the preceding observations, it is now clear that all the formal analytic procedures which we studied in terms of sinors—series and parallel combination of impedances, algebraic solutions of loop and node equations, tee-pi transformations, Thévenin's theorem, analysis by means of two-port parameters and signal flow graphs, and all the others—are directly applicable to the Laplace-transform solution, whenever the network is initially relaxed.

Referring to the list of steps at the beginning of this section it is clear that the first two are not performed separately but that we immediately write down the transformed equations, at least when the network is initially relaxed. As for the third step, we have spent most of our time in this book studying its various aspects. We can now say that we know how to perform this step. Only the fourth step remains to be studied.

One other circumstance remains, and that is the handling of the initial conditions. Glance back at the *V-I* relationships. Remember that the Laplace transform of a constant is that constant divided by s. Hence the quantity $v(0)/s$ appearing in the *V-I* relationship of the capacitor can be represented by a battery in series with the capacitor, as shown in Fig. 13-7(*a*). Similarly, from the last form of the *V-I*

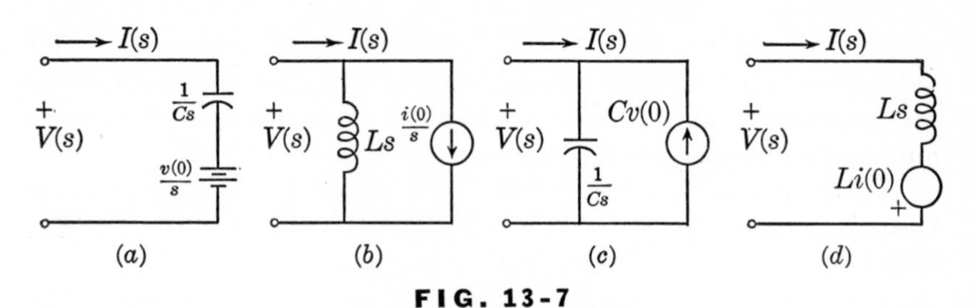

FIG. 13-7

Initial conditions as sources

relationship of the inductor, it is clear that the term $i(0)/s$ can be represented by a constant current source in parallel with the inductor, as shown in Fig. 13-7(*b*). (Note the references carefully.)

In view of the preceding discussion, we see that the initial conditions can be treated as sources, everything else in the analysis remaining unaltered. You will recall that, in order to define impedance, voltage gain, or any other network function, we specified that there should be no independent sources inside the network. Hence, when a network is not initially relaxed, we cannot use these functions. That is to say, if a voltage source is applied to an initially *live* (not *dead* or *relaxed*) network, and it is desired to find the input current, we cannot find an impedance with which to divide the voltage transform. We must, instead, use other methods such as Thévenin's theorem, loop equations, etc.

The circuits in Fig. 13-7(a) and (b) are not the only ones by which the initial conditions can be handled. Norton's theorem can be applied to the first one and Thévenin's theorem to the second one to yield the circuits shown in Fig. 13-7(c) and (d). Thus the initially charged capacitor can be replaced by an uncharged capacitor in parallel with a current source whose transform is equal to the capacitance times the initial voltage, $Cv(0)$. Similarly, an inductor with initial current can be replaced by an inductor without initial current in series with a voltage source whose transform is equal to the inductance times the initial current, $Li(0)$. (Note carefully the references of these sources.)

The question naturally arises as to what waveform it is that has a transform which is a constant. If we glance at our table of transforms, Table 13-1, we do not find such a pair. Observe, however, that the transform of t is $1/s^2$. Also, the transform of the unit step is $1/s$. But $1/s$ is s times $1/s^2$, and the unit step is the derivative (the right-handed derivative) of t. Now the constant 1 is s times $1/s$; hence we might extrapolate and say that the waveform whose transform is 1 is the derivative of the waveform whose transform is $1/s$, just as the waveform whose transform is $1/s$ is the derivative of the waveform whose transform is $1/s^2$. But the derivative of a unit step is zero everywhere except at the discontinuity, where it is not defined. You will recall that in Chap. 1 we defined this quantity to be the unit impulse $\delta(t)$. Analytically,

$$\int_{-\infty}^{t} \delta(x)\,dx = u(t) \qquad (13\text{-}47)$$

The answer to our question, then, is that the waveform whose transform is any constant K is an impulse whose strength is K. Actually, our reasoning in the last paragraph is pretty shaky. You will recall how we introduced the impulse function in Chap. 1 as the limit of a square pulse when the pulse width is reduced to zero, its height going to infinity. Perhaps a somewhat more appealing demonstration concerning the Laplace transform of an impulse can be given by finding the transform of the square pulse and then taking the limit as the pulse width goes to zero. The Laplace transform of the impulse, however, should be found by taking the limit *first* and then evaluating the Laplace integral. Is it justified to evaluate the Laplace integral first and *then* take the limit? Remember that earlier we answered a similar question in the affirmative, because the integral there was uniformly convergent, whereas in the present case it is not. Hence this interchange is not justified.

Nevertheless, there is a mathematical basis for the impulse function on a much higher level than we can discuss here. For our purposes we shall accept the unit impulse as a useful tool, and we shall take its Laplace transform to be 1. Actually, in using Fig. 13-7, it is not necessary to know what the waveform is that corresponds to a transform which is constant, since we shall be using these circuits in the transformed network anyway.

You have probably already observed a point of possible confusion in Fig. 13-7. In part (*c*) the current source has a value $Cv(0)$. Dimensionally, capacitance times voltage is not current but charge. Similarly, in part (*d*) the source has a voltage $Li(0)$ which is dimensionally not voltage but flux. Keep in mind, however, that these are transformed quantities. If you check the dimensions of the Laplace integral, you will see that the transform of $f(t)$ has the dimensions of f times time. Thus, dimensionally, the transform of a current is indeed charge, and the transform of a voltage is indeed flux.

In summary, then, when solving a problem by means of the Laplace transform, it is possible to proceed according to the steps outlined at the beginning of this section. We can proceed in an alternate way, however, by first replacing initial conditions with appropriate sources. Following this, we can write the transformed loop or node equations and solve them, or we can apply any of the other techniques such as Thévenin's theorem or signal flow graphs. Of course, the last step in the over-all procedure remains the same.

Let us now illustrate by means of an example all but the last step of the transform method of solution. Consider the network shown in Fig. 13-8. The left-hand loop has been active for a long time, when, at $t = 0$, the switch S is closed. It is desired to find the current in the inductor following the closing of the switch.

Let us first check the initial conditions. Since the left-hand loop has been operating for a long time, we can assume that any transients which appeared when the loop was first closed in the past have since disappeared. There will be a d-c current in the loop given by $24/(2 + 10) = 2$. Hence the current in the inductor just before the switch is closed is $i(0) = 2$, and this will be the current just after the switch is closed, as a result of the flux-linkage principle.

Now let us replace the initial condition by an appropriate source. In this case it would be better to have a source in series with L, because this source can then be combined with the battery which is already there. Hence, using Fig. 13-7(d) in which $Li(0)$ is here $2 \times 2 = 4$, and using transformed quantities throughout, the transformed network after the closing of the switch takes the form shown in Fig. 13-8(b). The

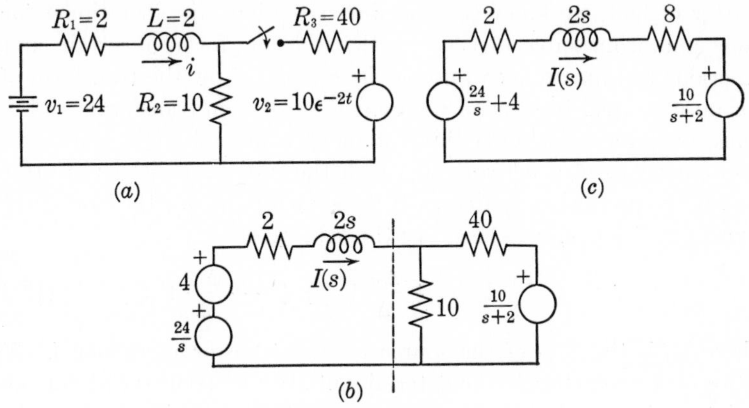

FIG. 13-8

Illustrative example

network has two loops, and it might be tempting to start writing loop equations. Let us, instead, replace the part of the network to the right of the dashed line by its Thévenin equivalent. The result is shown in Fig. 13-8(c). There is now a single loop, and the corresponding equation is

$$(2 + 2s + 8)I(s) = \frac{24}{s} + 4 - \frac{2}{s+2}$$

or

$$I(s) = \frac{12}{s(s+5)} + \frac{2}{s+5} - \frac{1}{(s+2)(s+5)} = \frac{2s^2 + 15s + 24}{s(s+2)(s+5)} \tag{13-48}$$

This is the transform of the desired current. At the extreme right the function was combined over a common denominator. In the next section we shall discuss procedures for finding the inverse transform of such functions. Note that the initial condition is already incorporated in the solution, and no additional calculation is needed after the inverse of $I(s)$ is found.

13.6 PARTIAL-FRACTION EXPANSIONS

The transform current which we calculated in the last section is seen to be a ratio of two polynomials in s, a rational function. As a matter of fact, we know that network functions are all rational functions, and, from Table 13-1, we see that the transforms of certain simple functions of time are also rational functions. Now the transform of a response is equal to sums of products of excitation transforms and network functions. To verify this statement, contemplate the transformed loop equations of a network in which the initial conditions are represented by appropriate sources. If we solve for any one the loop-current transforms by Cramer's rule, say the kth, we get

$$I_k(s) = \frac{\Delta_{1k}}{\Delta} V_1(s) + \frac{\Delta_{2k}}{\Delta} V_2(s) + \cdots \frac{\Delta_{nk}}{\Delta} V_n(s) \qquad (13\text{-}49)$$

where V_j is the sum of the source-voltage transforms on loop j. The ratios of the cofactors to the determinant are admittances (driving point and transfer), and these are rational. This expression has the form claimed.

We see that, if the external sources have rational transforms, then the transform of a response will also be rational. We shall, therefore, discuss a procedure for finding the inverse transform of a rational function.

A clue to the procedure has already been given in the illustrative example in Sec. 13-4. We know that the inverse transform of the function $1/(s - a)$ is the exponential ϵ^{at}. We also know that a polynomial can be factored and written as a product of its factors. If it could only be arranged to break up a rational function into a sum of terms each one of which had only one of the factors of the denominator, then the inverse transform could be written as a sum of exponentials. Such a decomposition of a rational function is a partial-fraction expansion.

Let us illustrate the procedure with the $I(s)$ function we found in Eq. (13-48). Remember that the roots of the denominator are called the poles of the function. In the present case there are three poles, at $s = 0$, $s = -2$, and $s = -5$. The partial-fraction expansion is written

$$I(s) = \frac{2s^2 + 15s + 24}{s(s + 2)(s + 5)} = \frac{K_0}{s} + \frac{K_1}{s + 2} + \frac{K_2}{s + 5} \qquad (13\text{-}50)$$

You have undoubtedly evaluated such partial fractions in the past.

One method for finding the constants K_0, K_1, and K_2 is to clear the fraction on both sides of the equation and then equate coefficients of like powers of s. Thus

$$\begin{aligned} 2s^2 + 15s + 24 &= K_0(s + 2)(s + 5) + K_1 s(s + 5) + K_2 s(s + 2) \\ &= (K_0 + K_1 + K_2)s^2 + (7K_0 + 5K_1 + 2K_2)s + 10K_0 \end{aligned} \tag{13-51}$$

which leads to

$$\begin{aligned} K_0 + K_1 + K_2 &= 2 \\ 7K_0 + 5K_1 + 2K_2 &= 15 \\ 10K_0 &= 24 \end{aligned} \tag{13-52}$$

Here there are three equations in three unknowns, which can be solved to yield

$$K_0 = \frac{12}{5}; \quad K_1 = -\frac{1}{3}; \quad K_2 = -\frac{1}{15}$$

With these values, Eq. (13-50) becomes

$$I(s) = \frac{12/5}{s} - \frac{1/3}{s + 2} - \frac{1/15}{s + 5} \tag{13-53}$$

Let us digress temporarily to complete the solution of this example which was started in Fig. 13-8. From the last equation we can now immediately write down the inverse transform of each of these terms by inspection (or by looking at the table). Hence the solution of the problem started in the last section is

$$i(t) = \mathcal{L}^{-1}[I(s)] = \left(\frac{12}{5} - \frac{1}{3}\,\epsilon^{-2t} - \frac{1}{15}\,\epsilon^{-at}\right) u(t) \tag{13-54}$$

The appearance of the step function reminds us that the solution is valid only for positive values of t. Now back to the partial-fraction expansion.

The procedure just used for evaluating the constants of the partial-fraction expansion is generally valid. Even in the present relatively simple problem, however, we see that a set of three equations must be solved simultaneously. In general, there will be as many simultaneous equations as there are terms in the expansion. It is certainly worthwhile to seek a way of avoiding this drudgery.

In going from Eq. (13-50) to Eq. (13-51), we multiplied both sides by the entire denominator of the left side. Suppose that, instead, we multiply by only one of the denominator factors, say $s + 5$. The result will be

$$\frac{2s^2 + 15s + 24}{s(s + 2)} = (s + 5)\left(\frac{K_0}{s} + \frac{K_1}{s + 2}\right) + K_2 \tag{13-55}$$

We observe that the constant K_2 is isolated on the right side. We further observe that, if we now substitute for s the value $s = -5$, everything on the right *except* K_2 will vanish. Hence the result will give an expression for K_2. Putting $s = -5$ on the left, we get the value $-1/15$, just as we found before. Thus we see that K_2 can be found by writing

$$K_2 = \left.\frac{2s^2 + 15s + 24}{s(s + 2)}\right|_{s=-5} = (s + 5)I(s)\Big|_{s=-5} \tag{13-56}$$

In a similar way, if both sides of Eq. (13-50) are multiplied by $s + 2$, the result will be

$$\frac{2s^2 + 15s + 24}{s(s + 5)} = (s + 2)\left(\frac{K_0}{s} + \frac{K_2}{s + 5}\right) + K_1 \tag{13-57}$$

Now if we give s the value $s = -2$, everything on the right except K_1 disappears. Hence

$$K_1 = \left.\frac{2s^2 + 15s + 24}{s(s + 5)}\right|_{s=-2} = (s + 2)I(s)\Big|_{s=-2} \tag{13-58}$$

which gives $K_1 = -1/3$, as before.

This procedure is perfectly general when all the poles of the transform function are simple, that is, when all the factors in the denominator are raised to the first power only. Thus, if $F(s)$ is a rational function and $(s - s_i)$ is a simple factor of the denominator [s_i is a simple pole of $F(s)$], then the corresponding constant K_i in the partial-fraction expansion, which is called the *residue* at the pole s_i, is given by

$$K_i = (s - s_i)F(s)|_{s=s_i} \tag{13-59}$$

That is, $F(s)$ is multiplied by the factor $(s - s_i)$, and the result is evaluated for $s = s_i$. In this way, each of the residues is evaluated individually, and the need for solving many simultaneous equations is removed.

To consolidate our ideas let us examine another example. Let

$$F(s) = \frac{s^2 + 5s + 8}{(s + 1)(s + 2)(s + 3)} = \frac{K_1}{s + 1} + \frac{K_2}{s + 2} + \frac{K_3}{s + 3} \tag{13-60}$$

Using Eq. (13-59), the residues are found to be

$$K_1 = (s + 1)F(s)\Big|_{s=-1} = \left.\frac{s^2 + 5s + 8}{(s + 2)(s + 3)}\right|_{s=-1} = 2$$

$$K_2 = (s + 2)F(s)\Big|_{s=-2} = \left.\frac{s^2 + 5s + 8}{(s + 1)(s + 3)}\right|_{s=-2} = -2$$

$$K_3 = (s + 3)F(s)\Big|_{s=-3} = \left.\frac{s^2 + 5s + 8}{(s + 1)(s + 2)}\right|_{s=-3} = 1$$

You should also find the constants by the usual method of clearing and equating coefficients, in order to compare the relative ease of the present method.

When a function has multiple poles (denominator factors at higher powers than 1), this procedure must be modified somewhat. Consider the following example.

$$F(s) = \frac{s+5}{(s+1)(s+2)^2} = \frac{K_1}{s+1} + \frac{K_2}{s+2} + \frac{K_3}{(s+2)^2} \quad (13\text{-}61)$$

Here there is a double pole, so the partial fraction must contain a term having a denominator $(s+2)^2$, as well as one having denominator $(s+2)$. A higher-order pole would have additional terms with higher powers of $(s+2)$ in the denominator.

K_1 can be found as before. If we next multiply both sides by $(s+2)$ and then set $s = -2$, nothing but trouble will result, because there will remain a factor $(s+2)$ in the denominator on the left. To avoid this difficulty we must multiply by $(s+2)^2$ instead of $s+2$. Then

$$(s+2)^2F(s) = \frac{s+5}{s+1} = (s+2)^2\frac{K_1}{s+1} + K_2(s+2) + K_3 \quad (13\text{-}62)$$

Now if we set $s = -2$, the result will give K_3. But how can we find K_2? Suppose we differentiate, with respect to s, both sides of the last equation. The result will be

$$\frac{d}{ds}[(s+2)^2F(s)] = \frac{-4}{(s+1)^2} = \frac{K_1s(s+2)}{(s+1)^2} + K_2 \quad (13\text{-}63)$$

The differentiation has eliminated the constant K_3 and has isolated K_2. Substituting $s = -2$ now gives us what we want. Thus

$$K_2 = \frac{d}{ds}[(s+2)^2F(s)]\Big|_{s=-2} \quad (13\text{-}64)$$

For higher-order poles we would need to differentiate more often. This is a straightforward procedure, but writing the expressions in general form makes them look more formidable than they are, so we shall not do so.

Although we have been dealing with real poles only, the same procedures apply to poles which are imaginary or complex. The numerical computations, of course, will now become somewhat more laborious, since they will involve complex numbers.

As an illustration of a case with imaginary poles, consider the following:

$$F(s) = \frac{10s}{(s+1)(s^2+4)} = \frac{K_1}{s+1} + \frac{K_2}{s-j2} + \frac{K_3}{s+j2} \quad (13\text{-}65)$$

Applying Eq. (13-59), we find the residues to be

$$K_1 = \left.\frac{10s}{s^2+4}\right|_{s=-1} = -2 \quad (a)$$

$$K_2 = \left.\frac{10s}{(s+1)(s+j2)}\right|_{s=j2} = \frac{5}{1+j2} = 1 - j2 \quad (b) \quad (13\text{-}66)$$

$$K_3 = \left.\frac{10s}{(s+1)(s-j2)}\right|_{s=-j2} = \frac{5}{1-j2} = 1 + j2 \quad (c)$$

Notice that the residues at the pair of imaginary poles are complex. Furthermore, one is the conjugate of the other. This is not an accident but a general result which we shall state as follows: *The residue of a rational function with real coefficients at a complex pole is the conjugate of the residue at the conjugate pole.* The proof will be left as an exercise for you.

With these values of the residues, Eq. (13-65) can be written

$$F(s) = \frac{10s}{(s+1)(s^2+4)} = -\frac{2}{s+1} + \frac{1-j2}{s-j2} + \frac{1+j2}{s+j2} \quad (13\text{-}67)$$

In order to introduce an important point in the development, let us find the inverse transform of $F(s)$. The inverse transform of each term is an exponential. Hence the inverse transform of this function will be

$$f(t) = \mathfrak{L}^{-1}[F(s)] = -2\epsilon^{-t} + (1-j2)\epsilon^{j2t} + (1+j2)\epsilon^{-j2t}$$
$$= -2\epsilon^{-t} + 2\cos 2t + 4\sin 2t \quad (13\text{-}68)$$

The last line was obtained by expanding the exponentials by Euler's theorem and collecting terms. All j's have disappeared, and the final result appears in a real form.

Now let us return to Eq. (13-67) and group together the two terms containing the imaginary poles. The reslut can be written

$$F(s) = -\frac{2}{s+1} + \frac{2s+8}{s^2+4} = -\frac{2}{s+1} + 2\frac{s}{s^2+4} + 4\frac{2}{s^2+4} \quad (13\text{-}69)$$

On the far right the last term has been split in such a way that the inverse transforms can be recognized (or can be found from Table 13-1) as a cosine and a sine. Thus the last line of Eq. (13-68) results immediately upon taking the inverse transform.

This example suggests that when a transform function has imaginary poles or, more generally, complex poles, we should contrive to collect the two terms containing conjugate pairs of poles in the partial-fraction expansion. If we do not do this, the inverse transform will contain complex exponentials with complex coefficients, which we must combine anyway if we wish to write the result in a real form, as in Eq. (13-68).

In order to give the preceding comments some general validity, let us examine the two terms in a partial-fraction expansion corresponding to a pair of complex conjugate poles $s = -a \pm jb$. Since the residues are also complex conjugate, say $k_1 \pm jk_2$, these terms can be written

$$F_1(s) = \frac{k_1 + jk_2}{s + a + jb} + \frac{k_1 - jk_2}{s + a - jb} = \frac{2k_1(s + a) + 2bk_2}{(s + a)^2 + b^2} \tag{13-70}$$

Note that this expression is valid for purely imaginary poles also, in which case $a = 0$. Let $F_2(s)$ represent this function with imaginary poles. Then, setting $a = 0$ in Eq. (13-70), we get

$$F_2(s) = 2k_1 \frac{s}{s^2 + b^2} + 2k_2 \frac{b}{s^2 + b^2} \tag{13-71}$$

We immediately recognize the two terms as transforms of a sine and a cosine. Hence the inverse transform becomes

$$f_2(t) = 2k_1 \cos bt + 2k_2 \sin bt \tag{13-72}$$

When the poles are complex, we can still separate the right side of Eq. (13-70) into two terms, as follows:

$$F_1(s) = 2k_1 \frac{s + a}{(s + a)^2 + b^2} + 2k_2 \frac{b}{(s + a)^2 + b^2} \tag{13-73}$$

This expression looks like the one in Eq. (13-71), except that here we have $(s + a)$ wherever s appeared before. Suppose that, in the function $F_2(s)$, wherever we see an s, we substitute $(s + a)$; the result will be $F_1(s)$, as given in Eq. (13-73). Thus

$$F_1(s) = F_2(s + a) \tag{13-74}$$

It remains to determine what happens to $f_2(t)$ when in its transform we replace s by $s + a$. By the definition of a Laplace transform, we have

$$F_2(s) = \int_0^\infty f_2(t)\epsilon^{-st}\, dt \tag{13-75}$$

The argument of $F_2(s)$ is whatever multiplies $-t$ in the exponential.

Thus

$$F_2(s + a) = \int_0^\infty f_2(t)\epsilon^{-(s+a)t}\,dt$$
$$= \int_0^\infty [f_2(t)\epsilon^{-at}]\epsilon^{-st}\,dt = F_1(s) \qquad (13\text{-}76)$$

The last step follows from Eq. (13-74). Hence, by the definition of the Laplace transform, the function which appears within brackets in the last line multiplying the exponential ϵ^{-st} is the inverse transform of $F_1(s)$; that is,

$$f_1(t) = \mathcal{L}^{-1}[F_1(s)] = \epsilon^{-at}f_2(t)$$
$$= 2k_1\epsilon^{-at}\cos bt + 2k_2\epsilon^{-at}\sin bt \qquad (13\text{-}77)$$

A pair of complex conjugate poles thus leads to damped sinusoids.

Although the preceding development was carried out for a specific case, the result is quite general and very useful. We shall state it as follows:

Let $F(s)$ be the Laplace transform of $f(t)$. Then the Laplace transform of $\epsilon^{-at}f(t)$ is $F(s + a)$. The general proof is left to you. This result is called the *complex shifting theorem.* The reason for this name is clear from a consideration of Fig. 13-9. A given configuration of poles and

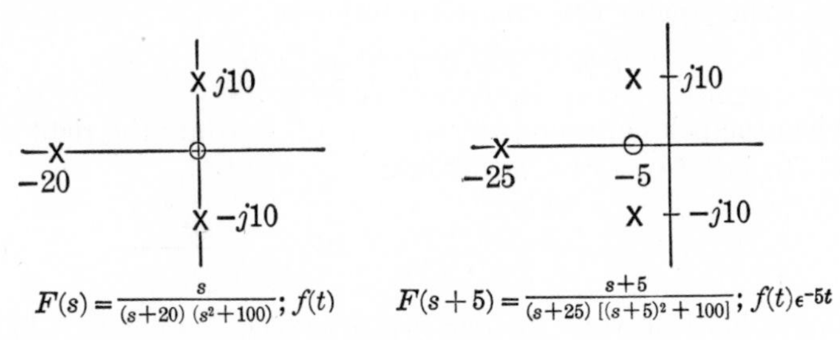

FIG. 13-9

Illustrating the shift of zeros and poles of the transform in the complex plane when $f(t)$ is multiplied by ϵ^{-at}

zeros of a transform function is translated or shifted bodily to the left a distance of five units when the inverse transform is multiplied by ϵ^{-5t}.

As an illustration of the use of the complex shifting theorem, consider the amplitude-modulated waveform

$$f_1(t) = f(t)\cos\omega_0 t = \frac{f(t)\epsilon^{j\omega_0 t}}{2} + \frac{f(t)\epsilon^{-j\omega_0 t}}{2} \qquad (13\text{-}78)$$

where $f(t)$ is the modulating function and ω_0 is the carrier frequency.

On the right side the cosine has been expressed in terms of exponentials. Let $F(s)$ be the Laplace transform of the modulating function. Then, applying the complex shifting theorem, we find the transform of the modulated waveform to be

$$F_1(s) = \mathcal{L}[f_1(t)] = \frac{1}{2}F(s - j\omega_0) + \frac{1}{2}F(s + j\omega_0) \qquad (13\text{-}79)$$

Remember that $F(s)$ is the "spectrum function" of the modulating function. Hence, by $F(s - j\omega_0)$, we understand the "spectrum" of the modulating function, but shifted up in the *vertical* direction in the complex plane a distance ω_0. Similarly, $F(s + j\omega_0)$ is the "spectrum" of the modulating function shifted vertically downward a distance ω_0. You should compare this discussion on a qualitative basis with that given in Chap. 12 on the frequency spectrum of an amplitude-modulated wave.

With the complex shifting theorem, and a knowledge of the transforms of a unit step function, a sine, a cosine, and a power of t, the partial-fraction expansion can be used effectively to find the inverse transform of any rational function, whether the poles are real or complex, simple or multiple.

To illustrate this last statement let us look back at Eq. (13-61) in which a second-order pole appeared. The first two terms we can recognize as transforms of exponentials, but the term involving $(s + 2)^2$ we do not recognize. If we contemplate using the complex translation theorem, however, we can write

$$F_1(s + 2) = \frac{K_3}{(s + 2)^2} \qquad (a)$$

Then (13-80)

$$F_1(s) = \frac{K_3}{s^2} \qquad (b)$$

The inverse transform of the last function is found from Table 13-1 to be

$$f_1(t) = \mathcal{L}^{-1}[F_1(s)] = K_3t$$

Finally, using the complex shifting theorem, we get

$$\frac{K_3}{(s + 2)^2} = \epsilon^{-2t}\mathcal{L}^{-1}\left(\frac{K_3}{s^2}\right) = K_3t\epsilon^{-2t} \qquad (13\text{-}81)$$

We see that a second-order pole leads not to a simple exponential but to an exponential multiplied by t.

As a final illustration of rational functions, let us take a relatively complicated example and obtain a partial-fraction expansion in considerable detail.

$$F(s) = \frac{4(s+4)(s+5)}{(s+1)^2(s+3)(s^2+4s+5)} = \frac{K_1}{(s+1)^2} + \frac{K_2}{s+1} + \frac{K_3}{s+3} + \frac{K_4 s + K_5}{(s+2)^2+1} \tag{13-82}$$

We have one simple real pole, one double real pole, and a pair of complex conjugates. K_1 and K_3 can be found most easily. $F(s)$ is multiplied by $(s+1)^2$, then evaluated at $s = -1$ to find K_1; it is multiplied by $(s+3)$ and evaluated at $s = -3$ to find K_3. To find K_2, $F(s)$ is multiplied by $(s+1)^2$; then it is differentiated before evaluating at $s = -1$, as in Eq. (13-64). Thus

$$K_1 = \left.\frac{4(s+4)(s+5)}{(s+3)(s^2+4s+5)}\right|_{s=-1} = 12$$

$$K_3 = \left.\frac{4(s+4)(s+5)}{(s+1)^2(s^2+4s+5)}\right|_{s=-3} = 1$$

$$K_2 = \frac{d}{ds}\left[\frac{4(s+4)(s+5)}{(s+3)(s^2+4s+5)}\right]_{s=-1}$$

$$= 4\left.\frac{(s+3)(s^2+4s+5)(2s+9) - (s+4)(s+5)(3s^2+14s+17)}{(s+3)^2(s^2+4s+5)^2}\right|_{s=-1} = -11$$

At this point we could go on with the complex poles, using Eq. (13-59) to find the residues. This promises, however, to be a lengthy computation involving complex numbers. Notice, though, from Eq. (13-82), that if we subtract from $F(s)$ the first three terms, which are already computed, the remainder will contain only the pair of complex poles. Hence

$$\frac{K_4 s + K_5}{(s+2)^2+1}$$

$$= \frac{4(s+4)(s+5)}{(s+1)^2(s+3)(s^2+4s+5)} - \frac{12}{(s+1)^2} + \frac{11}{s+1} - \frac{1}{s+3}$$

$$= \frac{1}{(s+1)^2(s+3)}\left[\frac{4(s+4)(s+5)}{s^2+4s+5} - 12(s+3) + 11(s+1)(s+3) - (s+1)^2\right]$$

$$= \frac{10}{(s+1)^2(s+3)}\left(\frac{s^4+7s^3+17s^2+17s+6}{s^2+4s+5}\right)$$

$$= \frac{10(s+2)}{(s+2)^2+1} \tag{13-83}$$

In the third line of this sequence of steps, the fractions are combined over a common denominator, and the numerator seems to be of the fourth degree. Since the poles at $s = -1$ and $s = -3$ have been subtracted from the original function, however, the remainder should not have these poles; that is, the numerator should have the factors $(s + 1)^2$ and $(s + 3)$, which will then cancel with the denominator, leaving the result shown in the fourth line. As a matter of fact, this step provides a numerical accuracy check on the preceding calculations, because, if the numerator does not have the factors $(s + 1)^2$ and $s + 3$, a computational error must have occurred.

The final expansion of the original function is

$$F(s) = \frac{12}{(s+1)^2} - \frac{11}{s+1} + \frac{1}{s+3} + 10\frac{s+2}{(s+2)^2+1} \tag{13-84}$$

The second and third terms can be recognized as transforms of exponentials. The inverses of the first and last terms can be evaluated with the aid of the complex translation theorem. Thus the inverse transform of $F(s)$ is

$$f(t) = \mathcal{L}^{-1}[F(s)] = \epsilon^{-t}(12t - 11) + \epsilon^{-3t} + 10\epsilon^{-2t}\cos t \tag{13-85}$$

There is another class of transform functions, besides the rational functions, whose inverse transforms can be found relatively easily. Recall that, if a function of t is shifted along the horizontal axis t_1 units, its transform becomes multiplied by $\epsilon^{-t_1 s}$. Hence, if a transform function is a rational function multiplied by an exponential $\epsilon^{-t_1 s}$, or a sum of such terms, then use of the real shifting theorem can lead to the inverse transform.

To illustrate the procedure consider the following function:

$$F(s) = \frac{2(s+2)\epsilon^{-4s}}{(s+1)(s+3)} \tag{13-86}$$

Leaving the exponential as a factor, the remainder of the function can be expanded in partial fractions, leading to the result

$$F(s) = \frac{\epsilon^{-4s}}{s+1} + \frac{\epsilon^{-4s}}{s+3} \tag{13-87}$$

Finally, use of the real shifting theorem leads to the inverse transform.

$$f(t) = \mathcal{L}^{-1}[F(s)] = [\epsilon^{-(t-4)} - \epsilon^{-3(t-4)}]u(t-4) \tag{13-88}$$

13.7 THE INITIAL-VALUE AND FINAL-VALUE THEOREMS

The Laplace integral associates with a given function in the time domain a particular function in the s domain, and conversely. To get the function of s, we must integrate the time function over all positive values of t. Hence we cannot, in general, pick out some point in time and say that this is related to some particular point or region in the complex s plane. Although such a point correspondence between the two domains does not exist in general, however, there are two specific points in time which correspond to two specific points in the complex plane.

Consider again the expression for the Laplace transform of the derivative of a function which is assumed to be of exponential order.

$$\mathcal{L}\left(\frac{df}{dt}\right) = \int_0^\infty \frac{df}{dt}\,\epsilon^{-st}\,dt = sF(s) - f(0) \tag{13-89}$$

Since df/dt is of exponential order, there will be some positive value of s for which $\epsilon^{-st}\,df/dt$ will approach zero as t goes to infinity. For any larger positive value of s, the same result will be true *á fortiori;* in particular, for $s \to +\infty$. Let us take the limit of both sides of Eq. (13-89) as $s \to \infty$. Thus

$$\lim_{s\to\infty} \int_0^\infty \frac{df}{dt}\,\epsilon^{-st}\,dt = \lim_{s\to\infty}\,[sF(s) - f(0)] \tag{13-90}$$

The integration and the limit-taking process can be interchanged on the left because the integral is uniformly convergent.† As just discussed, the limit of $\epsilon^{-st}df/dt$ will vanish as $s \to \infty$. Hence the left side of Eq. (13-90) will vanish, giving the final result that

$$\lim_{s\to\infty} sF(s) = f(0) \tag{13-91}$$

This result is called the *initial-value theorem*. It expresses the value of a function of time at $t = 0$ in terms of the limit approached by s times the Laplace transform of the function as s approaches infinity. If the transform of a function is known, the initial-value theorem permits us to determine the value of the function at $t = 0$ without actually finding the inverse transform. Furthermore, the theorem can be applied over and over again to find the initial slope of the function as well as initial values of higher derivatives.

† It is a theorem of Laplace-transform theory that the Laplace integral converges uniformly for those values of s for which it converges at all.

To illustrate, it is required to find the initial value and the initial slope of the function whose transform is

$$F(s) = \frac{(s+2)(s+4)}{(s+1)(s+3)(s+5)}$$

From the initial-value theorem we get

$$f(0) = \lim_{s\to\infty} sF(s) = \lim_{s\to\infty} \frac{s(s+2)(s+4)}{(s+1)(s+3)(s+5)} = 1$$

When going to the limit, the highest powers of s in the numerator and denominator dominate; hence the other powers can be neglected. Knowing the initial value of f, we can now find the Laplace transform of the derivative as

$$\mathcal{L}[f'(t)] = sF(s) - f(0) = \frac{s(s+2)(s+4)}{(s+1)(s+3)(s+5)} - 1$$

$$= \frac{-(3s^2+15s+15)}{(s+1)(s+3)(s+5)}$$

Now applying the initial-value theorem a second time, we get

$$f'(0) = \lim_{s\to\infty} s\mathcal{L}[f'(t)] = \lim_{s\to\infty} \frac{-s(3s^2+15s+15)}{(s+1)(s+3)(s+5)} = -3$$

Thus we find that the function $f(t)$ starts out at a positive value, but with a negative slope.

The initial-value theorem relates zero time and infinite s. This naturally suggests that we seek the converse relationship, relating infinite time and zero s, if one exists. Let us turn again to Eq. (13-89) and, this time, take the limit as s approaches zero. Thus

$$\lim_{s\to 0} \int_0^\infty \frac{df}{dt} \epsilon^{-st}\, dt = \lim_{s\to 0} [sF(s) - f(0)] \qquad (13\text{-}92)$$

Again we contemplate interchanging the integration and the limit taking, but here we run into difficulty. This interchange is permitted only if the value of s under consideration, namely $s = 0$, is a value for which the integral converges. For example, suppose $df/dt = \epsilon^{5t}$. Then, in order for the integral to converge, s must have a real part greater than 5. Clearly, in this case, we can not interchange the two operations and let s go to zero before integrating. We conclude that, in order to interchange the two operations, the given function $f(t)$ must be such that the smallest value of s for which its Laplace integral converges is less than zero.

Assuming that this condition is satisfied, we now interchange the operations and allow s to approach zero under the integral. Then

$$\lim_{s\to 0} \int_0^\infty \frac{df}{dt} \epsilon^{-st}\, dt = \int_0^\infty \frac{df}{dt}\, dt = \lim_{t\to\infty} f(t) - f(0) \qquad (13\text{-}93)$$

When this expression is inserted into Eq. (13-92), the final result becomes

$$\lim_{t\to\infty} f(t) = \lim_{s\to 0} sF(s) \qquad (13\text{-}94)$$

This result is called the *final-value theorem.* It gives the limit approached by a function of time for large t in terms of the limit approached by $sF(s)$ for zero s. Note that the conditions of applicability of this theorem are more restrictive than those of the initial-value theorem. It is actually easier to state the condition in terms of $F(s)$ than in terms of $f(t)$. This condition is that $F(s)$ *should have no poles in the right half plane, or on the imaginary axis.*

Thus the final-value theorem cannot be applied to such simple functions as sinusoids or increasing exponentials, because

$$\mathcal{L}(\sin \omega t) = \frac{\omega}{s^2 + \omega^2}$$

$$\mathcal{L}(\epsilon^{at}) = \frac{1}{s - a}; \quad a > 0$$

The first has poles on the imaginary axis, the second in the right half plane.

13.8 SOLVING NETWORK PROBLEMS

We are now ready to use the transform method in solving network problems. In this section we shall examine a number of specific examples, and perhaps we shall be led to some general conclusions from the results.

As a first example consider Fig. 13-10(a). A square pulse of voltage is applied to the low-pass RC circuit which is initially relaxed. It is desired to find the shape of the output voltage. Since the network is initially relaxed, the transform of the response is equal to the voltage-gain function times the transform of the excitation. Thus

$$V_2(s) = G_{21}(s) V_1(s) = \frac{1}{RCs + 1} V_1(s) \qquad (13\text{-}95)$$

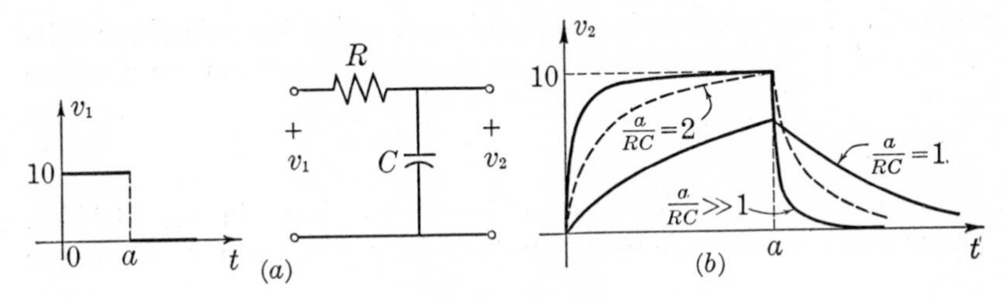

FIG. 13-10

Pulse response of low-pass network

The pulse waveform can be expressed as the difference of two step functions, one displaced in time. Thus

$$v_1(t) = 10u(t) - 10u(t-a) \tag{13-96}$$

It is necessary only to know the transform of a step function and the shifting theorem in order to find the transform of the pulse. It is

$$V_1(s) = \frac{10}{s} - \frac{10}{s}\epsilon^{-as} \tag{13-97}$$

Inserting this into Eq. (13-95), we get

$$V_2(s) = \frac{10}{RC}\left[\frac{1}{s(s+1/RC)} - \frac{\epsilon^{-as}}{s(s+1/RC)}\right]$$

$$= F(s) - F(s)\epsilon^{-as} \qquad (a)$$

where (13-98)

$$F(s) = \frac{10/RC}{s(s+1/RC)} = 10\left(\frac{1}{s} - \frac{1}{s+1/RC}\right) \qquad (b)$$

If we find the inverse transform of $F(s)$, then the inverse transform of $V_2(s)$ can be found by the shifting theorem. The inverse of $F(s)$ is easily recognized to be

$$f(t) = \mathcal{L}^{-1}[F(s)] = 10(1 - \epsilon^{-t/RC})u(t)$$

Hence the desired output voltage is

$$v_2(t) = \mathcal{L}^{-1}[V_2(s)]$$

$$= 10(1 - \epsilon^{-t/RC})u(t) - 10(1 - \epsilon^{-(t-a)/RC})u(t-a) \tag{13-99}$$

The exact waveform of the output will depend on the relative values of the pulse width a and the RC time constant. Figure 13-10(b) gives a sketch of the waveform when $a/RC \gg 1$, $a/RC = 2$, and $a/RC = 1$. The figure shows that the pulse is transmitted with relatively small distortion only when the pulse width is much greater than the time constant. For a given pulse width this means that the pole of the voltage gain (which is $-1/RC$) must be relatively far out on the negative real axis.

As the next example consider the network shown in Fig. 13-11(a). The network has been in operation for a long time when, at $t = 0$, the

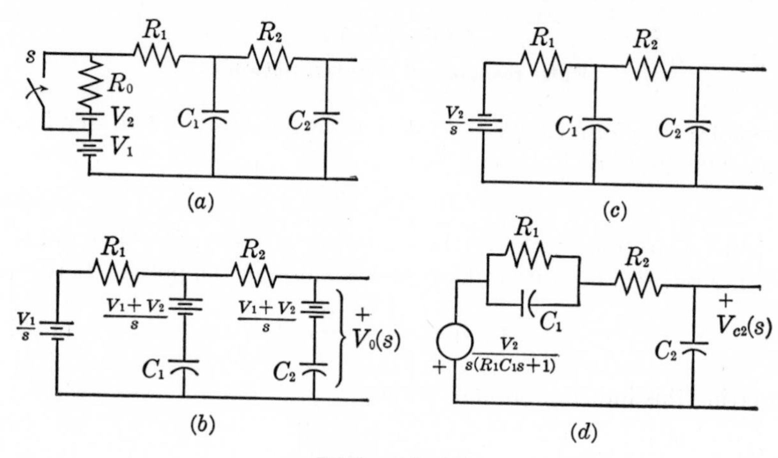

FIG. 13-11
Illustrative example

switch s is closed, thus removing V_2 and R_0 from the network. It is desired to find an expression for the voltage across the output capacitor following the closing of the switch.

Since the original network has been operating for a long time, any transients have long since disappeared. Hence, just before switching, the voltage on each capacitor is $V_1 + V_2$. This is also the initial voltage just after switching, by the principle of charge conservation. Hence the transformed network after switching takes the form shown in Fig. 13-11(b). The initial voltages are represented by sources.

Several alternative approaches are now available for finding the output voltage transform $V_0(s)$. The formal approach would be to write a set of loop or node equations and proceed from there. Let us, instead, try other approaches. One possibility is to use Thévenin's theorem by replacing the network to the left of R_2 by its equivalent. Another observation we can make is that the network is plagued with a multiplicity of sources. This situation can be remedied by the voltage-shift theorem. The battery in series with R_2 can be shifted through the node between R_1, R_2, and the center branch. It can then be combined with the sources already in these branches. If this is done, we see that the battery in series with C_1 is completely canceled by the shifted source, while that in series with R_1 is partially canceled. The resulting network takes the form shown in Fig. 13-11(c). Applying Thévenin's theorem to the network to the left of R_2, we now get the result shown in Fig. 13-11(d). The transform of the voltage across C_2 in this network can now be found by the voltage-divider relationship to be

$$V_{C_2}(s) = \frac{-V_2}{s(R_1C_1s + 1)} \frac{1/C_2s}{\dfrac{R_1}{R_1C_1s + 1} + R_2 + \dfrac{1}{C_2s}}$$

$$= \frac{-V_2}{s[R_1C_2s + (R_1C_1s + 1)(R_2C_2s + 1)]} \tag{13-100}$$

The quantities R_1C_1, R_2C_2, R_2C_1, and R_1C_2 are all dimensionally time constants. For convenience, let us assign them numerical values: $R_1C_1 = 1$, $R_2C_2 = 1/2$, $R_2C_1 = 2/3$. Note, however, that $V_{C_2}(s)$ is not the desired voltage $V_0(s)$. From Fig. 13-11(b) we see that $V_0(s)$ is the sum of $V_{C_2}(s)$ and $(V_1 + V_2)/s$. Thus

$$V_0(s) = \frac{V_1 + V_2}{s} + V_{C_2}(s)$$

$$= \frac{V_1 + V_2}{s} - \frac{V_2}{R_1C_1R_2C_2s\left[s^2 + s\left(\dfrac{1}{R_1C_1} + \dfrac{1}{R_2C_2} + \dfrac{1}{R_2C_1}\right) + \dfrac{1}{R_1C_1R_2C_2}\right]}$$

$$= \frac{V_1}{s} + \frac{V_2}{s}\left(1 - \frac{2}{s^2 + 3s + 2}\right) = \frac{V_1}{s} + \frac{V_2(s + 3)}{(s + 1)(s + 2)}$$

$$= \frac{V_1}{s} + \frac{2V_2}{s + 1} - \frac{V_2}{s + 2} \tag{13-101}$$

In the last step a partial-fraction expansion has been written, from which the inverse transform is readily found to be

$$v_0(t) = (V_1 + 2V_2\epsilon^{-t} - V_2\epsilon^{-2t})u(t) \tag{13-102}$$

Let us now examine this solution to see whether any known facts about the voltage are corroborated. For example, setting $t = 0$, we find that $v_0(0) = V_1 + V_2$, which is the correct initial voltage. We could have applied the initial-value theorem to $V_0(s)$ to determine this result also. From our classical ideas we know that the forced component of the voltage response should have the same waveform as the excitation, which, in this case, is the constant V_1. Also, we would expect this network to have two natural frequencies. Both of these facts are manifest in Eq. (13-102).

As a third example let us consider the transformer network shown in Fig. 13-12. It is required to find the voltage across the output

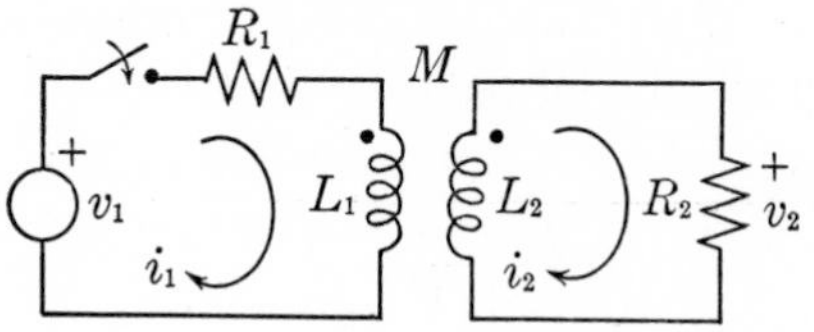

$v_1 = 10 \cos 2t$, $R_1 = 1$, $R_2 = 4$, $L_1 = 1$, $L_2 = 4$, $M = 3$

FIG. 13-12

Transformer network

resistor as a function of time after the closing of the switch, assuming initially relaxed conditions.

Although we can again use Thévenin's theorem or replace the transformer by a tee equivalent and proceed from there, let us, this time, write loop equations. With the loop currents shown in the figure, the transformed loop equations will be

$$(R_1 + L_1 s)I_1(s) - MsI_2(s) = V_1(s) = 10\frac{s}{s^2+4} \quad (a)$$
$$-MsI_1(s) + (R_2 + L_2 s)I_2(s) = 0 \quad (b) \qquad (13\text{-}103)$$

where M is a positive member. Solving for $I_2(s)$ and multiplying by R_2, we get

$$\begin{aligned} V_2(s) = R_2 I_2(s) &= R_2 \frac{\Delta_{12}}{\Delta} V_1(s) \\ &= \frac{R_2 Ms(10s/(s^2+4)}{(L_1L_2 - M^2)s^2 + (R_1L_2 + R_2L_1)s + R_1R_2} \\ &= \frac{40\sqrt{3}s^2}{(s^2+4)(s^2+8s+4)} \end{aligned} \qquad (13\text{-}104)$$

In the last step we inserted the numerical values given in the figure.

The next step is to find the partial-fraction expansion. The two real poles are at $s = -0.536$ and $s = -7.46$. The residues at these poles are found to be

$$K_1 = (s + 0.536)V_2(s)\Big|_{s=-0.536} = 0.671$$

$$K_2 = (s + 7.46)V_2(s)\Big|_{s=-7.46} = -9.34$$

If we now subtract from $V_2(s)$ the two terms in the partial-fraction expansion which represent the two real poles, the result will contain only the imaginary poles. Thus

$$V_2(s) - \frac{0.671}{s + 0.536} + \frac{9.34}{s + 7.46} = \frac{40\sqrt{3}s^2}{(s^2 + 4)(s^2 + 8s + 4)}$$

$$+ \frac{8.67s}{s^2 + 8s + 4} = \frac{s}{s^2 + 8s + 4}\left(\frac{40\sqrt{3}s}{s^2 + 4} + 8.67\right) = \frac{8.67s}{s^2 + 4}$$

The final expression for $V_2(s)$ now becomes

$$V_2(s) = \frac{0.671}{s + 0.536} - \frac{9.34}{s + 7.46} + \frac{8.67s}{s^2 + 4} \tag{13-105}$$

Finally, the inverse transform is

$$v_2(t) = \mathcal{L}^{-1}[V_2(s)]$$
$$= (0.671\epsilon^{-0.536t} - 9.34\epsilon^{-7.46t} + 8.67 \cos 2t)u(t) \tag{13-106}$$

This time we find a sinusoid, corresponding to the steady state, and two decaying exponentials corresponding to the transient. It is perhaps surprising that the steady-state output voltage is in phase with the input. We can check this by making a steady-state calculation, using Eq. (13-104) but with the Laplace transform $V_1(s)$ replaced by the appropriate sinor, which in this case is $10 \underline{/0°}$ (on an amplitude basis, rather than rms). Hence

$$V_2\ (\text{sinor}) = \left.\frac{40\sqrt{3}s}{s^2 + 8s + 4}\right|_{s=j2} = 5\sqrt{3} = 8.66 \tag{13-107}$$

which corroborates our previous result in Eq. (13-106).

As a final illustration let us find the step function response (the response to a unit step) of the amplifier network shown in Fig. 13-13. This is the model of a so-called "shunt-peaked" amplifier. Assuming

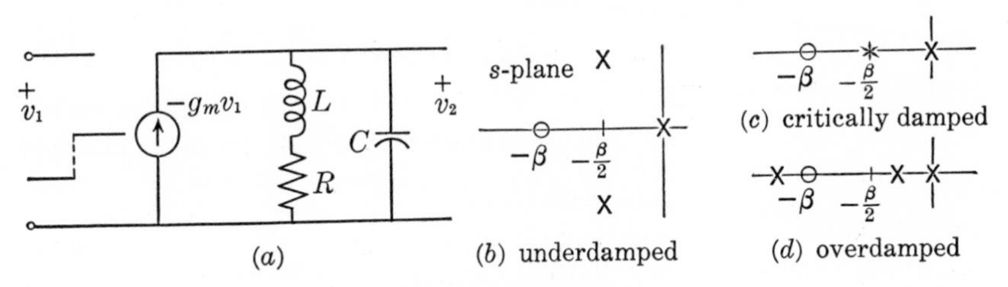

FIG. 13-13

"Shunt-peaked" amplifier model

initially relaxed conditions, the output voltage transform is equal to the impedance times the transform of the controlled source. Thus

$$V_2(s) = -g_m V_1(s) \frac{1}{sC + \dfrac{1}{sL + R}} = \frac{-g_m}{Cs} \frac{s + R/L}{s^2 + \dfrac{R}{L}s + \dfrac{1}{LC}}$$

$$= \frac{-g_m}{Cs} \frac{s + \beta}{s^2 + \beta s + \omega_0^2} = \frac{-g_m}{Cs} \frac{s + \beta}{(s + \beta/2)^2 + \omega_d^2} \qquad (13\text{-}108)$$

The last line is written in terms of the 3-db bandwidth β and the resonant frequency ω_0, or the damped resonant frequency ω_d. (See Chap. 6.) The specific form of the response will depend on the location of the poles. There are three possibilities: (1) the *underdamped* case in which the poles are complex, (2) the *overdamped* case in which the poles are real and distinct, and (3) the *critically* damped case in which the poles are real and equal. The pole-zero diagrams for these three cases are shown in Fig. 13-13(*b*), (*c*), and (*d*).

Let us assume that the element values are adjusted so that $\beta = 10$, $\omega_0 = 6.4$, and $g_m/C = 100$; the poles will be complex and will be located at $s = -5 \pm j4$. The residue K_0 at $s = 0$ is easily found to be

$$K_0 = sV_2(s)\bigg|_{s=0} = -100 \frac{(s + 10)}{(s + 5)^2 + 16}\bigg|_{s=0}$$

$$= -\frac{1{,}000}{41} = -24.4 \qquad (13\text{-}109)$$

Subtracting K_0/s from $V_2(s)$, we get

$$V_2(s) + \frac{24.4}{s} = \frac{1}{s}\left[\frac{-100(s + 10)}{(s + 5)^2 + 16} + 24.4\right] = \frac{24.4s + 144}{(s + 5)^2 + 16}$$

The final expression for $V_2(s)$ may now be written

$$V_2(s) = -\frac{24.4}{s} + 24.4 \frac{s + 5}{(s + 5)^2 + 16} + 5.5 \frac{4}{(s + 5)^2 + 16} \qquad (13\text{-}110)$$

The by-now-familiar complex shifting theorem permits the calculation of the inverse transforms of the last two terms. Thus, finally,

$$v_2(t) = [-24.4 + \epsilon^{-5t}(24.4 \cos 4t + 5.5 \sin 4t)]u(t) \qquad (13\text{-}111)$$

The cosine and sine can be written as a single sinusoid if desired.

A sketch of this result is given in Fig. 13-14. Note that the large damping constant causes the oscillations to die down quite rapidly. There is an *overshoot* where the response exceeds its eventual steady-state value, but this is quite small.

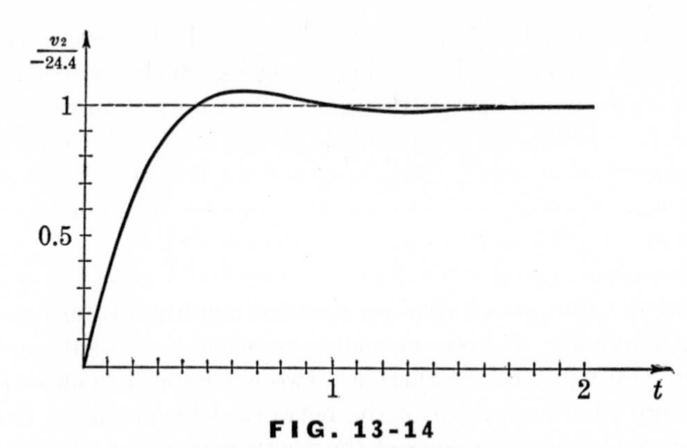

FIG. 13-14
Sketch of normalized response

When first discussing the topic of inverse transforms in Sec. 13-4, we introduced the idea of a table of transform pairs. When, in the process of solving a problem, we stated, a transform function is obtained, we can enter the table to find the corresponding inverse transform. In the preceding illustrations, however, we did not follow this precise procedure. Instead, we decomposed the transform function into simple terms each of whose inverse transform we could recognize. As a matter of fact, it is enough for us to know the transform of a step function, a power of t, a sine, a cosine, and an exponential, and to know the real and complex shifting theorems. With this meager arsenal and the partial-fraction expansion, we can solve almost any problem we are likely to meet.

13.9 CONCLUSION

In this chapter we arrived at the culmination of the objective which we originally set for ourselves, that of finding a voltage or current in an arbitrary network in response to an arbitrary excitation. The total response can be visualized as consisting of two components. One of these, the natural component, normally consists of decaying exponentials or exponentially damped sinusoids, and hence may also be called the transient. The other is the forced component; in case the excitation is periodic (or a constant) this component is also called the steady state.

We introduced the Laplace-transform method whereby the complete response is computed in a single process which also accounts for the influence of initial-energy storage. When initial conditions are relaxed, we found that all the procedures of analysis we have learned for the sinor method of steady-state solution can be applied here without change; it is necessary only to use the complex-frequency variable s and to deal with transforms rather than sinors.

Any response transform will have the general form of the expression in Eq. (13-49), that is, a sum of products of excitation transforms times network functions. If you consider expanding Eq. (13-49) in partial functions, you will see that there are two categories of poles—poles of the response which are zeros of the network determinant Δ, and poles of the response which are poles of the excitations. The latter poles will lead to terms in the time response having the same waveform as the corresponding excitation. The zeros of Δ are the natural frequencies, and these poles all contribute to the natural response. You can demonstrate these facts by referring to the illustrative examples. Thus we find that the transform method gives us a clear insight into the network behavior, showing exactly how each term in the response arises.

There is a tendency to look upon the transform method as a "crank-turning" technique, a sequence of steps which can be performed without much thought. This is far from the truth. The transform method, which includes additional concepts and techniques which we have not discussed,† sheds considerable light on network behavior and leads to many interesting relationships among the steady-state response and the transient response.

We shall cite one example of such understanding by considering the response at the terminals of an initially relaxed network to which a unit impulse voltage is applied. This is illustrated in Fig. 13-15. Since the network is initially relaxed, the transform of the input current will be the transform of the input voltage times the appropriate network function, which, in this case, is the admittance. Similarly, the transform of the output voltage will be the transform of the input voltage times the voltage-gain function. Thus

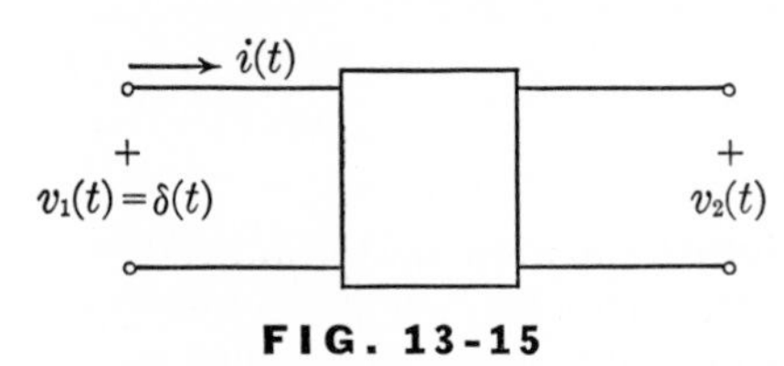

FIG. 13-15

Impulse response of network

† The convolution and superposition integrals are among these.

$$\begin{aligned} I(s) &= Y(s)V_1(s) = Y(s)\mathcal{L}[\delta(t)] = Y(s) \qquad (a) \\ V_2(s) &= G_{21}(s)V_1(s) = G_{21}(s) \qquad (b) \end{aligned} \qquad (13\text{-}112)$$

since the transform of a unit impulse is 1.

Let us use the term *impulse response* to signify a response as a function of time when the excitation is a unit impulse. Hence we see that a network function—driving point or transfer impedances and admittances, and voltage- or current-gain functions—is simply the Laplace transform of some particular impulse response, like the input current and output voltage in the preceding example.

The concept of impulse response plays an important role in our understanding of the behavior of linear networks. We have noted here its intimate relationship with a network function, but this is just a crack in the door. These relationships are made evident and are exploited through the agency of the Laplace transform. This brief discussion should help to dispel the notion that the Laplace transform is little more than a substitute for thought.

Within our context we have used the Laplace transform in the solution of electrical network problems. It should be clear, however, that this is a mathematical procedure which can be applied to the solution of any linear differential equation, or set of equations, no matter what the physical laws that lead to these equations and no matter what physical significance the variables may have. In our application the independent variable t has the significance of time. In other applications, such as in mechanical systems, however, the independent variable may be linear distance or angular displacement; the dependent variables may be velocity, acceleration, force, torque, and so on. A good grasp of the transform method, as applied to linear electrical networks, will be useful when you desire to solve problems in other engineering disciplines.

Some of the concepts introduced in this chapter may appear to be presented in an *ad hoc* manner, perhaps without adequate motivation and without adequate mathematical support. You may have a sense of uneasiness because you do not feel fully "at home" with the subtleties of the subject. A comment which is typically made under such circumstances is, "I can follow each step you perform, but I do not know why you are doing it."

It is not possible for us to give a completely satisfying and rigorous exposition with the mathematical background at our disposal. A thorough understanding of the Laplace transform and its properties, especially the inverse transform, requires a knowledge of complex-function theory and advanced calculus which most of you do not yet have at your command.

The reasons for seeking a transform process such as the one under discussion here are certainly clear. The steps in arriving at the desired mathematical tools are perhaps somewhat less clear. The use to which we put the mathematical tools, however, should have been amply demonstrated in the preceding sections.

It is hoped that the brief exposure given here to the transform concept in general, and the Laplace transform in particular, will serve as an enticement for you to pursue the subject in your later studies. Transform methods of solving engineering problems still constitute an area for research, and perhaps some of you will someday make a contribution to our understanding of such methods.

Problems

13-1 Find the Laplace transforms of the following functions.

(*a*) $f(t) = t\epsilon^{-2t}$

(*b*) $f(t) = t \sin 10t$

(*c*) $f(t) = 10 \cos (\omega t - \pi/3)$

(*d*) $f(t) = at^3 + bt^2 + ct + d$

(*e*) $f(t) = t\epsilon^{-2t}u(t-2)$

(*f*) $f(t) = (t-2)\epsilon^{-2(t-2)}u(t-2)$

(*g*) $f(t) = \sin 10t\, u(t-1)$

(*h*) $f(t) = \sin (10t - 10)u(t-1)$

13-2 Find the Laplace transforms of the functions shown in Fig. P13-2.

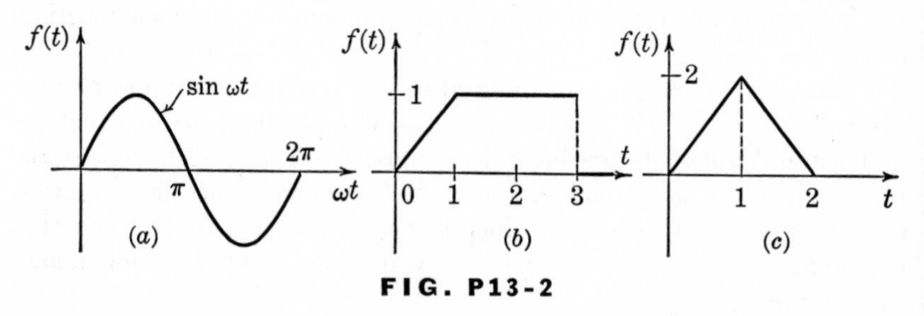

FIG. P13-2

13-3 Using only the derivative theorem, $\mathcal{L}\,(df/dt) = s\mathcal{L}[f(t)] - f(0)$, find the Laplace transforms of ϵ^{at}, $\sin \omega t$, $\cos \omega t$, $\sinh bt$, and $\cosh bt$.

13-4 (*a*) Using the derivative theorem and a knowledge of the transforms of $\sin \omega t$ and $\cos \omega t$, show that

$$\mathcal{L}(t \cos \omega t) = \frac{s^2 - \omega^2}{(s^2 + \omega^2)^2}$$

$$\mathcal{L}(t \sin \omega t) = \frac{2\omega s}{(s^2 + \omega^2)^2}$$

(*b*) Verify by direct calculation. (*c*) Also verify by expressing $\cos \omega t$ exponentially and using the complex shifting theorem.

13-5 From a knowledge of the Laplace transforms of the functions in the left-hand column, find the transforms of those on the right using the derivative theorem or integral theorem.

$$\mathcal{L}(t^2) = \frac{2}{s^3} \qquad f(t) = t^3$$

$$\mathcal{L}(t^4) = \frac{24}{s^5} \qquad f(t) = t^3$$

$$\mathcal{L}(\sin \omega t) = \frac{\omega}{s^2 + \omega^2} \qquad f(t) = \cos \omega t$$

13-6 Consider the periodic function $f(t)$ shown in Fig. P13-6(*a*). Let $f_1(t)$ be the function which is identical with $f(t)$ over the first period but zero elsewhere. $f(t)$ can be considered as made up of a summation of functions consisting of $f_1(t)$, $f_1(t)$ displaced to the right by one period, $f_1(t)$ displaced by two periods, etc. On this basis prove that

$$\mathcal{L}[f(t)] = \frac{\mathcal{L}[f_1(t)]}{1 - \epsilon^{-Ts}}$$

That is, to find the Laplace transform of $f(t)$, it is necessary to find the transform of its first cycle only.

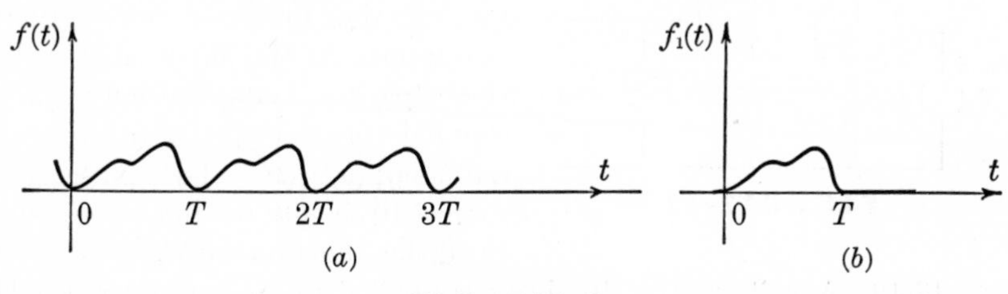

FIG. P13-6

13-7 Find the Laplace transforms of the following periodic functions: (*a*) a half-wave rectified sine function; (*b*) a full-wave rectified sine function; (*c*) a saw-tooth function; (*d*) a symmetrical square wave; (*e*) an iscosceles triangular wave.

13-8 Use Laplace-transform procedures to find the solution of the problems in Chap. 3.

13-9 Find the inverse transforms of the following functions. Use any simplification you can think of, including some of the theorems discussed in the text, rather than "bulling" through.

(*a*) $F(s) = \dfrac{3s + 1}{s^2 + 2s + 5}$

(*b*) $F(s) = \dfrac{2s + 5}{s(s^2 + 4s + 8)}$

(*c*) $F(s) = \dfrac{s(s + 1)}{(s + 2)(s + 5)}$

(*d*) $F(s) = \dfrac{3s + 2}{s(s + 3)(s + 4)}$

(*e*) $F(s) = \dfrac{s^2 + 2s + 2}{s(s + 1)^2}$

(*f*) $F(s) = \dfrac{2s\epsilon^{-2s}}{(s + 1)(s + 3)}$

(*g*) $F(s) = \dfrac{s(1 - \epsilon^{-2s})}{s^2 + 2s + 2}$

(*h*) $F(s) = \dfrac{(s + 3)\epsilon^{-s}}{(s^2 + 2s + 10)^2}$

(*i*) $F(s) = \dfrac{2s^2 + s + 3}{s^2(s^2 + 4)}$

(*j*) $F(s) = \dfrac{s(s + 2)\epsilon^{-3s}}{(s + 1)(s + 4)}$

13-10 Use the initial-value theorem to find the initial values and the initial slopes of the inverse transforms of the following functions. When possible, find also the final value.

(*a*) $F(s) = \dfrac{3s + 2}{s^2 + 3s + 6}$

(*b*) $F(s) = \dfrac{2s + 5}{(s + 1)(s^2 + 4)}$

(*c*) $F(s) = \dfrac{2s^2 + 10}{s^3 + 2s^2 + 5s}$

(*d*) $F(s) = \dfrac{(2s + 7)(1 - \epsilon^{-s})}{s^2 + 2s + 3}$

13-11 In Fig. P13-11 assume that the tube can be replaced by its small-signal low-frequency equivalent. Find the output voltage, assuming the following waveforms are applied to the input at $t = 0$, the capacitor being initially uncharged. Assume $r_p \to \infty$. (*a*) $v_1(t) =$ unit step; (*b*) $v_1(t) =$ 1-sec pulse; (*c*) $v_1(t) = 10 \sin 10t$; (*d*) full-wave rectified sine function with $\omega = 5$.

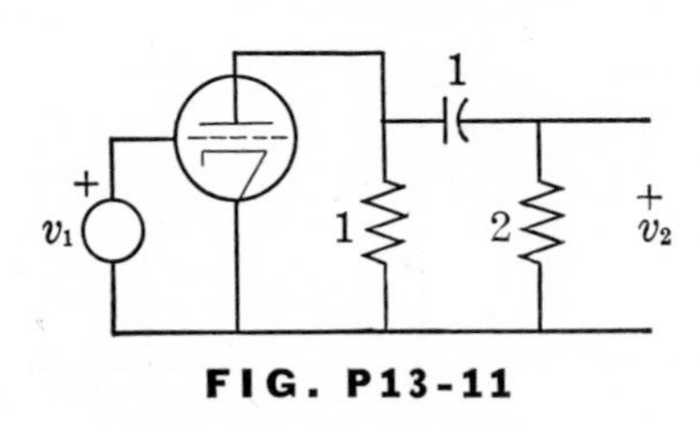

FIG. P13-11

13-12 A voltage $v_1 = 10 \sin 2t$ is applied to an initially relaxed network. The resulting current is $i(t) = 5\sqrt{2} \cos (2t - \pi/2) - 5\epsilon^{-t}$. Let the voltage applied to the same network be $v_2 = 5 \cos 2t$. Find the resulting current. (Hint: Use the derivative theorem.) Can you generalize this result to apply to any two signals, one of which is the derivative of the other?

APPENDIX

In this appendix is presented a discussion of some specific networks whose physical counterparts are widely used in electronic equipment. The networks have a number of similarities. They will be analyzed in terms of the locations of the poles and zeros of their transfer functions in the complex s plane. Their frequency response will be examined in the light of the pole locations.

Even though these particular networks are important, this reason is not the primary one for the appendix. The discussion here is an amplification and expansion of the ideas introduced in Chaps. 5 and 6, and it is intended to illustrate further the concepts of poles and zeros, the complex-frequency plane, and the frequency response.

In order to avoid becoming overburdened with algebra, many steps in the mathematical calculations will be omitted.

A.1 STAGGERED TUNED CIRCUITS

The filtering property or frequency selectivity exhibited by the tuned circuits discussed in Chap. 5 is a sought-after feature which it is desirable to enhance. In attempting to improve the bandwidth, it may be contemplated to connect two tuned circuits in parallel. This approach cannot succeed, since the combined network will have natural frequencies which are determined not by the parameters of the individual networks separately but by all of the parameters collectively, because the two circuits are not isolated. Isolation, however, can be achieved by amplifiers.

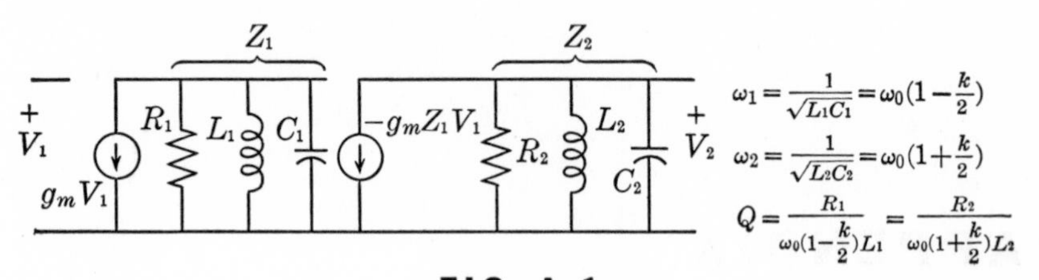

FIG. A-1

A staggered pair

With this discussion as background, consider the two-stage amplifier shown in Fig. A-1. Each stage has a parallel tuned circuit as a load. The voltage gain is easily found to be

$$G_{21} = \frac{V_2}{V_1} = g_m{}^2 Z_1 Z_2 \tag{A-1}$$

where Z_1 and Z_2 are

$$\begin{aligned} Z_1 &= \frac{L_1 s}{L_1C_1s^2 + \frac{L_1}{R_1}s + 1} \\ Z_2 &= \frac{L_2 s}{L_2C_2s^2 + \frac{L_2}{R_2}s + 1} \end{aligned} \tag{A-2}$$

Suppose that the two circuits have the same Q and are tuned to the same frequency. Then Z_1 and Z_2 will have the same poles and so the voltage-gain function will have double poles, as shown in Fig. A-2. If we contemplate sketching the magnitude of the voltage gain as a function of frequency, using the geometrical procedure of measuring line lengths from the poles and zeros to points on the $j\omega$ axis, we see that the response curve will resemble a simple resonance curve, but it will be more peaked.

Suppose the resonant frequencies are somewhat displaced from each other, however, one above and the other below ω_0: $\omega_1 = \omega_0(1 - k/2)$, $\omega_2 = \omega_0(1 + k/2)$. k is called the *staggering coefficient.* Assuming that the Q's are still the same, the poles of G_{21} will be located as shown in Fig. A-2(b). If the pole separations were great enough, we would ex-

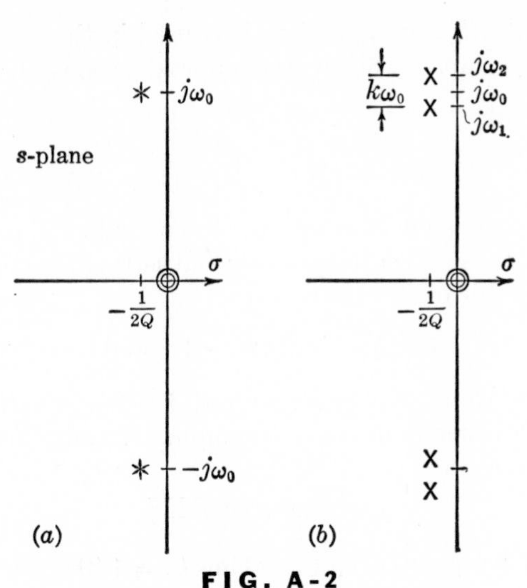

FIG. A-2

Poles of the voltage-gain function

pect the frequency-response curve to have two peaks, one opposite each of the poles in the upper half plane. At some intermediate point the character of the curve should change from a single-peaked one to a double-peaked one.

Let us examine the previous comments quantitatively. With the expressions given in Fig. A-1 substituted into the impedances, the voltage-gain function can be written

$$G_{21}(j\omega) = \frac{g_m{}^2R^2}{\left[1 + jQ\left(\frac{\omega/\omega_0}{1 - k/2} - \frac{1 - k/2}{\omega/\omega_0}\right)\right]\left[1 + jQ\left(\frac{\omega/\omega_0}{1 + k/2} - \frac{1 + k/2}{\omega/\omega_0}\right)\right]} \tag{A-3}$$

When the frequency is in the vicinity of ω_0, it is convenient to deal with the fractional frequency deviation δ, defined by

$$\delta = \frac{\omega}{\omega_0} - 1 \tag{A-4}$$

For high-Q circuits and small values of δ, the following approximations are valid.

$$\frac{\omega/\omega_0}{1 - k/2} - \frac{1 - k/2}{\omega/\omega_0} \doteq 2\left(\delta + \frac{k}{2}\right) \qquad (a)$$

$$\frac{\omega/\omega_0}{1 + k/2} - \frac{1 + k/2}{\omega/\omega_0} \doteq 2\left(\delta - \frac{k}{2}\right) \qquad (b) \tag{A-5}$$

Compare these with the approximations used in the case of the single tuned circuit in Chap. 5. With these approximations, the magnitude of G_{21} becomes

$$|G_{21}(j\omega)| = \frac{g_m{}^2R^2}{\sqrt{[1 + (2Q\delta + kQ)^2][1 + (2Q\delta - kQ)^2]}} \tag{A-6}$$

With this expression it is possible to compute the positions of the peaks of the response by differentiating and setting the derivative equal to zero. The result of this operation is

$$\delta[(2Q\delta)^2 - (Q^2k^2 - 1)] = 0 \tag{A-7}$$

Thus the derivative always has a zero at $\delta = 0(\omega = \omega_0)$. This corresponds to either a maximum or a minimum. In addition, it will have two other real zeros at

$$2Q\delta = \pm\sqrt{Q^2k^2 - 1} \tag{A-8}$$

provided that Q^2k^2 is greater than 1. Thus $k_c = 1/Q$ is a critical value of staggering. Below this value the curve has a single peak; above this value it has a double peak and an intervening valley at $\delta = 0$; it is *overstaggered.* At the critical value the curve is exceptionally flat at $\delta = 0$, since not only the first derivative but also the second and third derivatives are zero there.

The response curve is shown plotted in Fig. A-3 for two values of the staggering coefficient: the critical value $kQ = 1$, and $kQ = 2.414$. The heights of the response curve at the two maxima in the overstaggered

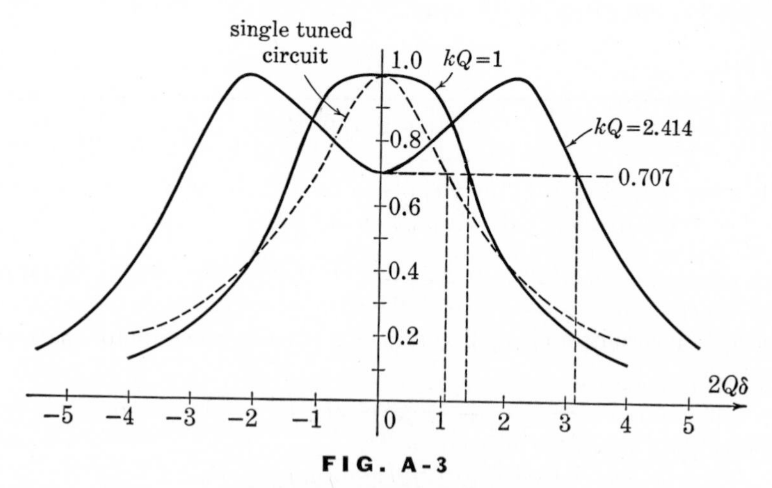

FIG. A-3

Frequency-response curves

case are computed by inserting Eq. (A-8) into Eq. (A-6). The result is

$$|G_{21}|_{\max} = \frac{g_m{}^2 R^2}{2kQ} \tag{A-9}$$

Thus we find that the two peak values are equal and are inversely proportional to the staggering coefficient.

At $\delta = 0$ the height of the curve at any value of k is found to be

$$|G_{21}|_{\delta=0} = \frac{g_m{}^2 R^2}{1 + k^2Q^2} = \frac{2kQ}{1 + k^2Q^2} |G_{21}|_{\max} \tag{A-10}$$

which, at critical staggering, is $g_m{}^2R^2/2$. Thus, although the two curves in Fig. A-3 are drawn with the same peak values, the scale of the ordinate should be divided by the corresponding value of kQ for each curve.

Let us now inquire into the bandwidth of the response. The 3-db points can be found by setting $|G_{21}|$ in Eq. (A-6) equal to $1/\sqrt{2}$ of $|G_{21}|_{\max}$ in Eq. (A-9). Four values of $2Q\delta$ are found at which the response is down 3 db. They are

$$2Q\delta = \pm\sqrt{k^2Q^2 - 1 + 2kQ} \qquad (a)$$
$$2Q\delta = \pm\sqrt{k^2Q^2 - 1 - 2kQ} \qquad (b) \tag{A-11}$$

For critical coupling the second of these expressions leads to imaginary values. In fact, the quantity under the radical is negative until kQ exceeds 2.414. At this point the value of $|G_{21}|$ at $\delta = 0$ is precisely 3 db down from the peak value, as seen from Eq. (A-10). For larger values of k the valley in the response curve will dip below the -3-db level.

The 3-db points on the outer skirts of the response curve are given by Eq. [A-11(a)]. Expressing δ in terms of ω, the bandwidth becomes

$$\beta = \omega_2 - \omega_1 = \omega_0(\delta_2 + 1) - \omega_0(\delta_1 + 1) = \omega_0(\delta_2 - \delta_1)$$

$$= \frac{\omega_0}{Q}\sqrt{k^2Q^2 - 1 + 2kQ} \tag{A-12}$$

This expression should be compared with that of the single tuned circuit which is $\beta = \omega_0/Q$. The bandwidth is improved by the factor $\sqrt{k^2Q^2 - 1 + 2kQ}$. At critical coupling the bandwidth is the same as that of a single tuned circuit. For $kQ = 2.414 = 1 + \sqrt{2}$, at which the valley in the curve has the same height as the 3-db band edges, the bandwidth is improved by a factor 2.7.

Thus, staggered tuned circuits lead to a flatter response over a wider band than the single tuned circuit.

A.2 THE DOUBLE TUNED SERIES CIRCUIT

A diagram of the double tuned series circuit is shown in Fig. A-4. We shall assume that the individual resonant frequencies of the two

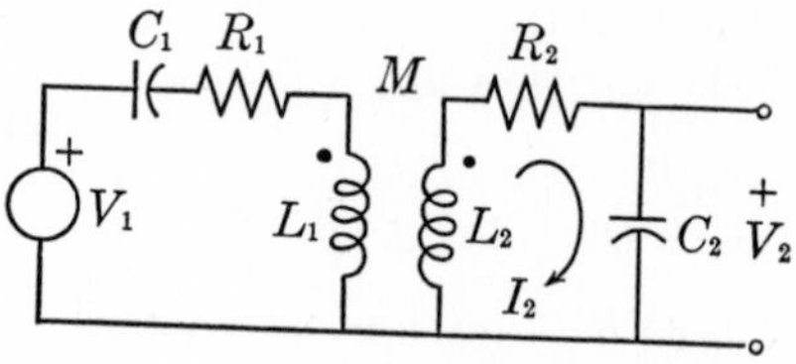

$$\omega_0^2 = \frac{1}{L_1C_1} = \frac{1}{L_2C_2}$$

$$Q = \frac{\omega_0 L_1}{R_1} = \frac{\omega_0 L_2}{R_2}$$

$$k = \frac{M}{\sqrt{L_1L_2}}$$

FIG. A-4

Series double tuned circuit

tuned circuits are the same, as are also the Q's. We shall also assume that the Q's are high. Using loop equations, the output voltage sinor can be found as

$$V_2 = \frac{1}{sC_2} I_2 = \frac{1}{sC_2}\frac{\Delta_{12}}{\Delta} V_1 = \frac{(M/C_2)V_1}{\Delta}$$

$$= \frac{(M/C_2)V_1}{\dfrac{(s^2L_1C_1 + R_1C_1s + 1)(s^2L_2C_2s + R_2C_2s + 1)}{C_1C_2s^2} - M^2s^2} \qquad \text{(A-13)}$$

Using the resonant frequency ω_0, the Q, and the coupling coefficient k as given in the figure, the voltage-gain function becomes

$$G_{21} = \frac{V_2}{V_1} = \frac{k\sqrt{\dfrac{C_1}{C_2}}\dfrac{s^2}{\omega_0^2}}{\left(\dfrac{s^2}{\omega_0^2} + \dfrac{s}{\omega_0 Q} + 1\right)^2 - k^2\dfrac{s^4}{\omega_0^4}}$$

$$= \frac{k\sqrt{\dfrac{C_1}{C_2}}\dfrac{s^2}{\omega_0^2}}{\left[(1-k)\dfrac{s^2}{\omega_0^2} + \dfrac{s}{\omega_0 Q} + 1\right]\left[(1+k)\dfrac{s^2}{\omega_0^2} + \dfrac{s}{\omega_0 Q} + 1\right]} \qquad \text{(A-14)}$$

The denominator of the first line is the difference of two squares and so can be factored to give the second line.

The voltage-gain function has a double zero at the origin and two pairs of poles whose locations are given by

$$\frac{s}{\omega_0} = -\frac{1}{2Q(1-k)} \pm j\frac{1}{\sqrt{1-k}}\sqrt{1-\frac{1}{4Q^2(1-k)}} \quad (a)$$
$$\frac{s}{\omega_0} = -\frac{1}{2Q(1+k)} \pm j\frac{1}{\sqrt{1+k}}\sqrt{1-\frac{1}{4Q^2(1+k)}} \quad (b) \qquad \text{(A-15)}$$

If we assume that $4Q^2(1-k) \gg 1$ [then $4Q^2(1+k) \gg 1$, *a fortiori*], the approximate location of the poles will be

$$\frac{s}{\omega_0} \doteq -\frac{1}{2Q(1 \pm k)} \pm j\frac{1}{\sqrt{1 \pm k}} \qquad \text{(A-16)}$$

These locations are shown in Fig. A-5 for a value of $k = 0.2$. The horizontal scale has been exaggerated. When there is no coupling

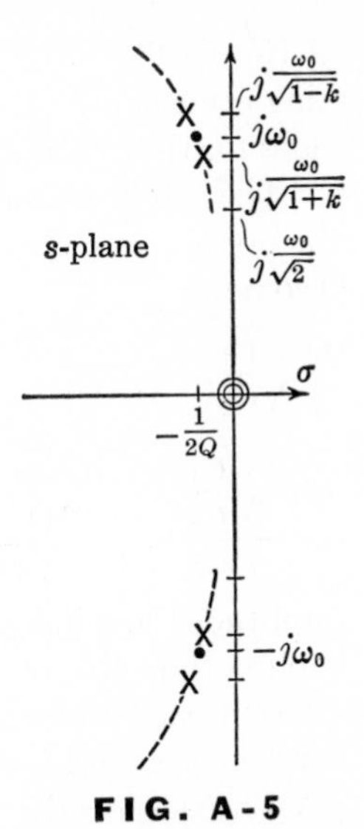

FIG. A-5

Pole-zero locations of double tuned series circuit

($k = 0$), the poles are coincident at $(-1/2Q \pm j1)\omega_0$. As k increases, the poles follow the dashed line shown in the figure. The movement of the upper pole in the upper quadrant is much greater than that of the lower one as k increases.

Let us now consider the frequency-response curve. Qualitatively, the same considerations enter into the determination of the shape of the curve as in the case of staggered tuning. Using the approximation

$$\frac{\omega}{\omega_0} - \frac{\omega_0}{\omega} \doteq 2\delta \tag{A-17}$$

the magnitude of Eq. (A-14) for $s = j\omega$ can be written

$$|G_{21}(j\omega)| = \frac{Q^2 k \sqrt{\frac{C_1}{C_2}}}{\sqrt{\left[1 + \left(2Q\delta - kQ\frac{\omega}{\omega_0}\right)^2\right]\left[1 + \left(2Q\delta + kQ\frac{\omega}{\omega_0}\right)^2\right]}} \tag{A-18}$$

Compare this with Eq. (A-6) which is the corresponding expression for staggered tuning.

Since the frequency range is limited to the region $\omega \doteq \omega_0$, the variation of ω/ω_0 will be small compared with the variation of δ. Hence let us neglect this variation and take $\omega/\omega_0 = 1$. The denominator then becomes identical with that of Eq. (A-6). The coupling coefficient plays the same role as the staggering coefficient. Hence Eq. (A-8) giving the locations of the maxima is valid for the present case also. The curves in Fig. A-3 are also applicable. The condition for which $kQ = 1$ is referred to as *critical coupling*. In the present case the expressions corresponding to Eqs. (A-9) and (A-10) become

$$|G_{21}|_{\max} = \frac{Q}{2}\sqrt{\frac{C_1}{C_2}} \qquad (a)$$

$$|G_{21}|_{\delta=0} = \frac{Q^2 k \sqrt{\frac{C_1}{C_2}}}{1 + k^2Q^2} = \frac{2kQ}{1 + k^2Q^2}|G_{21}|_{\max} \qquad (b) \tag{A-19}$$

Note that the peak amplitudes are independent of the coupling. Furthermore, for critical coupling the value of $|G_{21}|$ at $\delta = 0$ is the same as $|G_{21}|_{\max}$. Hence, in the present case, the scales for the two curves in Fig. A-3 are the same. For less than critical coupling ($kQ < 1$) the multiplier of $|G_{21}|_{\max}$ in Eq. [A-19(*b*)] is less than unity. Thus the single peak in the response will become smaller as the coupling is reduced from the critical value. This is in contrast with the staggered tuned case in which this peak becomes larger for decreasing k.

All considerations of bandwidth discussed before apply in the present case as well.

A.3 THE DOUBLE TUNED PARALLEL CIRCUIT

A diagram of this circuit is shown in Fig. A-6. Again we assume that the two individual resonant frequencies are the same, as are also

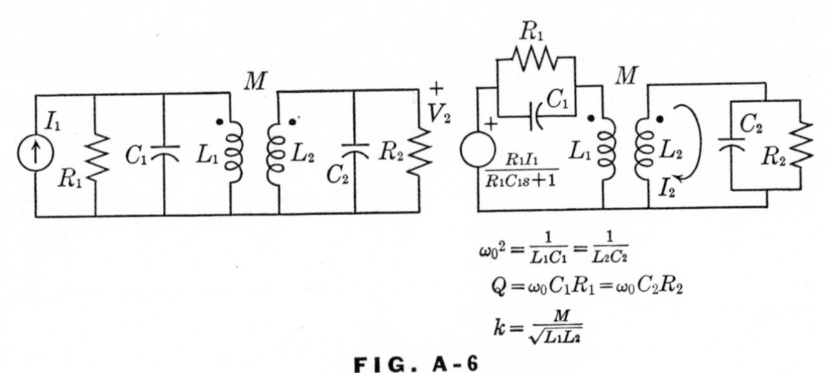

FIG. A-6

Double tuned parallel circuit

the Q's. Using Thévenin's theorem, the network can be converted to a two-loop network, as shown in Fig. A-6. Using loop equations, the output voltage sinor can be written

$$\begin{aligned} V_2 &= \frac{R_2}{R_2C_2s+1} I_2 = \frac{R_2}{R_2C_2s+1} \frac{\Delta_{12}}{\Delta} \frac{R_1I_1}{R_1C_1s+1} \\ &= \frac{R_1R_2MsI_1}{(R_1C_1s+1)(R_2C_2s+1)\left[\left(\frac{R_1}{R_1C_1s+1}+sL_1\right)\left(\frac{R_2}{R_2C_2s+1}+sL_2\right) - M^2s^2\right]} \\ &= \frac{R_1R_2MsI_1}{(R_1L_1C_1s^2+sL_1+R_1)(R_2L_2C_2s+sL_2+R_2) - k^2L_1L_2s^2(C_1R_1s+1)(C_2R_2s+1)} \end{aligned} \tag{A-20}$$

Using the definitions of ω_0, Q, and k, the transfer impedance becomes

$$\begin{aligned} Z_{21}(s) &= \frac{Ms}{\left(\frac{s^2}{\omega_0{}^2} + \frac{1}{Q}\frac{s}{\omega_0} + 1\right)^2 - k^2\left(\frac{s^2}{\omega_0{}^2} + \frac{1}{Q}\frac{s}{\omega_0}\right)^2} \\ &= \frac{k\frac{s^2}{\omega_0{}^2}\left(\frac{1}{\sqrt{C_1C_2}s}\right)}{\left[(1-k)\frac{s^2}{\omega_0{}^2} + \frac{1-k}{Q}\frac{s}{\omega_0} + 1\right]\left[(1+k)\frac{s^2}{\omega_0{}^2} + \frac{1+k}{Q}\frac{s}{\omega_0} + 1\right]} \end{aligned} \tag{A-21}$$

The denominator of the first line is the difference of two squares and so can be factored to give the second line. Compare this expression with Eq. (A-14). In the present case we have a simple zero at the origin and two pairs of poles whose locations are given by

$$\frac{s}{\omega_0} = -\frac{1}{2Q} \pm j\frac{1}{\sqrt{1 \pm k}}\sqrt{1 - \frac{1 \pm k}{4Q^2}} \doteq -\frac{1}{2Q} \pm j\frac{1}{\sqrt{1 \pm k}} \tag{A-22}$$

The approximate locations on the right are obtained under the assumption that $(1 + k)/4Q^2 \ll 1$, which is certainly satisfied with a Q as low as 5 for any k. The locations of the poles are shown in Fig. A-7 for a value of $k = 0.2$. In comparing the pole locations with those of the

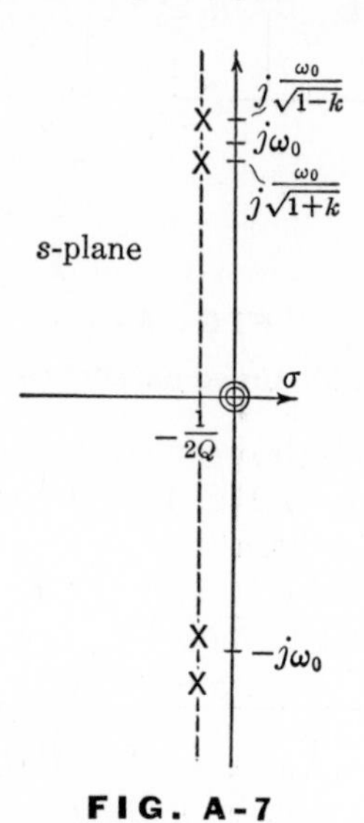

FIG. A-7

Pole-zero locations of double tuned parallel circuit

series circuit, note that the real part of the poles is now unaffected by the coupling coefficient. As k increases, the poles move out vertically from the points $\omega_0(1/2Q \pm j1)$, as in the staggered tuned case. Now, however, the distance above and below ω_0 is not the same. Nevertheless, for small values of k the pole separations will be approximately the same in the two cases. Thus

$$\omega_2 - \omega_1 = \frac{\omega_0}{\sqrt{1-k}} - \frac{\omega_0}{\sqrt{1+k}} = \omega_0\left(1 + \frac{k}{2} - \cdots\right) - \omega_0\left(1 - \frac{k}{2} + \cdots\right)$$
$$\doteq \omega_0 k \tag{A-23}$$

Here the square roots have been expanded by the binomial theorem. For small k only the first two terms need be retained. This is the same pole separation as shown in Fig. A-2.

The essential features of the frequency-response curve are the same as those of the previous case. Dividing numerator and denominator of Eq. (A-21) by s^2/ω_0^2, using the approximation of Eq. (A-17), and neglecting the variation of ω relative to that of δ, the magnitude of the transfer impedance can be written

$$|Z_{21}(j\omega)| = \frac{\dfrac{kQ^2}{\sqrt{C_1C_2}\omega_0}}{\sqrt{[(1-k)^2 + (2Q\delta - kQ)^2][(1+k)^2 + (2Q\delta + kQ)^2]}} \tag{A-24}$$

Aside from the multiplying constant in the numerator, this expression differs from the one in Eq. (A-18) only by the appearance of $(1 - k)^2$ and $(1 + k^2)$ in place of 1 in the factors of the denominator. The effect of this difference is to cause the peaks and valleys of the response curve to occur at slightly different values of δ. The previous quantitative values, however, are approximately valid for large values of Q and small values of the coupling coefficient.

A.4 TUNED COUPLED CIRCUIT

Consider the appearance of the double tuned parallel circuit if the transformer is replaced by a pi equivalent, as shown in Fig. A-8(*a*). There are two parallel tuned circuits coupled through an inductance L_c. This suggests another possibility, as shown in Fig. A-8(*b*), in which the two tuned circuits are coupled through a capacitor rather than an inductor.

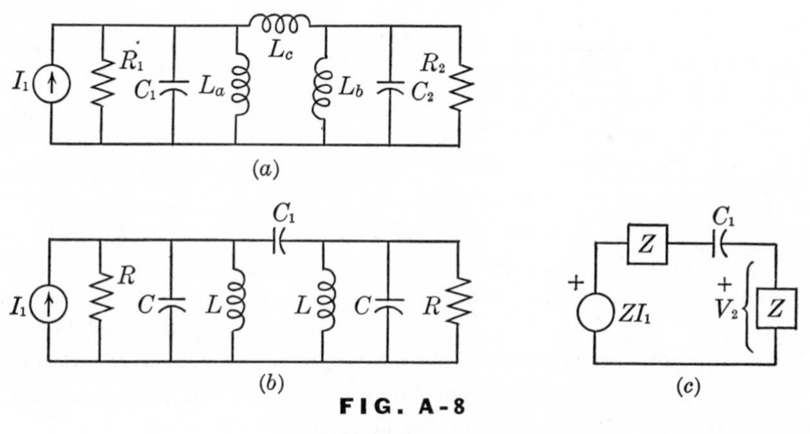

FIG. A-8

Tuned coupled circuits

Let us assume that the two tuned circuits are identical. Use of Thévenin's theorem to the left of C_1 leads to the result shown in Fig.

A-8(*c*), where Z is the impedance of R, L, and C in parallel. The voltage-divider relationship gives the following expression for the output voltage.

$$V_2 = \frac{Z}{2Z + \dfrac{1}{sC_1}} ZI_1 = \frac{sC_1}{2C_1s + Y}\frac{I_1}{Y}$$

$$= \frac{C_1sI_1}{[(Y + C_1s) + C_1s][(Y + C_1s) - C_1s]} \qquad \text{(A-25)}$$

where $Y = 1/Z$ and the two factors of the denominator, $(2C_1s + Y)$ and Y, have been rewritten in the form shown on the right. The quantity $Y + C_1s$ is the admittance of the tuned circuit in parallel with the coupling capacitor C_1. It can be written as

$$Y + C_1s = Cs + \frac{1}{R} + \frac{1}{Ls} + C_1s = \frac{1}{Ls}\left[(C + C_1)Ls^2 + \frac{Ls}{R} + 1\right] \qquad \text{(A-26)}$$

Let us now define ω_0 to be the resonant frequency of L and the parallel combination of C and C_1, and let us define Q correspondingly. That is,

$$\omega_0^2 = \frac{1}{L(C + C_1)} \qquad \text{(A-27)}$$

$$Q = R\sqrt{\frac{C + C_1}{L}} = \omega_0 R(C + C_1)$$

Using the last two equations in Eq. (A-25), the transfer impedance becomes

$$Z_{21} = \frac{C_1s}{\left[\dfrac{L(C + C_1)s^2 + Ls/R + 1}{Ls} + C_1s\right]\left[\dfrac{L(C + C_1)s^2 + Ls/R + 1}{Ls} - C_1s\right]}$$

$$= \frac{\dfrac{R}{Q}\dfrac{C_1}{C + C_1}\dfrac{s^3}{\omega_0^2}}{\left(\dfrac{s^2}{\omega_0^2} + \dfrac{1}{Q}\dfrac{s}{\omega_0} + 1 + \dfrac{C_1}{C + C_1}\dfrac{s^2}{\omega_0^2}\right)\left(\dfrac{s^2}{\omega_0^2} + \dfrac{1}{Q}\dfrac{s}{\omega_0} + 1 - \dfrac{C_1}{C + C_1}\dfrac{s^2}{\omega_0^2}\right)} \qquad \text{(A-28)}$$

On comparing this expression with the voltage gain of the double tuned series circuit in Eq. (A-14), we find that the denominators are identical if we let $C_1/(C + C_1)$ play the role of the coupling coefficient k. In the present case there is a triple zero at the origin instead of a double one. Thus the pole-zero pattern shown in Fig. A-5 will apply here also, except that an additional zero must be added at the origin.

The entire discussion pertaining to the frequency-response curve applies here also, again assuming that in the region near ω_0 the variation of ω can be neglected relative to that of δ, and ω can be replaced by ω_0.

INDEX